Plant Genetic Resources and Climate Change

FSC
www.fsc.org
MIX
Paper from responsible sources
FSC® C013604

CABI Climate Change Series

Climate change is a major environmental challenge to the world today, with significant threats to ecosystems, food security, water resources and economic stability overall. In order to understand and research ways to alleviate the effects of climate change, scientists need access to information that not only provides an overview of and background to the field, but also keeps them up to date with the latest research findings.

This series addresses many topics relating to climate change, including strategies to develop sustainable systems that minimize impact on climate and/or mitigate the effects of human activity on climate change. Coverage will encompass all areas of environmental and agricultural sciences. Aimed at researchers, upper level students and policy makers, titles in the series provide international coverage of topics related to climate change, including both a synthesis of facts and discussions of future research perspectives and possible solutions.

Titles Available

1. Climate Change and Crop Production
   Edited by Matthew P. Reynolds
2. Crop Stress Management and Global Climate Change
   Edited by José L. Araus and Gustavo A. Slafer
3. Temperature Adaptation in a Changing Climate: Nature at Risk
   Edited by Kenneth B. Storey and Karen K. Tanino
4. Plant Genetic Resources and Climate Change
   Edited by Michael Jackson, Brian Ford-Lloyd and Martin Parry

Plant Genetic Resources and Climate Change

Edited by

Michael Jackson

Formerly International Rice Research Institute (IRRI), the Philippines

Brian Ford-Lloyd

University of Birmingham, UK

and

Martin Parry

Imperial College London, UK

www.cabi.org

CABI is a trading name of CAB International

CABI
Nosworthy Way
Wallingford
Oxfordshire OX10 8DE
UK

Tel: +44 (0)1491 832111
Fax: +44 (0)1491 833508
E-mail: info@cabi.org
Website: www.cabi.org

CABI
38 Chauncey Street
Suite 1002
Boston, MA 02111
USA

T: +1 800 552 3083 (toll free)
T: +1 617 395 4051
E-mail: cabi-nao@cabi.org

A catalogue record for this book is available from the British Library, London, UK.

Library of Congress Cataloging-in-Publication Data

Plant genetic resources and climate change / edited by Michael Jackson, Brian Ford-Lloyd and Martin Parry.
p. cm. -- (CABI climate change series ; 4)
Includes bibliographical references and index.
ISBN 978-1-78064-197-3 (alk. paper)
1. Crops--Germplasm resources. 2. Food crops--Germplasm resources. 3. Crops and climate. 4. Germplasm resources, Plant. 5. Climatic changes. I. Jackson, Michael, 1948- II. Ford-Lloyd, Brian. III. Parry, M. L. (Martin L.) IV. Series: CABI climate change series ; 4.

SB123.3.P6315 2013
571.8'2--dc23

2013021989

ISBN-13: 978 1 78064 197 3

Cover photo: Flood-tolerant rice varieties with the *Sub1* gene thrive - in contrast to their non-*Sub1* but otherwise identical counterparts - after 17 days of submergence in experimental plots at the headquarters of the International Rice Research Institute (IRRI) in the Philippines. Climate change may cause long-term flooding in various parts of the rice-growing world and these varieties are ready for it.
Photo: Gina Vergara/IRRI.

Commissioning editor: Victoria Bonham
Editorial assistant: Alexandra Lainsbury
Production editor: Simon Hill

Typeset by Columns Design XML Limited, Reading.
Printed and bound in the UK by CPI Group.

Contents

Contributors vii

Preface xi

1. **Food Security, Climate Change and Genetic Resources** 1
   Robert S. Zeigler

2. **Genetic Resources and Conservation Challenges under the Threat of Climate Change** 16
   Brian Ford-Lloyd, Johannes M.M. Engels and Michael Jackson

3. **Climate Projections** 38
   Richard A. Betts and Ed Hawkins

4. **Effects of Climate Change on Potential Food Production and Risk of Hunger** 61
   Martin Parry

5. **Regional Impacts of Climate Change on Agriculture and the Role of Adaptation** 78
   Pam Berry, Julian Ramirez-Villegas, Helen Bramley, Mary A. Mgonja and Samarendu Mohanty

6. **International Mechanisms for Conservation and Use of Genetic Resources** 98
   Gerald Moore and Geoffrey Hawtin

7. **Crop Wild Relatives and Climate Change** 114
   Nigel Maxted, Shelagh Kell and Joana Magos Brehm

8. **Climate Change and On-farm Conservation of Crop Landraces in Centres of Diversity** 137
   Mauricio R. Bellon and Jacob van Etten

9. **Germplasm Databases and Informatics** 151
   Helen Ougham and Ian D. Thomas

10. Exploring 'Omics' of Genetic Resources to Mitigate the Effects of Climate Change **166**
Kenneth L. McNally

11. Harnessing Meiotic Recombination for Improved Crop Varieties **190**
Susan J. Armstrong

12. High Temperature Stress **201**
Maduraimuthu Djanaguiraman and P.V. Vara Prasad

13. Drought **221**
Salvatore Ceccarelli

14. Salinity **236**
William Erskine, Hari D. Upadhyaya and Al Imran Malik

15. Response to Flooding: Submergence Tolerance in Rice **251**
Abdelbagi M. Ismail and David J. Mackill

16. Effects of Climate Change on Plant–Insect Interactions and Prospects for Resistance Breeding Using Genetic Resources **270**
Jeremy Pritchard, Colette Broekgaarden and Ben Vosman

Index **285**

Contributors

Susan J. Armstrong, Senior Lecturer, School of Biosciences, University of Birmingham, Edgbaston, Birmingham, B15 2TT, UK. E-mail: s.j.armstrong@bham.ac.uk

Mauricio R. Bellon, Principal Scientist, Bioversity International, Via dei Tre Denari 472/a, Maccarese, Rome, Italy. E-mail: m.bellon@cgiar.org

Pam Berry, Senior Research Fellow, Environmental Change Institute, University of Oxford, Dyson Perrins Building, South Parks Road, Oxford, OX1 3QY, UK. E-mail: pam.berry@eci.ox.ac.uk

Richard A. Betts, Professor and Head of the Climate Impacts, Met Office Hadley Centre, FitzRoy Road, Exeter, Devon EX1 3PB, UK. E-mail: richard.betts@metoffice.gov.uk

Helen Bramley, Research Associate, Institute of Agriculture, The University of Western Australia, 35 Stirling Highway, Crawley, WA 6009, Australia. E-mail: helen.bramley@uwa.edu.au

Joana Magos Brehm, Collaborator, Centre for Environmental Biology, University of Lisbon, Portugal and Research Assistant, School of Biosciences, University of Birmingham, Edgbaston, Birmingham, B15 2TT, UK. E-mail: joanabrehm@gmail.com

Colette Broekgaarden, Postdoctoral Fellow, Wageningen UR Plant Breeding, PO Box 16, 6700 AJ Wageningen, The Netherlands. E-mail: colette.broekgaarden@wur.nl

Salvatore Ceccarelli, Former Barley Breeder, International Center for Agricultural Research in the Dry Areas (ICARDA), Aleppo, Syria (now retired). E-mail: s.ceccarelli@cgiar.org

Maduraimuthu Djanaguiraman, Postdoctoral Research Associate, Department of Agronomy, 2004 Throckmorton Plant Science Center, Kansas State University, Manhattan, KS 66506, USA. E-mail: jani@ksu.edu

Johannes M.M. Engels, Honorary Research Fellow, Bioversity International, Via dei Tre Denari 472/a, Maccarese, Rome, Italy. E-mail: j.engels@cgiar.org

William Erskine, Professor and Director, International Centre for Plant Breeding Education and Research (ICPBER) and Centre for Legumes in Mediterranean Agriculture (CLIMA), The University of Western Australia, 35 Stirling Highway, Crawley WA 6009, Perth, Australia. E-mail: william.erskine@uwa.edu.au

Jacob van Etten, Theme Leader – Climate Change Adaptation, Bioversity International, Regional Office of the Americas, CIAT, Recta Cali - Palmira Km. 17, Palmira, Colombia. E-mail: j.vanetten@cgiar.org

Brian Ford-Lloyd, Emeritus Professor, School of Biosciences, University of Birmingham, Edgbaston, Birmingham, B15 2TT, UK. E-mail: b.ford-lloyd@bham.ac.uk

Ed Hawkins, NERC Advanced Research Fellow, National Centre for Atmospheric Science, Department of Meteorology, University of Reading, Earley Gate, PO Box 243, Reading, RG6 6BB, UK. E-mail: e.hawkins@reading.ac.uk

Geoffrey Hawtin, Former Director General, International Plant Genetic Resources Institute (IPGRI), Maccarese, Rome, Italy (now retired), geoffhawtin@mac.com

Abdelbagi M. Ismail, Principal Scientist – Plant Physiology, International Rice Research Institute (IRRI), DAPO 7777, Manila 1301, Philippines. E-mail: abdelbagi.ismail@cgiar.org

Michael Jackson, Former Head of the Genetic Resources Center and Director for Program Planning and Communications, International Rice Research Institute (IRRI), DAPO Box 7777, Manila 1301, Philippines (now retired). E-mail: gen-res@mike.jacksonmail.org

Shelagh Kell, Research Fellow, School of Biosciences, University of Birmingham, Edgbaston, Birmingham, B15 2TT, UK. E-mail: s.p.kell@bham.ac.uk

David J. Mackill, Adjunct Professor, Department of Plant Sciences, University of California, Davis, CA 95616, USA and former Principal Scientist – Rice Breeding, International Rice Research Institute (IRRI), DAPO 7777, Manila 1301, Philippines. E-mail: david.mackill@effem.com

Al Imran Malik, Research Associate, Centre for Legumes in Mediterranean Agriculture (CLIMA) and Institute of Agriculture, The University of Western Australia, 35 Stirling Highway, Crawley, WA 6009, Australia. E-mail: imran.malik@uwa.edu.au

Nigel Maxted, Senior Lecturer in Genetic Conservation, School of Biosciences, University of Birmingham, Edgbaston, Birmingham, B15 2TT, UK. E-mail: nigel.maxted@dial.pipex.com

Kenneth L. McNally, Senior Scientist II – Molecular Genetics and Computational Biology, International Rice Research Institute (IRRI), DAPO Box 7777, Manila 1301, Philippines. E-mail: k.mcnally@irri.org

Mary A. Mgonja, Principal Scientist and Program Leader (Genetic Resources Enhancement and Management), International Crops Research Institute for the Semi-Arid Tropics (ICRISAT), Regional Office for Eastern and Southern Africa, United Nations Avenue, World Agroforestry Centre, Gigiri PO Box 39063-00623, Nairobi, Kenya. E-mail: m.mgonja@cgiar.org

Samarendu Mohanty, Head, Social Sciences Division, International Rice Research Institute (IRRI), DAPO Box 7777, Manila 1301, Philippines. E-mail: s.mohanty@irri.org

Gerald Moore, Former Legal Counsel, Food and Agriculture Organization of the United Nations (FAO), Rome, Italy (now retired). E-mail: gerald.moore@pcg.it

Helen Ougham, Former Reader, Institute of Biological, Environmental and Rural Sciences, Aberystwyth University, Penglais, Aberystwyth, Ceredigion, SY23 3DA, UK (now retired). E-mail: hjo@aber.ac.uk

Martin Parry, Visiting Professor, Grantham Institute and Centre for Environmental Policy, Imperial College London, London, SW7 2AZ, UK. E-mail: martin@mlparry.com

P.V. Vara Prasad, Associate Professor and Director of K-State Center for Sorghum Improvement, Department of Agronomy, 2004 Throckmorton Plant Science Center, Kansas State University, Manhattan, KS 66506, USA. E-mail: vara@ksu.edu

Jeremy Pritchard, Senior Lecturer and Head of Education, School of Biosciences, University of Birmingham, Edgbaston, Birmingham, B15 2TT, UK. E-mail: j.pritchard@bham.ac.uk

Julian Ramirez-Villegas, Doctoral Researcher, Institute for Climatic and Atmospheric Science (ICAS), School of Earth and Environment, University of Leeds, Leeds, UK, CGIAR Research Program on Climate Change, Agriculture and Food Security (CCAFS), Cali, Colombia, and International Center for Tropical Agriculture (CIAT), Cali, Colombia. E-mail: j.r.villegas@cgiar.org

Ian D. Thomas, Research Scientist, Institute of Biological, Environmental and Rural Sciences, Aberystwyth University, Penglais, Aberystwyth, Ceredigion, SY23 3DA, UK. E-mail: idt@aber.ac.uk

Hari D. Upadhyaya, Principal Scientist, Assistant Research Program Director – Grain Legumes, and Head – Gene Bank, International Crops Research Institute for the Semi-Arid Tropics (ICRISAT), Patancheru 502 324, Andhra Pradesh, India. E-mail: h.upadhyaya@cgiar.org

Ben Vosman, Senior Scientist – Resistance Breeding, Wageningen UR Plant Breeding, PO Box 16, 6700 AJ Wageningen, The Netherlands. E-mail: ben.vosman@wur.nl

Robert S. Zeigler, Director General, International Rice Research Institute (IRRI), DAPO Box 7777, Manila 1301, Philippines. E-mail: r.zeigler@irri.org

Preface

Almost 25 years ago, in April 1989, we convened the Second International Workshop on Plant Genetic Resources at the University of Birmingham. A year later, *Climatic Change and Plant Genetic Resources* was published, with 11 chapters summarizing perspectives on climatic change and how it might affect plant populations, and the expected impact on agriculture around the world.

There is much less scepticism today about greenhouse gas-induced climate change and what its consequences might be, even though the full impacts of climate change cannot yet be predicted with certainty. On the other hand, the nature of weather variability – particularly in the northern hemisphere in recent years – has left some again questioning whether our climate really is warming. But the evidence is there for all to see, even as the sceptics refuse to accept the empirical data of increases in atmospheric CO_2, for example, or the unprecedented summer melting of sea ice in the Arctic and the retreat of glaciers in the Alps.

Over the past decade, the world has experienced a number of severe climate events – wake-up calls to what might be the normal pattern in the future under a changed climate – such as extreme drought in one region or unprecedented flooding in another. Even the 'normal' weather patterns of Western Europe appear to have become disrupted in recent years leading to increased stresses on agriculture.

Some of the same questions we asked in 1989 are still relevant. There are, however, some very important differences today from the situation then. Our understanding of what is happening to the climate has been refined significantly over the past two decades, as the efforts of the International Panel on Climate Change (IPCC) have brought climate scientists worldwide together to provide better predictions of how climate will change. Furthermore, governments are now taking the threat of climate change seriously, and international agreements such as the Kyoto Protocol to the United Nations Framework on Climate Change, which came into force in 2005, even with their limitations, have provided the basis for society and governments to take action to mitigate the effects of climate change.

It is in this context, therefore, that this new book *Plant Genetic Resources and Climate Change* was commissioned to bring together, in a single volume, some of the latest perspectives about how genetic resources can contribute to achieving food security under the challenge of a changing climate. We also wanted to highlight some key issues for plant genetic resources management, to demonstrate how perspectives have changed over two decades, and discuss some of the actual responses and developments.

In our 1990 book, we asked (in a final chapter) whether genetic resources could cope with climate change, and would plant breeders be able to access and utilize genetic resources as building blocks of new and better-adapted crops? We listed the ten consensus conclusions from the workshop on which the book was based:

- The importance of developing collection, conservation and utilization strategies for genetic resources in the light of climatic uncertainty should be recognized.
- There should be marked improvement in the accuracy of climate change predictions.
- There must be concern about sea level rises and their impact on coastal ecosystems and agriculture.
- Ecosystems should be preserved thereby allowing plant species – especially crop species and their wild relatives – the flexibility to respond to climate change.
- Research should be prioritized on tropical dry areas because these might be expected to be more severely affected by climate change.
- There should be a continuing need to characterize and evaluate germplasm that will provide adaptation to changed climates.
- There should be an increase in screening germplasm for drought, raised temperatures and salinity.
- Research on the physiology underlying C_3 and C_4 photosynthesis should merit further investigation with the aim of increasing the adaptation of C_3 crops.
- Better simulation models should drive a better understanding of plant responses to climate change.
- Plant breeders should become more aware of the environmental impacts of climate change, so that breeding programmes could be modified to accommodate these predicted changes.

So what has happened during the past two decades or so? In 1990, world population was under 6 billion, but today there are more than 1 billion additional mouths to feed. The World Food Program estimates that there are 870 million people in the world who do not get enough food to lead a normal and active life. Food insecurity remains a major concern. In an opening chapter, Robert Zeigler provides an overview on food security today, how problems of food production will be exacerbated by climate change, and how – in the case of one crop, rice – access to and use of genetic resources have already begun to address many of the challenges that climate change will bring.

Expanding on the plant genetic resources theme, Brian Ford-Lloyd and his co-authors provide (in Chapter 2) a broad overview of important issues concerning their conservation and use, including conservation approaches, strategies and responses that become more relevant under the threat of climate change.

In three chapters, Richard Betts and Ed Hawkins, Martin Parry, and Pam Berry and her co-authors describe scenarios for future projected climates (Chapter 3), the effects of climate change on food production and the risk of hunger (Chapter 4), and regional impacts of climate change on agriculture (Chapter 5), respectively. During the past two decades, development of the global circulation models now permits climate change prediction with greater certainty. And combining these with physiological modelling and geographical information systems (GIS) we now have a better opportunity to assess what the impacts of climate change might be on agriculture and where.

In the 1990s, we became more aware of the importance of biodiversity in general, and several international legal instruments such as the Convention on Biological Diversity (CBD) and the International Treaty on Plant Genetic Resources for Food and Agriculture were agreed among nations to govern access to and use of genetic resources for the benefit of society. A detailed discussion of these developments is provided by Gerald Moore and Geoffrey Hawtin in Chapter 6.

In Chapters 7 and 8, we explore the *in situ* conservation of crop genetic resources and their wild relatives. Nigel Maxted and his co-authors provide an analysis of the importance of crop wild relatives in plant breeding and the need for their comprehensive conservation. Mauricio Bellon and Jacob van Etten discuss the challenges for on-farm conservation in centres of crop diversity under climate change.

Discussing the data management aspects of germplasm collections, Helen Ougham and Ian Thomas describe in Chapter 9 several developments in genetic resources databases, and regional projects aimed at facilitating conservation and use. Two decades ago we had little idea of what would be the impact of molecular biology and its associated data today on the identification of useful crop diversity and its use in plant breeding. In Chapter 10, Kenneth McNally provides a comprehensive review of the present and future of how genomics and other molecular technologies – and associated informatics – are revolutionizing how we study and understand diversity in plant species. He also provides many examples of how responses to environmental stresses that can be expected as a result of climate change can be detected at the molecular level, opening up unforeseen opportunities for precise germplasm evaluation, identification, and use. Susan Armstrong (Chapter 11) describes how a deeper understanding of sexual reproduction in plants, specifically the processes of meiosis, should lead to better use of germplasm in crop breeding as a response to climate change.

In a final series of five chapters, responses to a range of abiotic and biotic stresses are documented: heat (by Maduraimuthu Djanaguiraman and Vara Prasad, Chapter 12); drought (Salvatore Ceccarelli, Chapter 13); salinity (including new domestications) by William Erskine and his co-authors in Chapter 14; submergence tolerance in rice as a response to flooding (Abdelbagi Ismail and David Mackill, Chapter 15); and finally plant–insect interactions and prospects for resistance breeding using genetic resources (by Jeremy Pritchard and co-authors, Chapter 16).

The climate change that has been predicted is an enormous challenge for society worldwide. Nevertheless, progress in the development of scenarios of climate change – especially the development of more reliable projections of changes in precipitation – now provide a much more sound basis for using genetic resources in plant breeding for future climates. Although important uncertainty remains about changes to variability of climate, especially to the frequency of extreme weather events, enough is now known about the range of possible changes (for example by using current analogues of future climate) to provide a basis for choosing genetic resources in breeding better-adapted crops. Even the challenge of turbo-charging the photosynthesis of a C_3 crop like rice has already been taken up by a consortium of scientists worldwide under the leadership of the International Rice Research Institute in the Philippines.

Unlike the situation in 1989, estimates of average sea-level rise, and consequent risks to low-lying land areas, are now characterized by less uncertainty than most climate projections, and indicate the location and scale of the challenges posed by inundation, by soil waterlogging and by land salinization. Responses to all of these challenges and the progress achieved are spelt out in detail in several chapters in this volume.

We remain confident that research will continue to demonstrate just what is needed to mitigate the worst effects of climate change; that germplasm access and use frameworks – despite their flaws – facilitate breeders to choose and use genetic resources; and that, ultimately, genetic resources will be used successfully in crop breeding for climate change thereby enhancing food security.

Michael Jackson
Brian Ford-Lloyd
Martin Parry

April 2013

1 Food Security, Climate Change and Genetic Resources

Robert S. Zeigler
International Rice Research Institute (IRRI), Los Baños, Philippines

1.1 Introduction

In the last two decades of the 20th century, and well into the first decade of this century, food security was essentially absent from any discussions on global development priorities. Hunger, although being mentioned in the first Millennium Development Goal (United Nations, 2010), was seen as a consequence of poverty rather than an absolute supply issue. This was in stark contrast to the high priority food production was given following the end of the Second World War. This shift in attitude was the result of spectacular gains in food production, especially in developing countries that just a few decades earlier were facing persistent famine. The success, often referred to as the 'Green Revolution', was fuelled by increasing productivity in staples such as rice and wheat, particularly in Asia and Latin America. The rapid yield growth these cereals enjoyed during the 1970s and 1980s was built on a solid foundation of systematic development of genetic resources.

By the early 1990s, the World Bank, the Asian Development Bank and Inter-American Development Bank all saw very large drops in their lending for agriculture-related programmes compared with their previous major investments in infrastructure and institutions (Zeigler and Mohanty, 2010). Following the United Nations Conference on Environment and Development, commonly referred to as the Rio, or Earth, Summit in 1992, many overseas development assistance agencies shifted their emphasis from agriculture and food security to the environment. Agriculture, it seemed, was the antithesis of a healthy environment and the two could not coexist. Although an obvious fallacy, this attitude remains in some quarters to this day.

This perspective began to change – and change rapidly – with the food price crisis in 2007 and 2008. The prices of rice, maize and wheat spiked to levels not seen for decades, global trade was disrupted, and civil unrest spread across Asia and Africa. Rice, which is the staple food for most of the poorest people in the world and in fact the staple for half the world's population, was particularly hard hit. High bread prices were among the causes of the rioting that eventually led to the 'Arab Spring'. Prices quickly subsided but to higher levels than before the crisis. Today, the price of internationally traded rice is 70% higher than in 2005. Because food can make up 50% or more of total household expenditures for hundreds of millions of the very poor, millions of people were forced deeper into poverty and hunger became even more widespread. Food prices rose again in 2010–2011, and 2012 saw severe drought in the North American grain belt, Russia and the Ukraine. In the five years since 2008, these droughts, plus extreme flooding in Thailand and Pakistan, two 'delayed' monsoon seasons in South Asia and catastrophic storm surges in the Irrawaddy Delta of Burma (Myanmar), have caused many to speculate that global climate change is threatening agricultural productivity gains.

The interplay between food security and climate change and the means by which genetic resources may contribute to dampening the negative impacts that will be experienced are extraordinarily broad topics. Several authors in this volume directly address the range of changes that are being projected for our global climate. Others deal with how these changes may affect global food production, while in closing specific examples of the traits needed to counteract global climate change are presented. This opening chapter will necessarily be a scan of the issues but, because of my particular expertise in rice, I will draw specific examples from work ongoing in rice genetic resources to serve as the first illustrations of why I believe we will be able to adapt to whatever weather future climate has in store for us.

There are several reasons for my optimism. First, we are already experiencing today whatever severe weather events we will experience in the future. We can just expect that they may be more frequent, more widespread, at times more severe and possibly occurring in places where they did not occur before. That is, the abnormal events today may well be considered normal in the future, but by and large they will not be unprecedented (though I will draw a distinction for sea-level rise and rising night-time temperatures). Second, we have clear demonstrations that problems associated with extreme weather events previously thought to be insurmountable, such as serious flash flooding, can actually be solved. Third, the collections of genetic resources that our predecessors so painstakingly gathered and conserved are intact and without doubt contain many traits that will help our major crops cope with more extreme weather. Finally, enormous strides have been made in generating DNA sequences, relating DNA sequence to plant performance, handling large-scale data storage and analysis, doing crop modelling and making inter-specific crosses. These breakthroughs will enable us to systematically understand and use the vast treasure of traits that the world's public gene banks hold.

1.2 Food Security

For such a seemingly straightforward concept, the definition of 'food security' has been evolving since it became a widely used term in the early 1970s. Initially viewed as globally sufficient and available food supplies adequate to prevent price volatility, the term now encompasses availability, affordability, nutritional value, safety and desirability adequate for an active and productive life. Thus, the definition of food security now encompasses the household and individuals (Clay, 2002).

The Global Hunger Index (GHI) is an annual global assessment of food security, broken down by region and country (von Grebmer *et al.*, 2012). It is based on the extent of undernourishment, underweight children and child mortality in the area analysed. Although the GHI has dropped by more than 26% since 1990 and all regions have made progress over that time, it is not surprising that there is considerable heterogeneity among regions. The GHI in Southeast Asia declined by 46%, and in Latin America dropped by 44%. The decline in South Asia and sub-Saharan Africa (SSA) has been uneven. South Asia, after a good drop in the early 1990s, has a persistently high GHI of around 22 for the last 15 years, and SSA, while remaining high during the 1990s, has shown a promising decline over the last decade. The high GHI values of 22.5 for South Asia and 20.7 for SSA are not very different, but the enormous differences in total population between the two regions mean that the number of hungry people in South Asia comes close to the entire population of SSA. There is an expected link between economic growth and GHI: as per capita gross domestic income increases, the GHI for the country drops exponentially.

These numbers mask some appalling details. In India, for example, more than 43% of children under five years of age are underweight. Taken as a proxy for adequate nutrition, this is particularly alarming given that early childhood nutrition has been clearly shown to directly affect cognitive abilities, success in school, later health outcomes and adult productivity in the workforce (Victora *et al.* 2008). Considering the importance of rice as the main staple over much of South Asia, particularly for the poor, it is essential to have ample and affordable supplies of rice to meet demand. As overall nutrition is the determinant of a food-secure population, balanced diets are needed; but, for the very poor, these are simply not an option. Thus, improving staples such as rice for their nutritional content, particularly iron, zinc and vitamins, can be an important way to improve nutrition in poor populations.

Rice remains the most important staple for most of the world's poor (Gulati and Narayanan, 2002). After growing by 35% between 1960 and 1990, global per capita consumption has remained steady at 65 kg year^{-1} (calculated by dividing total rice consumption data from the United States Department of Agriculture (USDA) PSD online database by population data from the Food and Agriculture Organization of the United Nations database FAOSTAT; S. Mohanty, Los Baños, 2012, personal communication). In many Asian countries, however, consumption far exceeds 100 kg per person per year (Chapagain and Hoekstra, 2010). Even in rapidly growing economies such as India and China, per capita consumption has dropped only slightly and appears to be stable. Thus, population growth alone will drive demand for rice over the foreseeable future. Likewise, rice consumption is growing fastest, albeit from a fairly low base, in SSA, followed by Latin America. The International Rice Research Institute (IRRI) estimates that the world will need at least an additional 116 million tons of milled rice by 2035, or roughly 25% more than today's production.

Half of the world's rice-growing area is irrigated – typically flooded for most of the growing season. Yet, this land produces around three-quarters of the world's rice supply. Situated in low-lying deltas and coastal regions or large inland river valleys, these areas are susceptible to flooding and storm surges, despite sometimes elaborate canal and drainage systems. Large dams constructed from the 1950s to the 1990s provide irrigation water for large areas of rice production across Asia. Many are dependent on melt water from glaciers. It is noteworthy that seven of the great river systems in Asia have their headwaters in the Tibetan Plateau and derive much of their water from annual melting of snow pack and now, apparently, from rapidly melting glaciers. This could translate into an excess of water over several decades, followed by lower runoff rates much later.

The other half of the rice area depends exclusively on rainfall for its water needs and is therefore much more susceptible to variations in rainfall patterns. Rice is particularly sensitive to drought, moderately sensitive to salinity and, like most other crop plants, dies when fully submerged for just a few days. Not surprisingly, rainfed rice produces only 25% of global rice. Yet, it provides livelihoods for hundreds of millions of mostly very poor farmers and their families.

It should be clear that changes in total rainfall and its distribution could wreak havoc on global rice supplies. Storm surges will cause flooding and seawater intrusion to rice paddies. Higher temperatures and sea-level rise will also be detrimental to rice crops. However, given the very low resolution of our climate models, it is impossible to predict with any degree of precision where these changes will occur, how severe these changes will be and over what time frame. This I do not see as a problem. Many of the detrimental changes predicted in future climate scenarios already adversely affect poor rice farmers, and all are amenable to solutions that in part require access to genetic resources. I find comfort in this 'convenient convergence' of research and development objectives targeting the immediate needs of poor rice farmers and preparing the world's rice production for our future, but unknown, climate.

1.3 Available Genetic Resources

Some controversy exists over when and where rice (*Oryza sativa*) was domesticated (Sweeney and McCouch, 2007; Huang *et al.*, 2012). It is fairly safe to say that rice was being cultivated at least 10,000 years ago and that it was domesticated from *Oryza rufipogon* (Khush, 1997; Cheng *et al.*, 2003). Two major subgroups of rice, *indica* and *japonica*, led rice genetic resource specialists to conclude that there were two centres of origin. One was thought to be in the tropical regions of South Asia where *indica* rice varieties dominated and the other near central China where *japonica* rice dominated (cf. Londo *et al.*, 2006; Vaughan *et al.*, 2003). With the discovery that there are tropical *japonica*-like traditional varieties and another major group in South Asia called *aus* rice, things have become somewhat less clear. A recent study based on extensive DNA sequence analysis of *indica* and *japonica* varieties has concluded that there may have been only one centre of domestication and that all rice radiated from the Yellow River region (Huang *et al.*, 2012). Because rice was domesticated from *O. rufipogon*, however, the authors suggest that repeated crosses occurred between increasingly domesticated rice and its wild ancestor, resulting, along with natural mutations, in increasing genetic diversity in rice up until modern times. As early farmers and traders took rice from outside its centre of origin, it could continue to intercross with *O. rufipogon*, with farmers retaining those new traits that were of interest to them (Kovach *et al.*, 2007). This outcrossing continues today, with 'weedy rice' being an irritant to most farmers at one time or another. Thus, the genetic bottleneck of domestication from wild species that restricts genetic diversity in many species was nowhere near as narrow in rice as it was in wheat or groundnut, for example (Huang *et al.*, 2012). This probably explains the extraordinary diversity that we see in the International Rice Genebank at IRRI. This diversity, the sharp distinctions among subgroups of *O. sativa* and especially the differences among accessions collected from different regions, is beautifully illustrated in Plate 1.

Many collections of genetic resources exist for the world's principal food crops. Genebanks hold >7 million accessions of crops and their relatives in more than 1700 collections worldwide (FAO, 2010). IRRI holds well over 117,000 rice accessions, mostly landraces, in its genebank, and including almost 4500 accessions of all the wild *Oryza* species. When IRRI was created in the 1960s to develop new tropical rice varieties and production practices that would transform Asian agriculture, it was readily apparent to its rice breeders that there was simply not enough genetic diversity at hand to work with. So, they began to collect popular traditional rice varieties from around the region (IRRI, 2007). However, once the potential magnitude of their early breakthroughs became clear – the possibility, since realized, that the many thousands of low-yielding landraces would be replaced by a few very high yielding varieties – they set about to systematically collect, characterize and conserve these genetic resources before they were lost forever.

The first widely grown modern semi-dwarf tropical rice variety, IR8, was derived from three traditional varieties (Fig. 1.1). As breeders were compelled to address deficiencies in the early modern varieties, they increasingly turned to traditional varieties as sources for pest, disease and stress tolerance. Occasionally, they were also able to tap closely related *Oryza* species. The schematic of the pedigree of IR72, a later modern variety released in 1988, clearly shows how frequently breeders tapped traditional varieties (Fig. 1.1). The genetic background for the later varieties is impressive but, typically, only a tiny part of any particular parent can make it into the final variety. This use notwithstanding, and although it was known that some traditional varieties carried tolerance of the major abiotic constraints, efforts to transfer this tolerance to modern varieties generally met with only limited success. Likewise, because crossing a very productive and highly appreciated rice variety with a landrace tends to produce mostly very undesirable progeny owing to linkage drag, breeders are reluctant to make such crosses unless there is no alternative.

Modern rice varieties spread rapidly across Asia, with a number reaching the status of mega-varieties. One of the most popular varieties, IR64, covered as much as 35 million hectares and is still widely grown today. So, areas that once supported hundreds and perhaps thousands of

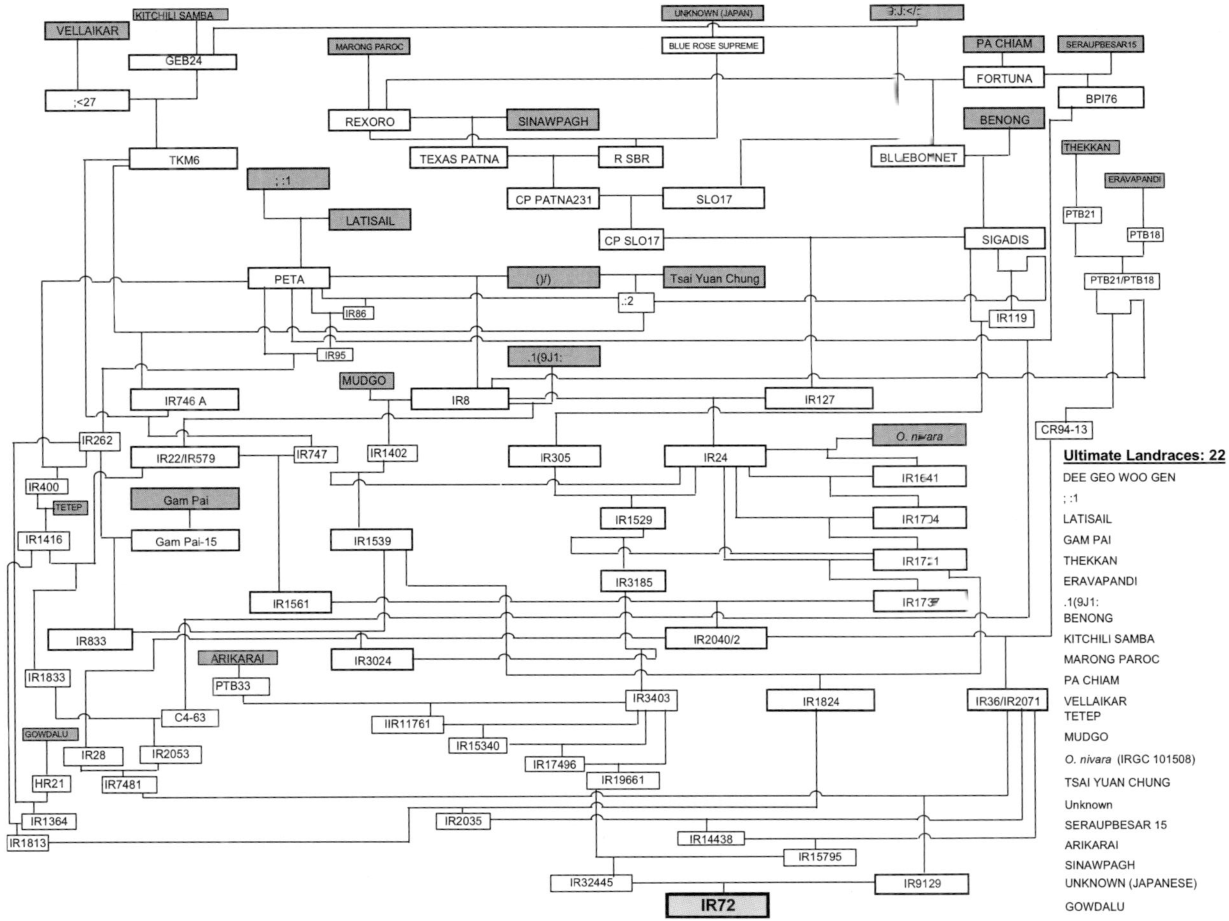

Fig. 1.1. Pedigree of rice variety IR72 showing 22 landraces (boxes with bold lines) in its pedigree. In contrast, IR8, the first widely grown modern semi-dwarf variety (indicated by the arrow), had only three landraces in its pedigree.

traditional varieties came to be covered by a single variety. This spread was accompanied by the development of large irrigation schemes that in turn were justified by the much larger returns from rice farming created by the modern varieties. This 'chicken and egg' relationship (which came first?), with mega-environments being created to support mega-varieties, masked an underlying reality of the rice sector. No vibrant seed sector existed that could support the production of more than a few varieties.

Another problem with the mega-varieties was that they were highly susceptible to drought, floods and salinity. Most pest and pathogen problems were tackled fairly early in the breeding process, but progress on the abiotic stresses was essentially non-existent before the turn of this century. Thus, the adoption of modern varieties was limited to irrigated areas with good water control and drainage. Farmers living in areas where only rainfed rice could be used were largely marginalized from the benefits conferred by modern varieties and crop management practices. It is noteworthy that wheat followed a trajectory similar to rice until recently, but why the difference?

Simply put, rice is a much easier crop to work with genetically. A diploid, it has by far the smallest genome of the world's cereals; in contrast, bread wheat, a hexaploid, has a genome 40 times larger than rice (International Rice Genome Sequencing Project, 2005). The combination of tractability (the small size and relative simplicity of its genome) and the extraordinary importance of rice as a food for the poor led the Rockefeller Foundation to create the Rice Biotechnology Program. The Foundation invested US$ 100 million over 15 years beginning in 1984 with the express purpose of applying the tools of molecular genetics and biotechnology to meet the needs of the world's poor. By investing in degree and non-degree training for young scientists from developing countries, this program also created a global scientific community focused on rice research. A specific research target was to create the tools to overcome the impasse breeders were facing with abiotic stresses. Significant investments were also made in supporting work to create viable progeny from crosses between species comprising the ten genomes that make up the genus *Oryza* as they diverged over the last 10 million years (Ge *et al.*, 1999), with the exception of *Oryza schlechteri*, which never flowers. A view of some of the traits found in these species immediately reveals why these crosses would be of such interest (Table 1.1).

As a direct result of the Rockefeller Foundation's investments, IRRI, along with many partners around the world, embarked on an ambitious project to sequence the DNA of a representative sample of the genetic diversity of rice and, eventually, its wild relatives (Yu *et al.*, 2002; International Rice Genome Sequencing Project, 2005; Huang *et al.*, 2012). Already, the DNA sequences of 3000 rice lines from the International Rice Genebank have been obtained. What is most important is that the loose consortium of institutions involved is also working on the means to do extensive, thorough and precise phenotypic analysis of the lines that have been sequenced. This will allow geneticists to apply the tools of association genetics that have been so successful in elucidating gene function in humans (Venter *et al.*, 2001). In addition, the rice community already has extensive mutant and mapping populations of lines that have been or will be sequenced. Finally, within *O. sativa*, a number of Multi-parent Advanced Generation Inter-Cross (MAGIC) populations are being created among the most popular and productive rice varieties in the world.

The most exciting prospect in tapping genetic resources to prepare rice for changing weather patterns is, in my opinion, the ability to cross *O. sativa* with all of the other species of *Oryza* and recover viable and fertile progeny. This has required the creation of special tetraploid genetic 'bridges' and the application of extremely labour-intensive and tedious embryo rescue techniques. This breakthrough means that rice breeders will now have at their disposal the products of 10 million years of natural selection and evolution that have taken place in a very wide range of environments,

Table 1.1. Useful traits of wild species of *Oryza* for adaptation to climate change.

Wild species	Genome	Chromosome no.	Traits
O. rufipogon Griff.	AA	24	Drought (seedling stage) tolerance, stem elongation ability under deep water, soil toxicity tolerance, anaerobic germination ability
O. longistaminata A. Chev. & Roehrich.	AA	24	Drought avoidance (seedling stage), out-crossing ability, submergence tolerance, biological nitrogen fixation ability
O. nivara S.D. Sharma et Shastry	AA	24	Drought tolerance (seedling stage), anaerobic germination
O. barthii A. Chev.	AA	24	Drought avoidance (seedling stage), anaerobic germination
O. glumaepatula Steud.	AA	24	Deep-water elongation ability, drought tolerance (reproductive stage), anaerobic germination (reproductive stage), anaerobic germination
O. meridionalis Ng	AA	24	Deep-water elongation ability, drought avoidance
O. punctata Kotschy ex Steud.	BB	24	Weed competitiveness, moderate tolerance of heat and salinity
O. officinalis Wall. ex G. Watt	CC	24	Heat tolerance (early flowering), drought and cold tolerance
O. eichingeri Peter	CC	24	Heat tolerance
O. rhizomatis D. A. Vaughan	CC	24	Drought (seedling stage) avoidance, heat, cold and salinity (seedling stage) tolerance
O. minuta J.S. Presl. ex K.B. Presl.	BBCC	48	Drought (seedling stage) avoidance, heat and cold tolerance
O. malampuzhaensis Krishnasw. & Chandrasekh	BBCC	48	None determined
O. schweinfurthiana Prodoehl	BBCC	48	None determined
O. latifolia Desv.	CCDD	48	Drought tolerance, submergence tolerance
O. alta Swallen.	CCDD	48	Drought tolerance, submergence tolerance
O. grandiglumis (Doell) Prodoehl.	CCDD	48	Drought tolerance, submergence tolerance
O. australiensis Domin.	EE	24	Drought (seedling and reproductive stages) tolerance, salinity and cold tolerance (seedling stage), heat tolerance (evening flowering)
O. brachyantha A. Chev. & Roehrich	FF	24	Submergence and cold tolerance, tolerance of problem soils
O. granulata Nees & Arn. ex G. Watt	GG	24	Tolerance of low light intensity (shade), adaptation to aerobic soil, drought and cold tolerance (all stages)
O. meyeriana (Zoll. & Moritzi) Baill.	GG	24	Tolerance of low light intensity (shade), adaptation to aerobic soil, drought and cold tolerance
O. neocaledonica Morat	GG	24	None determined
O. ridleyi Hook. f.	HHJJ	48	Drought tolerance and anaerobic germination ability
O. longiglumis Jansen	HHJJ	48	Drought (reproductive stage) tolerance
O. schlechteri Pilg.	Unknown	48	Drought (seedling stage) tolerance, stoloniferous
O. coarctata Roxb	KKLL	48	Salinity tolerance (all growth stages), heat tolerance and night flowering ability

including brackish water, very low light intensities and much higher temperatures than cultivated rice normally experiences (Khush, 1997; Jena and Khush, 2000).

1.4 Adapting to Climate Change

Views diverge on whether or not it makes sense to specifically target traits related to expected changes in weather patterns, given that we cannot predict exactly where and when they will occur. Sceptics further argue that the time frame over which the changes will occur is longer than normal breeding cycles of 10–15 years. They conclude that, by simply selecting the best performing plants under a gradually changing environment, breeders will automatically be selecting plants adapted to the new environments (Ziska *et al.*, 2012). As I will show later, this may be true for more global trends such as rising temperatures or increasing cloud cover. Plant breeding facilities are, however, generally protected from the drought, floods and salinity that will surely plague farmers of the future. Specific facilities will have to be developed to expose plants to these stresses under highly controlled conditions. And, as I argued earlier, we need these traits in any likely weather scenario, including today's weather, if we are to meet the needs of the poorest farmers who until now have been left behind.

I will only briefly introduce each area in which genetic resources can make a contribution because later chapters will deal with these in much more detail and for many more crops.

1.4.1 Heat

Analysis of the impact of high temperature on rice has been divided into high daytime temperatures and increasing night-time temperatures.

High daytime temperatures

Almost all rice varieties normally flower late in the morning until around noon. Temperatures above a critical threshold of 33°C at flowering result in unfilled grains (Nakagawa *et al.*, 2002; Jagadish *et al.*, 2007), although this threshold could vary depending on relative humidity. High temperatures combined with high relative humidity are more devastating than in combination with low relative humidity (Weerakoon *et al.*, 2008). There are two critical development stages: microspore development and anther dehiscence, with the latter being most sensitive. Post-pollination high temperatures have little direct effect on grain filling, assuming water and nutrients are adequate. Fortunately, some varieties identified in the International Rice Genebank flower at dawn. Additionally, some distantly related wild species also flower at dawn (Table 1.1). This simple avoidance measure can allow plants to flower when temperatures are well below the critical threshold for pollen sterility. It will be simply a matter of crossing these into commercially acceptable varieties.

High night-time temperatures

Analysis of yields in IRRI's Long-Term Continuous Cropping Experiment between 1967 and 1999 (Peng *et al.*, 1999) revealed an alarming trend. Yields of the same varieties were shown to be declining at about 10% for every 1°C rise in night-time temperatures. A similar relationship has been observed for wheat (Pradhan *et al.*, 2012). Analyses of rice yields across South and South-east Asia also showed a decline in yields when night-time temperatures surpassed 24°C. Experiments comparing the performance of different varieties selected over the decades at the IRRI experiment station in Los Baños (located at 14.145211, 121.261465) under controlled night-time temperatures in specially designed field plots revealed that newer varieties showed less of a yield decline than those developed in the 1960s (Peng *et al.*, 2004). It would appear then that selection under gradually changing conditions that affect the entire breeding population uniformly may allow breeders to unconsciously select for adaptation to a slowly changing environment. But, this may be one of the few examples for which such a *laissez faire* approach would be likely to succeed.

1.4.2 Overcoming stresses

Drought

This is a very complex phenomenon that will be dealt with extensively in a later chapter (see Ceccarelli, Chapter 13, this volume). Rice breeders and physiologists worked for decades with only limited success to improve the crop's notorious sensitivity to water deficit. Much work focused on identifying mechanisms of drought tolerance known to work in other crop species. Root architecture, leaf morphology and anatomy, stomatal behaviour and physiological processes such as osmotic adjustment were explored in some detail but little of practical application was found. The consensus after decades of work was that drought tolerance was rare in rice, largely limited to tropical *japonica* and *aus* varieties, and was under the control of a large number of genes, each with a small effect (Fukai *et al.*, 1999; Price and Courtois, 1999; Pantuwan *et al.*, 2002).

Despite this assessment, it was felt that the trait was too important to abandon work completely. A fresh approach was adopted to focus on whole-plant performance under precisely controlled water-stress conditions, using yield as the ultimate measure of tolerance. Tolerance was indeed identified in tropical *japonica* and *aus* backgrounds and crossed into *indica* backgrounds. Not surprisingly, many loci that appeared promising in one background did not perform well in *indica* backgrounds; however, others did perform well in a range of different genetic backgrounds. What was even more surprising was that several of the loci of interest had large-effect quantitative trait loci (QTLs) that determined drought tolerance (Bernier *et al.*, 2007; Kumar *et al.*, 2007; Venuprasad *et al.*, 2009). This discovery is quite novel and runs contrary to conventional wisdom based on experiences with wheat, maize and barley (Araus *et al.*, 2002). At least three large-effect QTLs have been combined and there seems to be little, if any, additive effect. The implications of this are still not clear.

Flash flooding

This causes severe losses over millions of hectares of rice land across Asia each year. It is difficult to determine whether these flood events are becoming more common or severe; however, several models have predicted flood incidence to increase due to increases in the intensity of tropical storms, precipitation extremes and sea-level rise (Coumou and Rahmstorf, 2012). The unprecedented flooding in Pakistan and Thailand over the past few years suggests that areas not previously affected by flash flooding may be at risk in the future. How this is linked to much heavier rainfall in areas that drain into the large river basins and the involvement of glacier melting is still in the realm of speculation. A source of flood tolerance (FR13A) has been known from India since the late 1970s. Since then, research at IRRI and elsewhere succeeded in transferring the trait into only a handful of varieties, but most of them were not widely accepted by farmers. The donor parent was low yielding and susceptible to lodging, and had poor eating quality. As with drought tolerance, it was concluded that the reason it was difficult to transfer the trait was because it, too, was controlled by a large number of QTLs, each with a small effect.

Subsequent research led to the identification and fine mapping of a single, large-effect QTL that conferred tolerance of complete submergence for at least 14 days. The gene, named *SUB1A-1* (Xu *et al.*, 2006), appears to control multiple traits that affect submergence tolerance. It has been shown to suppress gibberellin-mediated shoot elongation through down-regulating ethylene-dependent gene expression, and to stimulate anaerobic respiration and cytokinin-mediated delay in leaf senescence via inhibiting ethylene accumulation and sensitivity during submergence (Bailey-Serres *et al.*, 2010; Jung *et al.*, 2010). The *SUB1* locus is now successfully being deployed to farmers using current popular varieties, with considerable yield advantages in areas affected by flash floods (Mackill *et al.*, 2012; Ismail and Mackill, Chapter 15, this volume).

Analysis of submergence tolerance in wild species in the C genome (see Table 1.1) has shown that there are other sources of tolerance of complete submergence, independent of that conferred by *SUB1A* (Niroula *et al.*, 2012), leading to prospects of combining multiple sources of tolerance for higher and more stable submergence tolerance even for severe conditions, when the floodwater is very turbid or when submergence persists for longer than two weeks. Tolerance of other types of floods also needs to be combined with *SUB1*, such as partial stagnant flooding where deep water stays in the fields, sometimes for months, adversely affecting the rice crop (Singh *et al.*, 2011). Tolerance of floods during germination (anaerobic germination) is another feature that could be very important as rice farming systems in both rainfed and irrigated areas change because of labour shortages and other factors, and crop establishment shifts from transplanting rice seedlings into mud to sowing dry seed in soil that could be flooded by torrential rainfall, killing the seed before it can germinate. Anaerobic germination is also important in irrigated areas as a safeguard, when lands are not well levelled or if it rains after irrigating fields that are directly seeded. It can also be effectively used for weed management through shallow flooding after seeding, as weeds are a major problem for the large-scale adoption of direct seeding (Ismail *et al.*, 2009, 2012 and Chapter 15, this volume).

Salinity tolerance

Salt can adversely affect rice production in coastal areas, where sea water intrudes into rice paddies in river deltas. This can occur as freshwater that is perched on lower, denser saltwater near the sea coast, and from groundwater when all the fresh water is pumped from an aquifer, leaving only brackish water. In all river deltas of consequence in Asia, the rising and falling of the tides acts as an oscillating 'dam' of sorts. At high tide, rivers back up and farmers use this as a means to trap freshwater in the irrigation and drainage canal systems to supplement irrigation of their rice paddies. During the dry season, which may coincide with winter in the Himalayas, river flow decreases and saltwater moves further upstream, sometimes as much as 100 km in southern Bangladesh, for example. With unusually high tides or storm surges, seawater can move very far inland. More frequent and severe storms would exacerbate seawater intrusion. It is documented that global temperatures are rising and that sea levels are rising. The currently observed sea-level rise is greater than that predicted by even the most pessimistic scenarios of the IPCC just a decade ago (Kemp *et al.*, 2011). Presumably, this rise comes from thermal expansion and melting of glaciers. So, it is certain that salinity and flood events driven by the tides and/or storm surges will become more serious in the future.

Inland irrigated rice areas may also be adversely affected by increasing salt in the soil. Poor management of irrigation can result in high accumulations of salt in fields, making them unsuitable for any crop production. Higher temperatures, higher evapotranspiration and efforts to use less water or water with poor quality for irrigation may conspire and lead to salt accumulation on existing rice lands.

In all cases, salt-tolerant varieties will be increasingly needed. A moderate to high degree of salt tolerance is available in traditional rice varieties such as Pokkali (Thomson *et al.*, 2010; Platten *et al.*, 2013), which has been effectively exploited through breeding, and numerous salt-tolerant varieties have been released in India and Bangladesh, and some other countries. These varieties showed considerable yield benefits over farmers' current varieties in salt-affected areas, sometimes exceeding 2 t ha^{-1}. Several major QTLs associated with tolerance of salt stress were also identified, and one of them – *Saltol* on chromosome 1 – was transferred into several varieties, showing moderate but significant improvement of their salt tolerance (Thomson *et al.*, 2010). A few other major QTLs on different chromosomes are now being targeted for combination with *Saltol* for higher tolerance. The greatest promise comes, however, from the wild relative *Oryza coarctata* that thrives in brackish water (Jena, 1994, as *Porteresia coarctata*). The significance of recovering fertile, highly salt-tolerant progenies from a cross between this KKLL genome tetraploid and *O. sativa* (with the AA genome) is difficult to overstate.

Combined drought, submergence and salt tolerance

In less favourable crop production areas, multiple stresses usually affect farmers' fields, such as flooding and salinity in coastal areas and drought and floods at different times within the same season in rainfed lowlands. With the identification of large-effect QTLs controlling drought, submergence and salt tolerance, it is now possible to imagine rice varieties that will combine tolerance of all these constraints and their various manifestations. In no case is there a negative yield penalty, so they all appear to be inducible traits. Knowledge of their location on the rice chromosomes and their modes of action suggests that there is no reason to suspect that they will be mutually exclusive. IRRI breeders have already combined salt tolerance with submergence tolerance, and drought tolerance with submergence tolerance, with no observed negative effects on plant performance under normal growing conditions. The effectiveness of *SUB1A-1* has not been adversely affected when placed under controlled flood conditions. At least in terms of tolerance of these three abiotic stresses, the convenient convergence of multiple stress tolerance for today's poor farmers and future climates seems to be occurring.

Changing distribution of rice production and emerging pests

Evidence already exists that rice cultivation is moving farther north in China (Hijmans, 2007). So, it is reasonable to expect that the pest and disease complex will move as well. But, how will higher night-time temperatures affect pathogen growth and development? Will additional generations of fast-reproducing insect pests such as the planthopper complexes result in more frequent and severe pest outbreaks? How will temperature affect pathogens? Whatever the answer, it will do no harm to continue to evaluate traditional varieties for resistance to and tolerance of pathogens and pests. Renewed emphasis should be placed on the evaluation of wild relatives as well. These issues are discussed by Pritchard *et al.* in relation to insect pests (Chapter 16, this volume)

With these many new traits becoming available and all having a reasonable claim to being considered a high priority, is the science of rice breeding up to the task? Can a world of mega-varieties and mega-environments accommodate the many different trait combinations that will be required in the future?

1.5 The Future of Rice Breeding

The widespread adoption of modern semi-dwarf varieties that occurred beginning in the late 1960s was a result of a combination of factors. First, they generally outperformed available varieties by a wide margin, even though they may have had less than desirable quality. Second, this higher performance was seen across very large areas of the tropics when access to irrigation and fertilizer was good. This wide adaptation was specifically a result of low to absent photoperiod sensitivity and weak seed dormancy. This meant that varieties could be planted in any latitude and produce a crop in 130 days or so and the turnaround time between crops could be rapid, easily making possible two or even three crops per year. Third, governments and development agencies were sharply focused on achieving rice production targets, and were much less concerned about grain quality, growth duration and, to a large extent, pest management if they believed pesticides could keep pests in check. So, if a variety could deliver reliably high yields, that was reason enough to promote it over large areas. Fourth, the private rice seed industry was virtually non-existent and public seed systems (production and distribution) were typically centralized and bureaucratic, meaning that they could effectively manage only a few varieties. This extended to the capacity of national research and extension systems to create and then rigorously evaluate the large numbers of segregating populations that would be needed to have a highly diverse suite of varieties available to farmers. Finally, national seed laws were developed that tended to discourage the approval of large numbers of new varieties. So, the spectacular adoption of mega-varieties such as IR8, IR36 and IR64, though great improvements over what farmers previously had available, was by no means perfect.

The low interest in the seed industry by the private sector was driven by rice's inbreeding means of reproduction. This means that farmers could save seed from each crop for the next planting and, once having purchased a variety, never return for fresh seed. The perception that certified seed conferred an adequate advantage for farmers to pay a premium for it never really took hold in most rice-growing countries of the tropics. Tropical hybrid rice until very recently did not show the yield gains experienced in subtropical and temperate rice-producing areas of China, nor did their adoption have strong government backing.

The situation today, though, is changing. Private local and regional breeding programmes are being established all across Asia. Not only are multinational corporations seeing opportunity in the rice seed market, primarily in hybrid rice, but national and even local seed companies, including large non-governmental organizations (NGOs) such as BRAC in Bangladesh, are entering the rice seed business. Although hybrid rice seed looks to be the obvious business model for a large seed company (farmers must return to buy seed each year if they choose to continue growing hybrids), there is renewed interest in also developing a private inbred rice seed industry. The advances outlined earlier in this chapter, and that will be explored more fully in later chapters, will most likely turn out to be drivers that make an inbred rice seed industry possible.

IRRI's experience in disseminating mega-varieties such as IR64 and Swarna that carry desirable traits has clearly shown that farmers are willing to buy inbred seed if it carries these traits. The private sector has shown a strong interest in providing this seed and hundreds of small private seed enterprises have already begun producing *SUB1* seed. As new traits such as salinity tolerance, drought tolerance, tolerance of stagnant flooding and growth duration options, to name just a few, are added to the mega-varieties, at least two things will happen. First, as they will progressively be targeting narrower environments, they may well cease to be mega-varieties. But, more importantly, there will be a steady flow of new varieties with powerful traits that will attract farmers to purchase inbred rice seed. Perhaps not every year at the outset, but, with experience, farmers should quickly see the advantages of fresh high-quality seed. Thus, even regions and farmers that were effectively excluded from the mega-variety revolution will soon not only have access to vastly improved varieties derived from them, they will also play important roles in the evolution of mega-varieties.

These new traits are not the only way that a world of mega-varieties will be transformed. Ongoing breeding programmes, including anticipated results from the MAGIC populations, are increasingly targeting much finer markets from the outset. These markets are increasingly targeting more specific maturity groups (to enable much more flexibility in post-monsoon cropping alternatives) and, most importantly, grain quality types. These will include suitability for specific processed foods as well as nutritionally enhanced rice carrying beta-carotene and higher iron and zinc, and having a low glycaemic index.

Other trends will also drive a much more refined rice seed market. Consumers are demanding to have much more knowledge about the origins of even their staple foods. IRRI and a wide range of private and public partners are developing a framework that will result in a 'Sustainable Rice Production' (SRP) platform. Consumers purchasing SRP rice will know that their rice was produced using the most environmentally sensitive practices, that it meets clearly stated minimum quality criteria and that farmers received a fair price for their crop. This will be another outlet through which the myriad of traits becoming available through increased access to rice genetic resources will transform the rice industry at all levels. Supermarkets increasingly becoming the source of even staples in urban markets will drive a similar trend.

Finally, a number of countries will be sourcing their rice from beyond their borders and they are increasingly interested in controlling all means of production (*China Daily*, 2012). This, along with SRP-like drivers, will increasingly result in more and more farmers entering into a contractual relationship with a provider. This relationship will increasingly eliminate intermediate traders and encourage farmers to grow rice that will be in demand. This also aligns with farmers being less interested in small-scale rice farming, increasing their off-farm income and outsourcing increasingly labour-intensive and drudgery-intensive aspects of farm operations.

1.6 Conclusions

The 'convenient convergence' of meeting the needs of today's poorest farmers while tackling the range of challenges of future climate change will be made possible by the systematic tapping of genetic resources in rice and, eventually, other crops. The extraordinary genetic diversity available in our rice genebanks captures the wisdom and experience not only of those who collected the material but also of the countless generations of farmers who saw, nurtured and carried forward novelty as they encountered it. Incorporating new traits in excellent genetic backgrounds will allow some of the world's poorest farmers to benefit from some of the most advanced plant science. Generating novel variation from MAGIC populations of the world's elite rice varieties is a second stream through which genetic resources will be used to meet future challenges. Lastly, unravelling the 10 million years of natural selection and evolution that created the many genera of *Oryza* by crossing them into cultivated rice will allow rice breeders to weave an entirely new tapestry of rice for decades to come. This tapestry will allow us to face with confidence the vagaries of weather that any future climate gives us.

References

Araus, J.L., Slafer, G.A., Reynolds, M.P. and Royo, C. (2002) Plant breeding and drought in C_3 cereals: what should we breed for? *Annals of Botany* 89, 925–940.

Bailey-Serres, J., Fukao, T., Ronald, P.C., Ismail, A.M., Heuer, S. *et al.* (2010) Submergence tolerant rice: *SUB1*'s journey from landrace to modern cultivar. *Rice* 3, 138–147.

Bernier, J., Kumar, A., Ramaiah, V., Spaner, D. and Atlin, G. (2007) A large-effect QTL for grain yield under reproductive-stage drought stress in upland rice. *Crop Science* 47, 505–516.

Chapagain, A.K. and Hoekstra, H. (2010) The green, blue and grey water footprint of rice from both a production and consumption perspective. *Value of Water Research Report* Series No. 40, published by UNESCO-IHE Institute for Water Education, Delft, the Netherlands, in collaboration with University of Twente, Enschede, the Netherlands and Delft University of Technology, Delft, the Netherlands.

Cheng, C., Motohashi, R., Tsuchimoto, S., Fukuta, Y., Ohtsubo, H. *et al.* (2003) Polyphyletic origin of cultivated rice: based on the interspersion pattern of SINEs. *Molecular Biology and Evolution* 20, 67–75.

China Daily (2012) Rice plant helps grow Sino-Cambodian ties. Available at: http://europe.chinadaily.com.cn/world/2012-11/19/content_15939799.htm (accessed 18 December 2012).

Clay, E. (2002) Food security: concepts and measurement. A paper for the FAO Expert Consultation on Trade and Food Security: Conceptualising the Linkages, 11–12 July 2002. Food and Agriculture Organization of the United Nations (FAO), Rome, Italy.

Coumou, D. and Rahmstorf, S. (2012) A decade of weather extremes. *Nature Climate Change* 2, 491–496.

FAO (Food and Agriculture Organization of the United Nations) (2010) *The Second Report on the State of the World's Plant Genetic Resources for Food and Agriculture*. Available at: www.fao.org/docrep/013/i1500e/i1500e00.htm (accessed 18 December 2012).

Fukai, S., Pantuwan, G., Jongdee, B. and Cooper, M. (1999) Screening for drought resistance in rainfed lowland rice. *Field Crops Research* 64, 61–74.

Garris, A.J., Tai, T.H., Coburn, J., Kresovich, S. and McCouch, S. (2005) Genetic structure and diversity in *Oryza sativa L. Genetics* 169, 1631–1638.

Ge, S., Sang, T., Lu, B.R. and Hong, D.Y. (1999) Phylogeny of rice genomes with emphasis on origins of allotetraploid species. *Proceedings of the National Academy of Sciences USA* 96, 14400–14405.

von Grebmer, K., Ringler, C., Rosegrant, M.W., Olofinbiyi, T., Weismann, D. *et al.* (2012) *2012 Global Hunger Index: The Challenge of Hunger: Ensuring Sustainable Food Security under Land, Water, and Energy Stresses.* Deutsche Welthungerhilfe, Bonn, Germany, International Food Policy Research Institute, Washington, DC, USA and Concern Worldwide Dublin, Ireland. Available at: www.ifpri.org/sites/default/files/publications/ghi12.pdf (accessed 18 December 2012).

Gulati, A. and Narayanan, S. (2002) *Rice Trade Liberalization and Poverty.* MSSD Discussion Paper No. 51. International Food Policy Research Institute, Washington, DC, USA.

Hijmans, R. (2007) Relocating rice production in China. *Rice Today* 6, 26–27.

Huang, X., Kurata, N., Wei, X., Wang, Z.X., Wang, A. *et al.* (2012) A map of rice genome variation reveals the origin of cultivated rice. *Nature* 490, 497–501.

International Rice Genome Sequencing Project (2005) The map-based sequence of rice genome. *Nature* 436, 793–800.

IRRI (2007) Luck is the residue of design. Pioneer Interview with IRRI's first rice breeder, Peter Jennings, on 20 July 2007. Available at: http://archive.irri.org/publications/today/Jennings.asp (accessed 18 December 2012).

Ismail, A.M., Ella, E.S., Vergara, G.V. and Mackill, D.J. (2009) Mechanisms associated with tolerance to flooding during germination and early seedling growth in rice (*Oryza sativa*). *Annals of Botany* 103, 197–209.

Ismail, A.M., Johnson, D.E., Ella, E.S., Vergara, G.V. and Baltazar, A.M. (2012) Adaptation to flooding during emergence and seedling growth in rice and weeds, and implications for crop establishment. *AoB PLANTS* 2012: pls019 doi:10.1093/aobpla/pls019.

Jagadish, S.V.K., Craufurd, P.Q. and Wheeler, T.R. (2007) High temperature stress and spikelet fertility in rice (*Oryza sativa* L.). *Journal of Experimental Botany* 58, 1627–1635.

Jena, K.K. (1994) Production of intergeneric hybrid between *Oryza sativa* L. and *Porteresia coarctata* T. *Current Science* 67, 744–748.

Jena, K.K. and Khush, G.S. (2000) Exploitation of alien species in rice improvement: achievements and future challenges. In: Primlani, M. (ed.) *Rice Breeding and Genetics – Research Priorities and Challenges.* Oxford and IBH Publishing Co., Pvt. Ltd, New Delhi. pp. 269–285.

Jung, K.-H., Seo, Y.-S., Walia, H., Cao, P., Fukao, T. *et al.* (2010) The submergence tolerance regulator *Sub1A* mediates stress-responsive expression of *AP2/ERF* transcription factors. *Plant Physiology* 152, 1674–1692.

Kemp, A.C., Horton, B.P., Donnelly, J.P., Mann, M.E., Vermeer, M. *et al.* (2011) Climate related sea-level variations over the past two millennia. *Proceedings of the National Academy of Sciences USA* 108, 11017–11022.

Khush, G.S. (1997) Origin, dispersal, cultivation and variation of rice. *Plant Molecular Biology* 35, 25–34.

Kovach, M.J., Sweeney, M.T. and McCouch, S.R. (2007) New insights into the history of rice domestication. *Trends in Genetics* 23, 578–587.

Kumar, R., Venuprasad, R. and Atlin, G.N. (2007) Genetic analysis of rainfed lowland rice drought tolerance under naturally-occurring stress in eastern India: heritability and QTL effects. *Field Crops Research* 103, 42–52.

Londo, J.P., Chiang, Y.C., Hung, K.H., Chiang, T.Y. and Schaal, B.A. (2006) Phylogeography of Asian wild rice, *Oryza rufipogon*, reveals multiple independent domestications of cultivated rice, *Oryza sativa. Proceedings of the National Academy of Sciences USA* 103, 9578–9583.

Mackill, D.J., Ismail, A.M., Singh, U.S., Labios, R.V. and Paris, T.R. (2012) Development and rapid adoption of submergence-tolerant (Sub1) rice varieties. *Advances in Agronomy* 115, 303–356.

Nakagawa, H., Horie, T. and Matsui, T. (2002) Effects of climate change on rice production and adaptive technologies. In: Mew, T.W., Brar, D.S., Peng, S., Dawe, D. and Hardy, B. (eds) *Rice Science: Innovations and Impact for Livelihood.* Proceedings of the International Rice Research Conference, 16–19 September 2002, Beijing, China. International Rice Research Institute, Chinese Academy of Engineering, and Chinese Academy of Agricultural Sciences, Beijing, China. pp. 635–657.

Niroula, R.K., Pucciariello, C., Ho, V.T., Novi, G., Fukao, T. *et al.* (2012) SUB1A-dependent and -independent mechanisms are involved in the flooding tolerance of wild rice species. *Plant Journal* 72, 282–293.

Pantuwan, G., Fukai, S., Cooper, M., Rajatasereekul, S. and O'Toole, J.C. (2002) Yield response of rice (*Oryza sativa* L.) to drought under rainfed lowlands 3. Plant factors contributing to drought resistance. *Field Crops Research* 73, 181–200.

Peng, S., Cassman, K.G., Virmani, S.S., Sheehy, J. and Khush, G.S. (1999) Yield potential trends of tropical rice since the release of IR8 and the challenge of increasing rice yield potential. *Crop Science* 39, 1552–1559.

Peng, S., Huang, J., Sheehy, J.E., Laza, R.C., Visperas, R.M. *et al.* (2004) Rice yields decline with higher night temperature from global warming. *Proceedings of the National Academy of Sciences USA* 101, 9971–9975.

Platten, J.D., Egdane, J.A. and Ismail, A.M. (2013) Salinity tolerance, Na^+ exclusion and allele mining of *HKT1;5* in *Oryza sativa* and *O. glaberrima*: many sources, many genes, one mechanism? *BMC Plant Biology* 13, 32.

Pradhan, G., Vara Prasad, P.V., Fritz, A.K., Kirkham, M.B. and Gill, B.S. (2012) Effect of drought and high temperature stress on synthetic hexaploid wheat. *Functional Plant Biology* 39, 190–198.

Price, A.H. and Courtois, B. (1999) Mapping QTLs associated with drought resistance in rice: progress, problems and prospects. *Plant Growth Regulation* 29, 123–133.

Singh, S., Mackill, D.J. and Ismail, A.M. (2011) Tolerance of longer-term partial stagnant flooding is independent of the *SUB1* locus in rice. *Field Crops Research* 121, 311–323.

Sweeney, M. and McCouch, S. (2007) The complex history of the domestication of rice. *Annals of Botany* 100, 951–957.

Thomson, M.J., de Ocampo, M., Egdane, J., Rahman, M.A., Sajise, A.G. *et al.* (2010) Characterizing the *Saltol* quantitative trait locus for salinity tolerance in rice. *Rice* 3, 148–160.

United Nations (2010) *We Can End Poverty 2015 – Millennium Development Goals.* A Gateway to the UN System's Work on the MDGs. Available at: www.un.org/millenniumgoals (accessed 18 December 2012).

Vaughan, D.A., Morishima, H. and Kadowaki, K. (2003) Diversity in the *Oryza* genus. *Current Opinion in Plant Biology* 6, 139–146.

Venter, J.C., Adams, M.D., Myers, E.W., Li, P.W., Mural, R.J. *et al.* (2001) The sequence of the human genome. *Science* 291, 1304–1351.

Venuprasad, R., Dalid, C.O., Del Valle, M., Bool, M.E., Zhao, L. *et al.* (2009) Identification and characterization of large-effect quantitative trait loci (QTL) for grain yield under lowland drought stress and aerobic conditions in rice using bulk-segregant analysis (BSA). *Theoretical and Applied Genetics* 120, 177–190.

Victora, C.G., Adair, L., Fall, C., Hallal, P.C., Martorell, R. *et al.* (2008) Maternal and child undernutrition: consequences for adult health and human capital. *The Lancet* 371, 340–357.

Weerakoon, W.M.W., Maruyama, A. and Ohba, K. (2008) Impact of humidity on temperature-induced grain sterility in rice (*Oryza sativa* L.). *Journal of Agronomy* & *Crop Science* 194, 135–140.

Xu, K., Xia, X., Fukao, T., Canlas, P., Maghirang-Rodriguez, R. *et al.* (2006) *Sub1A* is an ethylene response factor-like gene that confers submergence tolerance to rice. *Nature* 442, 705–708.

Yu, J., Hu, S., Wang, J., Wong, G.K., Li, S. *et al.* (2002) A draft sequence of the rice genome (*Oryza sativa* L. ssp. *indica*). *Science* 296, 79–92.

Zeigler, R. and Mohanty, S. (2010) Support for international agricultural research: current status and future challenges. *New Biotechnology* 27, 565–572.

Ziska, L.H., Bunce, J.A., Shimono, H., Gealy, D.R., Baker, J.T. *et al.* (2012) Food security and climate change: on the potential to adapt global crop production by active selection for rising atmospheric carbon dioxide. *Proceedings of the Royal Society B* 279, 4097–4105.

2 Genetic Resources and Conservation Challenges under the Threat of Climate Change

Brian Ford-Lloyd[1], Johannes M.M. Engels[2] and Michael Jackson[3]

[1]*School of Biosciences, University of Birmingham, Birmingham, UK;* [2]*Bioversity International, Rome, Italy;* [3]*formerly International Rice Research Institute (IRRI), Los Baños, Philippines (now retired)*

2.1 Introduction

Since the publication just over two decades ago of *Climatic Change and Plant Genetic Resources* (Jackson *et al.*, 1990) there have been a number of important developments regarding the conservation of and access to plant genetic resources for food and agriculture (hereafter PGRFA). Furthermore, there is today considerably less uncertainty about the nature of and evidence for global climate change, as several chapters in this book document (see Betts and Hawkins, Chapter 3; Parry, Chapter 4; and Berry *et al.*, Chapter 5, this volume).

During these two decades, our understanding of conservation issues broadened and several important books were published bringing together the latest perspectives on a wide range of genetic conservation topics: *ex situ* and *in situ* conservation approaches (Maxted *et al.*, 1997; Hawkes *et al.*, 2000); crop wild relatives (Maxted *et al.*, 2008); the value and costs of conserving crop germplasm in genebanks (Fuccillo *et al.*, 1997; Koo *et al.*, 2004); core collections (Hodgkin *et al.*, 1995); the application of biotechnology (Callow *et al.*, 1997); and managing crop diversity (Engels *et al.*, 2002).

With the threat of global climate change looming, it is imperative to ensure that plant breeders have access and permission to use the germplasm already conserved in genebanks around the world, and to determine if and whether it might be necessary to mount collecting expeditions for novel diversity not already in a genebank. In this chapter we describe and discuss some of the initiatives and developments that make the world better prepared to respond to climate change in terms of plant genetic resources, as well as some of the conservation challenges under a changing climate.

2.2 The Biodiversity Decade

The 1990s were the 'biodiversity decade' during which the profile of environmental issues such as conservation was significantly raised. The Convention on Biological Diversity (CBD) was agreed at the Earth Summit of June 1992 in Rio de Janeiro, and entered into force in December 1993. Concurrent with the ratification of the CBD, a number of initiatives were undertaken by the Food and Agriculture Organization of the United Nations (FAO), through its Commission on Genetic Resources for Food and Agriculture (hereafter the Commission). The CBD had paid scant attention to biodiversity important for agriculture. At the same time, the development of different international legal instruments was taking place in an environment of growing concern about possible privatization of biodiversity and intellectual property rights. The implications (and politics) of these legal instruments, and the policies in general around PGRFA, are discussed later (see Moore and Hawtin, Chapter 6, this volume).

The Commission decided it was necessary not only to document what was already being conserved around the world, but also to develop internationally agreed plans to safeguard and use this diversity. This led to two important reports (discussed below and on p. 18) on the State of the World's Plant Genetic Resources (first published in 1996), and the Global Plan of Action that stemmed from this report. The Commission was also encouraging the adoption of agreed standards for germplasm conservation in genebanks, as well as their consistent documentation for conservation, evaluation and use.

The negotiations in the Commission finally led to the adoption in 2001 of the International Treaty on Plant Genetic Resources for Food and Agriculture (hereafter the Treaty). It is important to emphasize these points because the CBD and the Treaty provide the legal frameworks that will facilitate the use of germplasm as a response and adaptation to climate change.

2.3 State of the World's Plant Genetic Resources

Two reports on the State of the World's Plant Genetic Resources for Food and Agriculture have been published by FAO, the most recent in 2010 (hereafter referred to as SOW2) (FAO, 2010). SOW2 is based on information gathered from more than 100 countries, as well as from regional and international research and support organizations and academic programmes. The report documents the current status of plant genetic resources diversity, conservation and use, as well as the extent and role of national, regional and international efforts that underpin the contributions of PGRFA to food security. It highlights the most significant changes that have occurred in the sector since 1996, when the first report was produced by FAO (FAO, 1996a), as well as the gaps and needs that remain for setting future priorities. SOW2 provided the basis for the updating of the Global Plan of Action for the Conservation and Sustainable Utilization of Plant Genetic Resources for Food and Agriculture (hereafter referred to as GPA2).

These are the core messages in the SOW2 report:

- PGRFA are essential raw materials for helping farmers to respond to climate change. Plant breeding efforts in general need to be strengthened and breeding programmes have to be expanded to develop varieties with traits needed to meet this challenge.
- The loss of PGRFA has reduced options for the agricultural sector. Land clearing, human population pressures, overgrazing, environmental degradation and changing agricultural practices were identified as the main causes of genetic erosion.
- Local PGRFA diversity found in farmers' fields is still to a large extent inadequately documented and managed. The importance of this diversity and the contributions it makes to local food security is increasingly being recognized.
- Further progress has been achieved in securing PGRFA diversity in a large number of (national) genebanks. However, much diversity of crop wild relatives (CWRs) and of underused species important for food and agriculture still needs to be conserved properly for present and future uses.
- Significant and fast technological advances, especially in information technology and molecular biology have introduced new techniques for PGRFA conservation and use. Their wider application offers opportunities to strengthen the interface between conservation and use and thus to increase the efficiency of the conservation–production value chain.
- Significant policy developments have changed the landscape of PGRFA management over the past 15 years or so. Many more countries have established

national (plant) genetic resources programmes, laws and regulations for biodiversity following the adoption of the CBD and the Treaty.
- Improved communication, collaboration and partnerships are needed among institutions that deal with PGRFA management through to plant breeding and the support of seed systems. These are key factors for an integrated conservation and utilization strategy and for delivering sustainable solutions to build a world without hunger.

The issue of climate change has been addressed throughout the SOW2 report, especially in a newly added chapter on contributions of PGRFA to food security and sustainable agriculture. It concludes that some of the most densely populated regions of the world might be most affected by climate change and that the adaptation of agriculture to the new conditions will require a shift to more drought-tolerant or heat-tolerant varieties, or even to different crops. Drastic changes in pest and disease patterns are foreseen and will require new resistant or tolerant varieties. The anticipated and already observed less predictable weather patterns may also require the development of varieties with a broader adaptation, and the development of varieties that are able to mitigate climate changes, such as:

- Those that root deeper and, coupled with appropriate agricultural practices, allow capturing of more carbon in the soil.
- The production of feed and forage varieties for less methane emission by ruminants; and varieties that use nitrogen more efficiently and that need less fertilizers, hence less total energy.

In order to achieve such developments it will be critically important to increase efforts for conserving landraces and CWRs with traits likely to become more important in the future. This must be coupled with enhanced capacity building in traditional and modern crop improvement techniques to take advantage of these genetic resources through plant breeding, especially since the shortage of trained plant breeders is now a recognized constraint. This problem will be greatest in developing countries that are also likely to be hardest hit by climate change because they have least ability to respond to change.

2.4 The Global Plan of Action

The preparation of the GPA2 was sponsored by the FAO Commission and was adopted by the FAO Council in November 2011 (FAO, 2012). GPA2 is an update of the first plan that was adopted during the Fourth International Technical Conference on Plant Genetic Resources held in 1996 in Leipzig, Germany (FAO, 1996b). GPA2 responds to the needs and priorities that were identified in SOW2 and was prepared through a series of regional consultation meetings, with the participation of 131 countries, including representatives of the public and private sectors, the international research community, and civil society. The Global Plan of Action was a key element for the Commission to fulfil its mandate with respect to PGRFA. GPA1 was the main reference document for national, regional and global efforts to conserve and use PGRFA sustainably and to share the benefits deriving from their use in an equitable and fair manner. GPA2 continues these efforts. The Global Plan of Action set the stage for the successful completion of the negotiation of the Treaty, and also assisted governments with formulation of national policies and strategies on PGRFA. It has been used by the international community to: define priorities at the global level; reorient research and development agendas; improve the coordination of conservation and use efforts; and generate synergies among the genetic resources stakeholders and activities.

GPA2 makes reference to major challenges the world is facing, including an increasing food insecurity reflected in, among other aspects, highly volatile food prices, climate change, increasing urbanization, the need for more sustainable agriculture as well as the need to safeguard the plant genetic diversity and to minimize genetic erosion. At the same time the Global Plan mentions new opportunities for the improvement of the management of PGRFA that impressive advancements in communication and information technologies provide as well as those in biotechnology and the development of bioproducts derived from agriculture.

In addition to the technological advances, we have also seen significant changes in the policy environment, in particular the entry into force of the Treaty, the Cartagena Protocol on Biosafety, the adoption by the Conference of the Parties of the CBD of the Strategic Plan for Biodiversity 2011–2020 and the Nagoya Protocol on Access to Genetic Resources and the Fair and Equitable Sharing of Benefits Arising from their Utilization, although several of these instruments have yet to enter into force (see Moore and Hawtin, Chapter 6, this volume.)

While SOW2 identified the needs and priorities related to the conservation and use of PGRFA, the Global Plan is a well-defined and coordinated response; GPA2 specifies 18 priority activities (from the original 20). Greater visibility is given to plant breeding in this second plan. Climate change is also recognized by the Plan as an immediate and unprecedented threat to livelihoods and food security and that it might well be a major barrier to achieving the 60% increase in food production that is needed by 2050 to feed all people. Several strategic elements are proposed to safeguard PGRFA and use them optimally for adaptation to climate change:

- A greater emphasis on *in situ* conservation of genetically diverse populations, especially CWRs, to allow evolution to continue and thus to facilitate the continued generation of adaptive traits.
- A significantly expanded *ex situ* conservation programme, especially of CWRs, to ensure the maintenance of diversity of species, populations and varieties, in particular those adapted to the extreme conditions and those from areas that are expected to be highly affected by climate change.
- Increased research and improved availability of information on the characteristics of genetic resources material that are conserved *ex situ* and that will become useful under new climate conditions.
- Increased support for access to and the movement of PGRFA to meet the greater interdependence of countries that result from these new environmental conditions.
- Increased support to build capacity in plant breeding and seed-systems management in order to facilitate the effective and sustainable use of PGRFA.
- Increased and targeted involvement of farmers and farming communities in national and local crop improvement activities, including support for participatory research and plant breeding.

GPA2 reflects the new challenges and opportunities that have emerged with respect to the conservation and use of plant genetic resources for food and agriculture since 1996, including climate change, as well as new developments in the policy environment, in particular the entry into force of the Treaty. It puts greater emphasis on the sustainable use of PGRFA as a means to enhance food security in a changing environment.

2.5 The Molecular Biology Revolution

The application of molecular biology has opened a window on the diversity of genebank collections that was only a dream just a few short years ago, coupled with advances in computational power to handle huge data sets and high-speed telecommunications that permit researchers all over the world to collaborate on the same data in real time. The resolution of the genomes of many of the most important crop species in agriculture, beginning with rice in 2002, and followed by wheat and maize later on, as well as the identified homology or synteny between cereal genomes, for example (Devos and Gale, 1997) now permits breeders to assess diversity more comprehensively and determine the genetic basis of traits more accurately using molecular markers (see

Chapters 13–17 on the use of germplasm, this volume). The development of association genetics, first demonstrated in rice by Virk *et al.* (1996) and now the application of genomics to genebank collections (McNally *et al.*, 2006; McCouch *et al.*, 2012), will rapidly facilitate identification and use of germplasm as plant breeders strive to adapt crops to climate change (see McNally, Chapter 10, this volume, and Chapters 12–16).

2.6 *Ex Situ* Conservation

Germplasm of seed crops and their wild relatives are safe in *ex situ* genebanks – whatever the climate changes – provided that seed samples are handled properly following collecting, and during drying and storing to maximize long-term viability.

Nevertheless, several issues are particularly relevant in terms of *ex situ* conservation under the threat of global climate change:

- The scope of germplasm collections in terms of species genepools and how representative the germplasm is.
- Management of conserved germplasm in terms of multiplication and regeneration, and the conservation of vegetatively propagated crops, core collections and genebank standards.
- Documentation and information systems that enhance and facilitate both the conservation and use of crop diversity.
- Opportunities for collecting and ensuring the long-term safety of genetic diversity that will permit adaptation to climate change.

2.6.1 What is in the collections?

In terms of existing genebank collections, it is really not possible, in general, to determine their genetic completeness, although this is possibly more straightforward for vegetatively propagated crops like potato, for example. In agricultural systems in the Andes, potatoes are maintained by farming communities as clearly identified clones. If all the known clones in a designated area have been collected, there is a high probability that a collection does indeed represent a very high proportion of known genetic variation. New variation is, however, frequently introduced into these farming systems through cycles of sexual reproduction (since new clones do appear from time-to-time) such as in the Cuyo Cuyo valley in southern Peru (Jackson *et al.*, 1980); and this is also the case for other vegetatively propagated crops.

But the situation for seed conservation is very different – for both self- and cross-pollinated species. It is virtually impossible that all genetic combinations could ever be collected. Instead, what is important, and of increasing urgency as climate change begins to threaten agriculture and crop varieties become less adapted to changing conditions, is to ensure the broadest genetic representation as possible in germplasm collections and to conserve the processes that lead to new diversity. And we can estimate genetic representativeness by analysing the geographical distribution of conserved germplasm and determining where any gaps occur.

This is an approach taken for decades but as climate change begins to impact on agriculture and on the varieties farmers can grow, it becomes imperative to enhance more strategically the coverage of genetic diversity in the bank, more so now because for many crops and wild species this diversity is increasingly threatened with loss. However, uncertainty during the 1990s over ownership of and access to genetic resources, the implementation of 'bioprospecting' legislation in several countries (such as Executive Order 147 in the Philippines), and the entering into force of the CBD on 29 December 1993 (see Moore and Hawtin, Chapter 6, this volume) did impact negatively on germplasm collecting. This situation has not been adequately resolved. The completeness of *ex situ* germplasm collections is recognized as a key component of response to climate change.

2.6.2 Management of seed collections

It is fortunate that the seeds of the most important food crops – and many others – can be safely stored at low moisture contents, and at low temperatures, typically around −18°C for long periods. These are the so-called 'orthodox' species. The increasing focus on and attention to acceptable genebank standards (see p. 24) has, however, highlighted the need for genebanks to take into consideration a whole range of factors that can, and often do, affect the long-term survival of germplasm. Insufficient attention to these aspects – during regeneration and in harvest, drying and storage conditions – unfortunately leads to seeds being placed in storage with less than optimum long-term viability. With the advent of climate change, the impact of which at different genebank locations cannot yet be estimated precisely, the opportunities for successful germplasm multiplication and regeneration may be compromised.

Let us provide one important example, namely the case of the International Rice Genebank (IRG) at the International Rice Research Institute (IRRI) in the Philippines. As an international genebank, the IRG holds germplasm samples from all over the world, mainly tropical locations but also from some temperate areas, in line with the provisions of the Treaty because rice is one of its Annex 1 species (Jackson, 1997). The vast majority of rice accessions can be successfully grown at IRRI's Los Baños research centre (located at 14.145211, 121.261465) south of Manila. It can, however, be a struggle to multiply or regenerate some accessions that are less well adapted there. That is why a number of studies – among the most detailed regarding seed production for rice germplasm – were carried out to match germplasm multiplication and regeneration with the seed production environment to produce seeds of the highest quality from different varieties (Ellis *et al.* 1993; Ellis and Jackson, 1995; Kameswara Rao and Jackson, 1996a,b,c, 1997; Ellis, 2011).

Although these studies are specific for rice they do have implications for other crops. Why? There are two issues that we need to explore further. The first concerns the changing environment(s) where a genebank multiplies or regenerates its germplasm. And the second, in terms of an international genebank like the IRG, relates to plant quarantine issues.

The extent to which climate change will impact on the seed production environment should not be underestimated because just a degree or two temperature change may negatively affect fertility. This relationship has been clearly demonstrated in rice where elevated temperatures – well within the expected impact of predicted climate change – affected pollen viability (Prasad *et al.*, 2006; see Djanaguiraman and Vara Prasad, Chapter 12, this volume) and thus seed set, possibly as a result of meiotic disruptions during pollen development. So it is important to consider and monitor carefully changes in the seed production environment.

In terms of temperate rice accessions, such as the japonica rices that are less easy to grow in Los Baños (Kameswara Rao and Jackson, 1996b; Ellis, 2011), one option often considered is to choose an environment better suited to these rices, even if it meant a location outside the Philippines. There are, however, two important constraints that are likely to apply to other international genebanks as well. First, the newly chosen locations are also going to be affected by climate change and perhaps more adversely than more tropical locations. Second, plant quarantine regulations for the import and export of germplasm may require, in any event, a local multiplication and certification of seeds, as is the case with the IRG in the Philippines. Even if an external location is chosen for certain accessions, they have to be field evaluated in the Philippines before germplasm distribution is subsequently permitted – thus negating any advantages of an external regeneration of germplasm. In any case, we must also be aware of other plant health issues that will emerge under a changing climate (see Pritchard *et al.*, Chapter 16, this volume in relation to insect pests).

Of course there are differences between inbreeding and outbreeding species.

Outbreeding is common in CWRs, even though the crops themselves may be inbreeding. It therefore leads to high levels of genetic heterogeneity within and between genebank accessions, affecting the most significant trait for seed conservation showing seed-to-seed variation, namely longevity (Hay *et al.*, 2010). Given this additional complexity it is essential to understand how a crop species or CWR responds in a given environment to ensure adequate seed production and safe long-term conservation of high-quality seeds.

Clearly these are yet more examples of thoroughly understanding the behaviour of germplasm for long-term conservation, which will become more challenging under climate change. For a multi-species genebank this may be an insurmountable challenge, with compromises necessary. The agreed genebank standards (see p. 24) allow genebank managers to understand the scope – and limitations – of the conservation measures they adopt.

2.6.3 The need for *in vitro* conservation

The long-term cold-storage of viable plant material has been acknowledged as being a challenge worth confronting for many years. Arguably there can be no better demonstration of its potential than the discovery recently that plant material can survive for thousands of years in the Siberian permafrost (Yashina *et al.*, 2012). It has proven possible to grow and reproduce plants of *Silene* from maternal fruit tissue discovered in fossil squirrel burrows in permafrost and estimated to be 30,000 years old. Two points are pertinent: one is that the tissue has been effectively stored at around –7°C for a very long time; and the other is that it required knowledge of plant tissue culture techniques to regenerate whole reproductive plants – *in vitro* conservation in a nutshell!

Plant tissue culture has actually been used for some time for conserving vegetatively propagated crops, crop species that have 'recalcitrant' seeds (seeds that will not withstand low temperature or moisture conditions), and those where it is desirable to conserve specific genotypes that would otherwise be lost when propagating by seed. Target species for *in vitro* conservation include coconut, cocoa, coffee, oil palm (with recalcitrant or partially recalcitrant seeds), and potato, cassava, yam, sweet potato, sugar cane, bananas/plantains, garlic and apples (for which vegetatively propagated and/or specific genotypes needed to be preserved). It is possible to conserve such material as entire plants in so-called field genebanks or as dormant vegetative propagules for short periods. In the case of field genebanks, cost and space are issues when needing to maintain hundreds or thousands of actively growing tree species over many years for instance. Risk of loss is relatively high, and indeed may be increasing under climate change because of more frequent and extreme weather conditions such as drought, floods and typhoons and high winds, leaving aside changes in normal annual weather patterns. These issues are the same as those for the regeneration of seed crops, yet the consequences for recalcitrant or vegetatively propagated species are more detrimental.

Consequently, slow (minimal) growth storage and cryopreservation are the two major categories of conservation involving tissue culture, where minimal growth storage can only be regarded as useful for short- to medium-term conservation, whereas cryopreservation is appropriate to store germplasm for very long periods of time. Minimal growth storage is not ideal as it only allows storage for the medium term because of the constant need to transfer the plant material to new storage medium at regular intervals. Cryopreservation, in contrast, uses temperatures as low as that of liquid nitrogen and broadly ensures conservation for hundreds of years or more (see *Silene* above). But while the general methodologies have not substantially changed over recent years, considerable research has been needed to identify the correct precise protocols necessary for *in vitro* conservation, because in many cases these protocols have to be developed specifically for individual species and in a number of cases even for specific varieties/genotypes. Much recent research and development effort has gone into developing guidelines and quality control procedures following those for biorepositories in general. They take on board, *inter alia*, the need to ensure purity and freedom from contamination, authenticity in terms of ensuring the correct identity, and stability as in trueness to type (Benson *et al.*, 2011).

A good example of *in vitro* conservation is that of the bananas and plantains of the genus *Musa* (Panis, 2009). Bioversity International maintains the world collection of *Musa* germplasm at its International Transit Centre hosted by the Katholieke Universiteit Leuven in Belgium. Because bananas and plantains are clonally propagated, and cultivars do not generally produce seeds, besides maintenance in field genebanks only an *in vitro* approach offers a suitable conservation opportunity. Over 1000 accessions are maintained in an 'active collection' under low light and reduced temperature to slow the growth of the cultures. Even so, on average it is necessary to subculture each accession once a year. This is not only labour intensive, but germplasm is prone to infection and also to genetic instability. So the collection is now being introduced into cryopreservation – storage in liquid nitrogen where each of these problems can be countered. Several protocols have been developed and have become routine, serving as a good model for the cryopreservation of many other species (Panis, 2009).

2.6.4 Core collections – searching for useful diversity

Whether agrobiodiversity is fully and effectively used for crop improvement is still an ongoing issue (PGR Secure, 2012). In 1984, Sir Otto Frankel (Frankel and Brown, 1984) questioned how easy it is to select germplasm for use when there are so many accessions being stored in genebank collections that are inadequately documented and evaluated. Frankel's view was that we needed to develop a better understanding of how and what diversity is contained within collections, particularly with respect to quantitative traits. To achieve this he proposed that small numbers of genetically representative samples taken from large germplasm collections be established for intensive study and deeper evaluation, and this gave rise to the 'core collection' concept: 'a sub-sample of a larger germplasm collection that contains, with a minimum of repetitiveness, the maximum possible genetic diversity of the species in question'. Initially there was considerable debate on several issues (Hawkes *et al.*, 2000):

- How should material for these core collections be chosen?
- How large should core collections be?
- Would core collections be a threat to the continued maintenance of the large germplasm collections themselves?
- Would potentially valuable 'rare' alleles not be captured and therefore overlooked?
- Would core collections actually be of any use?

There have been many publications over the last 10 years or so addressing the development of core collections for different crops (Gouesnard *et al.*, 2001; Chandra *et al.*, 2002; Li *et al.*, 2004; McKhann *et al.*, 2004; Jansen and van Hintum, 2007; Wang *et al.*, 2007; Le Cunff *et al.*, 2008; Odong *et al.*, 2011; Jing *et al.*, 2012). Many have focused on the optimal size of core collections, whereas others assessed the levels of genetic diversity captured using different statistical sampling procedures or the combined use of traditional and molecular data in the selection process of individual accessions. Given the discussion surrounding the development of core collections, it is not unreasonable to ask whether core collections have been put to good use – the extent to which they have actually enhanced the use of genetic resources for crop improvement. There is now a substantial number of core collections across a range of different crops, including the major crops such as rice and wheat, and

Table 2.1. Examples of existing core or mini-core collections that have been developed for a range of crops with an indication of their use.

Core details	Use	Publication
Wild barley	Aluminium tolerance	Dai *et al.*, 2011
Hordeum – USDA-ARS National Small Grains Collection	Aphid resistance	Dahleen *et al.*, 2012
Beta	Allele mining	Reeves *et al.*, 2012
Chickpea mini-core	Salt tolerance	Krishnamurthy *et al.*, 2011
Global chickpea mini-core	Nitrogen fixation capacity	Biabani *et al.*, 2011
Finger millet	Grain nutrients	Upadhyaya *et al.*, 2011
Groundnut reference mini-core	Drought tolerance	Hamidou *et al.*, 2012
Sour orange core	*Citrus tristeza* virus resistance	Bernet *et al.*, 2008
Pisum – USDA	Seed nutrients	Kwon *et al.*, 2012
US peanut mini-core	*Sclerotinia minor* resistance	Chamberlin *et al.*, 2010
Pigeonpea – mini-core collection	Waterlogging response	Krishnamurthy *et al.*, 2012
Pigeonpea – mini-core collection	*Fusarium* wilt and sterility mosaic disease	Sharma *et al.*, 2012
Oryza sativa – USDA-ARS National Small Grains Collection	Rice blast disease – Pib gene	RoyChowdhury *et al.*, 2012
International rice core	Blast resistance gene Pi-ta	Wang *et al.*, 2010
Rice core subset	Rice blast resistance – Pi-z gene	RoyChowdhury *et al.*, 2012
Rice USDA core	Sheath blight resistance	Jia *et al.*, 2011
Rice USDA mini-core	Grain yield QTL	Li *et al.*, 2004
Oryza sativa	Grain quality traits	Redestig *et al.*, 2011
Sorghum mini-core	Grain mould and downy mildew resistance	Sharma *et al.*, 2012
Winter wheat USDA core subset	Bacterial leaf streak resistance	Adhikari *et al.*, 2011
Winter wheat core	Grain yield under drought	Dodig *et al.*, 2012
Chinese wheat micro-core	Grain dormancy	Chang *et al.*, 2010

they are actually being put to considerable use (Table 2.1).

The use broadly falls into two categories: first, as a convenient source of easily accessible diverse material for sourcing markers for use in marker assisted selection (Kwon *et al.*, 2012), allele mining (Zhao *et al.*, 2010; Reeves *et al.*, 2012) or SNP discovery (Le Cunff *et al.*, 2008); and second, for more direct screening and evaluation for useful traits, genes or quantitative trait loci (QTLs). Such use is now being extended to identify material for metabolomic studies of quality traits (Redestig *et al.*, 2011) and nitrogen-fixing potential (Biabani *et al.*, 2011). Perhaps the most frequent use, however, is to identify sources of resistance to biotic stresses (Jia *et al.*, 2011; Dahleen *et al.*, 2012; RoyChowdhury *et al.*, 2012; Sharma *et al.*, 2012) and abiotic stresses (Dai *et al.*, 2011; Dodig *et al.*, 2012; Krishnamurthy *et al.*, 2012; Hamidou *et al.*, 2012).

2.6.5 Changes to genebank standards

In 1994, the FAO and the International Plant Genetic Resources Institute (IPGRI – now Bioversity International) published a set of genebank standards, based on the scientific findings from genebanks and scientists conserving threatened crop genetic resources over the previous three decades or so. They were intended to provide guidance and targets, such as the seed drying procedures referred to earlier, to the global network of genebanks conserving PGRFA in base and active collections as part of the Global System, as well as for the emerging regional and crop-specific genetic resources networks (Engels and Visser, 2003).

The standards are underpinned by a set of broad underlying principles that provide the overarching framework for effective and efficient management of genebanks. The key principles that are at the heart of genebank operations are:

- Preservation of the genetic identity of accessions.
- Maintenance of their viability and genetic integrity.
- Germplasm health.
- Physical security that ensures their availability.
- Easy access to associated information to facilitate the use of the material.

This last principle emphasizes the need for curators to manage genebanks while adhering to legal and regulatory frameworks at national and international levels, as they relate to access, availability and distribution of materials, and plant and seed health. The genebank standards provide specificity that aids adherence to these underlying principles.

The main reason to develop such standards had been 'to minimize the loss of genetic integrity in seed accessions during storage and regeneration' (FAO and IPGRI, 1994). They should be viewed as targets for developing an efficient, effective, rational and transparent global system of *ex situ* conservation that provides optimal maintenance of seed, plant and tissue viability, and genetic integrity in genebanks, thereby ensuring access to, and use of, high-quality seeds, tissues or other propagules of conserved plant genetic resources. The standards define the level of performance of a routine genebank operation below which there is a high risk of losing genetic integrity. There are compelling scientific reasons why a genebank should aim to meet the proposed higher or 'preferred standards'. Where resources are limited, however, the genebank standards provide for an 'acceptable' level – a pragmatic compromise – such that even under less-than-ideal operating conditions a collection would not be placed in jeopardy.

The current Genebank Standards have been recently revised at the request of the Commission in light of changes in the global policy landscape and advances in science and technology (FAO, 2013). On the scientific front, advances in seed and tissue storage technology, biotechnology, and information and communication technology have added new dimensions to plant germplasm conservation. The revised Genebank Standards cover both orthodox and non-orthodox seed species as well as clonally propagated plants using *in vitro* culture, cryopreservation and field genebanks (Engels and Visser, 2003).

2.6.6 Crop descriptors and documentation

In the early 1980s IBPGR (now Bioversity International) started to develop descriptors for the management and evaluation of plant genetic resources and thus to assist countries and scientists to improve their capacity to describe, store, manage and share information about their plant resources, whether stored in genebanks or growing in their natural environments.

Over the past 30 years crop descriptor lists for 100 crop species were developed and published (Bioversity International, 2012). In addition to crop descriptors there are several others:

- Key access and utilization descriptors for crop genetic resources.
- FAO/Bioversity International multi-crop passport descriptors.
- Descriptors for genetic marker technologies.
- Descriptors for farmers' knowledge of plants.

The development of standards for documentation is a crucial issue because it enables information sharing that in turn facilitates the international exchange and use of plant resources, uniformity in data description, collection and record-keeping (see Ougham and Thomas, Chapter 9, this volume).

New genebank information management systems have been developed and are widely being used, such as GRIN-Global (2012, described on p. 28) and GENESYS (2012). The latter is a gateway to plant genetic resources with the aim of providing users of germplasm with a single access point to accession-specific characterization and evaluation information.

2.6.7 Effect of policies on collecting, access to and use of germplasm

Whereas the Treaty has been able to develop a unique and potentially powerful political and technical mechanism for the access to plant genetic resources and the equitable sharing of the benefits obtained from their sustainable use, it also has demonstrated some real constraints. The concept of establishing a multilateral system for access (and benefit-sharing) to PGRFA is arguably the only adequate solution for genetic resources that have been in 'the public domain' since their creation and domestication as these processes only function if the required genetic diversity is readily available (see Moore and Hawtin, Chapter 6, this volume for more specific aspects). With the political and legal constraints created through intellectual property rights, through technologies and recently also through political frameworks this 'ready access' has been jeopardized. The Treaty has so far not been able to adequately respond to these developments. In fact, it started creating its 'own' drawbacks to ready access by establishing a limited list of crops/species that are part of the multilateral system (i.e. Annex I is a list of crops predominantly of global economic importance and only focuses on genetic resources in the public domain); defining agricultural uses has proved an issue (in a non-comprehensive and sometimes rather confusing manner, such as for direct use or for biofuels, among others). Also, the interface between *in situ* and *ex situ* conservation efforts is not well defined.

Even within Europe's regional collaborative programme (ECPGR) for example, as well as at the global level, not all countries have ratified the CBD and/or the Treaty, and this is yet another weakness in making adequate progress with activities that build on, and indeed require, such membership for them to operate comprehensively (see also Chapter 6, this volume). In the case of the ECPGR and A European Genebank Integrated System (AEGIS, see p. 28) some pragmatic decisions have been made by accepting declarations of intention to ratify or to accept the legal conditions as set out in these instruments. In the case of access to plant genetic resources it has been agreed within the AEGIS context that non-Annex I crops/species will be exchanged with a 'footnoted' Standard Material Transfer Agreement (SMTA), stating that such resources will be treated in the same way as Annex I crops/species. Although the Treaty has been reported as functioning well in this respect, the CBD related instruments still cause uncertainty, especially at the national and local levels as well as within the scientific community (Ruiz and Vernooy, 2012).

One weakness of agreements like the CBD and the Treaty is that they need national legislation and regulatory frameworks to become effective. Whereas the execution of laws and regulations might still work, their implementation and even more their enforcement are very weak or even completely lacking in many countries. Within a 'United Nations' system there are hardly ever any 'sanctions' or even reprimands given to countries that do not follow the agreed principles and actions whereas they continue to have full access to the genetic resources placed in the Multilateral System. This therefore contributes to the non-functionality of the system. The situation is certainly not conducive for problem-free collecting of genetic resources, impacts negatively on access to the resources and thus often impacts negatively on the eventual use of the conserved resources.

2.7 Establishing Rational Conservation Approaches

International collaboration is essential to realize the full potential of germplasm collections, and even germplasm *in situ*. Here we highlight a few initiatives aimed at the long-term conservation and use of crop germplasm including CWRs.

2.7.1 The Global Crop Diversity Trust and the Svalbard Global Seed Vault

The Trust is an initiative of the CGIAR (formerly the Consultative Group on International Agricultural Research) centres and FAO that began in the late 1990s to place the long-term genetic conservation of the major food crops on a guaranteed financial and operational footing, for posterity. Having started as an entity hosted by Bioversity International in Rome, the Trust was formally established in October 2004 as an independent international organization operating from offices in FAO, but moved to Bonn, Germany in January 2013.

The policy framework for the Trust is provided by the principles and strategies of the Treaty as well as by the Global Plan of Action. The Trust's mission is to ensure the conservation and availability of crop diversity for food security worldwide because crop diversity is disappearing. The Trust is possibly the only dedicated worldwide funding organization for genetic conservation, and is operating in a unique political window of opportunity following the entering-into-force of the Treaty; it has partnered with all the important organizations in this field. The Trust offers a unique opportunity to put in place a rational and cost-effective system for the conservation of genetic resources. Sustainable funding has been guaranteed through the establishment of an endowment, the interest from which is sufficient to support effective conservation, and, crucially, the ready availability to those who wish to use it of the biological basis of all agriculture. It is the reliability of funding that is so fundamental to conserving germplasm because even short-term gaps in funding could lead to cutbacks in basic maintenance and the loss of unique genetic material. Crucially, with no secure or guaranteed funding, many of the world's genebanks do not know accurately what they are supposed to conserve, or even whether the germplasm is alive or dead. The Trust's response is to support a range of activities that will ensure these collections are better understood, more useful, more easily available, more safely conserved and more securely funded.

The Trust aims, ultimately, to ensure the conservation of the diversity within all crops of importance. For the time being, however, it focuses on a number of priority crops that are particularly important for food security and that coincide with the 23 or so major food crops included in Annex I of the Treaty, and some of their wild relatives. The critical importance of CWRs as a vital source of untapped genetic diversity useful for adaptation to climate change, but now at risk of disappearing just as it is most needed, has spurred the Trust to assign a high priority to their collection, evaluation and long-term conservation, as well as supporting pre-breeding.

The initiative undertaken by the Trust to generate sustainable funding for the long-term conservation of agriculturally important plant genetic resources is strategic and important. There is inevitably a strong *ex situ* focus. Furthermore, it can be argued that the strong (political) linkage with the Treaty is unnecessarily limiting the scope of targeted genetic resources. One consequence is the prioritization of genetic resources collections that are usually large in size and, in general, well-managed and more financially solvent, neglecting many genebanks that are in most need of support.

The Trust sees the Svalbard Global Seed Vault as an essential element of a rational and secure global system for the conservation of crop diversity. The Trust is therefore committed to supporting the ongoing operational costs of the Vault, and is assisting developing countries with preparing, packaging and transporting samples of unique accessions from their genebanks to the Arctic. The Vault currently holds more than 700,000 samples, originating from almost every country in the world, and the strategically important germplasm collections managed by the CGIAR centres. It should be pointed out, however, that the storage of safety duplicates in the Global Seed Vault is not meant to replace the agreed safety duplication of all genetic resources accessions in another genebank in another country and, whenever possible, on another continent, according to the revised FAO Genebank Standards (FAO, 2013). The Global Seed Vault is an added safety level. Unfortunately, there are no data readily available to show how much of the germplasm in Svalbard has already been safety-duplicated in another genebank.

2.7.2 Genetic conservation initiatives in Europe

The ECPGR established AEGIS (2012) across 43 countries. According to the ECPGR Strategic Framework Policy Guide (ECPGR, 2009) *ex situ* conservation will be carried out according to common, agreed quality standards, wherever the germplasm is physically located, and to facilitate close linkages with *in situ* conservation, as well as the use of, and research into, conserved germplasm. AEGIS focuses primarily on the *ex situ* conservation and use of genetically unique and important accessions of PGRFA, medicinal and ornamental species, and their wild relatives of European origin. It also includes other important species introduced and maintained in genebanks of the AEGIS member countries. AEGIS has been developed within the existing legal framework of the Treaty.

A number of important benefits are expected from AEGIS:

- Improved security of germplasm through long-term commitment and systematic safety duplication of the European accessions.
- Facilitated access to and availability of high-quality germplasm and related information.
- Improved and widely agreed minimum quality standards of conserved material.
- Cost-efficient conservation activities.
- Reduced duplication of germplasm material within and across collections.
- Improved sharing of knowledge and information.
- Shared responsibilities among European countries and their institutions.
- A direct contribution to the implementation of the Treaty at the national as well as at the regional level.
- An established European reference system for the effective and efficient conservation and the facilitated use in Europe.

EURISCO (2012) is a web-based catalogue that provides information about *ex situ* plant collections maintained in Europe, sponsored by the ECPGR. It is a one-stop shop window using international standards for information on these collections that enables users to search and access information on food crops, forages and wild-and-weedy species, including cultivars, landraces, farmers' varieties, breeding lines, genetic stocks and research material. The Catalogue currently contains passport data on over 1 million samples representing the genetic diversity of more than 5500 genera and 36,300 species from 43 countries (in May 2012). These data on crop diversity represent more than half of the *ex situ* accessions maintained in Europe and roughly 16% of total worldwide holdings. Knowledge of the scope and nature of these collections is essential to promote their use for adaptation to climate change.

2.7.3 GRIN-Global

The GRIN-Global project (a joint effort with the Global Crop Diversity Trust, Bioversity International and the Agricultural Research Service (ARS) of the US Department of Agriculture) aims to provide the world's crop genebanks with a powerful, flexible, easy-to-use global plant genetic resource information management system, based on the existing ARS Germplasm Resources Information Network (GRIN, 2012). This will constitute the keystone for an efficient and effective global network of genebanks to permanently safeguard plant genetic resources vital to global food security, and to encourage the use of these resources by researchers, breeders and farmer–producers. By improving the capability of genebanks to provide data to a global accession-level information system (i.e. GENESYS), it will be possible to assess more accurately the 'State of the World' for plant genetic resources and to identify the priority global needs for plant genetic resource conservation under the threat of climate change.

2.8 *In Situ* Conservation

In situ conservation has been defined as 'the conservation of ecosystems and natural habitats and the maintenance and recovery of viable populations of species in their natural surroundings and, in the case of domesticates or cultivated species, in the surroundings where they have developed their distinctive properties' (United Nations, 1992). For wild species related to crops, conservation will be enabled through genetic reserves or protected areas, whereas traditional varieties or landraces of crops will be maintained 'on-farm' (Maxted *et al.*, 2012; Maxted *et al.*, Chapter 7, and Bellon and Van Etten, Chapter 8, this volume).

In situ conservation within protected areas, specifically genetic reserves of CWRs, offers a complementary approach to *ex situ* genebank conservation. The potential risk of genetic diversity loss through climate change is, however, an important issue that may well need to be assessed reserve by reserve. There is already available sound knowledge of *ex situ* conservation techniques and protocols; substantial progress has also been made in drafting the necessary protocols for *in situ* CWR conservation (Iriondo *et al.*, 2008), as well as in identifying high priority species and sites for the conservation of wild relatives of globally important food crops. The bigger issue is in terms of how much progress has been made in actually formally establishing *in situ* reserves, with only few examples to quote such as the reserves established in Madeira, Mexico, Turkey, Vietnam and Peru.

For landrace conservation on-farm (see Bellon and Van Etten, Chapter 8, this volume), political, economic and sociological factors are recognized as being important and were recently built in to a European 'Strategic Approach' (Vetelainen *et al.*, 2012) to their conservation. However, the inadequacy of current landrace conservation was recognized along with the existence of an information gap in terms of their diversity. Nevertheless, examples of on-farm conservation exist in several European countries, including forage species in Norway, maize and other crops in Portugal, and several other crop species in Italy.

2.8.1 Adaptive evolution under rapid climate change

Climate change must inevitably dominate discussion regarding the value and effectiveness of different approaches to conservation of plant genetic resources. If plant species or their component populations, particularly CWRs, are in any danger of becoming extinct then *ex situ* conservation is imperative, if indeed germplasm can be stored in a genebank. On the other hand, if, as appears to be the case, plant species and populations can genetically adapt and evolve rapidly enough to survive rapid climate change, then there is an equal imperative to maintain populations in their natural environments to allow them to do so.

One of the challenges for *in situ* conservation in its two principal forms has been to determine the extent to which climate change is likely to either invalidate or support the use of such conservation techniques, and the answers may be different for each case. It had been considered that one of the benefits of *in situ* conservation would be to allow for 'evolution' to continue in the form of local adaptation either of natural populations or local landraces. In the face of rapid climate change serious questions have been raised as to whether such adaptive evolution could take place over a short time span and whether *in situ* reserves of genetic resources could survive for any length of time, making *in situ* conservation inappropriate. There is now increasing evidence that in certain situations rapid genetic adaptation can and does occur, hence proving *in situ* conservation to be important (see section 2.8.2).

2.8.2 Genetic reserves, adaptation and climate change

Natural populations in genetic reserves can respond to climate change in four different ways by demonstrating phenotypic plasticity, genetic adaptation/evolution, migration/gene flow or extinction (Kramer and Havens, 2009; Mercer and Perales, 2010). If extinction is the outcome then clearly *in situ* conservation of populations in genetic reserves will not be appropriate. The likelihood may be uncertain, particularly given the debate over the extent to which genetics plays a part in the extinction process. However, leaving aside cause and effect, from various meta-analyses it does seem to be the case that significantly lower genetic variation can be found among rare plant species compared with common species; estimates of genetic diversity using neutral markers may indeed be valuable in determining the likelihood of extinction (Kramer and Havens, 2009). One other finding is that the rapidity of the current climate change may not be exceptional; climate change during the late Pleistocene was even more rapid, but seemingly did not lead to a substantial level of species extinctions (Hof *et al.*, 2011).

Phenotypic plasticity means that plants may change their phenotype in response to changes in the environment. This plasticity can be adaptive or non-adaptive and it is the adaptive form that is of interest when considering the likely survival of species under climate change. We know that plastic changes in seed longevity, phenology, leaf lifespan and the temperature responses of metabolic processes do occur in response to climate change factors (Nicotra *et al.*, 2010). It may well be that phenotypic plasticity can provide a short-term buffer against the negative effects of climate change, in which case a better understanding of phenotypic plasticity in terms of its extent in different species and how much evolutionary fitness might be affected during the plastic response is called for (Chevin *et al.*, 2010). Evidently, we do know that adaptive phenotypic plasticity evolves itself in heterogeneous environments and this should lead us to better design of genetic reserves for conservation. We are also building up an understanding of the molecular basis of plasticity and which genes are involved (Nicotra *et al.*, 2010). But plasticity will have its limits in most species, and to what extent it will act against genetic adaptation in populations assuming that there is scope for such genetic change to occur is also unknown.

To what extent can genetic evolutionary adaptation occur over the short timescales predicted for climate change? There is now increasing evidence that it can and does occur but this may be species and life history dependent because the potential to adapt may be related to lifespan, with short-lived annuals being most likely to be able to respond (Jump and Peñuelas, 2005). The capacity to adapt may also vary among populations of a species, making it difficult to consider vulnerability to climate change at a species distribution level (Jay *et al.*, 2012). But, nevertheless, rapid evolution in response to climate change may be widespread in natural populations, importantly being underpinned by existing genetic variation (Jump *et al.*, 2008). As an example, adaptive evolution for drought escape is demonstrated to occur just within a few generations of an annual species (Franks *et al.*, 2007). The fact that the molecular basis for local adaptation and candidate loci controlling fitness can be identified among *Arabidopsis* ecotypes is starting to point the way to how populations are able to respond to rapid changes in their environmental conditions and could inform *in situ* conservation planning in the future (Fournier-Level *et al.*, 2011).

A prediction of the effect of climate change is the movement of the distributions of plant species to higher elevations or latitudes and there is evidence that this is occurring (Jump and Peñuelas, 2005). Some plant species, however, may not be able to migrate or shift their ranges fast enough to keep up with climate change. Gene flow, however, may have positive effects on adaptation. Although gene flow will act to smooth out allele frequencies spatially and therefore act against local genetic adaptation, there may equally be positive effects by way of the spread of adaptive genes on which selection can act. This may apply particularly to the long-distance dispersal of genes that is known to occur in tree species (Kremer *et al.*, 2012) and could support migration.

2.8.3 Landraces, adaptation and climate change

As with CWRs, under climate change landraces will be confronted with the same four response challenges or opportunities of plasticity, genetic adaptation/evolution, migration/gene flow and extinction. *Ex situ* seed conservation was promoted in the 1960s as an urgent measure to prevent the 'extinction' or loss of landraces in the face of the Green Revolution and their replacement by modern varieties (Hawkes *et al.*, 2000). The same need for *ex situ* conservation exists now under the threat of climate change, at least for where landraces are still grown.

In 1968, Erna Bennett made the proposition '... the difficulty is to find the borderline between "adaptive change" and "genetic erosion"' (Bennett, 1968). She saw 'no advantage in the steady state – the purpose of conservation is not to capture the present moment in evolutionary time, in which there is no special virtue, but to conserve material so that it will continue to evolve'. This raises the major issue affecting landrace conservation, the dilemma being how to determine where the line is drawn between adaptive change and genetic erosion.

The principal form of genetic erosion in landraces will be extinction – the complete loss of individual landraces. So what are the alternatives to this (apart from *ex situ* conservation)? Plasticity may, in some cases, allow landraces to maintain their genetic fitness as the environment changes, so plasticity may act to buffer landraces against climate change and against genetic change, allowing local farmers to maintain their crop yields and feed their families (Mercer and Perales, 2010). But if, in contrast to natural selection, local farmer selection reduces genetic fitness, then this could lead to genetic erosion. Farmer selection of sufficiently fit and well-yielding genotypes has, however, taken place over thousands of years quite successfully so the only question would be whether this can continue given the rapidity of current climate change.

Although Bellon and Van Etten address the issues of on-farm conservation in detail (see Chapter 8, this volume), some specific comments are relevant here. Landraces maintained on-farm will be confronted with selective pressures resulting from climate change, but also with farmer-imposed selection, and the two may act in opposition. However, neither will be effective in maintaining fitness and landrace survival if there is insufficient genetic variation within the landraces. Mercer and Perales (2010) refer to 'antagonistic' correlations that could result in landraces with sub-optimal phenotypes and reduced fitness. Notwithstanding these issues, Bennett's view could still mean that landraces maintained *in situ* will fare better under climate change than modern varieties because of their tendency to hold greater levels of genetic variation and this may be the key to the success of future on-farm conservation.

Such genetic variation could be maintained by migration and gene flow but whether these processes continue would be down to farmers' practices (Bellon and Brush, 1994). So if farmers maintain more than one landrace, or different landraces are grown by neighbouring farmers, there will always be the prospect of genetic variation being enhanced owing to gene flow, but dependent of course upon the breeding system of the crop species. The equivalent of migration in the on-farm context could be viewed as seed exchange within farmers' markets, which again could help maintain genetic variation within landraces (Louette and Smale, 2000; Barnaud *et al.*, 2007).

As with natural populations of CWRs, it is not clear as to the extent to which conservation objectives can be achieved on-farm; success is likely to vary regionally and by crop species. This is specifically demonstrated by way of landraces of maize in Mexico, which can suffer contrasting levels of genetic erosion depending upon altitude (Mercer and Perales, 2010). This once again points to the need for complementary conservation strategies, and despite genebanks maintaining the 'steady state', they are actually needed to capture the genetic variation that exists at any one point in time as a safeguard against unpredictable levels of genetic erosion resulting from rapid climate change *in situ* and to make this variation available to users in a targeted manner. The International Rice Genebank in the Philippines maintains many thousands of landraces of rice, an unknown number of which still thrive and continue to evolve on subsistence farms across south and South-east Asia. We know that the genetic variation captured in the genebank over a 30-year period has not significantly diminished, reflecting the situation *in situ*, but we cannot predict for how long this will prove to be the case (Ford-Lloyd *et al.*, 2008) so seed conservation must be supported.

2.9 New Technologies to Enhance the Use of Conserved Germplasm

There is now good evidence that core collections and their diminutives (mini- and micro-cores) can be invaluable for enhancing the use of major germplasm collections. Much of the use of these core collections has been to provide realistically sized sets of germplasm, making molecular genetic dissection feasible. Advances in molecular biology have opened up many new possibilities for identifying genes and genotypes of interest for crop improvement (see McNally, Chapter 10, this volume), although success has been limited for poorly understood, complex traits such as drought tolerance (see Ceccarelli, Chapter 13, this volume). The application of new molecular technologies has, however, been less towards identifying patterns of genetic diversity *in situ* and planning for *in situ* conservation, where both traditional population genetic and landscape genetic approaches could benefit from high-throughput molecular genetic data if such approaches were not held back by the need for massive computational power.

Increasing capability with reduced time and cost of whole-genome sequencing, together with massive capability for identifying and employing single nucleotide polymorphisms (SNPs) is enabling technology for a range of more or less traditional and novel applications. These can include identifying meaningful SNP variation in genes that are expressed (gene mining) (Edwards and Batley, 2010). SNPs can be used for producing high density genetic maps, for fine mapping of traits and for identifying marker-trait associations by way of 'association genetics' (Waugh *et al.*, 2009). The Beijing Genomics Institute is currently sequencing 10,000 rice genotypes aiming for high-throughput screening of phenotypic traits, especially those that are relevant to improve rice production or quality. At the same time, all biological samples being sequenced are kept in a bio-repository system that is part of the larger China GeneBank project. Phenotypic characters, together with genomics sequences, are deposited in GeneBank's database in a systematic way so that the combined dataset will further help to improve the overall understanding of the species and its valuable traits (Xin Zhou, 2013, personal communication).

Second-generation sequencing technologies may soon lead to third-generation single-molecule sequencing, providing even greater speed of analysis at lower cost. This will enhance gene discovery for locating genes underpinning traits of interest and greater capacity to following these genes through generations of breeding programmes will speed up the plant-breeding process (Stamp and Visser, 2012). Inevitably, new problems will then be evident in terms of the masses of data generated. The issue will not just be the ease with which data analysis can be undertaken, but also how to store such quantities of data. Perhaps one solution to the storage problem will be if actual sequencing becomes so inexpensive it may be easier to re-sequence rather than to store data! As well as for data mining, this genetic information could be used to plan both *in situ* and *ex situ* conservation. Simple population genetic analysis done on a scale that is an order of magnitude greater than that which we are used to could be a very powerful tool for use in *in situ* conservation.

There are increasingly good examples of the use of these advancing technologies. Although focused on *Arabidopsis*, genome-wide association studies (GWAS) have been used to identify candidate loci for lifetime fitness and local adaptation (Fournier-Level *et al.*, 2011). Climatic signals of local adaptation were particularly evident, suggesting a way forward for a greater understanding of how plants will or will not be able to respond to changes in climate.

CWRs, even those that are currently poorly described, may be put to greater use with a better understanding of their involvement in domestication processes. This has been achieved by high-throughput sequencing of whole genomes (Malory *et al.*, 2011). 'Massively parallel sequencing' has provided the opportunity to characterize homologues of known genes of rice in one of its distant relatives, *Microlaena stipoides*. Of the 18 genes where homologues were identified, several were related to heading date, which could have significance in terms of drought avoidance.

Studies of whole-genome expression, either by what are now the more traditional methods using Affymetrix chips or other forms of arrays (Agilent), or by the emerging RNA-Seq approach involving next-generation sequencing, have yielded valuable results already. The protein kinase gene *Pstol1* (phosphorus-starvation tolerance 1), discovered in the rice landrace Kasalath, conferring tolerance to phosphorus deficiency, has in part been identified using studies of whole-genome expression and is a clear example of the value of conserving and exploring traditional germplasm (Gamuyao *et al.*, 2012). RNA-Seq can also be used to identify the molecular mechanisms for adaptation to both biotic and abiotic stresses, and its advantage is that it can be applied to any species, not just those whose genomes have currently been sequenced. Studies of the adaptation of eucalyptus to water deficit conditions has painted a complex picture, but one that will be valuable in combating climate change (Villar *et al.*, 2011) in the future.

2.10 Conclusions

An enormous amount of crop germplasm and wild relatives is conserved in genebanks around the world, as well as *in situ*. Although not all this conserved germplasm is equally characterized and evaluated, there is sufficient documented germplasm available to plant breeders that can be used to tackle the growing threat of global climate change. Over the past decade there has been a better-defined appreciation of the potential scope and scale of climate change, as discussed in several chapters later in this volume. Not only are we better equipped to define the parameters for adaptation to climate change, we have a suite of molecular tools that will help to find and/or dissect the traits in conserved germplasm. And the rules for access to and use of germplasm are clear, even though not all these issues have been completely solved by any stretch of the imagination. Nevertheless, humanity is in a good position to exploit germplasm more effectively than ever before and ensure that food security for all under the threat of climate change becomes a reality not just wishful thinking.

References

Adhikari, T.B., Hansen, J.M., Gurung, S. and Bonman, J.M. (2011) Identification of new sources of resistance in winter wheat to multiple strains of *Xanthomonas translucens* pv. *undulosa*. *Plant Disease* 95, 582–588.

AEGIS (2012) A European Genebank Integrated System. Available at: www.aegis.cgiar.org/ (accessed 1 December 2012).

Barnaud, A., Deu, M., Garine, E., McKey, D. and Joly, H.I. (2007) Local genetic diversity of sorghum in a village in northern Cameroon: structure and dynamics of landraces. *Theoretical and Applied Genetics* 114, 237–248.

Bellon, M.R. and Brush, S.B. (1994) Keepers of maize in Chiapas, Mexico. *Economic Botany* 48, 196–209.

Bennett, E. (1968) Record of the FAO/IBP technical conference on the exploration, utilization, and conservation of plant genetic resources, 1967. FAO, Rome, Italy.

Benson, E.E., Harding, K., Debouck, D., Dumet, D., Escobar, R. *et al.* (2011) Refinement and standardization of storage procedures for clonal crops – Global Public Goods Phase 2: Part III. Multicrop guidelines for developing *in vitro* conservation best practices for clonal crops. System-wide Genetic Resources Programme, Rome, Italy.

Bernet, G.P., Gorris, M.T., Carbonell, E.A., Cambra, M. and Asins, M.J. (2008) Citrus tristeza virus resistance in a core collection of sour orange based on a diversity study of three germplasm collections using QTL-linked markers. *Plant Breeding* 127, 398–406.

Biabani, A., Carpenter-Boggs, L., Coyne, C.J., Taylor, L., Smith, J.L. *et al.* (2011) Nitrogen fixation potential in global chickpea mini-core collection. *Biology and Fertility of Soils* 47, 679–685.

Bioversity International (2012) Descriptor lists and derived standards. Available at: www.bioversityinternational.org/?id=3737 (accessed 30 November 2012).

Callow, J.A., Ford-Lloyd, B.V. and Newbury, H.J. (eds) (1997) *Biotechnology and Plant Genetic Resources.* Biotechnology in Agriculture Series, No. 19. CAB International, Wallingford, UK.

Chamberlin, K.D.C., Melouk, H.A. and Payton, M.E. (2010) Evaluation of the US peanut mini core collection using a molecular marker for resistance to *Sclerotinia minor* Jagger. *Euphytica* 172, 109–115.

Chandra, S., Huaman, Z., Krishna, S.H. and Ortiz, R. (2002) Optimal sampling strategy and core collection size of Andean tetraploid potato based on isozyme data – a simulation study. *Theoretical and Applied Genetics* 104, 1325–1334.

Chang, C., Zhang, H.P., Feng, J.M., Yin, B., Si, H.Q. *et al.* (2010) Identifying alleles of Viviparous-1B associated with pre-harvest sprouting in micro-core collections of Chinese wheat germplasm. *Molecular Breeding* 25, 481–490.

Chevin, L., Lande, R. and Mace, G.M. (2010) Adaptation, plasticity, and extinction in a changing environment: towards a predictive theory. *PLoS Biology* 8, e1000357.

Dahleen, L.S., Bregitzer, P., Mornhinweg, D. and Jackson, E.W. (2012) Association mapping of russian wheat aphid resistance in barley as a method to identify diversity in the national small grains collection. *Crop Science* 52, 1651–1662.

Dai, H., Shan, W., Zhao, J., Zhang, G., Li, C. *et al.* (2011) Difference in response to aluminum stress among Tibetan wild barley genotypes. *Journal of Plant Nutrition and Soil Science* 174, 952–960.

Devos, K.M. and Gale, M.D. (1997) Comparative genetics in the grasses. *Plant Molecular Biology* 35, 3–15.

Dodig, D., Zoric, M., Kobiljski, B., Savic, J., Kandic, V. *et al.* (2012) Genetic and association mapping study of wheat agronomic traits under contrasting water regimes. *International Journal of Molecular Sciences* 13, 6167–6188.

ECPGR (2009) *A Strategic Framework for the Implementation of a European Genebank Integrated System (AEGIS). A Policy Guide.* European Cooperative Programme for Plant Genetic Resources, Bioversity International, Rome, Italy.

Edwards, D. and Batley, J. (2010) Plant genome sequencing: applications for crop improvement. *Plant Biotechnology Journal* 8, 2–9.

Ellis, R.H. (2011) Rice seed quality development and temperature during late development and maturation. *Seed Science Research* 21, 95–101.

Ellis, R.H. and Jackson, M.T. (1995) Accession regeneration in genebanks: seed production environment and the potential longevity of seed accessions. *Plant Genetic Resources Newsletter* 102, 26–28.

Ellis, R.H., Hong, T.D. and Jackson, M.T. (1993) Seed production environment, time of harvest, and the potential longevity of seeds of three cultivars of rice (*Oryza sativa* L.). *Annals of Botany* 72, 583–590.

Engels, J.M.M. and Visser L. (eds) (2003) *A Guide to Effective Management of Germplasm Collections.* IPGRI Handbooks for Genebanks No. 6. IPGRI, Rome, Italy.

Engels, J.M.M., Ramanatha Rao, V., Brown, A.H.D. and Jackson, M.T. (eds) (2002) *Managing Plant Genetic Diversity.* CAB International, Wallingford, UK.

EURISCO (2012) Finding seeds for the future. Available at: www.ecpgr.cgiar.org/germplasm_databases.html (accessed 1 December 2012).

FAO (1996a) *Report on the State of the World's Plant Genetic Resources for Food and Agriculture.* Food and Agriculture Organization of the United Nations, Rome, Italy.

FAO (1996b) *Global Plan of Action for the Conservation and Sustainable Utilization of Plant Genetic Resources for Food and Agriculture.* Food and Agriculture Organization of the United Nations, Rome, Italy.

FAO (2010) *The Second Report on the State of the World's Plant Genetic Resources for Food and*

Agriculture. Food and Agriculture Organization of the United Nations, Rome, Italy.

FAO (2012) *Second Global Plan of Action for Plant Genetic Resources for Food and Agriculture*. Food and Agriculture Organization of the United Nations, Rome, Italy.

FAO (2013) Draft revised genebank standards for the conservation of orthodox seeds. Available at: www.fao.org/fileadmin/templates/agphome/documents/PGR/genebank/Genebankstandards-Nov2011english_FINAL.pdf (accessed 25 January 2013).

FAO and IPGRI (1994) *Genebank Standards*. Food and Agriculture Organization of the United Nations and International Plant Genetic Resources Institute, Rome, Italy.

Ford-Lloyd, B.V., Brar, D., Khush, G.S., Jackson, M.T. and Virk, P.S. (2008) Genetic erosion over time of rice landrace agrobiodiversity. *Plant Genetic Resources: Characterization and Utilization* 7, 163–168.

Fournier-Level, A., Korte, A., Cooper, M.D., Nordborg, M., Schmitt, J. *et al.* (2011) A map of local adaptation in *Arabidopsis thaliana*. *Science* 334, 86–89.

Frankel, O.H. and Brown, A.H.D. (1984). Plant genetic resources today: a critical appraisal. In: Holden, H.W. and Williams, J.T. (eds) *Crop Genetic Resources: Conservation and Evaluation*. George Allen and Unwin, London, pp. 249–257.

Franks, S.J., Sim, S. and Weis, A.E. (2007) Rapid evolution of flowering time by an annual plant in response to a climate fluctuation. *Proceedings of the National Academy of Sciences USA* 104, 1278–1282.

Fuccillo, D., Sears, L. and Stapleton, P. (eds) (1997) *Biodiversity in Trust – Conservation and Use of Plant Genetic Resources in CGIAR Centres*. Cambridge University Press, Cambridge.

Gamuyao, R., Chin, J.H., Pariasca-Tanaka, J., Pesaresi, P., Catausan, S. *et al.* (2012) The protein kinase Pstol1 from traditional rice confers tolerance of phosphorus deficiency. *Nature* 488, 535–539.

GENESYS (2012) Gateway to genetic resources. Available at: www.genesys-pgr.org/ (accessed 30 November 2012).

Gouesnard, B., Bataillon, T.M., Decoux, G., Rozale, C., Schoen, D.J. *et al.* (2001) MSTRAT: an algorithm for building germ plasm core collections by maximizing allelic or phenotypic richness. *Journal of Heredity* 92, 93–94.

GRIN (2012) Available at: www.ars-grin.gov/ (accessed 1 December 2012).

GRIN-Global (2012) Available at: www.grin-global.org (accessed 30 November 2012).

Hamidou, F., Ratnakumar, P., Halilou, O., Mponda, O., Kapewa, T. *et al.* (2012) Selection of intermittent drought tolerant lines across years and locations in the reference collection of groundnut (*Arachis hypogaea* L.). *Field Crops Research* 126, 189–199.

Hawkes, J.G., Maxted, N. and Ford-Lloyd, B.V. (eds) (2000) *The* Ex Situ *Conservation of Plant Genetic Resources*. Kluwer, Dordrecht, The Netherlands.

Hay, F.R., Smith, R.D., Ellis, R.H. and Butler, L.H. (2010) Developmental changes in the germinability, desiccation tolerance, hard-seededness, and longevity of individual seeds of *Trifolium ambiguum*. *Annals of Botany* 105, 1035–1052.

Hodgkin, T., Brown, A.H.D., van Hintum, Th.J.L. and Morales, E.A.V. (eds) (1995) *Core Collections of Plant Genetic Resources*. John Wiley & Sons, Chichester, West Sussex, UK.

Hof, C., Levinsky, I., Araújo, M.B. and Rahbek, C. (2011) Rethinking species' ability to cope with rapid climate change. *Global Change Biology* 17, 2987–2990.

Iriondo, J.M., Maxted, N. and Dulloo, M.E. (eds) (2008) *Conserving Plant Genetic Diversity in Protected Areas: Population Management of Crop Wild Relatives*. CAB International, Wallingford, UK.

Jackson, M.T. (1997) Conservation of rice genetic resources – the role of the International Rice Genebank at IRRI. *Plant Molecular Biology* 35, 61–67.

Jackson, M., Ford-Lloyd, B.V. and Parry, M.L. (1990) *Climatic Change and Plant Genetic Resources*. Belhaven Press, London.

Jackson, M.T., Hawkes, J.G. and Rowe, P.R. (1980) An ethnobotanical field study of primitive potato varieties in Peru. *Euphytica* 29, 107–113.

Jansen, J. and van Hintum, T. (2007) Genetic distance sampling: a novel sampling method for obtaining core collections using genetic distances with an application to cultivated lettuce. *Theoretical and Applied Genetics* 114, 421–428.

Jay, F., Manel, S., Alvarez, N.,Durand, E.Y., Thuiller, W. *et al.* (2012) Forecasting changes in population genetic structure of alpine plants in response to global warming. *Molecular Ecology* 21, 2354–2368.

Jia, L., Yan, W., Agrama, H.A., Yeater, K., Li, X. *et al.* (2011) Searching for germplasm resistant to sheath blight from the USDA rice core collection. *Crop Science* 51, 1507–1517.

Jing, R., Ambrose, M.A., Knox, M.R., Smykal, P., Hybl, M. *et al.* (2012) Genetic diversity in European *Pisum* germplasm collections. *Theoretical and Applied Genetics* 125, 367–380.

Jump, A.S. and Peñuelas, J. (2005) Running to stand still: adaptation and the response of plants to rapid climate change. *Ecology Letters* 8, 1010–1020.

Jump, A.S., Peñuelas, J., Rico, L., Ramallo, E., Estiarte, M. *et al.* (2008) Simulated climate change provokes rapid genetic change in the Mediterranean shrub *Fumana thymifolia*. *Global Change Biology* 14, 637–643.

Kameswara Rao, N. and Jackson, M.T. (1996a) Seed longevity of rice cultivars and strategies for their conservation in genebanks. *Annals of Botany* 77, 251–260.

Kameswara Rao, N. and Jackson, M.T. (1996b) Seed production environment and storage longevity of japonica rices (*Oryza sativa* L.). *Seed Science Research* 6, 17–21.

Kameswara Rao, N. and Jackson, M.T. (1996c) Effect of sowing date and harvest time on longevity of rice seeds. *Seed Science Research* 7, 13–20.

Kameswara Rao, N. and Jackson, M.T. (1997) Variation in seed longevity of rice cultivars belonging to different isozyme groups. *Genetic Resources and Crop Evolution* 44, 159–164.

Koo, B., Pardey, P.G., Wright, B.D., Bramel, P., Debouck, D. *et al.* (2004) *Saving Seeds – the Economics of Conserving Crop Genetic Resources* Ex Situ *in the Future Harvest Centres of the CGIAR*. CAB International, Wallingford, UK.

Kramer, A.T. and Havens, K. (2009) Plant conservation genetics in a changing world. *Trends in Plant Science* 14, 599–607.

Kremer, A., Ronce, O., Robledo-Arnuncio, J.J., Guillaume, F., Bohrer, G. *et al.* (2012) Long-distance gene flow and adaptation of forest trees to rapid climate change. *Ecology Letters* 15, 378–392.

Krishnamurthy, L., Upadhyaya, H.D., Saxena, K.B. and Vadez, V. (2012) Variation for temporary waterlogging response within the mini core pigeonpea germplasm. *Journal of Agricultural Science* 150, 357–364.

Kwon, S., Brown, A.F., Hu, J., McGee, R., Watt, C. *et al.* (2012) Genetic diversity, population structure and genome-wide marker-trait association analysis emphasizing seed nutrients of the USDA pea (*Pisum sativum* L.) core collection. *Genes & Genomics* 34, 305–320.

Le Cunff, L., Fournier-Level, A., Laucou, V., Vezzulli, S., Lacombe, T. *et al.* (2008) Construction of nested genetic core collections to optimize the exploitation of natural diversity in *Vitis vinifera* L. subsp *sativa*. *BMC Plant Biology* 8, 31 doi: 10.1186/1471-2229-8-31.

Li, C.T., Shi, C.H., Wu, J.G., Xu, H.M., Zhang, H.Z. *et al.* (2004) Methods of developing core collections based on the predicted genotypic value of rice (*Oryza sativa* L.). *Theoretical and Applied Genetics* 108, 1172–1176.

Louette, D. and Smale, M. (2000) Farmers' seed selection practices and traditional maize varieties in Cuzalapa, Mexico. *Euphytica* 113, 25–41.

Malory, S., Shapter, F.M., Elphinstone, M.S., Chivers, I.H. and Henry, R.J. (2011) Characterizing homologues of crop domestication genes in poorly described wild relatives by high-throughput sequencing of whole genomes. *Plant Biotechnology Journal* 9, 1131–1140.

Maxted, N., Ford-Lloyd, B.V. and Hawkes, J.G. (eds) (1997) *Plant Genetic Conservation – the* In Situ *Approach*. Chapman & Hall, London, UK.

Maxted, N., Ford-Lloyd, B.V., Kell, P.V., Iriondo, J.M., Dulloo, M.E. *et al.* (eds) (2008) *Crop Wild Relatives Conservation and Use*. CAB International, Wallingford, UK.

Maxted, N., Kell, S., Ford-Lloyd, B., Dulloo, E. and Toledo, A. (2012) Toward the systematic conservation of global crop wild relative diversity. *Crop Science* 52, 774–785.

McCouch, S.R., McNally, K.L., Wang, W. and Sackville Hamilton, R. (2012) Genomics of gene banks: a case study in rice. *American Journal of Botany* 99, 407–423.

McKhann, H.I., Camilleri, C., Berard, A., Bataillon, T., David, J.L. *et al.* (2004) Nested core collections maximizing genetic diversity in *Arabidopsis thaliana*. *Plant Journal* 38, 193–202.

McNally, K.L., Bruskiewich, R., Mackill, D., Buell, C.R., Leach, J.E. *et al.* (2006) Sequencing multiple and diverse rice varieties. Connecting whole-genome variation with phenotypes. *Plant Physiology* 141, 26–31.

Mercer, K.L. and Perales, H.R. (2010) Evolutionary response of landraces to climate change in centers of crop diversity. *Evolutionary Applications* 3, 480–493.

Nicotra, A.B., Atkin, O.K., Bonser, S.P., Davidson, A.M., Finnegan, E.J. *et al.* (2010) Plant phenotypic plasticity in a changing climate. *Trends in Plant Science* 15, 684–692.

Odong, T.L., van Heerwaarden, J., Jansen, J., van Hintum, T.J.L. and van Eeuwijk, F.A. (2011) Determination of genetic structure of germplasm collections: are traditional hierarchical clustering methods appropriate for molecular marker data? *Theoretical and Applied Genetics* 123, 195–205.

Panis, B. (2009) Cryopreservation of *Musa* germplasm: 2nd edition. Technical Guidelines No. 9. Engelmann, F. and Benson, E. (eds), Bioversity International, Montpellier, France.

PGR Secure (2012) Novel characterization of crop wild relative and landrace resources as a basis for improved crop breeding. Available at: www.pgrsecure.org/ (accessed 1 December 2012).

Prasad, P.V.V., Boote, K.J., Allen, L.H., Sheehy, J.E. and Thomas, J.M.G. (2006) Species, ecotype and cultivar differences in spikelet fertility and harvest index of rice in response to high temperature stress. *Field Crops Research* 95, 398–411.

Redestig, H., Kusano, M., Ebana, K., Kobayashi, M., Oikawa, A. *et al.* (2011) Exploring molecular backgrounds of quality traits in rice by predictive models based on high-coverage metabolomics. *BMC Systems Biology* 5 176 doi: 10.1186/1752-0509-5-176.

Reeves, P.A., Panella, L.W. and Richards, C.M. (2012) Retention of agronomically important variation in germplasm core collections: implications for allele mining. *Theoretical and Applied Genetics* 124, 1155–1171.

RoyChowdhury, M., Jia, Y., Jia, M.H., Fjellstrom, R. and Cartwright, R.D. (2012) Identification of the rice blast resistance gene *Pib* in the national small grains collection. *Phytopathology* 102, 700–706.

Ruiz, M. and Vernooy, R. (2012) *The Custodians of Biodiversity. Sharing Access to and Benefits of Genetic Resources*. Earthscan, Abingdon, UK.

Sharma, M., Rathore, A., Mangala, U.N., Ghosh, R., Sharma, S. *et al.* (2012) New sources of resistance to *Fusarium* wilt and sterility mosaic disease in a mini-core collection of pigeonpea germplasm. *European Journal of Plant Pathology* 133, 707–714.

Stamp, P. and Visser, R. (2012) The twenty-first century, the century of plant breeding. *Euphytica* 186, 585–591.

United Nations (1992) Convention on Bilogical Diversity. Available at: www.cbd.int/doc/legal/cbd-en.pdf (accessed on 25 January 2013).

Upadhyaya, H.D., Ravishankar, C.R., Narasimhudu, Y., Sarma, N.D.R.K., Singh, S.K. *et al.* (2011) Identification of trait-specific germplasm and developing a mini core collection for efficient use of foxtail millet genetic resources in crop improvement. *Field Crops Research* 124, 459–467.

Vetelainen, M., Negri, V. and Maxted, N. (2012) A second look at the European strategic approach to conserving crop landraces. In: Maxted, N., Dulloo, M.E., Ford-Lloyd, B.V., Frese, L., Iriondo, J.M. *et al.* (eds) *Agrobiodiversity Conservation: Securing the Diversity of Crop Wild Relatives and Landraces*. CAB International, Wallingford, UK, pp.181–185.

Villar, E., Klopp, C., Noirot, C.,Novaes, E., Kirst, M. *et al.* (2011) RNA-Seq reveals genotype-specific molecular responses to water deficit in eucalyptus. *BMC Genomics* 12 538 doi: 10.1186/1471-2164-12-538.

Virk, P.S., Ford-Lloyd, B.V., Jackson, M.T., Pooni, H.S., Clemeno, T.P. *et al.* (1996) Predicting quantitative variation within rice using molecular markers. *Heredity* 76, 296–304.

Wang, J.C., Hu, J., Xu, H.M. and Zhang, S. (2007) A strategy on constructing core collections by least distance stepwise sampling. *Theoretical and Applied Genetics* 115, 1–8.

Waugh, R., Jannink, J., Muehlbauer, G.J. and Ramsay, L. (2009) The emergence of whole genome association scans in barley. *Current Opinion in Plant Biology* 12, 218–222.

Yashina, S., Gubin, S., Maksimovich, S., Yashina, A., Gakhova, E. *et al.* (2012) Regeneration of whole fertile plants from 30,000-y-old fruit tissue buried in Siberian permafrost. *Proceedings of the National Academy of Sciences USA* 109, 4008–4013.

Zhao, J., Artemyeva, A., Del Carpio, D.P., Basnet, R.K., Zhang, N. *et al.* (2010) Design of a *Brassica rapa* core collection for association mapping studies. *Genome* 53, 884–898.

3 Climate Projections

Richard A. Betts[1] and Ed Hawkins[2]

[1]*Met Office Hadley Centre, Exeter, UK;* [2]*National Centre for Atmospheric Science, Department of Meteorology, University of Reading, Reading, UK*

3.1 Introduction

The Earth's climate system comprises the atmosphere, oceans, land surface and life. The atmosphere and oceans are both fluids moving on the surface of a rotating sphere, with energy being received from the sun in the form of solar (shortwave) radiation, and other energy being lost to space in the form of terrestrial (longwave) radiation. At any particular point on the Earth's surface, the intensity of solar radiation depends on the angle of the ground in relation to the sun – therefore the energy input varies across the day and throughout the year. The rotation of the Earth and the ongoing changes in the energy received from the sun in different locations drive the circulations of the atmosphere and oceans. There are some consistent, large-scale features of these circulations that mean that the general character of the climate can be reasonably predicted on the basis of past experience – the ocean current known as the North Atlantic Drift brings relatively warm water to western Europe, and the Hadley Circulation in the tropics is associated with convective activity and frequent thunderstorms around the equator and descending air masses with very little rain 1000–2000 km either side of the equator. Within this general structure, however, there is far more detail that is less easily predicted, behaving according to mathematical chaos, such as individual thunderstorms and weather systems.

The absorption of solar energy also depends on how much is reflected back to space by clouds or by the surface of the land or ocean – bright surfaces (such as snow, ice and clouds) reflect a large proportion back to space, whereas dark surfaces (ocean waters and forests) absorb more.

Some gases such as water vapour, carbon dioxide, methane and nitrous oxide are known as greenhouse gases, and they absorb and re-emit some of the heat radiation given off by the Earth's surface and hence warming the lower atmosphere. These gases occur naturally in the atmosphere and without their warming presence in the atmosphere the Earth's average surface temperature would be around −19°C as opposed to the current value of 14°C. In terms of its contribution to the natural greenhouse effect, the most important is water vapour, followed by carbon dioxide (CO_2).

Changes in the atmospheric concentration of CO_2 are a consequence of shifts in the balance of very large flows of carbon into and out of the atmosphere. In the natural system, CO_2 is produced through the chemical combination of carbon with oxygen in the process of respiration, most of which is a natural part of biological processes in soils, vegetation, plankton and animals, such as breathing and decay. Some CO_2 is also produced by combustion of vegetation – in other words, forest fires. These are natural sources of CO_2. Opposing these are the processes which take up CO_2 from the atmosphere, such as photosynthesis by plants, which combines CO_2 with water to produce sugars that store the energy absorbed from the sun during the process. Further uptake of CO_2 involves the dissolving of CO_2 in water (mostly in the oceans) from which it can then be extracted by plankton. These latter processes provide sinks of CO_2. All life on earth contains carbon, and indeed the vegetation, soil microbes, plankton and higher life forms collectively together store more carbon than is present in the atmosphere. Very large quantities of carbon are also held as dissolved CO_2 in sea water. So the 'carbon cycle' as described above consists of flows of carbon between large stores in the atmosphere, ocean and life.

The rates of release and removal of CO_2 by these processes are affected by environmental conditions, such as the concentration of CO_2 in the atmosphere, the temperature and the availability of moisture for plant growth. Hypothetically, if environmental conditions were unchanging, the natural flows of CO_2 into and out of the atmosphere would balance out and the atmospheric CO_2 concentration would not be changed. Even though the flows in each direction are large, in a balanced system they could in theory cancel each other out.

3.1.1 Climate variability and climate change

Both climate change and climate variability refer to timescales that are longer than individual weather events or the annual cycle of seasons, but the exact difference between climate variability and climate change is somewhat open to interpretation. Indeed there seem to be two ways in which these terms are distinguished. One definition could be regarded as centring on timescales, with shorter-term fluctuations that are later reversed being considered 'climate variability', whereas longer-term, more 'permanent' shifts being thought of as 'climate change'. This is still problematic, however, because such fluctuations occur on a wide range of timescales from years to millennia and longer. Nevertheless, the term 'climate change' would not generally be used for timescales of a decade or less. A further definition distinguishes between natural and human influences, with 'climate variability' being viewed as 'natural', whereas 'climate change' is specifically attributed to human causes. When viewed alongside the previous definition this would tend to imply that human influences on climate are by definition long term or permanent; however, should current human influences be reversed in the longer term then the resulting fluctuations would also be considered 'variability' as opposed to 'change'. For clarity, the term 'natural climate variability' is often used to distinguish such variability from human-induced climate change.

It is important to note that natural climate variability itself takes two forms – 'externally forced natural variability' and 'internal variability'. The former refers to fluctuations driven by external but natural causes, such as variations in the energy output of the sun, or the energy received by the Earth due to variations in the Earth's orbit or its axis of rotation. The latter refers to natural oscillations within the system, which are largely a property of the character of the atmosphere and oceans as fluids; an example is the El Niño/Southern Oscillation that arises from interactions between the atmosphere and oceans in the region of the equatorial Pacific.

3.1.2 Drivers of climate change and variability

Solar irradiance

One source of external forcing of the climate system is a change in the output of energy from the sun, or changes in the Earth's orbit or the tilt of its axis, which affect how much of the sun's energy is received by the Earth and the distribution of this across the Earth's surface.

Changes in greenhouse gases

Since the Industrial Revolution, a further source of CO_2 to the atmosphere has been added – the combustion of fossil fuels, namely coal, oil and natural gas. These substances are the remains of vegetation and plankton from millions of years in the past, which were buried underground or beneath the ocean floor. The carbon stored within them, as described above, becomes coal and oil. The energy that was stored as sugars as a result of photosynthesis is held within these substances and their concentrated form means that a small quantity of coal or oil can hold considerable quantities of energy that can then be released when the coal or oil are burnt – hence their value as fuels. Burning of these fuels, however, rapidly releases CO_2 that was previously removed over millions of years and held away from the atmosphere underground. This is therefore an extra release of CO_2 that tips the balance between sinks and sources.

Combustion produces CO_2, and humans have been burning fossil fuels such as coal, oil and natural gas at ever-increasing rates over the past 200–300 years. From international records of energy consumption, the quantity of CO_2 that has been produced by burning fossil fuels can be calculated; by 2000 this had reached nearly 30 billion tonnes of CO_2 per year, having risen by over 1% per year over the previous 20 years. In total, human activities have emitted more than a trillion tonnes of CO_2 since fossil fuel burning began (which is more than the amount of CO_2 increase in the atmosphere). Humans have also been clearing the world's forests to make way for farmland or degrading them through logging, and, because forests lock up carbon within their biomass, their removal (which again often involves combustion) inevitably leads to a further release of CO_2 to the atmosphere.

The rate of rise of CO_2 in the atmosphere is only about 60% of the rate of emission from fossil fuel burning; some of the additional CO_2 is absorbed through an increase in photosynthesis by vegetation and an increased dissolving of CO_2 in ocean waters. It is therefore clear that the observed increase in CO_2 in the atmosphere has resulted from a combination of burning fossil fuels and deforestation. It is less easy to determine the exact contribution of each of these sources, but the current best estimate is that fossil fuel burning has contributed approximately three-quarters of the current excess of CO_2 above pre-industrial levels and deforestation has provided approximately the remaining quarter. Cement production also makes a small contribution.

The removal of CO_2 from the atmosphere by vegetation growth or dissolving in ocean waters are slow processes, so a large proportion of a given quantity of emissions remains in the atmosphere for decades to centuries. The concentration of CO_2 in the atmosphere at any time depends not only on the recent emissions but also the history of emissions and removals over many previous decades.

Most methane emissions and one-third of nitrous oxide emissions are from human activities, largely agriculture. Humans have also introduced new greenhouse gases of their own, the 'halocarbons' such as chlorofluorocarbons or CFCs (which, incidentally, have also damaged the ozone layer in the stratosphere). Although CO_2, methane, nitrous oxide and the halocarbons by themselves do not contribute as much to the overall greenhouse effect as water vapour, they are increasing because of human activity. However, water vapour is not directly affected by humans to any appreciable degree. The fact that CO_2 and the other human-affected greenhouse gases are increasing leads to an enhancement of the greenhouse effect and hence a warming influence on climate.

The concentrations of CO_2, methane and nitrous oxide have shown a rapid increase during the past 200 years and this increase is accelerating. Further records of air bubbles in ancient ice show us that CO_2 is now at its highest concentration for more than 650,000 years. Methane and nitrous oxide, two other greenhouse gases, are also at record high levels.

Natural and human-made aerosols

Particles in the atmosphere ('aerosols' – not to be confused with the spray-cans of the same name) also influence climate by altering the Earth's radiation budget. As with greenhouse gases, some kinds of aerosols occur naturally, examples being dust, volcanic ash, sea salt, spora from plants and dimethylsulphide (DMS) from plankton. Aerosols are also produced by human activity, mostly from burning fossil fuels. Burning wood also produces aerosols so forest clearance and its use as fuel both increase aerosol concentrations. Additionally, desertification can increase the release of dust.

Aerosols have complex effects on climate. One effect is a cooling because many of the particles cause some of the sun's energy to be reflected back into space, either directly through their own brightness or indirectly by increasing the brightness or lifetime of clouds. But aerosols can also absorb some of the sun's energy or some of that given off by the Earth's surface, with both processes giving rise to a warming effect.

Large volcanic eruptions can inject very large quantities of aerosol high into the atmosphere, where they can spread around the globe and cool the Earth by blocking solar radiation.

Land-cover change

Vegetation cover can affect climate through its influence on the physical properties of the land surface. Surface albedo (reflectivity) is affected by the presence and nature of vegetation, especially if the underlying surface is relatively bright and reflective such as in snow-covered conditions. A forest has a low surface albedo so absorbs more of the sun's energy than brighter, more reflective open land. Vegetation also affects surface evaporation and heat flows, as plants extract water from the soil and release it to the atmosphere via microscopic pores in their leaf surfaces; this exerts a cooling influence. Increased vegetation cover therefore generally increases evaporation and cools the land surface and over large scales can also return moisture to the atmosphere for precipitation. For both albedo and evaporation effects, however, responses in the atmosphere such as changes in circulation can exert further effects both locally and further afield.

3.2 Estimating Future Climate Variations and Changes: Climate Models

3.2.1 General circulation models

Many aspects of the natural world, such as the motion of fluids or the energy changes involved in freezing of water, are understood well enough for them to be represented with mathematical equations. These processes can therefore be predicted in advance through application of such equations, often with great accuracy if the focus is in a single, relatively simple process. In theory, the combination of equations for many different processes can then allow certain aspects of even highly complex systems such as the Earth's atmosphere to be predicted if the equations can be solved sufficiently quickly, although there do seem to be limits to exactly what can be predicted owing to complex systems also exhibiting 'chaotic' behaviour as described above. This can be done using computer models, which essentially are tools for solving many different equations all together much more efficiently than could be done by a human mathematician. Computer models of the weather and climate therefore represent the application of theoretical understanding of the behaviour of the Earth's atmosphere, oceans and ecosystems. They can therefore be used to examine the extent to which this theoretical understanding can explain the behaviour of the system as measured in the real world, and for making predictions about future behaviour that can then act as a test of our understanding. If such tests demonstrate that the model predictions provide useful prior knowledge of the future, these predictions may also be used to inform decisions on how to respond to changes or variations in weather and climate.

Such computer models are used with a high degree of success for weather forecasts over the timescale of a few days, with useful predictions now regularly provided by the latest models. Over longer timescales, prediction of individual weather events is not possible because of the chaotic nature of the atmosphere, although in some parts of the world the general character of the weather over a season can be estimated with some degree of accuracy if there happens to be a strong influence of slower and more predictable processes such as variations in ocean temperature. Over the timescale of many years or decades, long-term changes in very large-scale quantities such as global or continental mean temperature can be reproduced and predicted, although changes in precipitation currently remain more challenging to model with accuracy.

Climate models bring together a vast array of understanding of the Earth's atmosphere, oceans and life, and represent the physical, chemical and biological processes in the form of mathematical equations solved by a computer program. These models are called 'General Circulation Models' (GCMs) because they simulate the general circulation of the atmosphere and ocean. GCMs simulate the global patterns of wind, the flows and changes of water, energy and chemical compounds between atmosphere, land and ocean, and the biological processes that affect these. They simulate the cycling of water through precipitation on the land and ocean, the flow of rivers to the ocean, evaporation back to the atmosphere from land and ocean, the condensation of water vapour back to liquid water ready for precipitation again, and the freezing and melting of ice at various points in the water cycle. The models simulate the energy that the planet receives from the sun, the proportions that are reflected back to space by clouds or that reach the surface, and the proportions of the latter that are absorbed by the surface or reflected. They also simulate the emission of energy from the Earth's surface and the proportion that is absorbed in the atmosphere through the 'greenhouse effect' – and again, clouds are also important here because they also reduce the loss of energy to space. These processes are affected by the chemical composition of the atmosphere, particularly the concentrations of greenhouse gases such as water vapour, CO_2 and methane, and again this chemical composition is simulated by the models. Large-scale ecosystem changes such as deforestation are also included in many models, and these affect the absorption of energy at the surface, the evaporation of water and the release and uptake of CO_2 and some other greenhouse gases.

The equations themselves have been established through careful observations, measurements and experiments both in laboratories and in the outside world, by several generations of scientists over the last few centuries. They are central to other aspects of physics, chemistry and biology as well as being the building blocks of computer models of climate. The climate models bring these equations together to provide an integrated view of the workings of the planet, and once again the models are tested and refined by comparison against observations.

GCMs solve the equations at thousands of points spaced out across a grid the globe, up through the atmosphere and down into the ocean. For models that represent the whole globe, the spacing of these grid points is typically approximately 100 km in current models, although as computing power increases this is enabling models to use higher-resolution grids with smaller spaces between points. Models can also be used to focus on more limited regions and hence use the available computing power for higher resolution – projections with higher resolution regional climate models (RCMs) are currently available at around 10 km to 25 km resolution, and these add increased precision to a forecast from a global model by taking the general large-scale climate changes simulated by a GCM and showing the implications of such changes for more local conditions once finer-scale processes such as the effects of mountains have been taken into consideration. It is important to note, however, the distinction between precision and accuracy, and the context of high uncertainty in regional-scale climate change patterns should be borne in mind. If the GCM simulation is basically unrealistic then the added precision of the RCM will not make it more realistic, it will just translate the unrealistic simulation into a higher resolution.

3.2.2 Simple climate models

In addition to the complex GCMs that simulate the atmospheric circulation, there are also 'Simple Climate Models' (SCMs) that focus on the global energy balance. These are far less computationally expensive than GCMs and hence can be used for many more simulations examining a wider range of scenarios. These models can be set up to approximate the behaviour of the more complex models in terms of the global mean temperature response to a particular change in greenhouse gas concentrations, so can be used to estimate the changes that the GCM would have simulated for a particular scenario.

3.2.3 Using models for forecasts and projections

Challenges in producing climate projections

There are a number of major uncertainties in predictions of future climate change and variability, which can be broadly categorized as:

- Natural internal variability.
- External drivers of climate.
- Response of the system to the external forcing.

External drivers can include natural and human drivers of change. The latter includes the change in the Earth's radiation balance owing to changes in the concentrations of greenhouse gases and aerosols, and in land cover. The climate can also be directly affected by other changes that do not directly affect the Earth's radiation balance but that nevertheless involve the movement of energy within the system, such as changes in evaporation at the land surface owing to land-cover change.

Although changes in greenhouse gas concentrations are currently driven primarily by emissions of these gases by human activity, the extent to which these emissions are translated in changes in concentration in the atmosphere is itself affected by climate because it also depends on source and sinks of carbon in the biosphere and oceans, which themselves depend on climate.

Butterflies and Pooh Sticks: forecasting climate variability

A key difficulty for meteorologists and climate scientists arises from the inherently chaotic nature of the atmosphere and oceans. Although the processes in the atmosphere and oceans are governed by the laws of physics, these laws include those of 'chaos theory' in which very small differences at the start of a process can lead to enormous differences in the final outcome. This was illustrated by Ed Lorenz with the classic 'butterfly effect', asking 'Does the flap of a butterfly's wings in Brazil set off a tornado in Texas?' Tiny changes in wind, such as would be caused by the butterfly's wings, can influence other, larger, wind flows, which in turn influence still larger winds until some major event is affected. Although a tornado could be 'set off' from such a small start, equally another tornado could be prevented from happening. The reason this is relevant to weather and climate modelling is that it is impossible to measure and model the atmosphere to infinite precision, and even the small approximations necessary in the measurements and models can lead to differences in the outcome. Minuscule changes in the numbers put into a model lead to different results, so this poses a limit on what is predictable, at least in terms of day-to-day weather. This is why weather forecasts for individual days are currently only given up to about a week ahead – any further ahead and the effects of today's butterflies are magnified so much as to make precise forecasting impossible. Because we do not include butterflies in our measurements of today's weather, we cannot hope to capture their effects when they become large enough to matter. And even if we could know where all the butterflies are and how hard they are flapping, there would still be some further level of detail we cannot capture.

But while day-to-day local weather cannot be forecast very far ahead, average or likely conditions over longer times and larger scales can be more predictable when large, slow oceanic changes or large drivers of change from outside are involved. A good example is the game of 'Pooh Sticks' – we throw several sticks into a river and know that they will all be carried downstream with the current, and so run to the other side of the bridge to watch them come out, but it is much harder (if not impossible) to predict which one will win the race because that depends on the individual swirlings in the water. Similarly, we can be confident that trapping more heat radiation from the Earth by increased greenhouse gas concentrations will cause a general warming on average over the coming decades, and can make projections of the general nature of changes in the atmospheric circulation. The local details of weather depend, however, on smaller-scale processes that again are much harder to predict even on average, let alone for particular years or months in the future. Understanding these processes and using this to improve predictability is a major focus for climate research.

Scenarios of human drivers of climate change: greenhouse gas and aerosol emissions, land-use change

A key factor for future climate change will be the quantity of greenhouse gas emissions. These will depend on the population, their lifestyle, and the way this is supported by the production of energy and the use of the land. A large population whose lifestyle demands high energy consumption and the farming of large areas of land, in a world with its main energy source being fossil fuel consumption, will inevitably produce more greenhouse gas emissions than a smaller population requiring less land and energy and deriving the latter from non-fossil sources. These factors could vary in a multitude of ways; the international community is already examining how energy demand and production can be modified to cause lower emissions but the implementation of this will depend on both the international political process and the actions of individuals. Even if no specific action is taken to reduce emissions, the future rates of emissions are uncertain because the future changes in population, technology and economic state are difficult if not impossible to forecast. Therefore, rather than make predictions of future emissions, climate science examines a range of plausible scenarios in order to examine the implications of each scenario and inform decisions on reducing emissions and/or dealing with their consequences.

The IPCC's climate models have generally used a set of scenarios known as 'SRES' (Special Report on Emissions Scenarios: Nakićenović *et al.*, 2000). These scenarios were generated by Integrated Assessment Models (IAMs) grounded in scenarios of the human socio-economic future, with differences in economy, technology and population but no explicit inclusion of emission reduction policies. Developed in the mid-1990s, these scenarios extended out to 2100 and varied widely in their projected emissions by that time even though none of them included climate policy.

Uncertainties in translating emissions to concentrations: importance of a weakening natural carbon sink

A key uncertainty in translating emissions into concentrations arises from uncertainty in the strength of natural land and ocean carbon sinks (Booth *et al.*, 2012) as described earlier about changes in greenhouse gases. Because these sinks currently slow the rate of CO_2 rise by partly offsetting anthropogenic emissions, the rise in concentrations would be more rapid if the natural carbon re-absorption service is weakened.

The land sink may weaken through a combination of three processes: photosynthesis ceasing to increase as CO_2 rises, some areas of forest dying as a result of a drying climate, and plant respiration and rates of decay in the soil increasing more rapidly with warming. At current levels, an increase in CO_2 concentration leads to an increase in photosynthesis, resulting in an uptake of CO_2 from the atmosphere by plants across the world. Experimental studies demonstrate, however, that this process has a limit, and the increase in photosynthesis becomes smaller and smaller as CO_2 concentrations become higher. The ecosystem service of an uptake of CO_2 therefore becomes weaker. Moreover, experimental studies suggest that the process of decay in the soil becomes more rapid under warming temperatures. Decay involves a release of CO_2 to the atmosphere and because the world's soils currently contain more than twice as much carbon as the atmosphere, such an increase in decay worldwide could potentially be a very large feedback on the CO_2 rise and climate warming. Coupled climate–carbon cycle models agree on projecting a decrease in the natural land carbon sink over the 21st century as a result of these processes.

CO_2 is also exchanged between the atmosphere and ocean waters, and models agree on the ocean carbon sink becoming weaker under a warmer climate owing to the reduced solubility of CO_2 in warmer water, reduced mixing of water near the surface and reduced salinity. However, the direct effect of increased atmospheric CO_2 concentrations is to enhance the uptake of CO_2 by ocean waters so there are competing effects between the direct effects of CO_2 concentration and the indirect effects via climate change. The net result of these competing effects varies between models. However, in models that project a net uptake of CO_2 by the oceans, this arises because the atmospheric CO_2 is already rising rapidly because of land carbon cycle feedbacks and so the overall effect of changes in the land and ocean sinks is still a net weakening.

Using the current scientific literature, the Intergovernmental Panel on Climate Change (IPCC) 4th Assessment Report (AR4) projections (Meehl *et al.*, 2007) take the possibility of these feedbacks into account as best they can, but the magnitude of these feedbacks is uncertain. This is one of the reasons for the large range of possible rates of warming.

Uncertainties in global climate sensitivity

Another source of uncertainty in projecting future climate change arises from difficulties in establishing the response of global temperatures to a given rise in greenhouse gas concentrations. This may seem like an easy problem to solve because greenhouse gas concentrations and temperatures can be measured well enough to be very confident that both are increasing. In theory, the problem can be expressed as a simple equation with three terms: radiative forcing, climate sensitivity and global mean temperature change. In this conceptual model, radiative forcing is the driving force of change, in the form of a change imposed on the Earth's energy balance. Climate sensitivity is the response of global mean temperature to a given radiative forcing, and is conventionally defined as the long-term equilibrium response in global average temperature to the radiative forcing associated with a doubling of CO_2 concentrations, representing how responsive the climate is to an imposed change. A related measure is transient climate response, which is the warming at the time of doubling CO_2 during an ongoing rise in CO_2 of 1% per year – this is smaller than climate sensitivity because longer-term feedback processes have not had time to fully take effect. However, climate sensitivity (and transient climate response) cannot be obtained from observations with a high degree of precision. Although global mean temperature change can be estimated reasonably well from observations and the radiative forcing due to greenhouse gases can be established quite precisely, the problem is made more difficult by the other contributions to the overall radiative forcing such as aerosols. As described above, the extent of the cooling influence of aerosols has not yet been established with much precision, which means that the net radiative forcing of greenhouse gases and aerosols together is also not very well known. Therefore, although the overall climate response can be established from temperature measurements, the forcing causing this response is not well known and therefore climate sensitivity cannot be quantified precisely either. Moreover, because the global observational record is only about 150 years long and greenhouse gas concentrations are still increasing, there are no observations of the full warming that would occur due to a doubling of CO_2, either at the time of doubling or for a long-term change reaching equilibrium.

Different climate models give different estimates of climate sensitivity, often related to the strength of cloud feedbacks. Clouds are a particular challenge for climate models, partly because the relevant processes largely take place on scales smaller than those resolved by the models and partly because the microphysical processes of changes between water vapour, liquid water and ice can be difficult to observe and measure *in situ*. Models rely on approximation techniques known as parametrizations in order to represent the overall effects of small-scale processes at the larger scale of the models, such as the net effect of several thunderstorm clouds on the overall total cloud cover, atmospheric motion and precipitation in a 100 km^2 model grid square.

A key issue arising from this is that the extent to which aerosols are masking the true effect of greenhouse gases is not known. If the current aerosol cooling is strong, this implies that the greenhouse warming that it is partly counteracting must also be strong. This is an issue because aerosol emissions can be reduced more easily than greenhouse gases and also aerosol particles are washed out of the atmosphere within days, whereas CO_2 is gradually taken up by the ocean and land ecosystems over decades and centuries. Therefore it would be relatively easy to reduce the aerosol cooling effect, and therefore unveil the full warming effect of greenhouse gases.

Uncertainties in regional climate responses

The response of climate in a particular region to an external forcing is not always easy to predict because it depends on changes in atmospheric circulation, which are complex and interact with many other factors. Some general principles are reasonably well understood conceptually, such as seasonal warming of land masses being an important component of monsoon circulations due to the warmer surface causing ascending air that draws in wind flows at lower levels. However, the details of how, say, a particular change in surface temperature in one region affects the overlying atmospheric circulation by driving convection also depend on more complex processes, which, although may be understood in general, may not be quantifiable in detail with confidence. Some of the key processes such as the formation of clouds take place at scales smaller than those explicitly represented in models and therefore need to be approximated in the equations applied at the model grid point scale. Moreover, with the atmosphere circulating across the whole globe, changes in one part of the world can have knock-on effects in other parts of the world, so even a good representation of processes in one region may not be adequate for prediction if there are complex effects involving interactions from other regions. The interactions between the atmosphere and oceans can be particularly challenging to measure, understand and represent in models.

A further issue is that different patterns of external forcing can give rise to different regional climate responses, although greenhouse gases are generally well mixed and therefore exert a reasonably uniform forcing across the globe.

Regional climate projections should therefore be interpreted as physically plausible scenarios of future change, rather than forecasts. Such projections may therefore be useful in scenario planning, preferably with a number of projections being used in order to explore a range of possible future changes. A comparison between the projections from several different models can give some indication of the robustness of the projections, provided this is also backed up with insight into the mechanisms involved in particular changes.

3.2.4 Testing climate model projections

Observations of climate change

Records from a worldwide network of weather stations on land and observations made on board ships can be used to establish change in global average temperature for roughly the past 150 years. These show that global average temperature has risen by approximately 0.8°C since the start of the 20th century. The rise has not been steady – before 1940 there was a warming of around 0.3°C, and then there was a cooling of approximately 0.2°C until 1950, followed by a renewed warming of 0.13°C per decade since then. This warming is observed over the oceans as well as over land, suggesting that it is a truly global phenomenon and not a conglomeration of 'local' warmings caused by some small-scale process such as the urban heat island effect.

Comparing climate models with observations

Early versions of climate models were developed in the 1960s and used to estimate the change in global temperature that would be expected from an increase in CO_2 concentration. An early model suggested that a doubling of CO_2 concentration should lead to a global warming of 2.4°C (Manabe and Wetherald, 1967). Around that time, it was predicted that CO_2 concentrations should increase by between 17% and 25% by the year 2000, and in 1972 a paper published by the Met Office (Sawyer, 1972) suggested that a 25% increase in CO_2 should increase global temperatures by approximately 0.6°C. The actual rise in CO_2 between the early 1970s and 2000 was approximately 17% and global warming approximately 0.5°C.

Projections issued by the IPCC in its First Assessment Report in 1990 (Houghton *et al.*, 1990) indicated a warming of 0.55°C between 1990 and 2010, and the observed warming was 0.35°C in the HadCRUT3 dataset and 0.39°C in the GISTEMP dataset. The IPCC (Houghton *et al.*, 1990) projection did not account for natural variability; if this is taken into account then the projected range of warming would be 0.28 to 0.81°C, or 0.27 to 0.83°C if a different method is used. The observed climate change comes within this range, suggesting that the observations are consistent with the 1990 projections once natural variability is accounted for (Frame and Stone, 2012).

The models used in the IPCC Third Assessment Report (TAR; Cubasch *et al.*, 2001) and Fourth Assessment Report (AR4; Meehl *et al.*, 2007) were used to simulate the 1990s as well as the 2000s and beyond, using a number of GCMs and also SCMs calibrated against the GCMs and (in AR4) carbon cycle models. Prior to 2000 the models were driven with observed concentrations of greenhouse gases and aerosols, including volcanic aerosols associated with the 1991 eruption of Mount Pinatubo in the Philippines. After 2000, the models are driven with the SRES emissions scenarios. Comparison of both sets of projections against observations is useful because the GCMs include natural variability so the importance of this can be seen, whereas the SCMs show the long-term trend more clearly. The average warming simulated by the SCMs was more rapid than observed from 1990 to 2006 (Rahmstorf *et al.*, 2007) but less rapid than observed over the full period of 1990–2011 (Rahmstorf *et al.*, 2012) (Fig. 3.1a). Nevertheless the observed change was within the range of model projections from the whole ensemble. Again, if natural variability is taken into account by removing solar and volcanic forcings and El Niño/Southern Oscillation variability (Foster and Rahmstorf, 2011) the agreement between models and adjusted observations becomes closer still (Rahmstorf *et al.*, 2012). The adjusted and unadjusted observational trends also fall within the range of trends simulation by the early part of the projections from the AR4 SCM models (Meehl *et al.*, 2007 (Fig. 3.1). When the wider range of changes arising from natural variability in the GCM is considered, it is clear that year-to-year global mean temperature changes also fall within the range simulated by the GCM (Fig. 3.1b). Moreover, a large instance of externally forced natural variability – a temporary cooling following the 1991 Mount Pinatubo eruption – is also captured by the GCM. These comparisons of observations with model projections from the 1970s, 1990s and 2000s suggest that the earlier models were capable of projecting the global warming trend in advance with some credibility, and more recent models can also

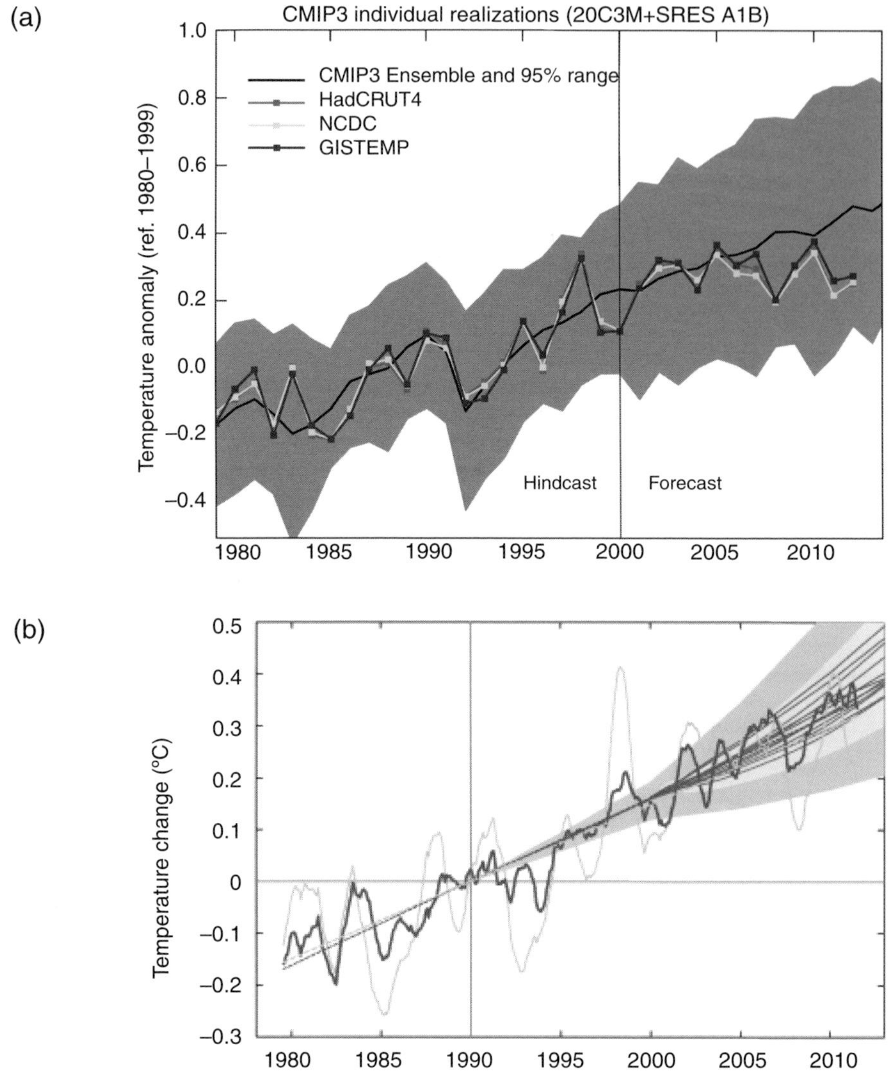

Fig. 3.1. Comparison of projections of global mean temperature change. (a) GCM projections from the IPCC 4th Assessment Report (black line and grey plume) against three observational datasets. (b) Simple model projections from the IPCC Third Assessment Report and 4th Assessment Report with observations and observations adjusted to remove the largest estimated effects of natural variability. Trends in adjusted and unadjusted observations are shown in dashed lines. The projections are from simple climate models calibrated against general circulation models, and represent the long-term average trend arising from anthropogenic forcing, without natural variability, hence the smooth nature of the lines.

reproduce the observed trend particularly if natural variability is taken into account.

The main uncertainty in climate for the next few years to decades arises from natural variability, because the year-to-year fluctuations are larger than the relatively gradual trend in long-term average conditions owing to the build-up of greenhouse gases. The details of these fluctuations are mostly unpredictable at present but a range of possibilities can be examined. Nevertheless, given that global greenhouse gas emissions will continue to some extent for the foreseeable future, this can be confidently expected to introduce systematic changes in regional climates of some kind on the medium to long term. Exactly what these systematic changes will be remains a difficult question, although a range of estimates can be made. By the middle of the century, the uncertainty in the regional climate responses to ongoing greenhouse gas build ups become more important in comparison with the uncertainty due to natural variability. The exact quantity of greenhouse gas emissions is important in the longer term, but this does not become a significant contribution to uncertainty until near the end of the 21st century owing to long timescales of response. For precipitation, the difference between the usual IPCC emissions scenarios remains relatively unimportant in comparison with uncertainties in the regional climate response and natural variability.

3.3 Long-term Projections

3.3.1 Temperature

The IPCC SRES emissions scenarios have been widely used as input to climate models to project further global warming over the coming century; this includes simulations from GCMs for some emissions scenarios and SCMs for other scenarios. In the IPCC AR4, up to 23 GCMs were used to simulate climate change for the B1, A1B and A2 emissions scenarios (Plate 2). The SCMs were calibrated against the GCMs for these scenarios, and then used to provide estimates of how the GCMs would have responded for the remaining scenario B2, A1T and A1FI (Plate 3). The SCMs were used to estimate the additional uncertainty in translating emissions scenarios to trajectories of greenhouse gas concentrations; the GCMs used in AR4 did not include climate-carbon cycle feedbacks and hence were driven by prescribed scenarios of CO_2 concentrations. The SCMs were calibrated against a number of different coupled climate–carbon cycle models from the Coupled Climate–Carbon Cycle Model Intercomparison Project (C4MIP) (Friedlingstein *et al.*, 2006) in order to capture the range of strengths of climate–carbon cycle feedback and the implications of this uncertainty for translating emissions into concentrations. So although the range of warming simulated by the set GCM for any particular emissions scenario did not take into account the uncertainties in climate–carbon cycle feedbacks, these additional uncertainties were represented through the SCM.

The SCM projected warming of between 1.1 and 6.4°C by the end of the 21st century, relative to the 1980–1999 average (Meehl *et al.*, 2007). Future warming projections are often expressed relative to the pre-industrial state, which was approximately 0.5°C cooler than the 1980–2000 average, so the above projections suggest a warming of between 1.6°C and 6.9°C by the end of the century relative to pre-industrial.

This wide range of projected warming by 2100 arises from differences in the emissions scenarios and also the range of possible warming under any particular scenario. For the lowest emissions scenario (B1), the warming projected by 2100 ranged from 1.1°C to 2.9°C, whereas for the highest emissions scenario (A1FI) the range was 2.4°C to 6.4°C. The ranges of projected warming from the highest and lowest AR4 scenarios therefore overlap slightly, illustrating the importance of the uncertainties in long-term climate response.

If the upper end of the AR4 range of projected warming is followed, which is less likely but still physically plausible, global warming relative to pre-industrial could reach 2°C relative to pre-industrial in the 2040s, and 4°C by approximately 2060. If the high emissions scenario (A1FI) were followed but the climate system is less responsive, the central estimate of model projections suggests reaching 2°C around approximately 2050 and 4°C in the 2070s.

Lower emissions scenarios are less likely to pass these thresholds, but most still pass 2°C in their central estimate and suggest some possibility of reaching 4°C if climate sensitivity is high enough.

Note that even the lowest bound of these business-as-usual projections (B1 emissions scenario and lowest climate sensitivity) suggests that warming would exceed 1.5°C relative to pre-industrial by the end of the century.

These changes would not be spread uniformly around the world. Some regions will warm faster than the global average, whereas others will warm less. Faster warming is expected near the poles, partly because the melting snow and sea-ice exposes the darker underlying land and ocean surfaces, which absorb more of the sun's radiation instead of reflecting it back to space as bright ice and snow do (Plate 4).

3.3.2 Precipitation

Changes in precipitation over the 21st century are also projected to vary from place to place, with large uncertainties. The uncertainties in projected precipitation change are generally larger than those for temperature and in many cases this includes uncertainty in even the sign of change. Although for temperature it is clear that the long-term trend is in one direction – warming – for precipitation there could be either an increase or a decrease depending on how the local climate regime responds. In some cases the projection may also be for no change in regional precipitation.

Much of the uncertainty in long-term precipitation projections arises from differences between models as opposed to different emissions scenarios – indeed the differences between the model responses to a given emissions scenario are generally larger than the differences due to emissions scenario alone (Fig. 3.2). This fact often seems to be overlooked in impacts studies – great care is often given to defining which emissions scenario is used, but less recognition is given to the large uncertainties in the regional climate response to that scenario. If only a single climate model projection is used, this gives an illustration of the impacts of one potential climate future, but care is needed in order to avoid giving the impression that this is a robust prediction unless there is an accompanying assessment of confidence in the climate change projection.

Because the main uncertainty concerns regional climate responses rather than emissions scenarios, the IPCC AR4 focused on one particular emissions scenario (A1B) in its discussion of regional climate projections and their uncertainties. The following discussion takes the same approach, summarizing key points from the AR4 assessment of multi-model projections of regional precipitation change for the SRES A1B scenario (Plate 5), with a prescribed scenario of CO_2 concentrations as opposed to including climate–carbon cycle feedbacks.

The projected changes and the level of agreement between models both vary from place to place and from season to season. Although in some regions of the world the models give a consistent projection of whether seasonal precipitation increases or decreases by 2100, in many regions the agreement between models is low so there is a lack of clarity over whether average precipitation is a projected increase or decrease in these regions. It is also important to realize that agreement between models does not by itself mean that the changes should be regarded with high confidence – an understanding of the processes involved in the changes is also required.

One key issue that was arguably not addressed well in the IPCC AR4 presentation (Plate 5) is that of projections of no change in precipitation in some regions. Tebaldi *et al.* (2011) point out that focusing only on consensus between models can give a misleading impression in regions where the model consensus appears to be small. There are two reasons why there may appear to be a low level model consensus (white areas in Plate 5): (i) a significant signal of change simulated in opposite directions by different models; and (ii) a low signal-to-noise ratio in most of the models. Tebaldi *et al.* (2011) suggest that confusion between these two reasons may lead to inappropriate adaptation responses by decision makers. If the science suggests that future change is likely

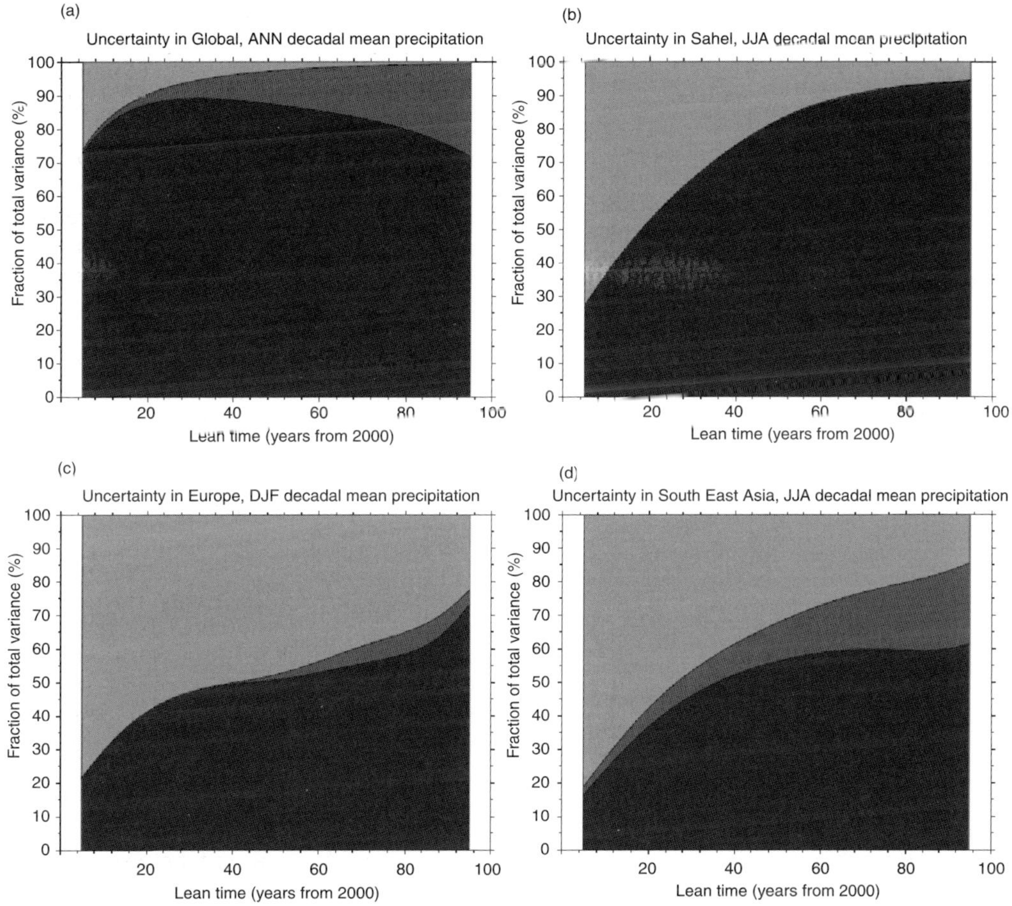

Fig. 3.2. Contribution to total uncertainty arising from differences in emissions scenarios (medium grey), model response to a given emissions scenario (dark grey) and internal climate variability (light grey), for decadal mean precipitation change in the global annual mean (a), the Sahel in June–July–August (b), Europe in December–January–February (c) and South-east Asia in June–July–August (d). Reproduced from Hawkins and Sutton (2010).

to be small then the appropriate response may be to take no adaptive action. However, if the lack of signal is misunderstood to indicate the possibility of large changes in either direction, then decision makers may unnecessarily take adaptive action to prepare for large but uncertain changes. To illustrate this, Tebaldi *et al.* (2011) reproduce the IPCC AR4 presentation of multi-model precipitation projections by 2080–2099, but they distinguish between areas where the signal-to-noise ratio is small and those where there are significant changes in some models but disagreement in the sign of change AR4 (Plate 6).

Christensen *et al.* (2007) took into account the CMIP3 multi-model projections and physical understanding in order to identify robust findings for future regional climate projections (Plate 7). The following discussion summarizes those findings.

The assessed likelihood of changes varied between regions, with highest likelihoods assigned to increased precipitation in northern latitudes and decreases in the Mediterranean.

The high-latitude regions are generally projected to become wetter, especially in winter: northern Europe, Canada, north-east USA and the Arctic were assessed in AR4 as being 'very likely' to undergo an increase in annual mean precipitation, and a winter precipitation increase is assessed as 'very likely' in Northern Asia and the Tibetan Plateau. In winter across the high latitudes, the multi-model average increase by 2100 being by 5% in the south of these regions and more than 20% further north. It was assessed as 'very likely' that the snow season will become shorter in most of Europe and North America. Increased risk of summer drought was assessed as 'likely' in central Europe and the Mediterranean.

Around the Mediterranean, in southern Europe, the far north of Africa and the Middle East, there is a high degree of model consensus on reduced precipitation in winter, summer and the annual mean; this drying signal was assessed as 'very likely'. Decreased precipitation is also projected in most models in southern Africa in June–July–August (JJA) and was assessed as 'likely' but little change in December–January–February (DJF; Plate 8). Increased precipitation was assessed as 'likely' in tropical and East Africa, with the multi-model mean increase being more than 30% by 2100 across Southern Sudan and Ethiopia. There were no other robust changes in Africa; either the signal-to-noise ratio was small (e.g. southern Africa in DJF) or a signal was seen in some models (Plate 6) and in the multi-model mean but not a sufficient majority of models (Plate 7, Plate 8).

Across eastern and southern Asia, increased mean precipitation and more intense precipitation are 'likely' in JJA, although in contrast, in India and western China a wetting is only actually seen in half of the models (Plate 9). In DJF, decreased precipitation is projected in two-thirds of the models over some parts of South Asia although there is little signal over much of Pakistan, India and southern China, and overall there is no robust change identified in this season and region.

In Australia, there is little signal of precipitation change across the centre of the continent in DJF. A drying in south-western Australia is 'very likely' in DJF, and is also seen in JJA but with lower model consensus and no identification as a robust finding. Decreased precipitation and increased drought are 'likely' across other parts of southern Australia in JJA. Over the South Island of New Zealand, increased precipitation and more intense precipitation are assessed as 'likely' but increased drought is 'likely' in the North Island.

There is some degree of consensus among the CMIP3 models on projecting reduced precipitation in JJA in north-eastern South America, although the extent of the drying varies between models and regions (Plate 10) and this was not identified by Christensen *et al.* (2007) as a 'robust finding'. In eastern Amazonia, two-thirds of the models project some degree of drying in this season, whereas 80–90% project drying in north-east Brazil. The multi-model mean precipitation reduction by 2100 is 30–50% over much of these regions. Simulated precipitation reductions in Amazonia seem to be linked to the extent of warming of sea surface temperatures simulated in the tropical north Atlantic (Good *et al.*, 2008) and equatorial east Pacific (Harris *et al.*, 2008). In DJF, the multi-model consensus is on increased precipitation.

3.3.3 Different patterns of regional climate change at specific global warming levels

The United Nations Framework Convention on Climate Change (UNFCCC) states its aim as to 'stabilize greenhouse gas concentrations at a level that would avoid dangerous climate change'. In association with this, there is a desire to identify what level of global mean temperature rise is 'dangerous'. Assessing the impacts of climate change at any particular level of warming is, however, difficult because of the large uncertainties in the associated patterns of regional climate change. This is not merely an issue of uncertainties in the ability of models to represent the real world – there are probably also many different patterns of regional temperature and precipitation change that could be associated with a single global mean temperature rise in the real world itself, either because of different spatial patterns of forcing or because of local feedback processes and natural internal variability in the climate system. The timing of reaching particular levels of warming is also important because rates of regional climate change can vary according to factors such as snow and ice feedbacks and the extent of the influence of the ocean, which warms more slowly than the land owing to the large heat capacity of water.

In the IPCC AR4 projections for the A1B scenario, while there are some regions where most models agree on precipitation changes associated with a given increase in global mean temperature, there are also many regions where there is low consensus between models (Plate 11).

Focusing on a specific region can illustrate how large the uncertainty can be in regional climate change for given levels of warming. For example, in eastern Amazonia, climate models produce widely varying projections of regional climate change for any given level of global warming and many of the AR4 models do not simulate changes that approach the critical threshold for rainforest even at 4°C global warming (Fig. 3.3). Therefore, robust identification of particular impacts with specific levels of global warming is challenging – it is often more appropriate to discuss the implications of warming in terms of risk, accompanied by appropriate assessment of uncertainty.

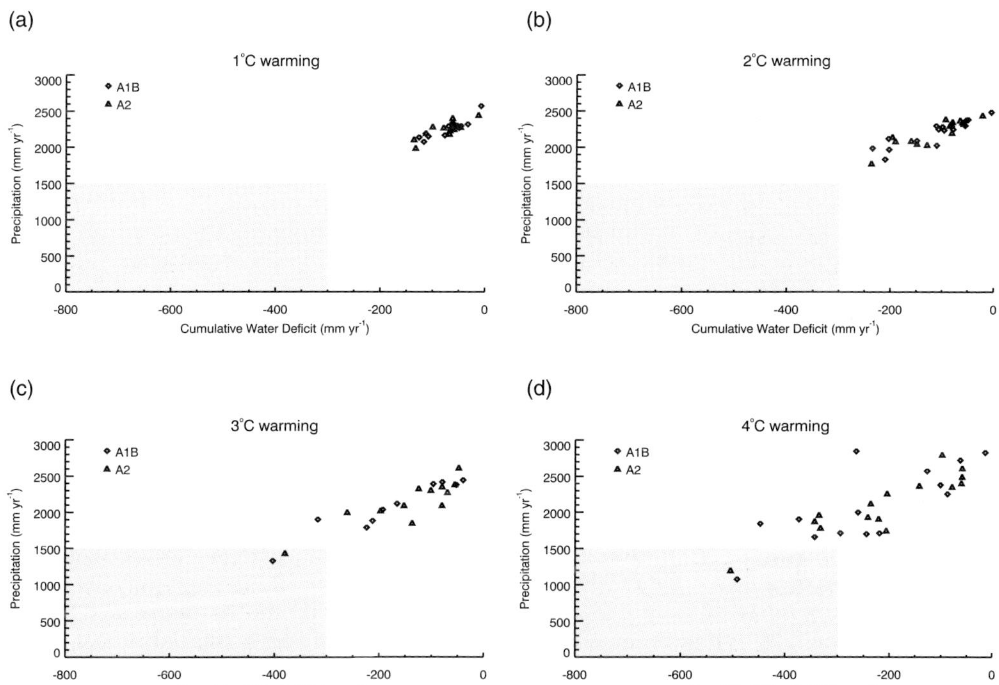

Fig. 3.3. Annual mean precipitation and cumulative water deficit projected over eastern Amazonia based on the CMIP3 GCM and HadCM3LC, for global warming at (a) 1°C, (b) 2°C, (c) 3°C and (d) 4°C, for the A1B and A2 emissions scenarios, in comparison with the thresholds identified as consistent with the presence of rainforest (Malhi *et al.*, 2009; shaded box). Values are derived by applying simulated anomalies to the observed present-day climatology. For 4°C warming, many of the points are derived through linear extrapolation for those models that do not simulate 4°C warming by 2100, which is the limit of the CMIP3 dataset.

3.3.4 Implications of uncertainties in CO_2 concentrations for a given level of global warming

One key issues of relevance to the impacts of CO_2-induced climate change on plants is the direct impact of CO_2 rise itself, and the importance of this relative to climate change. There is not a neat one-to-one correspondence between CO_2 concentration and global warming (climate sensitivity). This is important because in cases where the direct effects of CO_2 can offset or enhance the impacts of climate change, the overall impact of a given change in temperature (such as 2°C) on impacts such as ecosystem change and crop yield depends on the level of CO_2 concentrations that actually produced that warming. If climate sensitivity is high, a smaller CO_2 rise could lead to 2°C warming and the offsetting of warming impacts by CO_2 fertilization would be small. Conversely, if climate sensitivity is low, a higher CO_2 rise would be necessary to reach a 2°C warming, and the CO_2 fertilization would be larger and offset much more of the climate change impact.

For example, in the A2 scenario, the standard projected CO_2 concentration for the 2070s is 630 ppm (Plate 12b; Table 3.1), and the CMIP3 GCM project a range of warming of 2.0–2.7°C for this decade relative to 1971–2000 (Plate 12a; Table 3.1). In the B1 scenario for this time, the standard projected CO_2 concentration is 500 ppm and the GCMs project a warming of 1.0–2.0°C. The range of possible warming by the 2070s for each scenario includes 2°C even though there are different CO_2 concentrations in the two scenarios at this time. A warming of 2°C may be reached in the 2070s accompanied by a CO_2 concentration of 500 ppm if emissions follow the B1 scenario but climate sensitivity is high. Alternatively the 2°C warming could be accompanied by a CO_2 concentration of 650 ppm if emissions follow the A2 scenario but climate sensitivity is low. For these two scenarios of CO_2 at 2°C, the effects of CO_2 fertilization and increased water use efficiency would be less in the former case than the latter. This may have important implications for plant responses to a changing climate in combination with changing CO_2 concentrations.

Table 3.1. Ranges of projected global warming and CO_2 concentration for the 2070s in the A2 and B1 scenarios.

Scenario	Range of global warming relative to 1971–2000 (°C)	Approx. CO_2 concentration (ppm)
A2	2.0–2.7	650
B1	1.0–2.0	500

Similar issues will hold for different mixes of greenhouse gases (CO_2 and non-CO_2). Most non-CO_2 greenhouse gases do not impact directly on plants so the exact greenhouse gas mix will determine the overall ecosystem and hydrological response.

3.3.5 Changes in weather extremes

Rising temperatures are expected to act to enhance evaporation and hence exert a drying influence on soil moisture but to some extent this may be offset by increased efficiency of plants in their use of water under higher CO_2 concentrations.

The intensity of convective rainfall is expected to increase, which could increase the risk of river flooding and flash flooding. In regions and seasons where average precipitation is expected to increase, river flood risk could also be increased as a result of the higher likelihood of intense rain falling on saturated ground. The actual impact of these would depend on the existence of flood management infrastructure or practices.

IPCC AR4 predicts that a warmer future climate will bring an increased risk of more intense, more frequent and longer-lasting heatwaves. The frequency and intensity of heatwaves has already been seen to increase. Although an individual heatwave cannot be attributed to climate change because such events do occur naturally, studies with climate models suggest that a heatwave such as that in Europe in summer 2003 (which led to many thousands of excess deaths) is probably now twice as likely to occur than would have been the case in the absence of anthropogenic climate change.

The likelihood of such extremes is also projected to continue to increase over the 21st century as a result of ongoing greenhouse gas emissions. Generally it is not yet possible to forecast individual extreme events beyond the timescale of a few days, but the Met Office has shown that the overall level of tropical storm activity over a coming season can be successfully forecast in the North Atlantic.

3.4 Near-term Projections: Uncertainties in Natural Variability

The above discussion, and much of that by Christensen *et al.* (2007) in the IPCC AR4, focused on long-term changes by the end of the 21st century. This timescale is relevant to informing climate change mitigation decisions because it represents the potential future changes that may be avoided if large reductions in global greenhouse gas emissions take place. However, a certain amount of further global climate change is already inevitable in the coming decades, so adaptation to this will be necessary. Projection of regional climate changes in the next few decades involves an added dimension of complexity, because on these timescales the relative importance of internal climate variability is greater. The signal of anthropogenic climate change may be difficult to distinguish from the noise of internal variability, even for temperature, especially at local scales (Deser *et al.*, 2012). Moreover, the dominance of internal variability coupled with the relatively small effect of anthropogenic warming means that near-term changes may be temporarily opposite to those expected from anthropogenic effects in the long term. Hence, even in regions where the long-term climate change is regarded as robust, this may not emerge for some years or even decades due to internal variability (Plate 13).

In the CMIP3 ensemble projection for the A1B scenario, the signals of long-term change seen by 2080–2099 had begun to emerge by 2011–2030 in some places but in many regions there was no signal (Plate 14; Tebaldi *et al.*, 2011). Even in regions where the signal was emerging, multi-model consensus was lower in 2011–2030 than in 2080–2099; in all land regions, less than 80% of models showing a significant change agreed on the sign of the change by 2011–2030, whereas many regions showed more than 80% agreement (among models showing significant change) by 2080–2099 (Plate 6).

This means that it is extremely important not to assume that projections of long-term trends are relevant to planning and decision making on shorter timescales of the next few years or even decades. Although conditions are projected to continue to become warmer on average, some seasons may still be colder than average as a result of internal variability. Similarly, projections of a particular long-term trend in precipitation may not reflect short-term variations. For example, the UKCP09 projections (Murphy *et al.*, 2009) indicate reduced summer precipitation as the most likely outcome in the second half of the 21st century but the probability of increased summer precipitation still remains high in the first half of the century, particularly when considering local scales where the relative importance of variability is even higher (Fig. 3.4).

3.5 Beyond the AR4 Projections: Emerging Research

3.5.1 Abrupt climate change

An abrupt climate change may be defined as when 'the climate system is forced to cross some threshold, triggering a transition to a new state at a rate determined by the climate system itself and is faster than the cause'. This could refer to changes affecting the entire climate system, such as feedbacks between climate change and the carbon cycle that lead to an increase in the fraction of CO_2 emissions remaining in the atmosphere and hence an acceleration of global warming, or they could refer to more localized changes such as the initiation of an

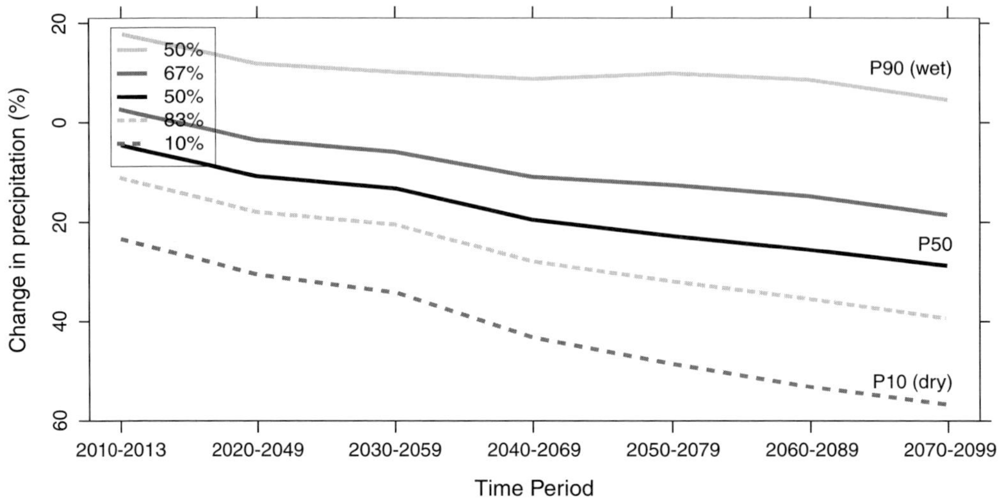

Fig. 3.4. Probabilistic projection of changes in summer mean precipitation from the 2020s to 2080s for South-east England under the A1FI scenario, reflecting uncertainties in both natural variability and long-term response to global warming for a single 25 km grid-box in central London. Although the long-term trend is for increased likelihood of reduced precipitation, the 10th–90th percentile range continues to include the possibility of increased precipitation throughout the 21st century. Reproduced from Murphy *et al.* (2009).

unstoppable collapse of an ice sheet or ecosystem.

These changes might be driven by strong destabilizing feedbacks, leading to nonlinear responses to external forcing. There is evidence of such changes in the past, in response to natural climate variability. For example, 8200 years ago the North Atlantic region appears to have undergone rapid cooling despite global temperatures being relatively stable around that time. It is suggested that this was due to a major change in ocean circulation, effectively equivalent to a collapse of the Gulf Stream, caused by sudden outflow of cold, freshwater from major lakes in North America previously held at bay by a major ice sheet. Potential collapse of the Gulf Stream has been suggested as one possible example of an abrupt climate change that might result from ongoing global warming, owing to increased input of cold, freshwater into the North Atlantic by the melting of the Greenland ice sheet. Although this mechanism is physically plausible, no current models show this occurring within this century, however.

Ice sheet melt is one example of where a ‘tipping point’ could be crossed leading to unstoppable, irreversible change. For example, as the Greenland ice sheet melts, its surface becomes lower in altitude and hence subject to warmer ambient air temperatures (because temperatures generally decrease with altitude). Therefore a point could be reached where warming causes melting that lowers the ice sheet surface, which exposes it to warmer temperatures leading to further melting, committing the ice sheet to a long-term but irreversible decline. Such a decline could take many centuries or even millennia, but once started it could be unstoppable.

Reduction in sea ice cover has the potential to occur abruptly, with melting ice exposing the darker underlying ocean surface, which absorbs a greater fraction of the incoming solar radiation and hence could enhance the warming. It is difficult, however, to find evidence for such effects in the palaeo record because sea ice does not leave a signature in proxy records – this potential candidate for an abrupt climate change mechanism therefore remains largely theoretical at present. Although a rapid decline in Arctic summer sea ice extent took place in 2007, there was a partial recovery in 2008 and 2009. Arctic sea ice extent still shows a declining trend but not as rapidly as might have been feared on the basis of the 2007 change alone.

Tipping points may also exist in large-scale ecosystems, especially where such ecosystems influence their own climate and/or are also impacted directly by human activity. One key example is the possible reduction in rainfall over Amazonia. About 1500 mm of rainfall per year are required to support a rainforest, and some models project the rainfall to fall below this level at higher levels of global warming (Fig. 3.3). Such a drying would increase the frequency of forest fires and prevent the subsequent regrowth of dense rainforest vegetation. The Amazon forest itself plays a part in maintaining precipitation in its local area, by returning moisture to the atmosphere through greater evaporation than would be the case if the forest were replaced by grassland, so loss of the forest could lead to further reductions in rainfall. Should climate change lead to a major reduction in rainfall in this region, this could therefore initiate an ongoing loss of forest. Direct deforestation itself could also act as the trigger for such 'die-back'. Again such rapid changes are projected by models but have not been seen in palaeo records, largely because of the lack of a climate state in the past that is both well recorded in palaeo records and analogous to projected future climate states.

A further tipping point may exist in the global exchange of CO_2 between the atmosphere and terrestrial ecosystems. As discussed at the beginning of this chapter, currently terrestrial ecosystems on the whole seem to be a sink of carbon from the atmosphere and hence help to buffer the atmospheric CO_2 concentration against the additional inputs from anthropogenic sources. Almost all of the models in the C4MIP simulate the terrestrial biosphere to be an increasingly strong sink of carbon over the 20th century and early part of the 21st century (Friedlingstein *et al.*, 2006), but many of the models project this increase to cease around the middle of the 21st century and several models project the sink strength to decline thereafter. Two models project the terrestrial biosphere to become a net source of carbon to the atmosphere by the end of the 21st century. This declining sink and potential shift to a source arises from varying combinations of a saturation of the fertilization of photosynthesis at higher CO_2 concentrations, an exponential increase in soil and/or plant respiration under rising temperatures, and, in some cases, severe regional impacts on some ecosystems such as the Amazon forest. These changes in atmosphere–land carbon exchange result in an increase in the airborne fraction of CO_2 emissions in all the C4MIP models over the 21st century. Importantly, however, none of the models simulates an increase in airborne fraction over the 20th century because the airborne fraction is dominated by emissions in previous years at that time (Betts *et al.*, 2010).

A further potential feedback on greenhouse gas concentrations involves the emission of methane. Although methane is produced from anthropogenic sources, there are also significant natural sources of methane that include wetlands, forest fires, oceans and termites. Wetlands are the largest natural source of methane and the presence of moisture is important because methane production requires decay in anaerobic conditions. A key aspect of climate change feedbacks involves changes in methane emissions from high-latitude peatlands due to permafrost thawing. Although warming is confidently predicted, uncertainties in the projected feedback strength are high due to the strong reliance on moisture content relating to changes in precipitation. Global model projections suggest approximately a 20% increase in methane emissions from wetlands under a 2°C global warming, whereas experimental studies at a limited number of sites are broadly consistent with this and suggest a potential 60% increase in methane emissions under a 4.5°C warming. Some models suggest larger increases. Observations suggest that methane emissions from high-latitude wetlands increased over the 20th century, correlating with soil temperature and the depth of the water table.

The probability of these changes is highly uncertain and perhaps small in comparison with other expected changes such as the general ongoing warming. The risk of a change in the mode of operation of a large or important Earth system is significant, however, owing to serious potential negative impacts on human and ecological systems and is likely to be increased by anthropogenic forcing of the climate system. Many abrupt climate changes are therefore considered to be 'low-probability, high-impact' events in that they are thought to be relatively unlikely to occur but could potentially lead to very severe impacts if they did occur.

3.5.2 Near-term forecasting (inter-annual to decadal)

Climate prediction in the near-term (years to a few decades) is made more difficult by the fact that natural variations are still relatively important on this timescale. Whereas on multi-decadal timescales we expect the increased greenhouse warming to lead to a warming that is greater than the natural year-to-year warming and cooling, in the nearer term natural cycles are likely to remain significant. This poses a challenge when attempting to forecast on these timescales in order to inform adaptation. This area of climate science is only just beginning to demonstrate predictive skill but nevertheless some skill is there. The key lies in good measurements of the current state of the climate (especially sub-surface ocean temperatures and salinity), including the direction of any trends in the system, in order to set off the forecast in the right direction. Measurements of ocean temperatures and salinities can help capture the natural variations within the climate system and allow them to be factored in to the models. Forecasts of global temperatures on inter-annual timescales have been shown to be credible, with techniques developed by Smith *et al.* (2007) in the Met Office Hadley Centre having been used to reproduce temperature changes in 5-year periods over recent decades, using only the information that would have been available to forecasters at the start of each of those periods (Fig. 3.5). The most recent forecast, issued in December 2012, predicted the average global temperature from 2013–2017 to be 0.43°C warmer than the baseline of 1971–2000 (Fig. 3.6) – for comparison, the warmest

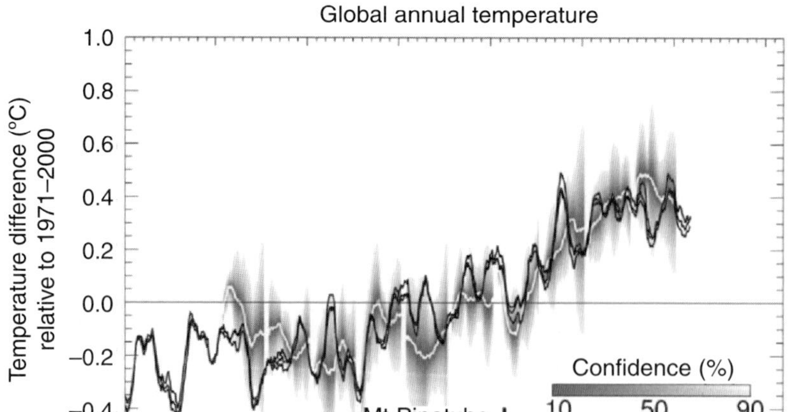

Fig. 3.5. 5-year hindcasts of global annual mean temperature (white curve with grey shading) with the HadGEM3 climate model initialized from observed data at the start of every 5-year period, compared against observations (black curve).

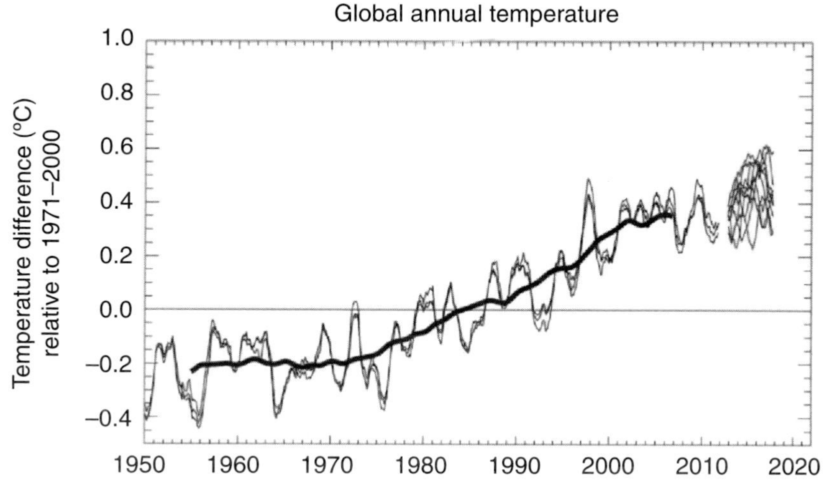

Fig. 3.6. Observed global annual mean temperature (thin grey curves) and running 5-year average (thick black curve), and 5-year forecast issued December 2012 (grey curves, far right) produced with the HadGEM3 climate model; those curves show 10 simulations, each started from slightly different initial conditions at the end of the observational record.

year on record to date was 1998, at 0.40°C above the 1971–2000 average.

Decadal forecasting is still rather experimental and significant challenges remain, especially for forecasting of local rather than global changes. Rapid progress is being made, however, and such forecasts may become useful to decision makers planning for adaptation to climate change and variability in the coming years.

References

Betts, R.A., Collins, M., Hemming, D.L., Jones, C.D., Lowe, J.A. *et al.* (2010) When could global warming reach 4°C? *Philosophical Transactions of the Royal Society. A* 2011 369, 67–84.

Booth, B.B.B., Jones, C.D., Collins, M., Totterdell, I.J., Cox, P.M. *et al.* (2012) High sensitivity of future global warming to land carbon cycle processes. *Environmental Research Letters* 7, 024002.

Christensen, J.H., Hewitson, B., Busuioc, A., Chen, A., Gao, X. *et al.* (2007) Regional climate projections. In: Solomon, S., Qin, D., Manning, M., Chen, Z., Marquis, M. *et al.* (eds) *Climate Change 2007: The Physical Science Basis.* Contribution of Working Group I to the Fourth Assessment Report of the Intergovernmental Panel on Climate Change. Cambridge University Press, Cambridge and New York, pp. 849–940.

Cubasch, U., Meehl, G.A., Boer, G.J., Stouffer, R.J., Dix, M. *et al.* (2001) Projections of future climate change. In: Houghton, J., Ding, Y., Griggs, D.J., Noguer, M., van der Linden, P.J. *et al.* (eds) *IPCC 2001: The Scientific Basis.* Contribution of Working Group 1 to to the Third Assessment Report of the Intergovernmental Panel on Climate Change. Cambridge University Press, Cambridge, UK.

Deser, C., Knutti, R., Solomon, S. and Phillips, A.S. (2012) Communication of the role of natural variability in future North American climate. *Nature Climate Change* 2, 775–779.

Foster, G. and Rahmstorf, S. (2011) Global temperature evolution 1979–2010. *Environmental Research Letters* 6, 044022.

Frame, D. and Stone, D. (2012) Assessment of the first consensus prediction on climate change. *Nature Climate Change* 3, 357–359.

Friedlingstein, P., Cox, P.M., Betts, R.A., Bopp, L., von Bloh, W. *et al.* (2006) Climate carbon-cycle feedback analysis: results from the C4MIP model intercomparison. *Journal of Climate* 19, 3337–3353.

Good, P., Lowe, J.A., Collins, M. and Moufouma-Okia, W. (2008) An objective tropical Atlantic sea surface temperature gradient index for studies of south Amazon dry-season climate variability and change. *Philosophical Transactions of the Royal Society B* 363, 1761–1766.

Harris, P.P., Huntingford, C. and Cox, P.M. (2008) Amazon Basin climate under global warming: the role of the sea surface temperature. *Philosophical Transactions of the Royal Society B* 363, 1753–1759.

Hawkins, E. and Sutton, R. (2009) The potential to narrow uncertainty in regional climate predictions. *Bulletin of the American Meteorological Society* 90, 1095–1107.

Hawkins, E. and Sutton, R. (2010) The potential to narrow uncertainty in projections of regional precipitation change. *Climate Dynamics* 37, 407–418.

Houghton, J.T., Jenkins, G.J. and Ephraums, J.J. (eds) (1990) *Climate Change: The IPCC Scientific Assessment*. Cambridge University Press, Cambridge.

IPCC (2007) Summary for policymakers. In: Solomon, S., Qin, D., Manning, M., Chen, Z., Marquis, M. *et al.* (eds) *Climate Change 2007: The Physical Science Basis*. Contribution of Working Group I to the Fourth Assessment Report of the Intergovernmental Panel on Climate Change. Cambridge University Press, Cambridge and New York.

Knutti, R., Furrer, R., Tebaldi, C., Cermak, J. and Meehl, G.A. (2010) Challenges in combining projections from multiple climate models. *Journal of Climate* 23, 2739–2758.

Malhi, Y., Roberts, J.T., Betts, R.A., Killeen, T.J., Li, W. *et al.* (2008) Climate change, deforestation, and the fate of the Amazon. *Science* 319, 169–172.

Malhi, Y., Aragão, L.E.O.C., Galbraith, D., Huntingford, C., Fisher, R. *et al.* (2009) Exploring the likelihood and mechanism of a climate-change-induced dieback of the Amazon rainforest. *Proceedings of the National Academy of Sciences USA* 106, 20610–20615.

Manabe, S. and Wetherald, R.T. (1967) Thermal equilibrium of the atmosphere with a given distribution of relative humidity. *Journal of Atmospheric Sciences* 24, 241–259.

Meehl, G.A., Stocker, T.F., Collins, W.D., Friedlingstein, P., Gaye, A.T. *et al.* (2007) Global climate projections. In: Solomon, S., Qin, D., Manning, M., Chen, Z., Marquis, M. *et al.* (eds) *Climate Change 2007: The Physical Science Basis*. Contribution of Working Group I to the Fourth Assessment Report of the Intergovernmental Panel on Climate Change. Cambridge University Press, Cambridge and New York.

Murphy, J.M. Sexton, D., Jenkins, G., Boorman, P., Booth, B. *et al.* (2009) *UKCP09 Climate Change Projections*. ISBN 978-1-906360-02-3.

Nakićenović, N., Alcamo, J., Davis, G., de Vries, B., Fenhann, J. *et al.* (2000) *IPCC Special Report on Emissions Scenarios*. Cambridge University Press, Cambridge and New York.

Rahmstorf, S., Cazanave, A., Church, J.A., Hansen, J.E., Keeling, R.F. *et al.* (2007) Recent climate observations compared to projections. *Science* 316, 709.

Rahmstorf, S., Foster, G. and Cazenave, A. (2012) Comparing climate projections to 1 observations up to 2011. *Environmental Research Letters* 7, 044035.

Sawyer, J.S. (1972) Man-made carbon dioxide and the 'greenhouse' effect. *Nature* 239, 23–26.

Smith, D.M., Cusack, S., Colman, A.W., Folland, C.K., Harris, G.R. *et al.* (2007) Improved surface temperature prediction for the coming decade from a global climate model. *Science* 317, 796–799.

Tebaldi, C., Arblaster, J.M. and Knutti, R. (2011) Mapping model agreement on future climate projections. *Geophysical Research Letters* 38, L23701.

4 Effects of Climate Change on Potential Food Production and Risk of Hunger

Martin Parry

Grantham Institute and Centre for Environmental Policy, Imperial College, London

4.1 Introduction

The Intergovernmental Panel on Climate Change (IPCC) concluded that even small amounts of warming at lower latitudes would tend to decrease agricultural yields overall; and beyond about two degrees of warming potential yields would be reduced in almost all parts of the world (IPCC, 2007). The implications for food security are substantial both at global, regional and local scales (see Zeigler, Chapter 1 and Berry *et al.*, Chapter 5, this volume). At present about 1 billion people in the developing world are estimated to be experiencing some form of shortage in food supply (FAO, 2009). Climate change will probably increase this risk. The extent of this change and its spatial distribution are the focus of what follows here. The approach taken is a chronological one and is largely an updated version of work undertaken for the World Food Programme (Parry *et al.*, 2009). The purpose of this chapter is to describe the scale of the challenge facing crop genetic management in the role it needs to play in confronting climate change. It summarizes the development of our knowledge about the risks posed by climate change to food production in the future and the consequent increases in risk of hunger.

4.2 First Analyses Using Low-resolution Climate Models with 2 × CO_2 Scenarios

The first model-based studies of the potential effects of climate change on global food supply were published in the early 1990s. The general conclusions of that work still broadly hold today: that climate change is likely to reduce global food potential and that risk of hunger will increase in the most marginalized economies (Rosenzweig and Parry, 1994). In these studies there were two main tasks:

- Effects on crop yields were modelled. These covered wheat, rice, maize and soybean, accounting for more than 85% of the world's traded grains and legumes (Rosenzweig and Iglesias, 1994; IBSNAT, 1989).
- Projected effects of yield changes on food production, prices and the number of people at risk of hunger were estimated by using estimated yield changes in a world food trade model (Fischer *et al.*, 1988).

The climate change scenarios for these early studies were 2 × CO_2, which are broadly equivalent to the continued high rates of unmitigated emissions of 2010–2013 (see

Betts and Hawkins, Chapter 3, this volume for updated climate projections). The enhanced effect on crop growth from elevated CO_2 was included in yield calculations. The non-climate scenarios, which are immensely important in determining the sensitivity of risk of hunger to climate change, were: a world population of 10.2 billion by 2060 (the UN median estimate); 50% trade liberalization in agriculture introduced gradually by 2020; moderate economic growth (ranging from 3.0% per year in 1980–2000 to 1.1% per year in 2040–2060); and crop yield for world total, developing and developed countries increasing annually by 0.7%, 0.9% and 0.6%, respectively.

4.2.1 Effects on yields and production

Under these scenarios of climate change and development, world cereal production is estimated to decrease between 1% and 7% depending on the General Circulation Model (GCM) climate scenario (Fig. 4.1). The largest negative changes occur in developing countries, averaging –9% to –11%. By contrast, production in developed countries is estimated to increase under all but the UK Meteorological Office scenario (+11% to –3%). The explanation for this difference in effect is that growing seasons in mid- and high-latitude regions, being currently often temperature-constrained, are extended by higher temperatures under conditions of climate change; whereas at lower latitudes growing seasons are more determined by water availability and therefore more likely to be shortened by increased evapotranspiration and reduced water availability.

Current disparities in crop production between the developed and developing countries are thus estimated to increase. Decreases in production are estimated to lead to increase in prices (by 25–150%) and increases in hunger (by 10–60%, or 60 to 350 million) (Fig. 4.1).

4.2.2 Effect under different levels of agricultural adaptation

The study tested the efficacy of two levels of adaptation:

- Level 1 adaptation included shifts in planting date that do not imply major changes in the crop calendar, additional application of irrigation water to crops already under irrigation and changes in crop variety to currently available varieties better adapted to the projected climate.
- Level 2 adaptation included large shifts in planting date, increased fertilizer application, development of new varieties and installation of irrigation systems.

Level 1 adaptations largely offset the negative climate-change-induced effects in developed countries, improving their comparative advantage in world markets. In these regions, cereal production increased by 4% to 14% over the reference case. In developing countries, however, these minor adaptations appear insufficient to accommodate the more significant reductions in crop potential in these regions. Production here, under these adaptations, remains down (–9% to –12%). Averaged global production is altered between 0% and –5% from the reference case. As a consequence, world cereal prices are estimated to increase by 10–100% and the number of people at risk from hunger by 5–50% (an additional 50–300 million) (Fig. 4.2).

More extensive adaptation (Level 2) is projected to reduce impacts by a third and in some cases virtually eliminates them. However, a reduction in the comparative advantage of developing countries under these scenarios is projected to lead to a reduced area planted to cereals. Cereal production in developing countries still declines by around 5%. Globally, however, cereal prices increase by only 5% to 35%, and the additional number of people at risk from hunger is reduced to around 10–50 million (Fig. 4.2). This suggests that substantial

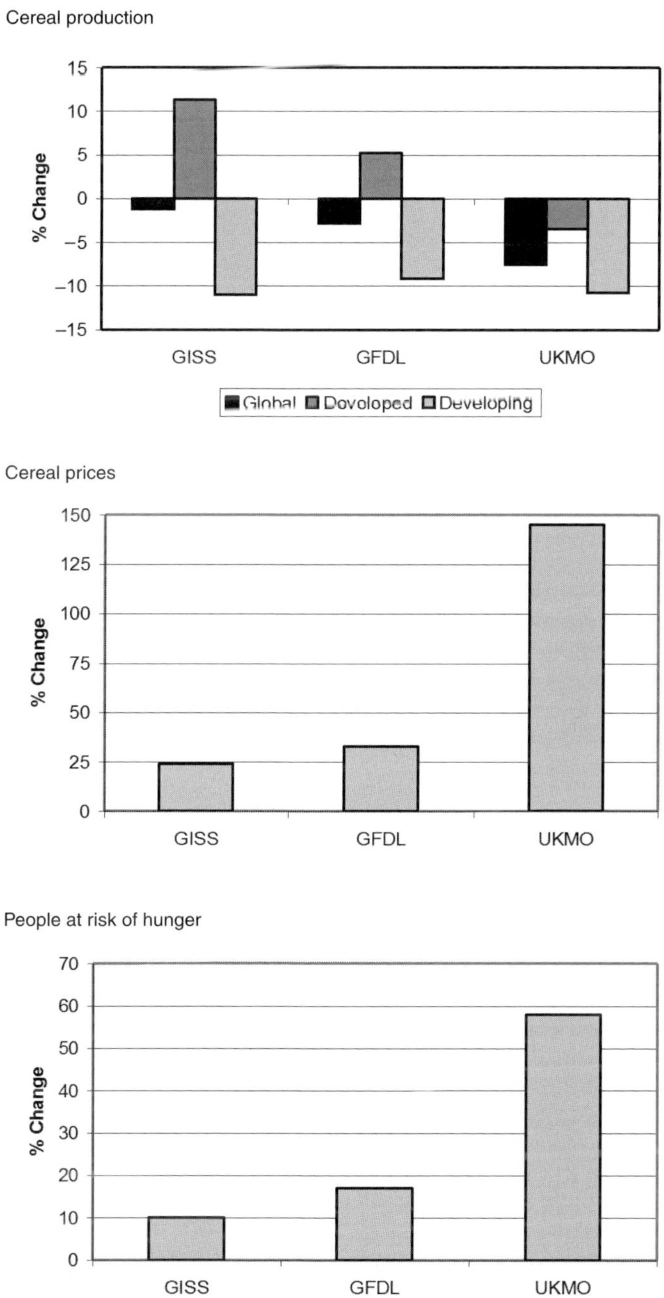

Fig. 4.1. Estimated effects of climate change on cereal production, cereal prices and people at risk of hunger in 2060. The reference case for 2060 assumes no climate change. The climate change is $2 \times CO_2$. Cereal production = global 3286 mmt, developed 1449 mmt, developing 1836 mmt; cereal prices 1970 = 100; and 641 million people at risk of hunger. GISS, Goddard institute for Space Studies; GFDL, Geophysical Fluid Dynamics Laboratory; UKMO, UK Meteorological Office (from Rosenzweig and Parry, 1994).

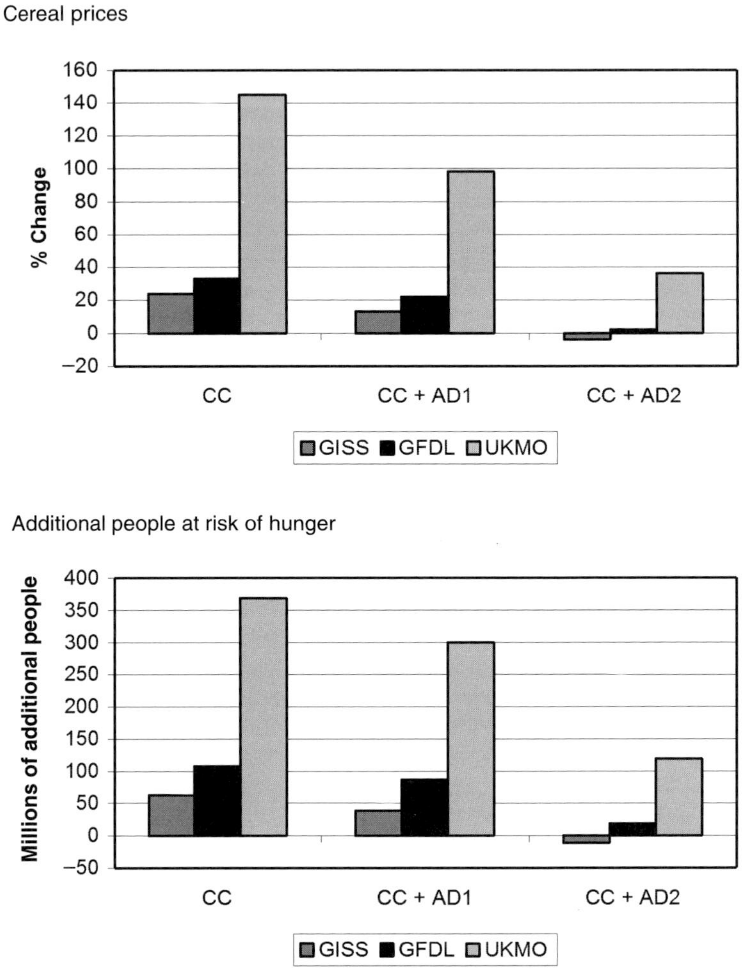

Fig. 4.2. Change in cereal prices and people at risk of hunger for 2 × CO_2 climate change scenarios (CC) and with Adaptation Levels 1 and 2 (AD1 and AD2). Scenarios and reference case are the same as in Fig. 4.1 (from Rosenzweig and Parry, 1994).

adaptations are required to mitigate the negative effects of climate change but these still do not eliminate them in developing countries.

4.2.3 Effects assuming full trade liberalization and lower economic and population growth rates

Assuming full trade (rather than 50%) liberalization in agriculture by 2020 provides for more efficient resource use and leads to 3.2% higher value added in agriculture globally and a 5.2% higher agriculture GDP in developing countries (excluding China) by 2060 compared with the reference case. This policy change results in almost 20% fewer people at risk from hunger.

Estimates were also made of impacts under a lower economic growth scenario (10% lower than reference), which leads to a tighter supply situation, higher prices and more people below the hunger threshold.

Prices are 10% higher and the number of people at risk from hunger is 20% greater.

Assuming the UN Low rather than Mid Estimate gives a population in 2060 of 7.3 billion instead of 5.9 billion, and leads to a model estimate of higher GDP/capita (about 10%) and 40% fewer people at risk from hunger compared with the reference scenario.

4.3 Subsequent Analyses Using Higher Resolution Climate Models and for Different Time Periods

The spatial resolution of models of the atmosphere (general circulation models or GCMs) increased and their simulation of air–ocean interactions and other feedback mechanisms improved substantially in the late 1990s. This enhanced the accuracy of their projections of climate change and several were now capable of producing time-dependent scenarios, thus enabling the evaluation of climate change impacts at different time horizons.

In the next suite of experiments (Parry *et al.*, 1999) the crop models were run for three future climate time periods (2020s, 2050s and 2080s) for the UK Hadley Centre's GCMs known as HadCM2 and HadCM3 (Mitchell *et al.*, 1995; Hulme *et al.*, 1999). All climate change scenarios assumed a 'business-as-usual' future of greenhouse gas emissions (termed IS92a by the Intergovernmental Panel on Climate Change), economic development and population growth, that is, ones that trended along recent historical lines.

Under the HadCM2 scenarios cereal prices increase by ~17% (+/– 4.5%) by the 2080s (Fig. 4.3). In turn these production and price changes were estimated to increase the number of people at risk of hunger by approximately 90 million (Fig. 4.3). The HadCM3 model, which projected slightly higher amounts of warming for the same emissions, gives climate conditions that lead to ~125 million additional people at risk of hunger by the 2080s.

The distribution of this additional risk of hunger is shown in the top part of Plate 16. Africa is the region most affected, with parts of South Asia and Central America also showing the most increases in risk owing to climate change.

4.4 Analyses for Different Development Pathways

More recently, the projected effects of climate change on global food supply have been considered under different pathways of future socio-economic development, expressed in terms of population and income level, which have been characterized by the Special Report on Emissions Scenarios (SRES) of the IPCC. Differing trajectories of population growth and economic development will affect the level of future climate change and, simultaneously, the responses of agriculture to changing climate conditions at regional and global scales. The goal of the study was to understand the nature of these complex interactions and how they affect people at risk of hunger in the coming decades (Parry *et al.*, 2004).

4.4.1 Effects on yield

In summary, the effects of climate change under these SRES scenarios result in similar crop yield decreases in developing countries and yield increases in developed countries as projected in the preceding studies (Table 4.1). The A1FI and A2 scenarios (characterizing a high temperature world) exhibit the greatest decreases both regionally and globally in yields, especially by the 2080s. Decreases are especially significant in Africa and parts of Asia with expected losses up to 30%. In these locations, effects of temperature and precipitation changes on crop yields are beyond the inflection point of the beneficial direct effects of CO_2. In North America, South-east Asia, South America and Australia, the effects of CO_2 on the crops partially compensate for the stress that the A1FI climate conditions impose on the crops and result in small yield increases. Climate change scenarios with smaller temperature increases (such as B1 and B2, which reflect

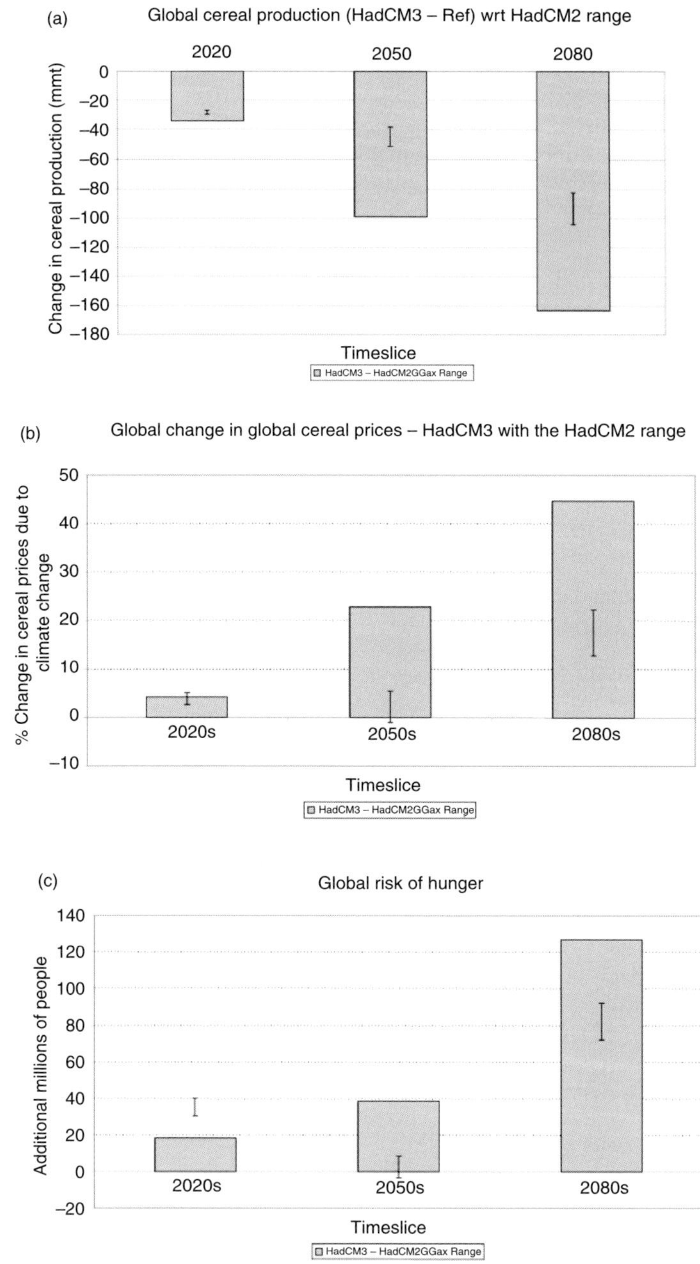

Fig. 4.3. (a) Changes in global cereal production (mmt). Blocks are the production change projected under the HadCM3 climate change scenario (compared with the reference case). Bars depict the range of change under the four HadCM2 ensemble simulations. (b) Percentage change in cereal prices. Blocks are the price changes projected under the HadCM3 climate change scenario (relative to the reference case). Bars depict the range of price change under the HadCM2 ensemble experiments. (c) Global estimates of the additional number of people at risk of hunger due to climate change compared with the reference case. HadCM3 estimates are represented by the blocks. Bars represent the range of results under the four HadCM2 ensemble simulations. Effects of elevated CO_2 on crop growth are included (from Parry *et al.*, 1999).

slower population growth but higher per capita income) lead, generally, to smaller cereal yield decreases. Table 4.1 illustrates the differing regional outcomes from scenarios of climate varying according to development pathway and mitigation policy (the S550 and S750 discussed below).

4.4.2 Effects on risk of hunger

The estimated number of additional millions at risk of hunger due to climate change is shown in Table 4.2 (see the DSSAT-BLS column in this table). The BLS is the basic linked system food model developed at the International Institute of Applied Systems Analysis (Fischer *et al.*, 1988). The main conclusion (for 2050) emerging here concerns the differential effect of development on impacts of climate change:

- For a pathway of continuing high population growth and regional disparities of income (A2 in the table) numbers at risk of hunger due to climate change are projected to be 10–20% greater than the number expected without climate change.
- For a pathway of lower population growth and more equitable income distribution (B1 and B2 in the table): additional numbers at risk are estimated to be 5% or less.

In these analyses, the conclusion remains that Africa is the region with most increases of hunger risk owing to climate change (about 65% of the global total is projected to occur here).

An important element of uncertainty stems from the role that elevated levels of ambient atmospheric CO_2 may have. Assuming a fully beneficial effect on C_3 crops such as wheat and rice, increased atmospheric CO_2 could elevate yields significantly but this estimation is drawn from laboratory experiments and may not be replicated in farmers' fields. At present, a reasonable guess is that roughly one half of the full effect might occur in reality, i.e. the mid-point between the numbers given in Table 4.2 for full effect (CC) and no effect (CC, no CO_2) is a best guess of the combined effect of climate change and elevated CO_2.

Table 4.2 shows broadly similar conclusions for risk of hunger based on different modelling methods (the AEZ-BLS, which is discussed later).

4.5 Comparable Recent Studies

In 2009, a study by IFPRI projected the responses of major crops, namely, rice, wheat and maize, to climatic variability under the SRES A2 scenario, using the US National Center for Atmospheric Research (NCAR) CCM3 and Australian Commonwealth and Industrial Research Organisation (CSIRO) Mk3.0 models (IFPRI, 2009; Nelson *et al.*, 2009). The results incorporate the biophysical effects of climate change on crop production; the changes in crop production including both biophysical and economic changes through higher prices induced by the initial biophysical shock; and the resulting impacts on calorie availability and child malnutrition. The following is taken

Table 4.1. Aggregated developing-developed country differences (percent) in average crop yield changes from baseline for the HadCM2 and HadCM3 scenarios (from Arnell *et al.* 2001 and Rosenzweig and Iglesias, 2006). See text for descriptions of the scenarios.

	HadCM3 – 2080s							HadCM2 – 2080s	
Scenario	A1F1	A2a	A2b	A2c	B1a	B2a	B2b	S550	S750
World	–5	0	0	–1	–3	–1	–2	–1	1
Developed	3	8	6	7	3	6	5	5	7
Developing	–7	–2	–2	–3	–4	–3	–5	–2	–1
Difference (%) Developed – Developing	10.4	9.8	8.4	10.2	7.0	8.7	9.3	6.6	7.7

Table 4.2. Estimated global number of people (in millions) at risk of hunger.

	2020		2050		2080	
	AEZ-BLS	DSSAT-BLS	AEZ-BLS	DSSAT-BLS	AEZ-BLS	DSSAT-BLS
Reference (millions)						
A1	663	663	208	208	108	108
A2	782	782	721	721	768	769
B1	749	749	239	240	91	90
B2	630	630	348	348	233	233
CC (millions)						
A1	666	687	219	210	136	136
A2	777	805	730	722	885	742
B1	739	771	242	242	99	102
B2	640	660	336	358	244	221
CC, no CO_2 (millions)						
A1	NA	726	NA	308	NA	370
A2	794	845	788	933	950	1320
B1	NA	792	NA	275	NA	125
B2	652	685	356	415	257	384

Reference, without climate change; CC, with climate change and full beneficial effects of carbon dioxide fertilization; CC, no CO_2, climate change without beneficial effects of carbon dioxide fertilization. AEZ, estimates from analysis of shift of agroecological zones; DSSAT, estimates from crop modelling analysis. Both analyses use the same global food model: BLS, the basic linked system food model (from Easterling and Aggarwal, 2007, based on Parry *et al.*, 2004 and Fischer *et al.*, 2002). A1, A2, B1 and B2 are socio-economic scenarios described in the text.

from a summary written by Dr Mark Rosegrant for the World Food Programme (Parry *et al.*, 2009).

Table 4.3 presents the results for the direct biophysical effects of the two climate change scenarios on production, with and without CO_2 fertilization, on the three crops modelled with DSSAT. For all regions, production declines in 2050 under climate change compared to no climate change when there is no CO_2 fertilization. Irrigated and rainfed wheat and irrigated rice are especially hard hit. In China, some crops fare reasonably well because higher future temperatures are favourable in locations where current temperatures are at the low end of the crop's optimal temperature. India and other parts of South Asia are particularly hard hit by climate change. With the CO_2 fertilization effect, the drop in production is

Table 4.3. Climate change effects on crop production, including biophysical effects.

	CSIRO No CF	CSIRO CF	NCAR No CF	NCAR CF
Maize, irrigated				
East Asia and the Pacific	–8.19	–7.75	–9.40	–8.81
Europe and Central Asia	–12.61	–12.58	–13.68	–13.63
Latin America and the Caribbean	–23.89	–23.47	–24.04	–23.65
Middle East and North Africa	–30.51	–30.85	–31.23	–31.32
South Asia	–25.99	–24.44	–25.33	–23.80
Sub-Saharan Africa	–39.67	–39.55	–39.53	–39.36
Developing countries	–15.67	–15.16	–16.39	–15.78
Developed countries	–3.22	–3.19	–10.55	–10.40
World	–8.86	–8.61	–13.19	–12.84

	CSIRO No CF	CSIRO CF	NCAR No CF	NCAR CF
Maize, rainfed				
East Asia and the Pacific	−19.03	−17.29	−13.70	−12.21
Europe and Central Asia	−24.40	−19.41	−74.80	−72.68
Latin America and the Caribbean	−11.24	−8.89	−12.96	−10.98
Middle East and North Africa	−24.54	−23.00	−57.40	−57.14
South Asia	−36.83	−34.78	−25.50	−22.95
Sub–Saharan Africa	−14.64	−14.80	−15.15	−15.32
Developing countries	−15.56	−13.90	−16.57	−15.20
Developed countries	−6.04	2.23	−14.67	−7.20
World	−12.73	−9.11	−16.01	−12.83
Rice, irrigated				
East Asia and the Pacific	−21.94	−6.33	−28.31	−11.64
Europe and Central Asia	−32.96	−19.80	−40.62	−26.06
Latin America and the Caribbean	−24.91	−20.74	−20.06	−13.80
Middle East and North Africa	−46.50	−37.18	−56.74	−47.46
South Asia	−19.79	−3.12	−23.69	−6.23
Sub-Saharan Africa	−39.43	−27.75	−41.28	−29.98
Developing countries	−22.61	−7.60	−27.26	−11.32
Developed countries	−17.27	−5.00	−19.00	−6.84
World	−22.30	−7.44	−26.78	−11.05
Rice, rainfed				
East Asia and the Pacific	−8.46	−0.88	−10.84	−2.60
Europe and Central Asia	−31.00	−25.70	−36.89	−6.34
Latin America and the Caribbean	−20.66	−15.14	−25.76	−19.56
Middle East and North Africa	−100.00	−100.00	−100.00	−100.00
South Asia	−16.41	−11.33	−12.62	−8.46
Sub-Saharan Africa	−11.48	−9.19	−10.05	−7.59
Developing countries	−13.09	−7.11	−12.70	−6.79
Developed countries	−30.50	−26.82	−34.59	−30.45
World	−13.13	−7.15	−12.75	−6.84
Wheat, irrigated				
East Asia and the Pacific	−19.22	−13.92	−24.42	−19.15
Europe and Central Asia	−46.77	−43.22	−53.39	−50.43
Latin America and the Caribbean	−21.47	−16.56	−26.11	−20.94
Middle East and North Africa	−30.13	−24.57	−35.49	−30.45
South Asia	−50.12	−41.84	−57.67	−50.28
Sub-Saharan Africa	−34.45	−30.14	−34.00	−28.61
Developing countries	−37.49	−30.96	−43.85	−37.85
Developed countries	−22.59	−18.94	−28.91	−25.34
World	−35.54	−29.39	−41.89	−36.21
Wheat, rainfed				
East Asia and the Pacific	−34.42	−27.18	−39.14	−34.13
Europe and Central Asia	−46.61	−41.74	−51.27	−46.55
Latin America and the Caribbean	−4.91	4.31	−3.06	3.99
Middle East and North Africa	−20.26	−10.86	−24.28	−16.01
South Asia	−54.67	−42.05	−57.73	−45.95
Sub-Saharan Africa	−32.82	−23.58	−35.69	−27.55
Developing countries	−30.99	−23.35	−33.78	−27.31
Developed countries	−17.11	−11.95	−22.11	−16.74
World	−24.52	−18.03	−28.34	−22.38

Percentage change in production compared with a reference scenario with no climate change, 2050 (from Nelson *et al.*, 2009). CF, carbon fertilization.

lower in all regions and rainfed maize has small improvement in production under CSIRO in developed countries, and rainfed wheat production increases in Latin America and the Caribbean under both CSIRO and NCAR.

The results in Table 4.3 broadly reinforce the findings of the earlier studies reported above but the negative impacts of climate change on production are larger, especially given that the results are for 2050, rather than 2080 for most of the earlier work. The effects of climate change on agricultural production are more negative in developing countries than in developed countries. Global impacts on production are negative in all cases, compared to the small positive impacts that were found in some of the earlier studies. Although results vary by crops, Africa and South Asia generally have the most negative impacts on production, consistent with earlier studies.

4.6 World Prices

The direct and indirect effects of climate change on agriculture play out through the economic system, altering prices, production, productivity investments, food demand, food consumption and ultimately human well-being.

World prices are the most useful single aggregate indicator of the effects of climate change on agriculture. Table 4.4 shows the price effects of various permutations of climate change, with and without the CO_2 fertilization effect. With no climate change, world prices for the most important cereal crops (rice, wheat and maize) will increase between 2000 and 2050, driven mainly by population and income growth and slowing productivity growth. Climate change results in much higher price increases in 2050 compared to the no climate change case, with prices in 2050 at 35–37% higher for rice, 99–102% higher for wheat and 58–62% higher for maize. A worse scenario is expected in 2050 where cereal price increase is projected at 25–150% using BLS as discussed in earlier in this chapter. In contrast, if CO_2 fertilization is effective in the field, these price increases owing to climate change are reduced, but still significantly higher in 2050 than the no climate change scenarios.

Table 4.5 reports effects of climate change on crop production, combining both rainfed and irrigated systems without CO_2 fertilization. It accounted for the biophysical effects of climate change on area and yield with the autonomous adjustments in these two variables due to price effects induced by the biophysical climate change shock and indirect effects from water stress in irrigated crops. Therefore, climate change negative effects on production shown in Table 4.4 are lower than in Table 4.2 as a result of inclusion of the autonomous economic response to the biophysical climate change shocks.

The rows show the difference in production between the GCM and no climate change scenarios in 2050. For instance, under CSIRO, maize production in South Asia is projected to fall by 23%, in comparison

Table 4.4. Projected world prices of selected crops and livestock products (USD/metric tonne) (from Nelson *et al.*, 2009).

		2050				
	2000	No climate change	NCAR No CF	CSIRO No CF	NCAR CF effect (%)	CSIRO CF effect (%)
Rice	190	305	419	414	−17.41	−16.36
Wheat	113	132	263	267	−10.39	−11.37
Maize	95	100	158	162	−11.41	−13.50

Prices are in year 2000 USD. The last two columns report the % difference between the price in 2050 with and without the CO_2 fertilization effect. For example, with the NCAR GCM, assuming CO_2 fertilization (CF) is effective in the field it results in a 17.41% decline in the world rice price in 2050 compared with the no carbon fertilization (no CF) price.

Table 4.5. Climate change effects on crop production, including biophysical and economic effects, no CO_2 fertilization, % change in production compared to a reference scenario with no climate change, 2050 (from Nelson *et al.*, 2009).

	South Asia	East Asia and the Pacific	Europe and Central Asia	Latin America and the Caribbean	Middle East and North Africa	Sub-Saharan Africa	Developed countries	Developing countries
Maize								
CSIRO (%)	–23	–12	–10	0	–26	–3	11	–9
NCAR (%)	–15	5	–25	–2	–22	–1	2	–2
Rice								
CSIRO (%)	–11	–12	–3	–20	–41	–16	–9	–13
NCAR (%)	–12	–13	0	–16	–44	–13	–8	–14
Wheat								
CSIRO (%)	–40	–13	–44	11	–14	–30	–6	–32
NCAR (%)	–40	–2	–46	15	–15	–35	–10	–31

with the 2050 reference scenario (no climate change) as compared to a drop of 26% (irrigated maize) or 37% (rainfed maize) in a scenario where yield and area responses to the increases in commodity prices induced by the biophysical climate change shock were not considered (shown in Table 4.3). For developing countries, under the CSIRO scenario (Table 4.5), maize, rice and wheat production will decline by 9%, 13% and 32%, respectively, values that are also less negative than the results in Table 4.3, due to autonomous adaptation to crop price changes.

4.7 Allowing for Large-scale Changes in Land Use

One of the drawbacks of the preceding studies is that they do not allow for changes in land use and assume that the crops that experience changes in yield will continue to be grown in the same areas in the future. In contrast, a logical response by farmers would be to change to crops that experience smaller reductions in yield and thus assume a comparative advantage over others. A study has examined the difference in outcome that might occur if such changes were considered (Cline, 2007).

The results differ only in degree from those of the previous studies. Global agricultural output is estimated to decrease by 16% assuming no carbon fertilization and by 3% with full carbon fertilization. The regional pattern shows quite strong adverse effects on yield in tropical areas, especially Africa, the Middle East and South Asia.

4.8 Analyses based on Changes in Agroecological Zones

A very different approach from point-based crop-growth modelling is the study of how zones of crop suitability, or agroecological zones, may shift location in response to changes of climate. When combined with modelling of the length of crop growing season, either due to changes in moisture or heat availability, this method enables evaluation of both changes in yield at any given place in combination with changes in extent of suitability (Fischer *et al.*, 2002).

The conclusions of this study with respect to effects on risk of hunger are given in Table 4.2 (here shown as AEZ-BLS to compare with DSSAT-BLS – data from the preceding studies based on crop modelling). Although both use the same global food model (the BLS), the two approaches adopt quite different methods in modelling altered crop yields. Yet their conclusions regarding the numbers of additional millions at risk of hunger owing to climate change are broadly similar and this gives greater confidence to estimates of ultimate effects on hunger.

4.9 Comparing the Range of Analyses

Other studies have either used a macroeconomic approach, or yield estimates from previous crop modelling as inputs to different economic models.

One uses six broad land classes around the world and analyses the change in extent of these owing to altered moisture and temperature, the yield results from which are very similar to the analyses from crop models (Darwin, 2004). But the economic assumptions in the approach, especially the land-use changes, eliminate three-quarters of the climate-induced reductions in production, at least until high levels of warming (above 3°C) start to reduce land suitability markedly.

A summary of the different approaches by the IPCC indicates that all conclude a reduction in output and consequent rise in prices, but vary in their conclusion regarding when (along a pathway of increasing global temperature) this decrease will occur (Easterling and Aggarwal, 2007). The crop modelling and agroecological analyses conclude that prices will rise with even small amounts of warming (1 to 2°C). The other analyses suggest that they will first increase (due to increased potential with extended growing seasons at higher latitudes) before decreasing when temperature increases exceed 2 or 3°C (Plate 15). This point of inflection from a positive to a negative effect on global food output, and whether it occurs at 1 or 2 or 3°C increase in global temperature, is central to the current debate as to whether global warming may, for the first few decades, have a beneficial effect. As we have seen, however, such a point of inflection is dependent on the very uncertain mix of the positive effects from higher CO_2 and the negative effects from higher temperature.

Moreover, in tropical and equatorial regions, the general pattern is of decreases in yield potential with even small amounts of climate change. The explanation for this is that in these regions crop plants are often currently being grown near their temperature and moisture limits, so that changes in heat and water availability can easily stress them. A consequence of this is that, even if global production potential were increased with small amounts of global warming (and this point remains unclear) before it declines with greater amounts of warming, in developing countries production potential is projected to fall with even small amounts of climate change. Risk of hunger at the regional level in poorer parts of the world is, therefore, likely to increase under all scenarios of climate change.

4.10 Reducing Impacts by Stabilizing CO_2 Concentrations at Lower Levels

This section explores the effect on risk of hunger of stabilizing CO_2 concentrations at defined levels (Arnell *et al.*, 2001). These stabilization scenarios are among the set defined by the IPCC (IPCC, 1997), stabilizing at 550 ppmv and 750 ppmv compared with the unmitigated emissions scenario (IS92a) (Mitchell *et al.*, 2000).

4.10.1 Effects on yield potential

Plate 16 shows the estimated changes in national potential grain yield by the 2080s, assuming no changes in crop cultivars, under the three emissions scenarios. Under unmitigated emissions, positive changes in mid and high latitudes are overshadowed by reductions in yield in the lower latitudes. These reductions are particularly substantial in Africa and the Indian subcontinent. Many of the mapped changes in yield are, however, small and indistinguishable from the effects of natural climate variability.

Stabilization at 550 ppmv produces far fewer reductions in yield than the unmitigated case, although there would still be reductions in the Indian subcontinent, most of the Pacific Islands, Central America and the majority of African nations. Stabilization at 750 ppmv to a large extent produces intermediate changes.

4.10.2 Implications for risk of hunger

Stabilization at 750 ppmv reduces the unmitigated impacts by around 75%, whereas stabilization at 550 ppmv achieves more modest reduction of around 50% (Tables 4.6 and 4.7). The assumption of full CO_2 effects seems to explain this, the beneficial effects of higher levels of CO_2 under 750 ppmv compensating for the adverse effects of higher temperatures. Under less optimal conditions of CO_2 effect, the reverse would be likely. Another study projects a 12–16% increase of people at risk of hunger under an unmitigated scenario (>800 ppmv) relative to a reference scenario, whereas under a 550 ppmv mitigated scenario, there is a reduction of 80–95% from the unmitigated scenario (Tubiello and Fischer, 2007).

Global figures, however, conceal considerable regional variations. The majority (~65%) of the additional people at risk of hunger in the future are in Africa. This partly reflects the greater-than-average reduction in yields in that continent owing to climate change, but is also due to higher levels of vulnerability there. It appears that the beneficial effects of stabilization are likewise less in this region. Under an S750 world the additional number of people at risk of hunger is reduced by ~30% while under an S550 future the reduction in the climate induced impact is only 20%. Because this conclusion assumes full CO_2 effects, more analysis is needed of the consequences of mitigation assuming partial effects of CO_2 fertilization.

Table 4.6. Average annual cereal production (million metric tonnes) (from Arnell *et al.*, 2001).

	No climate change	Unmitigated	S750	S550
1990	1800			
2020s	2700	2670–2674	2672	2676
2050s	3500	3475	3973	3477
2080s	4000	3927	3987	3949

The estimates assume no change in crop cultivar, and come from the Basic Linked System. The range in estimates for the unmitigated scenario represents the range between the four ensemble partners.

Table 4.7. Number of people at risk of hunger (millions) (from Arnell *et al.*, 2001).

	No climate change	Unmitigated	S750	S550
1990	521			
2020s	496	521–531	546	540
2050s	312	309–321	319	317
2080s	300	369–391	317	343

The range in estimates for the unmitigated scenario represents the range between the four ensemble partners.

Similar asymmetries of effect appear in a comparable study (Tubiello and Fischer, 2007). Some regions apparently become worse off with mitigation by not fully realizing positive agronomic impacts or economic benefits. However, the general and substantial effect of mitigation by stabilization of greenhouse gas concentrations is to reduce global additions to risk of hunger from climate change.

4.11 More Recent and Comprehensive Evaluations

In more recent analyses by IFPRI, the primary measures of the welfare effects of climate change are the change in calorie availability and the change in the number of malnourished children in 2050 without climate change and with various climate change scenarios (IFPRI, 2009; Nelson *et al.*, 2009). The following is taken from a summary written by Dr Mark Rosegrant for the World Food Programme (Parry *et al.*, 2009).

The diminishing consumption of cereals in particular translates into similarly large declines in calorie availability as the result of climate change. Results indicate that without climate change, calorie availability increases throughout the world between 2000 and 2050, except for a small decline in Latin America and the Caribbean (Table 4.8). The largest increase, of 12.6%, is in sub-Saharan Africa but Asian consumers also consume more, ranging from 3.5–7%.

With climate change, however, calorie availability is not only lower than the 'without climate change' scenario in 2050 (Plate 17). Worse, calorie availability actually declines relative to 2000 levels throughout the world. With CO_2 fertilization, the declines are 3–6% less severe but still large relative to the no climate change scenario. There is almost no difference in effect between the two climate scenarios.

Table 4.9 reports summary statistics for the child malnourishment indicator. With no climate change, only sub-Saharan Africa projects an increase in the number of malnourished children between 2000 and 2050. All other parts of the developing world have reductions in the number of malnourished children, driven by rapid income and agricultural productivity growth. Climate change eliminates much of that improvement. In East Asia and the Pacific, instead of 12 million malnourished children in 2050, results showed 16.5 million in both climate scenarios. In South Asia, instead of 52.3 million malnourished children in 2050, it is 58.1 million. In sub-Saharan Africa, climate change increases the number of malnourished children in 2050 by 10 million compared to the no-climate change case. Globally, climate change is projected to increase the number of malnourished children by 24 million in 2050 (about 20%) compared to the no climate change case. If CO_2 is in fact effective in the field, the negative effect of climate change on child malnutrition is reduced somewhat (to 16%).

The same model run for different assumptions about climate and development futures for 2050 arrived at lower levels of

Table 4.8. Daily per capita calorie availability with and without climate change (from Nelson *et al.*, 2009).

Region	2000	2050				
		No climate change	NCAR No CF	SIRO No CF	NCAR CF effects (%)	CSIRO CF effects (%)
South Asia	2381	2464	2089	2088	4	4
East Asia and the Pacific	2870	2979	2553	2555	4	4
Europe and Central Asia	3017	3231	2760	2759	3	3
Latin America and the Caribbean	2879	2862	2545	2546	3	3
Middle East and North Africa	2846	3112	2594	2595	3	3
Sub-Saharan Africa	2282	2570	2033	2017	6	6
Developed countries	3438	3606	3213	3213	2	2
Developing countries	2677	2750	2318	2315	4	4

Table 4.9. Projected number of malnourished children in 2000 and 2050 (thousands of children, under five years of age) (from Nelson *et al.*, 2009).

Region	2000	2050				
		No climate change	NCAR No CF	CSIRO No CF	NCAR CF effect (%)	NCAR CF effect (%)
South Asia	75,621	52,374	58,165	58,170	–3	–3
East Asia and Pacific	23,810	12,018	16,553	16,521	–8	–8
Europe and Central Asia	4,112	2,962	3,907	3,910	–4	–4
Latin America and the Caribbean	7,687	5,433	6,731	6,724	–4	–5
Middle East and North Africa	3,459	1,148	2,016	2,016	–10	–10
Sub-Saharan Africa	32,669	38,780	48,725	49,024	–5	–5
All developing countries	147,357	112,714	136,097	136,366	–4	–5

The last two columns in this table report the percentage difference between the number of malnourished children in 2050 with and without the CO_2 fertilization effect. For example, with the NCAR GCM, assuming CO_2 fertilization is effective in the field results in a 3% decline in the number of malnourished children in South Asia in 2050 relative to the climate change outcome without CO_2 fertilization.

impact (9–11% global increase in child malnourishment assuming CO_2 effect), but with a similar regional pattern (IFPRI 2010; Foresight. The Future of Food and Farming, 2011).

4.12 Conclusions about Projected Impacts on Risk of Hunger

In summary, if unmitigated by urgent and substantial reductions of greenhouse gases, climate change is projected to increase risk of hunger substantially, especially in the poorest parts of the world.

The scale of increased risk due to climate change varies with assumptions made about future development, especially future levels of poverty, but it is in the order of additional tens to hundreds of millions of people.

Most affected is expected to be Africa, especially its semi-arid regions north and south of the Equator, mainly because of projected increases in aridity due to climate change but also due to high vulnerability consequent on low levels of income. The poorest parts of South and South-east Asia are also likely to be substantially affected.

Initial studies concluded that decreases in yield of the major food grains (wheat, rice and maize), mainly due to increased heat and water stress resulting from climate change, would be greatest in developing countries, and estimated this decrease at

9 to 11% overall. The consequent decrease in global production of these grains was projected to be 25 to 150% and the related increase in risk of hunger was estimated to be 10 to 60%.

The above estimations are, however, essentially changes in potential and do not allow for adaptation in farming practice and institutions. With local small-scale adaptation, such as changes in planting times, it is estimated that increase in risk of hunger could be reduced to 5 to 50%; and with substantial adaptation, such as widespread additional irrigation, it could be reduced to 5%. Furthermore, the use of plant genetic resources in crop improvement will help address many of the constraints of climatic change on agriculture (see Djanaguiraman and Prasad, Chapter 12 on heat stress, Ceccarelli, Chapter 13 on drought, Erskine *et al.*, Chapter 14 on salinity, Ismail and Mackill, Chapter 15 on flooding tolerance in rice, and Pritchard *et al.*, Chapter 16 on pests, this volume).

Furthermore, a global response to climate change such as full liberalization of world food trade could, it is estimated, reduce the baseline numbers at risk of hunger by a fifth (i.e. the overall reference case in a world assumed not to be affected by climate change).

Subsequent analyses, using two different approaches, have both tended to confirm the previous estimates but with more detail about effects under different pathways of global development:

- For a development pathway of continuing high population growth and regional disparity of income (A2 in Table 4.2): numbers at risk of hunger due to climate change are projected to be 10 to 20% greater than the number expected without climate change (these data are for 2050).
- For a pathway of lower population growth and more equitable income distribution (B1 and B2 in Table 4.2): additional numbers at risk are estimated to be 5% or less (also for 2050).

In these analyses, the conclusion remains that Africa is the region with most increases of hunger risk owing to climate change (about 65% of the global total is projected to occur here).

Recent work by IFPRI confirms the magnitude and location of these production and food security impacts in developing regions, which are most severe in sub-Saharan Africa and South Asia. Critical impacts of climate change on food security include the effects in terms of calorie availability and the change in the number of malnourished children. Child malnourishment will be affected by climate change owing to impacts on production, food prices and consumption. The biggest increase in the number of malnourished children is in sub-Saharan Africa. Globally, climate change is projected to increase the number of malnourished children by about 10–20% in 2050 compared to the no climate change case, depending on the ultimate effects of CO_2 fertilization.

Actions to mitigate climate change by reducing greenhouse gas emissions would substantially reduce increases in risk of hunger. Stabilizing greenhouse gas concentrations to 550 ppm would reduce it by 50%. This illustrates the urgency of achieving international action to cut substantially global greenhouse gas emissions.

References

Arnell, N.W., Cannell, M.G.R., Hulme, M., Kovats, R.S., Mitchell, J.F.B. *et al.* (2001) The consequences of CO_2 stabilisation for the impacts of climate change. *Climatic Change* 12, 201–223.

Cline, W.R. (2007) *Global Warming and Agriculture: Impact estimates by country.* Center for Global Development, Washington, DC.

Darwin, R. (2004) Effects of greenhouse gas emissions on world agriculture, food consumption and economic welfare. *Climatic Change* 66,191–238.

Easterling, W. and Aggarwal, P. (2007) Food, fibre and forest products. In: Parry, M.L., Canzianai, O.F., Palutikof, J.P., van der Linden, P.J. and Hanson, C.E. (eds) *Climate Change 2007: Impacts, Adaptation and Vulnerability. Contribution of Working Group II to the Fourth Assessment Report of the Intergovernmental*

Panel on Climate Change. Cambridge University Press, Cambridge, UK, pp. 273–313.

FAO (2009) *The State of Food Insecurity in the World.* Rome.

Fischer, G., Frohberg, K., Keyzer, M.A. and Parikh, K.S. (1988) *Linked National Models: A Tool for International Food Policy Analysis.* Kluwer, Dordrecht, the Netherlands.

Fischer, G., Shah, M. and Van Velthuizen, H. (2002) *Climate Change and Agricultural Vulnerability.* IIASA Special report commissioned by the UN for the World Summit on Sustainable Development, Johannesburg 2002. International Institute for Aplied Systems Analysis, Laxenburg, Austria.

Foresight. The Future of Food and Farming (2011) *Final Project Report.* The Government Office for Science, London, UK.

Hulme, M., Mitchell, J., Ingram, W., Johns, T., New, M. *et al.* (1999) Climate change scenarios for global impacts studies. *Global Environmental Change* 9, 3–19.

IBSNAT (1989) *Decision Support System for Agrotechnology Transfer Version 2.1 (DSSAT V2.1).* Department of Agronomy and Soil Science, College of Tropical Agriculture and Human Resources, University of Hawaii, Honolulu, Hawaii.

IFPRI (2009) Addressing Climate Change in the Asia and Pacific Region: Building Resilience in the Agriculture Sector. TA 6479-REG. Short-term study funded by the Asian Development Bank. Unpublished report. Washington, DC, USA.

IFPRI, (2010) *Food Security, Farming, and Climate Change to 2050.* Washington, DC, USA.

IPCC (1997) *Stabilisation of Atmospheric Greenhouse Gases: Physical, Biological and Socio-economic Implications.* Intergovernmental Panel on Climate Change Technical Paper III. Intergovernmental Panel on Climate Change, Geneva, Switzerland.

IPCC (2007) *Climate Change 2007: Impacts, Adaptation and Vulnerability – Contribution of Working Group II to the IPCC Fourth Assessment. Climate Change 2007: Impacts, Adaptation and Vulnerability. Contribution of Working Group II to the Fourth Assessment Report of the Intergovernmental Panel on Climate Change.* Parry, M.L., Canzianai, O.F., Palutikof, J.P., van der Linden, P.J. and Hanson, C.E. (eds) Cambridge University Press, Cambridge.

Mitchell, J.F.B., Johns, T.C., Gregory, J.M. and Tett, S. (1995) Climate response to increasing levels of greenhouse gases and sulphate aerosols. *Nature* 376, 501–504.

Mitchell, J.F.B., Johns, T.C., Ingram, W.J. and Lowe, J.A. (2000) The effect of stabilising atmospheric carbon dioxide concentrations on global and regional climate change. *Geophysical Research Letters* 27, 2997–3100.

Nelson, G., Batka, M., Koo, J., Lee, D., Magalhaes, M. *et al.* (2009) *The Costs of Agricultural Adaptation to Climate Change.* Project report submitted to World Bank. World Bank, Washington, DC.

Parry, M.L., Fischer, C., Livermore, M., Rosenzweig, C. and Iglesias, A. (1999) Climate change and world food security: a new assessment. *Global Environmental Change* 9, 51–67.

Parry, M.L., Rosenzweig, C., Iglesias, A., Livermore, M. and Fischer, G. (2004) Effects of climate change on global food production under SRES emissions and socio-economic scenarios. *Global Environmental Change* 14, 53–67.

Parry. M.L., Evans, A., Rosegrant, M.W. and Wheeler, T. (2009) *Climate Change and Hunger. Responding to the Challenge.* World Food Programme, Rome, Italy.

Rosenzweig, C. and Iglesias, A. (1994) *Implications of Climate Change for International Agriculture: Crop Modeling Study.* US Environmental Protection Agency, Washington, DC.

Rosenzweig, C., and Iglesias, A. (2006) *Potential Impacts of Climate Change on World Food Supply: Data Sets from a Major Crop Modelling Study.* Goddard Institute for Space Studies, Columbia University, New York.

Rosenzweig, C. and Parry, M.L. (1994) Potential impacts of climate change on world food supply. *Nature* 367, 133–138.

Rosenzweig, C., Parry, M.L., Fischer, G., and Frohberg, K. (1993) *Climate Change and World Food Supply.* Research Report No. 3. Environmental Change Unit, University of Oxford, Oxford, UK.

Tubiello, F.N. and Fischer, G. (2007) Reducing climate change impacts on agriculture: Global and regional effects of mitigation, 2000–2080. *Technological Forecasting and Social Change* 74, 1030–1056.

5 Regional Impacts of Climate Change on Agriculture and the Role of Adaptation

Pam Berry[1], Julian Ramirez-Villegas[2], Helen Bramley[3], Mary A. Mgonja[4] and Samarendu Mohanty[5]

[1]Environmental Change Institute, University of Oxford, UK; [2]School of Earth and Environment, University of Leeds, UK and CGIAR Research Program on Climate Change, Agriculture and Food Security (CCAFS), Cali, Colombia; [3]Institute of Agriculture, The University of Western Australia, Crawley, Australia; [4]International Crops Research Institute for the Semi-Arid Tropics (ICRISAT), Nairobi, Kenya; and [5]International Rice Research Institute (IRRI), Los Baños, Philippines

5.1 Introduction

It has been estimated that during the last 300 years 42–68% of the Earth's land surface cover has changed, some multiple times, owing to cropping, pasture and wood harvesting, and nowadays 10% of the ice-free land is used for crop cultivation and 25% for pasture (Tubiello *et al.*, 2007). Increases in crop production have been met by extensification into new areas of primary or secondary vegetation or intensifying production through changing crop type or crop variety to produce more food per unit area (Foley *et al.*, 2011). It has been estimated that, while extensification is a future possibility for sub-Saharan Africa (27%) and Latin America and the Caribbean (33%), lack of available land will mean that intensification will be required in south and east Asia and the Near East/North Africa (Bruinsma, 2003). These land-use changes have consequences for the climate system but also they are affected by changes in climate.

These recent changes in climate provide an important means of understanding the impacts of climate change because they can be compared with observed changes in agriculture, such as type and timing of crops planted and yields. An analysis, using statistical models, of the effects of climate trends in temperature and precipitation from 1980 to 2008 on yields of maize, wheat, rice and soya beans showed that globally they had led to a net loss of 3.8% and 5.5% for maize and wheat, respectively, whereas the results were insignificant for the other two crops (Lobell *et al.*, 2011). There were important country-level variations, with Russia experiencing the largest losses, with almost a 15% decrease in wheat, whereas the largest producer – the USA – showed no significant effect because of a lack of a trend in climate. This study identified that the majority of the yields were driven by temperature, but that precipitation was an important driver of inter-annual variability. Lobell *et al.* (2011) suggest that this is consistent with modelled future projections, especially at the national and regional scales.

5.2 Impacts of Climate Change

Projections of future land-use changes using the latest Representative Concentration

Pathways (RCPs) in four different Integrated Assessment Models harmonized with inputs from historical data on land-cover changes have shown that in four focal case studies, three RCPs project cropland increases and three pasture decreases, but there was no consistency between them (Hurtt *et al.*, 2011). Only the RCP with the highest warming potential projected an increase in both, with variations in the geographical distribution and timings of these. This scenario projected that between 2000 and 2050 cultivated land could increase by 185 million ha and by a further 120 million ha by 2100 with a high density of cropland in the USA, Europe, and South-east Asia and of pasture in western USA, Eurasia, South Africa and Australia. This would be at the expense of natural or semi-natural vegetation with implications for biodiversity and other land uses.

There are several methods of projecting the impacts of climate change on agriculture, many of which involve modelling, either using statistical or semi-empirical parameterization of crop yield, climate scenario outputs to drive process-based crop models or process-based crop modelling (e.g. see Challinor *et al.*, 2009a for a review of different approaches; Lobell and Burke, 2010). It is important to note there are difficulties in matching spatial and temporal resolutions and most of these do not include key potential limiting factors, such as soil and water quality, pests and diseases (Easterling *et al.*, 2007) and management practices.

The modelling-based projections of the potential impacts of climate change on global agriculture display variations in the magnitude and direction of the impacts depending on the Global Climate Model(s) (GCM) and emissions scenarios used (see Betts and Hawkins, Chapter 3, this volume), crop and region under consideration (e.g. Challinor *et al.*, 2009b; Osborne *et al.*, 2013), although there are some broadly consistent trends.

This chapter analyses and discusses the impacts of future climate change for agricultural production and the projected regional differences in such impacts. First it reviews the aspects of climate change that can affect agricultural production, namely, increasing temperatures, changes in precipitation intensity and frequency, increasing atmospheric carbon dioxide (CO_2) concentrations, and changing patterns of pests and diseases. Case studies follow for three crops (wheat, rice and sorghum) that are particularly important in the human diet and in which modern plant breeding has been particularly successful (Ceccarelli *et al.*, 2010) in three representative, but contrasting, regions of the world (i.e. wheat in Western Australia, rice in Asia and sorghum in Africa). The chapter then illustrates how changes in climates could potentially result in climatic conditions not observed in past history, and discusses how these can affect future crop productivity. Each of the case studies highlights the main expected impacts, as well as the genetic traits that may be useful in countering such impacts. It concludes by discussing the needs of adaptation to climate change in agriculture and the potential benefits from improved plant genetic resources conservation, sharing and utilization.

Reviews of the global impacts of climate change on agriculture can be found in the IPCC 4th Assessment Report (Easterling *et al.*, 2007) and on crop productivity in Gornall *et al.* (2010). A synthesis of 69 studies showed a differential response across latitudes (Plate 18), with mid- to high latitudes generally showing slight increases in yields for temperature increases up to 3°C and with a range of rainfall changes and CO_2 concentrations. In low latitude regions, however, even slight temperature increases could have a negative impact on productivity of major crops because agriculture in these areas is more marginal and may be sensitive to even small changes (Gornall *et al.*, 2010). Changes in precipitation are more difficult to project and show great regional variations in the magnitude and sign of change (see Betts and Hawkins, Chapter 3, this volume), but are extremely important as more than 80% of agriculture is rainfed (Christensen *et al.*, 2007; Gornall *et al.*, 2010). More recent modelling (Nelson *et al.*, 2010) using climate scenarios from two GCMs that cover the range of the means of GCM ensembles – CSIRO (a dry relatively cool scenario) and MIROC (a warm, wet scenario) – compared the ratio of yields in 2000 with those for 2050. The study showed similar broad latitudinal outcomes, but some geographical and crop variations (Plate 19). In the tropics, the wetter MIROC scenarios lead to higher yields of rainfed crops, whereas the higher temperatures reduce rainfed wheat here and irrigated yields generally. Of particular note is the projected 33% reduction of average rainfed maize yields in eastern USA.

In addition, meta-analyses of climate change impacts exist for Africa (Müller *et al.*, 2011; Roudier *et al.*, 2011; Knox *et al.*, 2012) and South Asia (Knox *et al.*, 2012). These depict consistent pictures in terms of negatively impacted crops and regions, with rice and wheat showing the largest negative impacts across these two regions, and cassava, sorghum and millet standing as the least impacted crops (Lobell *et al.*, 2008; Jarvis *et al.*, 2012). Increases in temperatures are expected to reduce crop productivity directly through reductions in crop duration (Challinor *et al.*, 2009b; Singh *et al.*, 2012), heat stress during reproductive stages (Hawkins *et al.*, 2013; Challinor *et al.*, 2007b), increased senescence (Lobell *et al.*, 2012) and loss of efficiency in assimilate production (Zhu *et al.*, 2010); and indirectly through increases in evapo-transpirative demand in water-limited environments (Dai, 2012; Fraser *et al.*, 2013). Thus, increases in temperature will affect all cropland, but the warming is projected to be greater in mid-to-high latitudes, where temperatures are lower and more variable throughout the year (see Betts and Hawkins, Chapter 3, this volume).

The regional variations in climate change impacts on agriculture, therefore, are important and will be explored in more detail in the context of Europe and Africa, in order to provide contrasts in these by latitudinal location and type of agriculture. A review of the recent literature on climate change impacts on African agriculture has confirmed the importance of rainfall in affecting yields, and Africa is considered the region at the greatest risk from drought (Li *et al.*, 2009; Sheffield *et al.*, 2012). As with other regions, there are countries and crops that are projected increased yields, although many are negative (Challinor *et al.*, 2007a), but much also depends on the human dimensions. The median decreases in yields for maize, sorghum, millet, groundnut and cassava were –22%, –17%, –17%, –18% and –8%, respectively, but with country level variations (Schlenker and Lobell, 2010). Under 1°C, 3°C and 5°C projected rises in temperature, for example, maize yields are expected to decrease, but with a large range of responses, with Niger showing the smallest losses (–7.4%; –19.66% and –28.87%, respectively) and Zimbabwe the largest (–18.9%; –54.12% and –80.44%, respectively). These figures for the highest climate scenario are lower than those of Thornton *et al.* (2011) who found that the mean change in production relative to current conditions (2000s) by the 2090s with a 5°C increase in temperature was –24% for maize and –71% for *Phaseolus* beans. In both cases, the western region (Benin, Burkina Faso, Côte d'Ivoire, Gambia, Ghana, Guinea, Guinea Bissau, Liberia, Mali, Mauritania, Niger, Nigeria, Senegal, Sierra Leone and Togo) had the highest projected losses. Thornton *et al.* (2011) also showed the likelihood of crop failure to increase in all areas apart from central Africa and, in southern Africa, all rain-fed agriculture could fail every other year. They were, however, based on different GCMs, climate scenarios and crop models, once again emphasizing the difficulty of cross-study comparison.

For Europe, modelling has shown that by 2080 – depending on the scenario – the area of cropland could decrease by 28% to 47%, which could free up land for other uses including biofuels and nature conservation (Rounsevell *et al.*, 2005). In Europe, the impacts of climate change will also vary by region: higher increases in winter temperatures are projected for northern Europe but higher increases in summer temperatures in southern Europe, and annual rainfall is projected to decrease south of 55°N (Moriondo *et al.*, 2010). The increase in temperatures in northern Europe means that more temperature-sensitive crops, such as maize, sunflower and soybeans, will be able to be grown and yields could increase by as much as 30% by the 2050s, dependent on crop type (Olesen *et al.*, 2007), whereas other research using different scenarios and models suggested that yields of spring wheat could increase 7% and sunflower 8%, but soybean decrease 4% (Moriondo *et al.*, 2010). A regional analysis of changes in crop (wheat, maize and soybean) yields for 2071–2100 showed that the Boreal region (Finland and Sweden) could have the greatest gains (34–54%), depending on GCM and emissions scenario (Iglesias *et al.*, 2012). The same study showed that the greatest losses could be experienced in the Mediterranean and Atlantic South (a broad Atlantic coastal zone of France, Spain and Portugal) and other studies have also identified southern and south-eastern Europe as potentially similarly affected with possibly decreases of up to 30% by the 2050s (e.g. Olesen and Bindi, 2002; Maracchi *et al.*, 2005), although Moriondo *et al.* (2010) found lower decreases: soft wheat and sunflower (5%) and soybean (13%). These regional snapshots serve to show the range of variation in the magnitude of the changes in yields, although most studies agree on their direction.

5.2.1 Impacts of increased CO_2 concentrations

There is still uncertainty about the exact consequences of increased CO_2, but it will benefit C_3 crops, such as rice, wheat and the grain legumes (e.g. soybean, dry bean and groundnut) rather than C_4 ones, such as maize and sorghum. It has been estimated that during the past 50 years, rising CO_2 has led to an increase in yields of maize and soybean by 9% and 14%, respectively, under dry conditions (McGrath and Lobell, 2011). These findings approximate to those from free-air CO_2 enrichment (FACE) experiments, although there is a range of results with C_3 crops possibly showing up to a 73% increase in yield (Ainsworth and Long, 2005).

Modelling using multiple climate scenarios with CO_2 constant at 2000 levels (370 ppm) or allowing for a full CO_2 fertilization effect showed that if CO_2 increases are not taken into account, then globally crop yields could decrease by 0–13%, but its inclusion could lead to increases of 8–22% by 2050 (Müller *et al.*, 2010), whereas Rost *et al.* (2009) project global increases in crop net primary productivity of 28% by the 2050s. At the national and local scales, climate played a larger part in yield uncertainty, especially in Africa and Latin America (Müller *et al.*, 2010) and this may be partly a function of the relationship of CO_2 with water stress (Challinor and Wheeler, 2008). Although it is expected that water-stressed crops and similarly rain-fed cropping systems will benefit more from increased CO_2 because it increases water-use efficiency, the evidence is not consistent in models, chamber or FACE experiments (see Challinor and Wheeler, 2008; McGrath and Lobell, 2011).

5.2.2 Impacts of water availability

Water is critical to crop growth and it is thought that seasonal changes in precipitation may be more important that mean annual changes (Gornall *et al.*, 2010). In the past, drought events have led to large decreases in crop productivity (Sivakumar *et al.*, 2005) and, globally, drought is already increasingly affecting barley, maize, rice, sorghum, soya bean and wheat, with the area affected expanding from about 5–10% in the 1960s to 15–25% (Li *et al.*, 2009). A drought risk index (DRI) based on drought disaster frequencies, drought severity and levels of production, management and availability of irrigation showed Africa as having the highest current and future DRI (Li *et al.*, 2009) and, assuming that the linear relationship between the DRI and yield reduction rate continues into the future, they calculated that by the 2050s drought-related yield reductions for barley, maize, rice and wheat could be 17%, 21%, 7% and 18%, respectively, and by the end of the century, cropland drought-disaster risk could double.

Given the uncertainty in projecting precipitation, the amount, location and sign of changes in water availability show a high level of variation (e.g. Nelson *et al.*, 2009). Modelling of the impacts of climate change on water availability and CO_2 fertilization on agriculture using only the SRES A1B and A2 scenarios showed that by 2050s world food production could decrease by about 2.3% under both emissions scenarios (Calzadilla *et al.*, 2010). There would, however, be significant regional variations; just taking temperature, precipitation and CO_2 into account, production is expected to increase by the 2050s in the developed world, except Canada, the USA and the former Soviet Union, with possibly decreases of over 20% for the latter and decreases in developing countries, except China and sub-Saharan Africa. Scenarios combining temperature, precipitation and CO_2 fertilization, river flow and irrigation area, show that total production could decline by 0.5% in the 2020s and by 2.3% in the 2050s, with the same regional differentiation in the developed world, and decreases in developing countries being greatest in the Middle East (23.24%), South-east Asia (11.63%) and North Africa (8.9%) under the A1B scenario. Other studies have also highlighted the arid and semi-arid tropics in Africa as very vulnerable to drought risk and tropical Asia as sensitive to changes in the monsoon (Sivakumar *et al.*, 2005). In a study of the major agricultural regions in Argentina, Brazil, China, Hungary, Romania and the USA, Rosenzweig *et al.* (2004) found that mostly increases in water supply were projected, with north-east China likely to be most water-stressed. Given the importance of water, irrigation will be critical in some areas to maintaining yields or realizing projected increases, but it is not certain how much water will be available for agriculture given competing demands from industrial and domestic usage. Greater efficiency of agricultural water use may help offset some of the demand or crop yield losses, but it is unlikely to be sufficient on its own (Rost *et al.*, 2009). Other aspects of drought are considered by Ceccarelli (see Chapter 13, this volume).

5.2.3 Impacts of pests and diseases

Despite efforts to protect against pests and diseases, current crop losses in the field still lead to significant reductions in actual yields. In a review of actual losses, Oerke (2006) estimated that globally losses for soybean, wheat and cotton were 26–29%, and 31, 37 and 40% for maize, rice and potatoes, respectively, and without crop protection, losses for wheat, for example, could be about 50%. The main sources of global losses were weeds (34%), with animal pests and pathogens counting for 18% and 16%, respectively (Oerke, 2006). It is expected that these losses will reduce projected future yields, despite crop protection methods, such as increasing biodiversity, including species genetic diversity, with in-field genetic diversity being important in providing disease resistance (Frison *et al.*, 2011). Further discussion of the impacts of pests and diseases are discussed by Pritchard *et al.* (see Chapter 16, this volume).

5.3 Regional Impacts of Climate Change: Case Studies on Wheat, Rice and Millet

For each of these case studies, an analysis of climate shifts and climate similarity was undertaken and further discussion was based upon a selected region (south-western Australia for wheat, Asia for rice and eastern and southern Africa for sorghum).

The analyses presented here investigate the likely future shifts in crop-specific growing season climates by estimating the spatial variation of the overlap between current and future climates, as predicted by the Coupled Model Intercomparison Project phase 5 (CMIP5) GCM (Taylor *et al.*, 2012). Heat stress (i.e. maximum growing season temperature) was chosen as the variable of interest because it is one of the main drivers of crop yield loss (Challinor *et al.*, 2007b; Lobell *et al.*, 2012; Gourdji *et al.*, 2013) and because temperature changes are relatively well constrained in global model simulations (see Betts and Hawkins, Chapter 3, this volume). Historical and future (i.e. RCP 4.5) simulations of ten CMIP5 GCM[1] were used to investigate the extent to which changes in climates can induce heat-stress events never observed in recent history. Calculations were performed on a grid cell basis, with a grid cell size of 0.5 decimal degrees. The analysis focused on the harvested areas of each crop, as defined by Monfreda *et al.* (2008), and on the growing seasons defined in the dataset of Sacks *et al.* (2010). Using historical gridded observations from the Climatic Research Unit (CRU) climate dataset version 3.1 (New *et al.*, 2000; Mitchell and Jones, 2005), probability density functions (PDFs) of the 1–99% percentile of maximum growing season temperatures of the past 40 years (1966–2005) were first calculated (i.e. historical PDF). Projected future changes in monthly climatological means of maximum temperature (as predicted by each of the ten GCM simulations) for two 40-year periods centred on 2035 (2016–2055) and 2075 (2056–2095) were then added to the historical monthly observations of CRU (i.e. a procedure often called 'delta method'; also see Hawkins *et al.*, 2013). PDFs of future (2035 and 2075) maximum temperatures were constructed using this information. Using a similar approach to Burke *et al.* (2009) and Battisti and Naylor (2009), the overlap between the present and future distributions of each pixel was computed as a way of assessing the shift in climates between the present and the future. Uncertainties in the resulting climate overlaps were assessed by calculating relevant statistics (e.g. standard deviation, extremes of the distribution) using the multi-GCM ensemble.

Shifts in growing season climates by 2075 for wheat showed that the areas with the least similarities between current and future climates are located across the Andes, eastern USA, eastern Africa (central Kenya, southern Ethiopia and southern Uganda), and South-east Asia (Thailand, Myanmar, east India and Bangladesh). In these areas, overlap between current and future scenarios ranged between 10% and 40% (depending on location and GCM used) (Plate 20). Projected increases in temperatures are expected to be harmful, particularly for spring wheat growing areas where temperatures are already high (Gourdji *et al.*, 2013). In eastern China and central Russia, as well as some areas of eastern Europe (Ukraine, Romania and Poland), overlap between current and future climates was higher (40–60%).

Overlaps between observed and projected maximum temperatures for rice were generally lower than those of wheat, probably as a result of rice being grown more commonly during warmer periods. By 2075, roughly 5–30% (depending on the GCM) of the areas across the tropics showed less than 25% overlap between current and future maximum temperature distributions (Plate 20). Critically, the globally important producing areas of Bangladesh, Thailand, Malaysia in Asia, showed only 5–10% overlap (Plate 20). In addition, most growing areas in Latin America (eastern Andes, Brazil and Central America) and central Africa (eastern Tanzania, Democratic Republic of Congo) feature very low (<25% overlap) between future and historical climates in some areas.

Results for sorghum were somewhat similar to those of rice (Plate 20), although fewer areas showed low overlap (below 50%). Temperate areas in the USA as well as some areas of Europe could experience similar future climate overlaps (50–60%) to some areas in the tropics. Yet, the lowest overlaps (between 0% and 40% in most cases) were observed across the tropics. East and southern Africa show a spatially varying picture, with the subtropical areas of southern Africa depicting low proportions of novel climates (70–80% overlap). Conversely, East Africa (Tanzania, north of Mozambique, Uganda and Rwanda) showed a significantly larger proportion of non-analogue climates (overlap below 20%, Plate 20).

Changes in climates will thus probably increase temperatures beyond the observed variability in the growing areas (Plate 20; also see Burke *et al.*, 2009 for maize). For the three crops it is across the tropics where novel maximum temperatures are projected to occur much more frequently. This can be attributed to the lower temporal variation (both within the growing season and through years) that is observed across the tropics (Battisti and Naylor, 2009). Remarkably, regional variations are consistent across GCMs, although there is uncertainty in the values of projected climate overlap (Plate 20, right). This indicates that, regardless of GCM uncertainty, novel climates will be experienced throughout the world, and specifically in tropical areas where temperatures are high and show low temporal variability. This further agrees with previous research (Williams *et al.*, 2007; Battisti and Naylor, 2009). Projected temperature increases can reduce agricultural yields through a variety of mechanisms (Challinor *et al.*, 2010; Lobell *et al.*, 2012). It is very unlikely that enhanced CO_2 concentrations can offset the negative effects of higher temperatures in all growing areas (see Djanaguiraman and Prasad, Chapter 12, this volume), particularly if these are coupled with increased water stress (Battisti and Naylor, 2009; Sheffield *et al.*, 2012; Fraser *et al.*, 2013).

5.3.1 Climate change impacts on wheat production: a case study of south-west Western Australia

Wheat production in the south-west of Western Australia provides a key example of agricultural production under limited land resources and increasing pressure from climate change (WCWA, 2010; Sheffield *et al.*, 2012). Wheat is Australia's most important grain crop with an average of 26 million tonnes produced annually from half of the nation's cropping land and a gross production value of AUS$6 billion (ABS, 2006). It is the world's 9th largest wheat producer (FAO, 2010) with two-thirds of production being exported, mostly to Asia and the Middle East, making Australia the world's fourth largest wheat exporter. Despite hostile and nutrient-poor soils, increasing temperatures, climate variability and declining rainfall, wheat production steadily increased during the 20th century owing to improved agronomic practices and new varieties. However, the hotter and drier conditions predicted due to climate change, particularly the timing of these conditions during the growing season, threaten future production levels and reliability as an export commodity at a time when global demand for wheat is increasing. Research programmes that lead to a better understanding of wheat physiology and genetics are addressing these challenges to develop crops adapted to the future climate in targeted growing regions. This case study focuses on the region that produces the largest proportion of Australia's wheat: south-west Western Australia.

Growing conditions in south-west Western Australia

The agricultural region in south-west Western Australia has a Mediterranean-type climate, characterized by winter dominant rainfall and hot, dry summers. Growing mostly on sandy-loam soils with low water retention, wheat production in this region relies entirely on within season rainfall, which is highly variable even within the wettest months and has declined by approximately 7 mm per decade since 1900 (Australian Bureau of Meteorology, 2013). Wheat is sown in autumn, taking advantage of early season rains, and harvested in late spring/early summer when rainfall is decreasing and temperature and pan evaporation are increasing. Water deficit often occurs during the reproductive and grain-filling stages, called terminal drought, causing the greatest reduction in crop yields (Dracup *et al.*, 1998).

Climate change and impacts on wheat production

The decline in growing season rainfall since the 1970s has occurred mainly in June and July when rainfall exceeds crop demand for water so that wheat yields have apparently not been impacted (Ludwig *et al.*, 2009). Growing season rainfall, however, especially in spring, is predicted to further decline by 10 to 20% by 2030 (Pittock, 2003), which could have a significant impact on wheat yields because crop demand for water is high during flowering and grain-filling and daily temperatures are increasing. Water deficit during this critical period causes smaller grains in wheat owing to reduced carbon assimilation and increased leaf senescence (Palta *et al.*, 2004; Yang and Zhang, 2006). More late seasonal breaks for sowing in autumn are also expected, after which there is sufficient rainfall to sustain growth. Breaks late in the season will result in later sowing or risking water deficit during germination and crop establishment. Already, there is 50% chance that the first rains of the season are a false start (Armstrong *et al.*, 1996).

Mean temperatures in the agricultural zone of south-western Australia have increased by 0.9°C over the past 50 years (Hennessy *et al.*, 2008; CSIRO and Bureau of Meteorology, 2010), bringing with them more exceptionally hot years and days above 35°C but fewer frost risks (CSIRO, 2007; Hennessy *et al.*, 2008). According to models, every day above 34°C reduces wheat grain yields by 5%, mainly due to increased leaf senescence (Asseng *et al.*, 2011). Over the next 30 years, average daily temperatures in the region are expected to further increase by 1–2°C (Pittock, 2003), which is expected to shorten the length of the growing season because wheat phenology accelerates with increasing temperature. The duration of grain filling is critical in determining final grain yield of wheat (Loss and Siddique, 1994) and although more heat-tolerant cultivars tend to have greater rates of grain filling, the rate is insufficient to compensate for a shorter grain-filling duration (Wardlaw and Moncur, 1995; Zahedi and Jenner, 2003).

Crops are expected to respond positively to the increasing atmospheric CO_2 levels (Amthor, 2001; Kimball *et al.*, 2002), but yield gains will probably be only achieved in the cooler, wetter southern regions of Western Australia's wheat belt (Ludwig and Asseng, 2006). Moreover, yield gains will be dependent on genotype and the level of interaction between temperature, rainfall and CO_2 (Dias de Oliveira *et al.*, 2013). Average temperatures that exceed current ambient levels by more than 2°C will negate any beneficial effect of elevated CO_2.

Adaptation strategies for future climate change

Management strategies, such as conservation agriculture, have been adopted by most Australian wheat farmers. Sowing earlier is one way to avoid terminal drought but the late season breaks may make this impossible or, at the very least, genotypes with improved drought tolerance during emergence and establishment are needed.

Modern wheat varieties selected for Western Australia develop faster and flower earlier (Siddique *et al.*, 1989), which is also an advantage in avoiding terminal drought and heat stress conditions (van Ittersum *et al.*, 2003), but further improvements in phenology need to be linked with traits such as increased partitioning of assimilates to grains and rapid growth when rainfall is highest so that sufficient biomass is accumulated to support spike and grain development.

Modern varieties of wheat also have greater water use efficiency (WUE) compared with old varieties (Siddique *et al.*, 1990), that is they produce more grain per unit of water. Although high WUE is advantageous for crops growing on stored soil moisture, it is generally associated with lower yield potential and slower growth (Blum, 2005, 2009) so selection for more effective water use, such as early vigour, is needed for Western Australia.

Modern varieties also have fewer tillers, which are stronger and have greater survival rates than the numerous sub-tillers in old varieties (Siddique *et al.*, 1989); however, reduced tillering appears to limit the beneficial effect of elevated CO_2. Free-tillering genotypes had a yield advantage over reduced tillering genotypes mainly because they produced more ears per m^2 under elevated CO_2 combined with terminal drought and +2°C over ambient temperatures (Dias de Oliveira *et al.*, 2013).

Early root vigour is being targeted as a future trait for selection (Palta and Watt, 2009) because current wheat varieties have inadvertently smaller root systems (Siddique *et al.*, 1990). Root vigour, particularly increased branching, will enable more effective capture of water and nutrients in the sandy-loam soils. Predictions from simulation models suggest that early root vigour will improve yields under future climate conditions of elevated CO_2 and temperature, particularly in regions with greater rainfall (Ludwig and Asseng, 2010). However, the effect of elevated CO_2 and temperature on root growth needs investigation because preliminary studies indicate that vigour lines may have reduced root growth under elevated CO_2.

5.3.2 Climate change impacts on rice production: a case study of Asia

Rice, a staple for nearly half of the world's 7 billion population, is grown in more than 100 countries. More than 90% of rice is, however, produced and consumed in Asia where it constitutes the basic staple of the region's 560 million hungry people. The projected climate changes, with a greater frequency of extreme weather events (e.g. floods and droughts), and greater incidence of pest and diseases are likely to complicate further future rice production. Because rice is grown under diverse agroecological conditions, climate change is likely to have varying impacts on different rice-growing regions. According to Datta (2004), 20–70% of the potential yield in favourable ecosystems in the developing countries is already lost to various abiotic (flood, drought and salinity) and biotic stresses (pests and diseases). Among abiotic stresses, drought is the largest constraint, affecting nearly a third of the total rice area in Asia (Venuprasad *et al.*, 2008) and causing significant economic losses to poor rice producers in the region. According to Pandey and Bhandari (2007), average production loss due to drought in three regions, eastern India, north-east Thailand and southern China, adds up to US$200 million annually. Similarly, flooding irregularly causes severe rice yield damage on 16 million ha of flood-prone rice area in Asia (Johnson *et al.*, 2012). The annual loss in rice production in south Asia alone is estimated to be 4 million tons (IRRI, 2010). Salinity, although not as big a stress as drought and flood, still affects 10 million ha of coastal and inland areas (Johnson *et al.*, 2012).

Crop losses are likely to be greater in the future with changing climate that will make flood, drought and salinity more frequent and severe. In addition, rising sea levels are expected to spread the salinity problem to a wider coastal area where rice is currently grown under favourable conditions. The rice area in some deltas may also be affected by salinity. For example, the Mekong River delta, which accounts for nearly half of Vietnam's total rice production, is vulnerable to future sea-level rise. Salinity may also spread beyond the coastal areas and deltas to some groundwater irrigation agriculture in the interior regions, where excessive withdrawal could increase salinity in groundwater, making it unsuitable for use in rice production.

In addition to higher production losses caused by more frequent and more severe droughts, floods and salinity, rising temperatures will also negatively affect future yield and grain quality (Fitzgerald and Resurreccion, 2009), especially if temperature becomes higher during the reproductive phase, particularly at flowering (Jagadish *et al.*, 2010). In the past three decades, night-time minimum temperature has been rapidly rising and the current trend is expected to continue in the future (Fischer and Knutti, 2013). Studies by Welch *et al.* (2010) and Peng *et al.* (2004) estimated that a 1°C rise in night-time temperature above the critical temperature (>24°C) would reduce yield and biomass by 10%. The rise in frequency of drought and floods is also likely to make farmers in the stress-prone environments even more risk-averse, causing further decline in yields in normal years.

In addition to the above-mentioned abiotic stresses, many biotic stresses, such as pest and disease incidence and weed infestation, are also predicted to rise with climate change and would represent a major challenge for sustainable rice production. According to Sparks *et al.* (2012), average yield loss in Asia due to major pests and diseases now is a staggering mean of 37% with a range between 24% and 41%, depending on the production system. These losses would be greater with climate change that would lead to frequent extreme weather, water scarcity and uneven rainfall. This is particularly true for brown spot and blast, which are expected to intensify with water shortage and irregular rainfall (Savary *et al.*, 2005; Luck *et al.*, 2011).

Climate change adaptation for rice production

One of the most important adaptation strategies is to develop varieties that can withstand extreme weather conditions to enable them to perform better than current modern high-yielding varieties. In 2008, IRRI released submergence-tolerant rice varieties that were developed by transferring the submergence tolerance gene (*Sub1*) into widely grown modern rice varieties without affecting such preferred characteristics as high yield, good grain quality, and pest and disease resistance (see Ismail and Mackill, Chapter 15, this volume). These varieties can withstand submergence for up to 2 weeks, which is long enough to completely destroy the high yielding modern varieties that occupied 70% of Asian rice area in 1990 (IFPRI, 2002). So far, more than 15 popular varieties with the *Sub1* gene have been released in many South and South-east Asian countries. As expected, these flood-tolerant varieties have become very popular in South Asia, with preliminary adoption by nearly 1–2 million farmers in the past 4–5 years. Similarly, IRRI's first drought-tolerant variety, IR74371-70-1-1, which was recommended for release in South Asia in 2011, and a few more drought-tolerant varieties that are in the pipeline, if successful, could have an even greater impact on production than submergence-tolerant varieties. The rice varieties developed for salt tolerance, through collaborative research at IRRI and various national partners, are also starting to make an impact on millions of hectares of salt-affected rice.

A recent study by Mottaleb *et al.* (2012), using a multidisciplinary modeling framework, estimated the net economic benefits of developing and disseminating drought-tolerant varieties in South Asia. The study used the crop growth simulation model ORYZA2000 (Bouman *et al.*, 2001) to estimate average yield gain under different climatic scenarios and fed those yield changes into a partial equilibrium structural econometric model developed by Mohanty *et al.* (2011) to measure price and trade flow changes. The results suggest that if a drought-tolerant variety is released by 2016, then cumulative net economic benefits between 2016 and 2035 would amount to US$2 billion (assuming no climate change). Under different climate scenarios, however, total net benefits decline to US$1.5–1.9 billion because of lower yield gain. For example, the yield gain from the proposed drought-tolerant variety in Bangladesh and India could be 1.71% and 6.85%, respectively, under a no climate change scenario, whereas under an A1B climate scenario, the average yield gain could be 1.4% and 5%, respectively. Another important development by IRRI is varieties with multiple stress tolerance because many rainfed lowland areas are prone to the twin problems of drought and flood in the same season, the likelihood of which is expected to be higher in the future. IRRI scientists now have successfully introgressed a drought quantitative trait locus (QTL) into Swarna *Sub1* and results of initial trials on this variety under double stress are very promising.

Apart from planting stress-tolerant varieties, farmers need to modify/change their management practices to counter the negative effects of climate change on rice production. First and foremost, water-saving technologies, such as alternate wetting and drying, land levelling, improved tillage, bound preparation and direct seeding can make a significant difference in keeping the plant standing in the face of water shortage and irregular rainfall. In addition, farmers should be educated on pest dynamics and the benefits of diverse ecosystems to keep pests at a distance in order to minimize pesticide overuse. This cannot happen without the active participation of government agencies that can reform policy to prevent excessive use of pesticides.

5.3.3 Climate change impacts on sorghum production: a case study of eastern and southern Africa

Sorghum is cultivated in the drier areas of Africa, Asia, the Americas and Australia. Globally, it is the fifth most important cereal crop and is the dietary staple of over 500 million people in about 30 countries. Sorghum grain is mostly used for food purposes but also as a source of starch in industrial processing of animal and poultry feeds. The major producers in Africa are Nigeria and Sudan; others include Burkina Faso, Ethiopia, Tanzania, Mozambique and Somalia. In the eastern and southern Africa (ESA) region it is the third most important crop after maize and beans (ASARECA, 2005) and is mostly cultivated in zones with 500 to 1000 mm of rainfall. As a C_4, hot-season cereal crop, it has an extremely wide range of adaptation potential. Almost 85% of the sorghum produced in ESA is used for food and the non-food industrial uses are relatively underdeveloped but have the potential to increase.

Recent global trends in the major cereals show increases in both production and grain yield, partly as a result of the use of improved varieties and better crop management practices, as well as increased demand and higher world prices. Sorghum production in the ESA region is, however, characterized by declining yields. The regions where sorghum is grown are already towards the higher side of its temperature tolerance range, with the optimum temperature for reproductive growth being 25 to 28°C. High temperature stress during the reproductive phase can affect the crop substantially (Hammer and Broad, 2003), thus a small change could drastically reduce sorghum productivity. Smallholder agricultural productivity in ESA is already declining too and smallholder farmers across the region are concerned that shifting climatic conditions are a new problem that contributes to this falling productivity. Evidence suggests that climate conditions are changing and will continue to change (Boko *et al.*, 2007) with warming in sub-Saharan Africa (SSA) expected to be greater than the global average and rainfall declining in certain areas (Dimes *et al.*, 2008; Ringler *et al.*, 2010).

Ringler *et al.* (2010), on the basis of IFPRI IMPACT modelling projections, indicated that cereal production for a range of crops in SSA is projected to decline by 3.2% net in 2050 as a result of climate change. In Zimbabwe (Dimes *et al.*, 2008), the APSIM simulation model indicated potential crop yield reductions of 8–30% under climate change with increased temperature having the most significant impact on grain yield, with reductions of 16% for maize and sorghum and 31% for groundnut, but only 3% for pigeon pea. The combined effects of climate change could lead to a reduction in potential grain yield of 8% for pigeon pea, 30% for groundnuts, 25% for maize and 22% for sorghum. Sorghum will experience greater shortening of the vegetative phase (18%) with increased temperatures relative to the grain filling (14%) resulting in an increased Harvest Index (HI) while retaining a WUE of 6.7 kg ha^{-1} mm^{-1}. Results from another modelling case study in Botswana (Chipanshi *et al.*, 2003) indicated that climate change impacts could result in declining crop yields of 36% for maize and 31% for sorghum. For comparison, a simulation study in India found that a 1°C increase in average air temperature could decrease the yield of sorghum by from 4 to 8% in the rainy season and winter sorghum suffered yield losses of 8–15% with a 2°C rise in temperature (Srivastava *et al.*, 2010).

Overall these modelling results indicate that SSA will be more vulnerable to climate change because the low yield levels and limited agricultural inputs use are combined with a high dependence on rainfed agriculture and high poverty levels. Undoubtedly, increased production and productivity is a prerequisite to achieving food security, as well as commercialization for improved livelihoods and it is thought that the productivity of sorghum can be doubled or tripled from their current low levels of about 0.5 to 1.5 metric tons of grain per hectare through integrating improved cultivars with natural resource management, especially of water and soil fertility (Twomlow *et al.*, 2008). Nevertheless strategies are needed to sustain rainfed agriculture in SSA and facilitate adaptation to climate change.

Sorghum adaptation strategies for future climate change

There are many potential adaptation options available for marginal change of existing agricultural systems relevant to sorghum, including:

1. Targeted diversification of production systems and livelihoods. Because sorghum originated in Africa and is a more drought-resistant crop than other cereals, switching back to sorghum from other water-loving crops, such as maize, is a possible strategy and it is technologically feasible. Sorghum's climate resilience was highlighted during the devastating drought of 2012, especially in the Horn of Africa. There are ongoing activities in eastern Africa on introducing sorghum to areas that historically are not sorghum-producing areas, but that are becoming drought prone and where maize – a main staple – is no longer withstanding the climate. Participatory approaches of evaluation and promotion of integrated sorghum-based technologies for improved food security and income are being implemented in climate change pilot sites in eastern Kenya and Tanzania, with emphasis on productivity together with linking farmers to sorghum markets.

2. Technological developments in breeding, including developing variety types with high degrees of heterozygosity and genetic heterogeneity for adaptation traits to increase buffering capacity. Breeders are evaluating sorghum in representative and stratified test sites (Mgonja *et al.*, 2006) that include analogue sites to allow larger-scale, on-farm participatory testing and enable assessments of varietal performance under evolving climatic variability. Breeders also are using molecular markers in breeding platforms and these have led to the development and release of four *Striga*-resistant varieties of sorghum in the Sudan against the genetic backgrounds of the popular, but *Striga*-susceptible, improved sorghum varieties: Tabet, Wad Ahmed and AG-8 (Mugoya and Masiga, 2012). Reduced leaf senescence (stay-green) has been demonstrated to improve tolerance of post-flowering moisture stress in grain sorghum. A number of QTLs associated with stay-green have been identified in sorghum, to facilitate transfer of this trait into adapted genetic backgrounds. Other opportunities for the future include developing and deploying hybrids to increase yields for a wide range of production systems in Africa.

5.4 Possible Adaptation Strategies for Agriculture

Agricultural adaptation to climate change is expected to occur in various ways during the 21st century (Howden *et al.*, 2007), mostly as a result of two different, but complementary, processes: short-term coping strategies and longer-term planned system changes. Adapting a cropping system means that adjustments to the cropping system are done as a response to variations that are occurring or are expected to occur. Such adjustments can be implemented at a variety of temporal scales and, in some circumstances, may not be useful in achieving the desired objective (Ellis, 2000; IPCC, 2001; Moser and Ekstrom, 2010). It is due to such difference in temporal scales that the adaptation process is generally viewed as an iterative process.

In responding to climate change, it is critical to identify the magnitude of the risk involved in not responding, which varies depending on the lead time at which the impacts of climate change are analysed. At very short time scales (e.g. a year or less) adjustments (e.g. such as changes in sowing dates, timings and amounts of fertilizer, irrigation and fungicide) are done by farmers as a way to cope with conditions that are being or are expected to be detrimental. In this case, the cropping system evolves on the basis of these incremental short-term changes in a rather messy manner (Moser and Ekstrom, 2010). At longer lead times, where the degree of climate change is expected to be higher (Joshi *et al.*, 2011), adaptation to climate change may require changes to the cropping system as a whole, by for instance changing the crop rotations, increasing on-farm diversity (Bradshaw *et al.*, 2004), or bolstering the deployment and release of improved germplasm (Foley *et al.*, 2011; Gourdji *et al.*, 2013). These adjustments may be costly and/or require a degree of change in farmers' livelihoods (Rickards and Howden, 2012). Changes in climates can, however, make a cropping system no longer viable, hence requiring a complete system transformation. This often implies a new land use and/or livelihood to be adopted. One such case is the cropper–livestock keeper shift presented by Jones and Thornton (2009), or the coffee niche shrinking predicted by Laderach *et al.* (2011) across Central America.

As highlighted by the case studies presented here, of particular importance to the adaptation of an agricultural system is the availability of germplasm that responds well to future predicted conditions. The development of improved germplasm has proven the most successful strategy for enhancing crop yields and achieving better stability (Foley *et al.*, 2011; Gourdji *et al.*, 2013), yet can hold large potential for responding to future climate change (Butler and Huybers, 2012; Challinor *et al.*, 2010). For south-western Australian wheat, the development of germplasm with shorter-duration, better WUE, and faster and deeper roots constitutes an important strategy for adapting the crop to future climates. Modelling has suggested that net gains from developing climate-ready (i.e. waterlogging and drought tolerant) rice varieties could be in the order of US$1.5–1.9 billion in the next 20 years. Sorghum across eastern and southern Africa may warrant a different approach, with a much stronger focus on disease (i.e. *Striga*) resistance and reduced leaf senescence to enhance biomass accumulation during the grain filling period. In all cases, however, under the coalescence of increased temperatures (Joshi *et al.*, 2011) and increased frequency and intensity of extreme events (Orlowsky and Seneviratne, 2012), breeding will continue to be a priority need to reduce the risk of crop yield losses due to high temperatures while increasing food availability worldwide. This is particularly important because parts of the growing areas of the three crops are subjected to temperatures near the upper bound of the optimal temperature range for reproduction.

The availability of and access to germplasm that holds key traits for climate change adaptation is fundamental for achieving future breeding prospects (Guarino and Lobell, 2011). Both landraces and wild crop species have the potential of holding useful genetic variability for climate change adaptation (Lane and Jarvis, 2007; Burke *et al.*, 2009). Because similar genotypes can be found in areas where climate characteristics resemble each other (Ramirez-Villegas *et al.*, 2011, 2012), useful traits can also be investigated by looking at climate analogues between current and future climates of different sites (Ramirez-Villegas *et al.*, 2011; and the case studies of wheat, rice and sorghum in this chapter). An analysis was conducted to test whether current climates hold merit for climate change adaptation. For each pixel where the crop is reported to grow (i.e. reference site), similarity between the non-overlapping section of the future PDF ('novel' maximum temperature conditions) and the current climates of a number of neighbouring sites (target sites) was analysed. Target sites were searched within three domains: (a) a 250 km area around the site (i.e. pixel); (b) the country where the site is located; and (c) all crop growing areas. Similarity was calculated as the overlap between the novel climate of the reference site and the historical PDF of each target site.

The analysis indicated that in most cases, novel climates already exist at present for the three crops (hatched areas of left panels in Plate 20; also see Williams *et al.*, 2007; Ramirez-Villegas *et al.*, 2011). For wheat, 30–100% (depending on the GCM) of the locations where self-overlaps were below 50% had a representative climate within 250 km. For rice, 33–85% of locations where a significant proportion of novel climates are likely to appear are likely to have analogues in nearby areas. For sorghum, a larger number of cases (39–100 %) were found to have analogues within 250 km. These percentages almost doubled if all locations within a country are used as potential analogues. Analogue climates are also likely to exist in other areas of the globe. Only a very few locations (roughly 10% globally), mostly located in the Middle East, eastern Pakistan and in the Sahel, showed future climates with no current analogues.

This all suggests that traits to respond to certain conditions (high temperature in this case) may already currently exist for most locations where warming is expected to become a threat to crop production (also see Burke *et al.*, 2009). However, improved international exchange of genetic material and crop breeding will be required for incorporating these traits, while maintaining desirable local traits in current varieties (Reynolds *et al.*, 2011). It must be noted that, in some cases (Sahel, the Middle East and eastern Pakistan), climate similarity was low between reference sites and the global pool of temperature conditions, probably indicating that existing varietal diversity in terms of abiotic traits across growing areas is not enough for adaptation. In such cases, traits present in wild ancestors could, however, be incorporated into existing varieties (Hajjar and Hodgkin, 2007; Guarino and Lobell, 2011).

5.5 Conclusions

The above discussion has shown how the impacts of climate change on agriculture will depend on a number of factors, the first of which is the nature of the future climate, both means and extremes. These will vary according to the region under consideration, although some broad trends can be discerned. One of these is the differences in impacts between tropical regions, where many agricultural systems are close to their thermal limit, such that small increases in temperature, even of 1°C could lead to reduced yields, particularly of rice and wheat, while mid-to-high latitudes could benefit from increases of up to 3°C. In Europe, this would vary with many southern regions being sensitive to such increases in temperature, especially when coupled with projected decreases in rainfall. Rainfall is also critical in the tropics, where flooding could affect rice production in Asia, whereas Africa's food production is at particular risk from drought.

Climate change will impact differentially on the world's major crop types, depending on their sensitivity to particular climate parameters. Key crops, such as wheat and maize, are likely to be particularly affected and the case studies illustrated this for three important crops. In south-west Australia, for example, drought and heat stress could lead to decreased wheat yields, although some of this could be offset by the beneficial effects of enhanced CO_2, whereas extreme weather events (drought and floods), increasing salinity and frequency of pests and diseases could add to the impacts of mean climate change in Asian rice production. Sorghum production in ESA is already declining and climate change could exacerbate this, especially given the overall system's sensitivity to temperature increases.

Many areas and crops could experience a novel future climate where temperature is likely to increase beyond the range currently experienced. This is particularly the case for the tropical growing areas of rice and to a lesser extent sorghum. Both crops could have many more projected maximum temperatures that are novel for their production sites than wheat. Within the tropics certain areas were identified as having only a 5–10% overlap between historical and future climates (e.g. Bangladesh, Thailand and Malaysia), whereas for sorghum in East Africa there was less than 20% overlap.

The projected negative impacts of climate change and associated stresses on crop yields mean that adaptation is a critical response. There are a range of possible adaptation options to address crop sensitivity to climate, with different temporal and spatial applicability. As shown by the three case-study crops, the development of germplasm better adapted to certain climate conditions has been effectively used in the last century to increase food production per unit area. Similarly changing crop types and cropping practices can also be used to overcome future projected changes. Such adaptation strategies are vital, especially in marginal agricultural environments, to sustain human livelihoods and to meet the food demands of a growing population.

5.6 Acknowledgements

J.R.V. is supported by the CGIAR Research Program on Climate Change, Agriculture and Food Security. Authors thank the input from the Climate Impacts Group of the University of Leeds. J.R.V. thanks Prof. Piers Forster from the University of Leeds for providing local access to CMIP5 datasets. Authors acknowledge the World Climate Research Programme's Working Group on Coupled Modelling, which is responsible for CMIP, and we thank the climate modelling groups (listed in note 1 of this chapter) for producing and making available their model output. For CMIP the US Department of Energy's Program for Climate Model Diagnosis and Intercomparison provides coordinating support and led development of software infrastructure in partnership with the Global Organization for Earth System Science Portals.

Note

[1] Models were bcc-bcm1-1, bcc-bcm1-1-m, BNU-ESM, CanESM2, NCAR-CCSM4, CESM1-BGC, FGOALS-g2, HadGEM2-AO, INMCM4, and MIROC5. In all cases only one initial conditions ensemble member was used (see Taylor *et al.*, 2012).

References

ABS (2006) The Australian Wheat Industry, *Year Book Australia* 1301.0, No. 88, Australian Bureau of Statistics, Canberra, Australia, pp. 431–439.

Ainsworth, E.A. and Long, S.P. (2005) What have we learned from 15 years of free-air CO_2 enrichment (FACE)? A meta-analytic review of the responses of photosynthesis, canopy properties and plant production to rising CO_2. *New Phytologist* 165, 351–372.

Amthor, J.S. (2001) Effects of atmospheric CO_2 concentration on wheat yield: review of results from experiments using various approaches to control CO_2 concentration. *Field Crops Research* 73, 1–34.

Armstrong, L.J., Abrecht, D.G., Anderson, W.K. and Belford, R.K. (1996) The effect of non-lethal water deficits during establishment on the growth of wheat crops. In: Michalk, D.L. and Pratley, J.E. (eds) *Agronomy – Science with its Sleeves Rolled Up*. Proceedings of the 8th Australian Agronomy Conference, 30 January–2 February 1996, The University of Southern Queensland, Toowoomba, Queensland, Australia, pp. 80–83.

ASARECA (2005) *ASARECA Strategic Plan (2005 to 2015): Fighting Poverty, Reducing Hunger and Enhancing Resources through Regional Collective Action in Research for Development.* Available at: www.asareca.org/docs/ASARECA-Strategy.pdf (accessed 15 February 2013).

Asseng, S., Foster, I.A.N. and Turner, N.C. (2011) The impact of temperature variability on wheat yields. *Global Change Biology* 17, 997–1012.

Australian Bureau of Meteorology (2013) Australian climate variability & change – Time series graphs. Available at: www.bom.gov.au/cgi-bin/climate/change/timeseries.cgi (accessed 15 February 2013).

Battisti, D.S. and Naylor, R.L. (2009) Historical warnings of future food insecurity with unprecedented seasonal heat. *Science* 323, 240–244.

Blum, A. (2005) Drought resistance, water-use efficiency, and yield potential – are they compatible, dissonant, or mutually exclusive? *Australian Journal of Agricultural Research* 56, 1159–1168.

Blum, A. (2009) Effective use of water (EUW) and not water-use efficiency (WUE) is the target of crop yield improvement under drought stress. *Field Crops Research* 112, 119–123.

Boko, M., Niang, I., Nyong, A., Vogel, C., Githeko, A. *et al.* (2007) Africa. climate change 2007: Impacts, adaptation and vulnerability. In: Parry, M.L., Canziani, O., Palutikof, J., van der Linden, P.J. and Hanson, C.E. (eds) *Contribution of Working Group II to the Fourth Assessment Report of the Intergovernmental Panel on Climate Change.* Cambridge University Press, Cambridge, UK, pp. 433–467.

Bouman, B.A.M., Kropff, M.J., Tuong, T.P., Wopereis, M.C.S., Berge, H.F.M.t. *et al.* (2001) *ORYZA2000: Modeling Lowland Rice.* International Rice Research Institute, Los Baños, Philippines.

Bradshaw, B., Dolan, H. and Smit, B. (2004) Farm-level adaptation to climatic variability and change: crop diversification in the Canadian prairies. *Climatic Change* 67, 119–141.

Bruinsma, J. (2003) *World Agriculture: Towards 2015/2030; An FAO Perspective.* Earthscan Publications Ltd, London.

Burke, M.B., Lobell, D.B. and Guarino, L. (2009) Shifts in African crop climates by 2050, and the implications for crop improvement and genetic resources conservation. *Global Environmental Change* 19, 317–325.

Butler, E.E. and Huybers, P. (2012) Adaptation of US maize to temperature variations. *Nature Climate Change Letters* doi: 10.1038/NCLIMATE1585, 1–5.

Calzadilla, A., Rehdanz, K., Betts, R., Falloon, P., Wiltshire, A. *et al.* (2010) *Climate Change Impacts on Global Agriculture.* Working Paper FNU-185. Research Unit Sustainability and Global Change, Hamburg University, Hamburg, Germany.

Ceccarelli, S., Grando, S., Maatougui, M., Michael, M., Slash, M. *et al.* (2010) Plant breeding and climate changes. *Journal of Agricultural Science* 148, 627–637.

Challinor, A.J. and Wheeler, T.R. (2008) Use of a crop model ensemble to quantify CO_2 stimulation of water-stressed and well-watered crops. *Agricultural and Forest Meteorology* 148, 1062–1077.

Challinor, A., Wheeler, T., Garforth, C., Craufurd, P. and Kassam, A. (2007a) Assessing the vulnerability of food crop systems in Africa to climate change. *Climatic Change* 83, 381–399.

Challinor, A.J., Wheeler, T.R., Craufurd, P.Q., Ferro, C.A.T. and Stephenson, D.B. (2007b) Adaptation of crops to climate change through genotypic responses to mean and extreme temperatures. *Agriculture, Ecosystems & Environment* 119, 190–204.

Challinor, A.J., Osborne, T., Shaffrey, L., Weller, H., Morse, A. *et al.* (2009a) Methods and resources for climate impacts research. *Bulletin of the American Meteorological Society* 90, 836–848.

Challinor, A.J., Wheeler, T.R., Hemming, D. and Upadhyaya, H.D. (2009b) Ensemble yield simulations: crop and climate uncertainties, sensitivity to temperature and genotypic adaptation to climate change. *Climate Research* 38, 117–127.

Challinor, A.J., Simelton, E.S., Fraser, E.D.G., Hemming, D. and Collins, M. (2010) Increased crop failure due to climate change: assessing adaptation options using models and socio-economic data for wheat in China. *Environmental Research Letters* 5, 034012 doi: 10.1088/1748-9326/5/3/034012

Chipanshi, A.C., Chanda, R. and Totolo, O. (2003) Vulnerability assessment of the maize and sorghum crops to climate change in Botswana. *Climatic Change* 61, 339–360.

Christensen, J.H., Hewitson, B., Busuioc, A., Chen, A., Gao, X. *et al.* (2007) Chapter 11: Regional climate projections. In: Solomon, S., Qin, D., Manning, M., Chen, Z., Marquis, M. *et al.* (eds) *Climate Change 2007: The Physical Science Basis.* Contribution of Working Group I to the Fourth Assessment Report of the Intergovernmental Panel on Climate Change. Cambridge University Press, Cambridge and New York.

CSIRO (2007) Climate Change in Australia. Technical Report 2007. Available at: www.climatechangeinaustralia.gov.au (accessed 15 February 2013).

CSIRO and Bureau of Meteorology (2010) State of Climate. 11pp. Available at: http://www.bom.gov.au/announcements/media_releases/ho/state Climate2012.pdf (accessed 15 February 2013).

Dai, A. (2012) Increasing drought under global warming in observations and models. *Nature Climate Change* 3, 52–58.

Datta, S.K. (2004) Rice biotechnology: a need for developing countries. *AgBioForum* 7, 31–35.

Dias de Oliveira, E., Bramley, H., Siddique, K.H.M., Henty, S., Berger, J. *et al.* (2013) Can elevated CO_2 combined with high temperature ameliorate the effect of terminal drought in wheat? *Functional Plant Biology* 40, 160–171.

Dimes, J., Cooper, P. and Rao, K. (2008) Climate change impact on crop productivity in the semi-arid tropics of Zimbabwe in the 21st century. In: *Proceedings of the Workshop on Increasing Productivity and Sustainability of Rainfed Cropping System of Poor Smallholder Farmers, 22–23 September 2008.* Tamale, Ghana.

Dracup, M., Reader, M. and Palta, J.A. (1998) Timing of terminal drought is an important cause of yield variability in lupin grown on duplex soils in southern Australia. *Australian Journal of Agricultural Research* 49, 799–810.

Easterling, W.E., Aggarwal, P.K., Batima, P., Brander, K.M., Erda, L. *et al.* (2007) Food, fibre and forest products. In: Parry, M.L., Canziani, O., Palutikof, J., van der Linden, P.J. and Hanson, C.E. (eds) *Climate Change 2007: Impacts, Adaptation and Vulnerability.* Contribution of Working Group II to the Fourth Assessment Report of the Intergovernmental Panel on Climate Change. Cambridge University Press, Cambridge, pp. 273–313.

Ellis, F. (2000) The determinants of rural livelihood diversification in developing countries. *Journal of Agricultural Economics* 51, 289–302.

FAO (2010) Food and agricultural commodities production: countries by commodity. FAOSTAT, Rome, Italy. Available at: http://faostat.fao.org/site/339/default.aspx (accessed 9 January 2013).

Fischer, E.M. and Knutti, R. (2013) Robust projections of combined humidity and temperature extremes. *Nature Climate Change* 3, 126–130.

Fitzgerald, M.A. and Resurreccion, A.P. (2009) Maintaining the yield of edible rice in a warming world. *Functional Plant Biology* 36, 1037–1045.

Foley, J.A., Ramankutty, N., Brauman, K.A., Cassidy, E.S., Gerber, J.S. *et al.* (2011) Solutions for a cultivated planet. *Nature* 478, 337–342.

Fraser, E.D.G., Simelton, E., Termansen, M., Gosling, S.N. and South, A. (2013) 'Vulnerability hotspots': Integrating socio-economic and hydrological models to identify where cereal production may decline in the future due to climate change induced drought. *Agricultural and Forest Meteorology* 170, 195–205.

Frison, E.A., Cherfas, J. and Hodgkin, T. (2011) Agricultural biodiversity is essential for a sustainable improvement in food and nutrition security. *Sustainability* 3, 238–253.

Gornall, J., Betts, R., Burke, E., Clark, R., Camp, J. *et al.* (2010) Implications of climate change for agricultural productivity in the early twenty-first century. *Philosophical Transactions of the Royal Society B: Biological Sciences* 365, 2973–2989.

Gourdji, S.M., Mathews, K.L., Reynolds, M., Crossa, J. and Lobell, D.B. (2013) An assessment of wheat yield sensitivity and breeding gains in hot environments. *Proceedings of the Royal Society B: Biological Sciences* 280, 20122190, doi:10.1098/rspb.2012.2190.

Guarino, L. and Lobell, D.B. (2011) A walk on the wild side. *Nature Climate Change* 1, 374–375.

Hajjar, R. and Hodgkin, T. (2007) The use of wild relatives in crop improvement: a survey of developments over the last 20 years. *Euphytica* 156, 1–13.

Hammer, G.L. and Broad, I.J. (2003) Genotype and environment effects on dynamics of harvest index during grain filling in sorghum. *Agronomy Journal* 95, 199–206.

Hawkins, E., Fricker, T.E., Challinor, A.J., Ferro, C.A.T., Ho, C.K. *et al.* (2013) Increasing influence of heat stress on French maize yields from the 1960s to the 2030s. *Global Change Biology* 19, 937–947.

Hennessy, K., Fawcett, R., Kirono, D., Mpelasoka, F., Jones, D. *et al.* (2008) *An Assessment of the Impact of Climate Change on the Nature and Frequency of Exceptional Climatic Events.* Australian Government Bureau of Meteorology, Melbourne, CSIRO, Aspendale and Australian Government Department of Agriculture, Fisheries and Forestry, Canberra, Australia.

Howden, S.M., Soussana, J.-F., Tubiello, F.N., Chhetri, N., Dunlop, M. *et al.* (2007) Adapting agriculture to climate change. *Proceedings of the National Academy of Sciences USA* 104, 19691–19696.

Hurtt, G.C., Chini, L.P., Frolking, S., Betts, R.A., Feddema, J. *et al.* (2011) Harmonization of land-use scenarios for the period 1500–2100: 600 years of global gridded annual land-use transitions, wood harvest, and resulting secondary lands. *Climatic Change* 109, 117–161.

IFPRI (2002) *Green Revolution: Curse or Blessing?* IFPRI Brief. International Food Policy Research Institute, Washington, DC, USA.

Iglesias, A., Garrote, L., Quiroga, S. and Moneo, M. (2012) A regional comparison of the effects of climate change on agricultural crops in Europe. *Climatic Change* 112, 29–46.

IPCC (2001) *IPCC Third Assessment Report: Climate Change 2001 (TAR).* IPCC, Geneva, Switzerland.

IRRI (2010) Scuba rice: breeding flood-tolerance to Asia's local mega rice varieties. International Rice Research Institute, Los Baños, Philippines.

van Ittersum, M.K., Howden, S.M. and Asseng, S. (2003) Sensitivity of productivity and deep drainage of wheat cropping systems in a Mediterranean environment to changes in CO_2, temperature and precipitation. *Agriculture Ecosystems & Environment* 97, 255–273.

Jagadish, S.V.K., Muthurajan, R., Oane, R., Wheeler, T.R., Heuer, S. *et al.* (2010) Physiological and proteomic approaches to address heat tolerance during anthesis in rice (*Oryza sativa* L.). *Journal of Experimental Botany* 61, 143–156.

Jarvis, A., Ramirez-Villegas, J., Herrera Campo, B. and Navarro-Racines, C. (2012) Is cassava the answer to African climate change adaptation? *Tropical Plant Biology* 5, 9–29.

Johnson, D., Haefele, S. and Hardy, B. (2012) Responding to changing climate in unfavorable rice environments. IRRI Limited Proceedings 17, Los Baños, Philippines.

Jones, P.G. and Thornton, P.K. (2009) Croppers to livestock keepers: livelihood transitions to 2050 in Africa due to climate change. *Environmental Science & Policy* 12, 427–437.

Joshi, M., Hawkins, E., Sutton, R., Lowe, J. and Frame, D. (2011) Projections of when temperature change will exceed 2°C above pre-industrial levels. *Nature Climate Change* 1, 407–412.

Kimball, B.A., Kobayashi, K. and Bindi, M. (2002) Responses of agricultural crops to free-air CO2 enrichment. *Advances in Agronomy* 77, 293–368.

Knox, J., Hess, T., Daccache, A. and Wheeler, T. (2012) Climate change impacts on crop productivity in Africa and South Asia. *Environmental Research Letters* 7, 034032.

Laderach, P., Lundy, M., Jarvis, A., Ramirez, J., Portilla, E.P. *et al.* (2011) Predicted impact of climate change on coffee supply chains. In: Leal Filho, W. (ed.) *The Economic, Social and Political Elements of Climate Change.* Springer, Berlin and Heidelberg, Germany, pp. 703–723.

Lane, A. and Jarvis, A. (2007) Changes in climate will modify the geography of crop suitability: agricultural biodiversity can help with adaptation. *Journal of the Semi-Arid Tropics* 4, 1–12.

Li, Y., Ye, W., Wang, M. and Yan, X. (2009) Climate change and drought: a risk assessment of crop-yield impacts. *Climate Research* 39, 31–46.

Lobell, D.B. and Burke, M.B. (2010) On the use of statistical models to predict crop yield responses to climate change. *Agricultural and Forest Meteorology* 150, 1443–1452.

Lobell, D.B., Burke, M.B., Tebaldi, C., Mastrandrea, M.D., Falcon, W.P. *et al.* (2008) Prioritizing climate change adaptation needs for food security in 2030. *Science* 319, 607–610.

Lobell, D.B., Schlenker, W. and Costa-Roberts, J. (2011) Climate trends and global crop production since 1980. *Science* 333, 616–620.

Lobell, D.B., Sibley, A. and Ivan Ortiz-Monasterio, J. (2012) Extreme heat effects on wheat senescence in India. *Nature Climate Change* 2, 186–189.

Loss, S.P. and Siddique, K.H.M. (1994) Morphological and physiological traits associated with wheat yield increases in Mediterranean environments. *Advances in Agronomy* 52, 229–276.

Luck, J., Spackman, M., Freeman, A., Trebicki, P., Griffiths, W. *et al.* (2011) Climate change and diseases of food crops. *Plant Pathology* 60, 113–121.

Ludwig, F. and Asseng, S. (2006) Climate change impacts on wheat production in a Mediterranean environment in Western Australia. *Agricultural Systems* 90, 159–179.

Ludwig, F. and Asseng, S. (2010) Potential benefits of early vigor and changes in phenology in wheat to adapt to warmer and drier climates. *Agricultural Systems* 103, 127–136.

Ludwig, F., Milroy, S. and Asseng, S. (2009) Impacts of recent climate change on wheat production systems in Western Australia. *Climatic Change* 92, 495–517.

Maracchi, G., Sirotenko, O. and Bindi, M. (2005) Impacts of present and future climate variability on agriculture and forestry in the temperate regions: Europe. In: Salinger, J., Sivakumar, M.V.K. and Motha, R. (eds) *Increasing Climate Variability and Change.* Springer, the Netherlands, pp.117–135.

McGrath, J.M. and Lobell, D.B. (2011) An independent method of deriving the carbon dioxide fertilization effect in dry conditions using historical yield data from wet and dry years. *Global Change Biology* 17, 2689–2696.

Mgonja, M.A., Chandra, S., Monyo, E.S., Obilana, A.B., Chisi, M. *et al.* (2006) Stratification of SADC regional sorghum testing sites based on grain yield of varieties. *Field Crops Research* 96, 25–30.

Mitchell, T.D. and Jones, P.D. (2005) An improved method of constructing a database of monthly climate observations and associated high-resolution grids. *International Journal of Climatology* 25, 693–712.

Mohanty, S., Valera, H.G. and Molina, I. (2011) IGRM (IRRI Global Rice Model) tool: manual of the user interface for IGRM, beta version 1.0. International Rice Research Institute, Los Baños, Philippines.

Monfreda, C., Ramankutty, N. and Foley, J.A. (2008) Farming the planet: 2. Geographic distribution of crop areas, yields, physiological types, and net primary production in the year 2000. *Global Biogeochemical Cycles* 22, GB1022.

Moriondo, M., Bindi, M., Kundzewicz, Z., Szwed, M., Chorynski, A. *et al.* (2010) Impact and adaptation opportunities for European agriculture in response to climatic change and variability. *Mitigation and Adaptation Strategies for Global Change* 15, 657–679.

Moser, S.C. and Ekstrom, J.A. (2010) A framework to diagnose barriers to climate change adaptation. *Proceedings of the National Academy of Sciences USA* 107, 22026–22031.

Mottaleb, A.K., Rejesus, R.M., Mohanty, S., Murty, M.V.R., Li, T. *et al.* (2012) Ex ante impact assessment of a drought-tolerant rice variety in the presence of climate change. Paper presented at the 2012 Annual Meeting of the Agricultural and Applied Economics Conference, Seattle, Washington.

Mugoya, C. and Masiga, C.W. (2012) Striga resistant sorghum due for release soon. Available at: www.asareca.org/content/striga-resistant-sorghum-due-release-soon (accessed 16 January 2013).

Müller, C., Bondeau, A., Popp, A., Waha, K. and Fader, M. (2010) Climate change impacts on agricultural yields. Background note to the World Development Report 2010. Potsdam Institute for Climate Impact Research, Potsdam, Germany.

Müller, C., Cramer, W., Hare, W.L. and Lotze-Campen, H. (2011) Climate change risks for African agriculture. *Proceedings of the National Academy of Sciences, USA* 108, 4313–4315.

Nelson, G.C., Rosegrant, M.W., Koo, J., Robertson, R., Sulser, T. *et al.* (2009) *Climate Change Impact on Agriculture and Costs of Adaptation.* International Food Policy Research Institute, Washington, DC, USA.

Nelson, G.C., Rosegrant, M., Palazzo, A., Gray, I., Ingersoll, C. *et al.* (2010) *Food Security, Farming and Climate Change to 2050: Scenarios, Results and Policy Options.* International Food Policy Research Institute, Washington, DC, USA.

New, M., Hulme, M. and Jones, P. (2000) Representing twentieth-century space–time climate variability. Part II: Development of 1901–96 monthly grids of terrestrial surface climate. *Journal of Climate* 13, 2217–2238.

Oerke, E.C. (2006) Crop losses to pests. *Journal of Agricultural Science* 144, 31–43.

Olesen, J.E. and Bindi, M. (2002) Consequences of climate change for European agricultural productivity, land use and policy. *European Journal of Agronomy* 16, 239–262.

Olesen, J., Carter, T., Díaz-Ambrona, C., Fronzek, S., Heidmann, T. *et al.* (2007) Uncertainties in projected impacts of climate change on European agriculture and terrestrial ecosystems based on scenarios from regional climate models. *Climatic Change* 81, 123–143.

Orlowsky, B. and Seneviratne, S. (2012) Global changes in extreme events: regional and seasonal dimension. *Climatic Change* 110, 669–696.

Osborne, T., Rose, G. and Wheeler, T. (2013) Variation in the global-scale impacts of climate change on crop productivity due to climate model uncertainty and adaptation. *Agricultural and Forest Meteorology* 170, 183–194.

Palta, J.A. and Watt, M. (2009) Crop roots systems form and function: Improving the capture of water and nutrients with vigorous root systems. In: Sadras, V. and Calderini, D. (eds) *Crop Physiology: Applications for Genetic Improvement and Agronomy.* Academic Press, San Diego, California, pp. 309–325.

Palta, J.A., Turner, N.C. and French, R.J. (2004) The yield performance of lupin genotypes under terminal drought in a Mediterranean-type environment. *Australian Journal of Agricultural Research* 55, 449–459.

Pandey, S. and Bhandari, H. (2007) Drought: an overview. In: Pandey, S., Bhandari, H. and Hardy, B. (eds) *Economic Costs of Drought and Rice Farmers' Coping Mechanisms: a Cross-Country Comparative Analysis.* International Rice Research Institute, Los Baños, Philippines.

Peng, S., Huang, J., Sheehy, J.E., Laza, R.C., Visperas R.M. *et al.* (2004) Rice yields decline with higher night temperature from global warming. *Proceedings of the National Academy of Sciences USA* 101, 9971–9975.

Pittock, B. (2003) *Climate Change: An Australian Guide to the Science and Potential Impact.* Australian Greenhouse Office. Canberra, Australia.

Ramirez-Villegas, J., Lau, C., Köhler, A.K., Signer, J., Jarvis, A. *et al.* (2011) *Climate analogues: finding tomorrow's agriculture today.* CCAFS Working paper No. 12. CGIAR Research Program on Climate Change, Agriculture and Food Security (CCAFS), Cali, Colombia.

Ramirez-Villegas, J., Jarvis, A., Leibing, C. and Fujisaka, S. (2012) Crop and forage genetic resources: international interdependence in the face of climate change. In: Halewood, M., Lopez Noriega, I. and Louafi, S. (eds) *Crop Genetic Resources as Global Commons.* Routledge, Taylor and Francis Group, London and New York, pp. 78–98.

Reynolds, M.P., Hellin, J., Govaerts, B., Kosina, P., Sonder, K. *et al.* (2011) Global crop improvement networks to bridge technology gaps. *Journal of Experimental Botany* 63, 1–12.

Rickards, L. and Howden, S.M. (2012) Transformational adaptation: agriculture and climate change. *Crop and Pasture Science* 63, 240–250.

Ringler, C., Zhu, T., Cai, X., Koo, J. and Wang, D. (2010) *Climate Change Impacts on Food Security in Sub Sahara Africa. Insights from Comprehensive Climate Change Scenarios.* IFPRI Discussion Paper No. 1042. International Food Policy Research Institute, Washington, DC.

Rosenzweig, C., Strzepek, K.M., Major, D.C., Iglesias, A., Yates, D.N. *et al.* (2004) Water resources for agriculture in a changing climate: international case studies. *Global Environmental Change* 14, 345–360.

Rost, S., Gerten, D., Hoff, H., Lucht, W., Falkenmark, M. *et al.* (2009) Global potential to increase crop production through water management in rainfed agriculture. *Environmental Research Letters* 4, 044002.

Roudier, P., Sultan, B., Quirion, P. and Berg, A. (2011) The impact of future climate change on West African crop yields: What does the recent literature say? *Global Environmental Change* 21, 1073–1083.

Rounsevell, M.D.A., Ewert, F., Reginster, I., Leemans, R. and Carter, T.R. (2005) Future scenarios of European agricultural land use: II. Projecting changes in cropland and grassland. *Agriculture, Ecosystems & Environment* 107, 117–135.

Sacks, W.J., Deryng, D., Foley, J.A. and Ramankutty, N. (2010) Crop planting dates: an analysis of global patterns. *Global Ecology and Biogeography* 19, 607–620.

Savary, S., Castilla, N.P., Elazegui, F.A. and Teng, P.S. (2005) Multiple effects of two drivers of agricultural change, labour shortage and water scarcity, on rice pest profiles in tropical Asia. *Field Crops Research* 91, 263–271.

Schlenker, W. and Lobell, D.B. (2010) Robust negative impacts of climate change on African agriculture. *Environmental Research Letters* 5, 014010.

Sheffield, J., Wood, E.F. and Roderick, M.L. (2012) Little change in global drought over the past 60 years. *Nature* 491, 435–438.

Siddique, K., Belford, R., Perry, M. and Tennant, D. (1989) Growth, development and light interception of old and modern wheat cultivars in a Mediterranean-type environment. *Australian Journal of Agricultural Research* 40, 473–487.

Siddique, K., Tennant, D., Perry, M. and Belford, R. (1990) Water use and water use efficiency of old and modern wheat cultivars in a Mediterranean-type environment. *Australian Journal of Agricultural Research* 41, 431–447.

Singh, P., Boote, K.J., Kumar, U., Srinivas, K., Nigam, S.N. *et al.* (2012) Evaluation of genetic traits for improving productivity and adaptation of groundnut to climate change in India. *Journal of Agronomy and Crop Science* 198, 399–413.

Sivakumar, M.V.K., Das, H.P. and Brunini, O. (2005) Impacts of present and future climate variability and change on agriculture and forestry in the arid and semi-arid tropics: increasing climate variability and change. In: Salinger, J., Sivakumar, M.V.K. and Motha, R.P. (eds) *Increasing Climate Variability and Change – Reducing the Vulnerability of Agriculture and Forestry.* Springer, the Netherlands, pp. 31–72.

Sparks, A., Nelson, A. and Castilla, N. (2012) Where rice pests and diseases do the most damage. *Rice Today* 11, 26–27.

Srivastava, A., Naresh Kumar, S. and Aggarwal, P.K. (2010) Assessment on vulnerability of sorghum to climate change in India. *Agriculture, Ecosystems & Environment* 138, 160–169.

Taylor, K.E., Stouffer, R.J. and Meehl, G.A. (2012) An overview of CMIP5 and the Experiment Design. *Bulletin of the American Meteorological Society* 93, 485–498.

Thornton, P.K., Jones, P.G., Ericksen, P.J. and Challinor, A.J. (2011) Agriculture and food systems in sub-Saharan Africa in a 4°C+ world. *Philosophical Transactions of the Royal Society A: Mathematical, Physical and Engineering Sciences* 369, 117–136.

Tubiello, F.N., Soussana, J.-F. and Howden, S.M. (2007) Crop and pasture response to climate change. *Proceedings of the National Academy of Sciences USA* 104, 19686–19690.

Twomlow, S., Shiferaw, B., Cooper, P. and Keatinge, J.D.H. (2008) Integrating genetics and natural resource management for technology targeting and greater impact of agricultural research in the Semi-Arid Tropics. *Experimental Agriculture* 44, 235–256.

Venuprasad, R., Sta Cruz, M.T., Amante, M., Magbanua, R., Kumar, A. *et al.* (2008) Response to two cycles of divergent selection for grain yield under drought stress in four rice breeding populations. *Field Crops Research* 107, 232–244.

Wardlaw, I. and Moncur, L. (1995) The response of wheat to high temperature following anthesis. I. The rate and duration of kernel filling. *Functional Plant Biology,* 391–397.

WCWA (Water Corporation of Western Australia) (2010) Yearly streamflow for major surface water sources: impact on water availability – WA reduced inflows to dams. Available at: www.watercorporation.com.au/d/dams_streamflow.cfm (accessed 21 March 2013).

Welch, J.R., Vincent, J.R., Auffhammer, M., Moya, P.F., Dobermann, A. *et al.* (2010) Rice yields in tropical/subtropical Asia exhibit large but opposing sensitivities to minimum and maximum temperatures. *Proceedings of the National Academy of Sciences USA* 107, 14562–14567.

Williams, J.W., Jackson, S.T. and Kutzbach, J.E. (2007) Projected distributions of novel and disappearing climates by 2100 AD. *Proceedings of the National Academy of Sciences USA* 104, 5738–5742.

Yang, J. and Zhang, J. (2006) Grain filling of cereals under soil drying. *New Phytologist* 169, 223–236.

Zahedi, M. and Jenner, C.F. (2003) Analysis of effects in wheat of high temperature on grain filling attributes estimated from mathematical models of grain filling. *Journal of Agricultural Science* 141, 203–212.

Zhu, X.-G., Long, S.P. and Ort, D.R. (2010) Improving photosynthetic efficiency for greater yield. *Annual Review of Plant Biology* 61, 235–261.

6 International Mechanisms for Conservation and Use of Genetic Resources

Gerald Moore[1] and Geoffrey Hawtin[2]

[1]*Formerly Food and Agriculture Organization of the United Nations (FAO), Rome, Italy (now retired) and* [2]*formerly International Plant Genetic Resources Institute (IPGRI), Rome, Italy (now retired)*

6.1 Introduction

What we now know as plant genetic resources have been at the centre of controversy for several centuries. The early 19th century saw protracted struggles between Latin American and European countries over the genetic resources of the cinchona tree, for example, the bark of which was the sole source of quinine and the only known cure for malaria at the time. The Peruvian government banned the export of cinchona germplasm in an effort to maintain its monopoly, but the Dutch succeeded in smuggling out seeds and establishing their own plantations in Java, which by the 1930s were producing 97% of the world's supply of quinine. Historically however, plant genetic resources of crops used for food have presented a different picture, in that those resources have for the most part been freely exchanged throughout the world. The characteristics of the crops that form the basis of modern agriculture have been developed in many different countries and continents, with the result that today all countries and regions are dependent on plant genetic resources from other countries and continents for the maintenance and further development of their agriculture. Interdependence for access to and use of plant genetic resources for food and agriculture (PGRFA) is likely to become even more pronounced in a world faced with climate change and in which agroecosystems, and the varieties adapted to them, are likely to shift across national and even regional boundaries.

In recognition of this interdependence, the first international mechanisms, as typified in the International Undertaking on Plant Genetic Resources (FAO, 1983), were aimed at conserving those resources and ensuring their free availability as a global public good. However, the basic concepts of plant genetic resources as a public good and free availability have been subject to encroachment from two sides. On one side, the development of new technologies and globalization have resulted in attempts to appropriate the benefits of new advances in plant breeding through placing restrictions on the free availability of those resources through intellectual property rights. The various UPOV Conventions and the Agreement on Trade-related Aspects of Intellectual Property Rights (the TRIPS Agreement) are illustrative of this movement. On the other, and to a large extent as a reaction, states have laid increasing stress on their sovereign rights over their own genetic resources and their right to determine the conditions of access to those resources. The Convention on Biological Diversity (the CBD) represents an embodiment of this approach (United Nations, 1992).

The International Treaty on Plant Genetic Resources for Food and Agriculture (the Treaty) is an attempt to balance all of these competing forces and to create a legal regime for plant genetic resources for food agriculture that facilitates their exchange while recognizing the legitimate claims of innovators and the overall framework of national sovereignty over natural resources. The success of the Treaty is critical to the future of sustainable agriculture, especially in an age of climate change. If the Treaty fails, and the balance changes in favour of increasing restrictions on the exchange of PGRFA, then countries will find it increasingly difficult to access the genetic resources that are essential for developing the new varieties required to face the challenges of climate change.

It is too soon to assess fully the success or otherwise of the Treaty. Certainly the statistics on the exchange of materials by the centres of the Consultative Group on International Agricultural Research (CGIAR) are encouraging. But there are signs that a number of developing countries are hesitant in committing themselves fully to the implementation of the Treaty. In this chapter we will explore briefly the various international mechanisms for the conservation and use of genetic resources and how they have developed. Then we look more closely at the Treaty and explore ways in which its Contracting Parties, acting through the Governing Body of the Treaty, may be able to improve its implementation.

6.2 The First Formal International Attempts to Promote and Regulate Conservation and Use of PGRFA

The first international agricultural research centres[1] were established in response to concerns that food resources in developing countries would be insufficient to meet the needs of their growing populations and the need to seek improved varieties to increase food production. The work of the centres, and in particular that of plant breeders such as Norman Borlaug of CIMMYT and Gurdev Khush and Hank Beachell at IRRI, bore spectacular results, heralding the birth of the so-called 'Green Revolution'. At the same time, the introduction of the new improved varieties tended to supplant existing local varieties and threatened to erode the very biodiversity on which the original Green Revolution, and future crop improvements, depended. The response of the centres to this new threat was to increase their efforts to collect and conserve endangered PGRFA. In 1974, the CGIAR set up the International Board for Plant Genetic Resources (IBPGR)[2] hosted by FAO, with the task of coordinating an international plant genetic resources programme, including organizing collecting missions as well as building and expanding genebanks at the national, regional and international levels. Over the period 1974–1980, IBPGR supported the collection and conservation of more than 65,000 germplasm samples or accessions from over 70 countries. The material collected through these missions, which were for the most part carried out jointly with institutions in the countries concerned, were deposited for conservation in some 52 genebanks. These genebanks included both national or parastatal institutions[3] as well as international institutions such as the CGIAR centres,[4] AVRDC (The World Vegetable Center) in Taiwan and CATIE in Costa Rica.

The material was, in general, collected on the understanding that it would be made available to breeders and researchers worldwide without charge or restriction on how it could be used. The status of the material was, however, not entirely clear because agreements between the collectors and the countries concerned were often signed at the level of the individual institute in the host country, and it was not always clear that this institute had the authority to enter into such agreements on behalf of the government. This lack of clarity, as well as uncertainty concerning the status of the CGIAR centres' collections themselves, led to the first international initiative to regulate the status and availability of plant genetic resources that are important for food and agriculture. But before we launch into a description of the International Undertaking on Plant Genetic Resources, we should perhaps take a look at the international instruments establishing the legal basis for intellectual property over PGRFA.

6.3 International Instruments Establishing the Legal Basis for Intellectual Property over Innovations in the Area of Plant Genetic Resources

The main forms of intellectual property over innovations in the area of PGRFA are patents and plant breeders' rights.[5] Patents, which have the effect of preventing the use of protected innovations for research and further innovation, have always been viewed with a certain amount of misgiving by most breeders involved in the development of PGRFA and indeed have until recently been unavailable in the area of biological resources in general. In searching for ways that would allow for the protection of their investments in developing new varieties while still allowing for such new varieties to remain available for further research and genetic improvement, breeders came up with the concept of Plant Breeders' Rights, now encapsulated in the various conventions of the Union for the Protection of New Varieties of Plants (UPOV).

6.3.1 The UPOV Conventions

The First International Convention for the Protection of New Varieties of Plants, which also established UPOV as an intergovernmental organization, was adopted in Paris in 1961 (UPOV now has its headquarters at The World Intellectual Property Organization (WIPO) Headquarters in Geneva). The Convention was subsequently revised in 1972, 1978 and again in 1991 (UPOV, 1991). Under the UPOV Conventions, plant breeders are entitled to claim intellectual property protection, in the form of Plant Breeders' Rights over new varieties that they have developed, provided that the varieties are new, distinct, homogeneous and stable. Once the varieties are protected, prior authorization must be sought from the breeder for the production, offering for sale and marketing of the reproductive or vegetative propagating materials of the new variety. On the other hand, other breeders are freely entitled to take the new varieties, do research on them and further develop them into other new varieties without being required to seek the approval of the first breeder or make payments. Farmers who purchased the new varieties were also entitled to replant and exchange farm-saved seed. The term of protection is a minimum of 18 years for vines, fruit trees and their root-stock, and 15 years in the case of all other plants.

The subsequent revisions of the UPOV Convention have strengthened Breeders' Rights and made them closer to the protection offered by patents. In particular, the automatic exemption in respect of farm-saved seed has been removed, though member countries are allowed to provide for the exemption in national legislation in a more strictly limited form. Under the 1991 Convention, the research exemption, under which other breeders are entitled to conduct further research on the protected varieties and to use them to develop new varieties, does not extend to a new variety that is essentially derived from the protected variety.

To date some 69 countries plus the European Union are party to the UPOV Convention in one form or another, but new members are now only entitled to join the 1991 version of the UPOV Convention. The UPOV system of protection of Plant Breeders' rights has been given more prominence by the Agreement on Trade-Related Aspects of Intellectual Property Rights (the TRIPS Agreement), Article 27.3(b) of which provides that parties may exclude from patentability plants and animals, other than micro-organisms, and essentially biological processes for the production of plants and animals. However, they must provide for the protection of plant varieties either by patents or by an effective *sui generis* system or by a combination of both. Some 157 countries are now parties to the TRIPS Agreement.

6.3.2 The International Undertaking on Plant Genetic Resources

The International Undertaking on Plant Genetic Resources, which was a non-binding instrument, was adopted by the FAO Conference at its 22nd Session in 1983 (FAO, 1983). It was based on what was then considered, in Article 1, to be 'the universally accepted principle that plant genetic resources are a heritage of mankind and consequently should be made available without restriction'. Under Article 5 of the International Undertaking adhering governments and institutions agreed to allow access to samples of plant genetic resources under their control and to permit their export where the resources had been requested for the purpose of scientific research, plant breeding or genetic resource conservation. The samples were to be made available free of charge, on the basis of mutual exchange or on mutually agreed terms.

The International Undertaking also called for the development of 'an international network of base collections in gene banks under the auspices or jurisdiction of FAO, that have assumed the responsibility to hold, for the benefit of the international community and on the principles of unrestricted exchange, base or active collections of plant genetic resources of particular plant species'. A series of discussions was held in the FAO Commission on Plant Genetic Resources in the 1980s on the legal arrangements that would be most appropriate to establish the international network. At the time it was envisaged that the international network should cover both international and national collections.

In the event, the Commission decided to proceed only with the establishment of agreements with the CGIAR centres and the genebanks holding the COGENT (International Coconut Genetic Resources Network) coconut collections. We shall return to the in-trust agreements with the CGIAR centres later on in this chapter because they and the material transfer agreements adopted to implement them were influential in the design of the Multilateral System of Access and Benefit-sharing set up by the International Treaty on Plant Genetic Resources for Food and Agriculture.

The International Undertaking was never fully accepted by all member nations of FAO: a number of countries announced their reservations at the time the Undertaking was adopted, many of them on the grounds that the free availability of PGRFA was incompatible with plant breeders' rights. At the same time many developing countries had concerns about making their plant genetic resources available to developed countries unless those developed countries would undertake not to take out intellectual property rights over those resources. In fact, the concept of free availability that was the cornerstone of the International Undertaking began to be boxed in from both sides by intellectual property rights and the concept of national sovereignty over genetic resources.

These new considerations were recognized in a series of three 'Agreed Interpretations' of the International Undertaking that were adopted by the FAO Conference in 1989 and 1991.

The first agreed interpretation, adopted under Conference resolution 4/89, recognized that plant breeders' rights as provided for under the UPOV Convention were not incompatible with the International Undertaking. The Agreed Interpretation also recognized the rights farmers had as a result of their contributions over time to the conservation and development of plant genetic resources, which formed the basis for the concept of Farmers' Rights. The first agreed interpretation further recognized that free access did not necessarily mean free of charge (i.e. it was legitimate to charge for the cost of multiplying, preparing and distributing samples), and that the benefits to be derived under the International Undertaking were part of a reciprocal system and should thus be limited to countries adhering to the Undertaking.

The second agreed interpretation adopted by the FAO Conference under Resolution 5/89 further defined the concept of Farmers' Rights, while continuing to characterize plant genetic resources as a common heritage of mankind to be preserved and to be freely available for use for the benefit of present and future generations.

The third agreed interpretation, adopted two years later under Conference Resolution

3/91, recognized that the concept of the heritage of mankind, as applied in the International Undertaking, is subject to the sovereignty of the states over their plant genetic resources.

6.3.3 The FAO Commission on Genetic Resources for Food and Agriculture

At this point it is perhaps useful to take a look at the FAO Commission on Genetic Resources for Food and Agriculture. The FAO Council set up the Commission in 1983 as the FAO Commission on Plant Genetic Resources in implementation of a recommendation included in the International Undertaking. The Commission was tasked with monitoring the international arrangements set out in Article 7 of the International Undertaking and with overseeing the development of a Global System on Plant Genetic Resources for Food and Agriculture. In 1995, the FAO conference expanded the mandate of the Commission to include all genetic resources for food and agriculture and changed its name accordingly.

The FAO Commission was the only intergovernmental body that dealt with the complex legal issues of plant genetic resources, and later other elements of biodiversity, in so far as they affected food and agriculture. It was this body that oversaw the development of the first Report on the State of the World's Plant Genetic Resources and the adoption of the First Global Plan of Action by the Fourth International Technical Conference on Plant Genetic Resources, which took place in Leipzig in 1996. The Global Plan of Action listed some 20 priority areas of action and was later endorsed by both the FAO Commission and the FAO Conference. Both the report on the State of the World's Plant Genetic Resources and the Global Plan of Action have since been updated (FAO 2010, 2011a).

The FAO Commission also adopted a number of important voluntary codes and standards, such as the International Code of Conduct for Plant Germplasm Collecting and Transfer, and the Genebank Standards. The Commission is also responsible for overseeing the development of the various networks on plant genetic resources as well as the establishment of the World Information and Early Warning System on Plant Genetic Resources for Food and Agriculture (WIEWS).

Perhaps the most important activity carried out by the FAO Commission in recent years has been the negotiation of the International Treaty on Plant Genetic Resources for Food and Agriculture. We shall describe the contents and the importance of the Treaty later in this chapter. But first we should look at the developments in the general area of biodiversity and in particular the adoption of the Convention on Biological Diversity.

6.3.4 The Convention on Biological Diversity

In 1992, environmental concerns over the progressive deterioration of biodiversity in the world, including plant genetic resources, led to the adoption of the Convention on Biological Diversity (the CBD) at the Earth Summit in Rio de Janeiro.

The CBD takes as its starting point the recognition that states have sovereign rights over their own resources, including genetic resources, and can exploit them in accordance with their own environmental policies. The Contracting Parties, however, recognize the threat that exists to biological diversity and that the conservation of biological diversity is a common concern of mankind. The CBD is basically a framework treaty. In other words it sets down the basic principles that Parties are to follow in providing for the conservation and sustainable use of genetic resources, and in granting access to those resources. But it leaves it to each Contracting Party to implement those principles in its own territory in accordance with its own policies and legislation.

The Convention is oriented around three main themes:

- Conservation of biodiversity.
- Sustainable use of its components.
- Equitable sharing of benefits arising from the utilization of genetic resources.

The main commitments of the Parties are to develop national strategies, plans and programmes for the conservation and sustainable use of biological diversity, to identify and monitor the components of biodiversity, to promote *in situ* conservation through the use of protected areas and the protection of ecosystems, and to adopt measures for the *ex situ* conservation of biodiversity, predominantly for the purpose of complementing *in situ* measures. Contracting Parties commit themselves to integrating the consideration of the conservation and sustainable use of biological resources into their national decision making.

In recognition of the fact that states have sovereign rights over their genetic resources, the CBD stipulates that the authority to determine access to those resources rests with national governments and is subject to national legislation. Contracting Parties are to endeavour to create conditions to facilitate access to genetic resources by other Contracting Parties for environmentally sound uses. Access to genetic resources, where provided, is to be subject to the prior informed consent of the country of origin providing the resources and on the basis of mutually agreed terms. Although there is nothing in the CBD that requires that prior informed consent and mutually agreed terms be negotiated on a bilateral basis, this is the way in which the Convention has in practice been implemented, at least until the negotiation of the Treaty provided for multilateral approaches for a subset of biodiversity.

As noted above, the CBD sets down only the basic principles that Parties are to follow in providing for the conservation and sustainable use of genetic resources, and in granting access to those resources. More guidance, though of a non-binding nature, on possible terms and conditions for access and benefit-sharing, was provided to Parties through the Bonn Guidelines on Access to Genetic Resources and the Fair and Equitable Sharing of the Benefits arising from their Utilization, (the Bonn Guidelines), which were adopted by the CBD Conference of Parties at its 6th Session in 2002.

6.3.5 The Nagoya Protocol

In 2010, the Conference of Parties to the CBD at its 10th Meeting adopted a binding Protocol to the Convention that sets out in more detail the obligations of both Parties seeking access to genetic resources and those granting access with respect to the terms of access and benefit-sharing. The Protocol on Access to Genetic Resources and the Fair and Equitable Sharing of Benefits Arising from their Utilization (the Nagoya Protocol) draws heavily on the Bonn Guidelines: the list of possible monetary and non-monetary benefits set out in the Annex to the Nagoya Protocol is drawn directly from those guidelines. It deals in a parallel way with access to genetic resources and to traditional knowledge related to those resources: in both cases prior informed consent is required, and access and benefit-sharing is to be on mutually agreed terms. The prior informed consent of indigenous and local communities must be obtained where those communities have an established right to grant access to particular genetic resources and associated traditional knowledge, and access and benefit-sharing must be negotiated with them on mutually agreed terms. The Nagoya Protocol seeks to ensure legal certainty both with respect to the terms and procedures for access and benefit-sharing, and the status of genetic resources acquired under those procedures. It provides for the establishment of an Access and Benefit-sharing Clearing-House to act as a central information point for legislative, administrative and policy measures on access and benefit-sharing, including procedures, requirements and information on national focal points and national authorities, and for notifications of the issuance of permits or certificates of compliance with requirements for those accessing genetic resources.

The Nagoya Protocol recognizes the validity of existing special agreements. Article 4 provides that where a specialized international access and benefit-sharing instrument applies that is consistent with and does not run counter to the objectives of the CBD and the Nagoya Protocol, the Protocol will not apply for the Parties to that specialized instrument for the specific resources covered by the specialized instrument. This means that for Parties to the Treaty, access and benefit-sharing for PGRFA covered by the Multilateral System will be governed by the Treaty and not the Protocol. For access to PGRFA outside the Multilateral System, other than those accessed from the CGIAR centres, the terms and conditions of the Nagoya Protocol will apply.

As of September 2012, the CBD had 193 Parties, and the Nagoya Protocol had 92 signatures and 5 ratifications. The CBD entered into force in December 1993. The Protocol will enter into force 90 days after the deposit of the 50th instrument of ratification.

6.3.6 The in-trust agreements between the CGIAR centres and FAO

Soon after the adoption of the agreed interpretations of the International Undertaking and spurred on by the agreement on the CBD, the CGIAR centres and the FAO Secretariat started work on the implementation of the international institutional arrangements envisaged under Article 7 of the International Undertaking and the establishment of the international network of genebanks. Under the supervision of the FAO Commission, a series of agreements was drawn up between each of the 12 CGIAR centres holding *ex situ* collections and FAO, placing those collections under the auspices of FAO. For the first time, the agreements clarified the legal status of the CG collections that were recognized as being held in trust for the benefit of the international community. In exchange, the CGIAR centres agreed not to seek legal ownership over the material held in trust, nor to seek intellectual property rights over the material or related information. They also undertook to manage the in-trust germplasm in accordance with internationally accepted standards, and to make samples of the material available to users for the purpose of scientific research, plant breeding and genetic resources conservation without restriction. In such cases they undertook to ensure that the obligations to which they had agreed under the in-trust agreements would be passed on to the subsequent recipients. This they did through a Material Transfer Agreement (MTA). It was this use of a contractual instrument, the MTA, which formed the model for the Standard Material Transfer Agreement under which all material is transferred under the Multilateral System of the Treaty.

6.3.7 The International Treaty on Plant Genetic Resources for Food and Agriculture

The CBD left open a number of issues relating to PGRFA. One of these was the status of the *ex situ* collections acquired before the entry into force of the Convention, notably the *ex situ* collections held by the CGIAR centres. The second was the question of Farmers' Rights. Both of these questions were taken up by the FAO Conference in 1993, when it called for the renegotiation of the International Undertaking to bring it into harmony with the CBD. Although the mandate given by the CBD Conference was limited to the status of *ex situ* collections acquired before the entry into force of the Convention, the FAO Conference chose to expand that mandate to include consideration of the whole issue of access to plant genetic resources, including *ex situ* collections not addressed by the Convention, as well as for the issue of the realization of Farmers' Rights. The negotiations were carried out within the framework of the FAO Commission on Genetic Resources for Food and Agriculture, and a special contact group was established for that purpose. The negotiations took a total of 7 years.

The Treaty was finally adopted by the FAO Conference in November 2001 and entered into force in June 2004. The treaty now has some 127 parties.

The Treaty contains general provisions to promote the conservation and sustainable use of PGRFA. These provisions draw on the provisions set out in the Global Plan of Action as agreed at the Leipzig Technical Conference in 1996, and develop themes already set out in the CBD. Article 5 sets out the main tasks that Contracting Parties are required to carry out with respect to the conservation, exploration, collection, characterization, evaluation and documentation of PGRFA. Contracting Parties, subject to national legislation and in cooperation with other Contracting Parties where appropriate, are to promote an integrated approach to the exploration, conservation and sustainable use of PGRFA. They are also required to cooperate to promote the development of an efficient and sustainable system of *ex situ* conservation. One of the mechanisms through which this cooperation is in practice being carried out is the Global Crop Diversity Trust, which we will describe in more detail later in this chapter.

Article 6 of the Treaty sets out the main obligations of Contracting Parties regarding the sustainable use of PGRFA. They are required to develop and maintain appropriate policy and legal measures that promote the sustainable use of PGRFA. Such measures may include pursuing fair agricultural policies that promote diverse farming systems, strengthening research that enhances and conserves biological diversity, promoting plant breeding, broadening the genetic base of crops and increasing the range of genetic diversity available to farmers, and ensuring that regulations concerning variety release and seed distribution are appropriate.

Of particular significance are the provisions recognizing Farmers' Rights that appear here for the first time in an international agreement. These rights are themselves a recognition of the enormous contribution made by farmers in the conservation and development of PGRFA. Indeed it is through the innovations of farmers throughout the world that the crops that form the basis of modern agriculture have been developed, and it is these informal innovations that form the starting point for further advances through formal breeding and technology. Although the concept of Farmers' Rights is thus recognized in the Treaty, the responsibility for realizing those rights lies with national governments in accordance with their needs and priorities. Each Contracting Party should take measures to protect and promote Farmers' Rights, including the protection of relevant traditional knowledge, the right to participate in sharing benefits arising from the utilization of PGRFA, and the right to participate in decision making at the national level on matters relating to the conservation and sustainable use of PGRFA. The Treaty is neutral on the issue of the rights of farmers to save, use, exchange and sell farm-saved seed and propagating material. If those rights exist in national law, then the Treaty should not be interpreted as limiting them.

Perhaps the most critical provisions of the Treaty are those of Part IV establishing the Multilateral System of Access and Benefit-sharing. The CBD established the basic principles governing access to genetic resources in general. Access should be subject to the prior informed consent of the country granting access, and should be on mutually agreed terms that ensure the fair and equitable sharing of the benefits arising from the commercial utilization of genetic resources with the country that provided those resources. The underlying assumption of the CBD is that the mutually agreed terms of access and benefit-sharing would be negotiated on a bilateral basis. The underlying assumption of the Treaty is that the transaction costs of negotiating mutually agreed terms on a bilateral basis each time that access is required would be prohibitive. PGRFA have been shared by all countries over centuries and all countries are to a large degree dependent on others for the PGRFA that are essential to enable their agriculture to develop and to respond to new challenges posed, for example, by disease and climate change, as well as to produce the increased yields necessary to sustain growing populations. The approach taken by the Treaty to access and benefit-sharing is therefore a multilateral one.

Although the treaty itself covers all PGRFA, the Multilateral System established by the Treaty covers only a selection of plant genetic resources that are considered to be the most important for food security and sustainable agriculture, and on which countries are most inter-dependent. For those plant genetic resources, which are listed in Annex 1 to the Treaty, Contracting Parties agree to grant facilitated access for the purpose of research, breeding and training on the basis of standard terms that have been multilaterally agreed by the parties. These standard terms are set out in a Standard Material Transfer Agreement (SMTA), which was adopted by the Governing Body of the Treaty at its first session in 2006 (International Treaty on Plant Genetic Resources for Food and Agriculture, 2012a).

The main terms and conditions set out in the SMTA are:

- Facilitated access can be claimed only for the purpose of utilization and conservation for research, breeding and training for food and agriculture. This does not include chemical, pharmaceutical and/or non-food/feed industrial uses.
- Facilitated access must be provided expeditiously and free of charge, although an administrative fee may be charged that may not exceed the minimal cost involved.
- All available passport data and other non-confidential descriptive information must be made available with the material.
- Recipients must not claim intellectual property or other rights that would limit facilitated access to the PGRFA or their genetic parts or components in the form received from the Multilateral System.
- Access to PGRFA under Development is to be at the discretion of the developer.
- PGRFA accessed under the Multilateral System shall continue to be made available by recipients to the Multilateral System under the same terms and conditions, i.e. under the SMTA.

The approach to fair and equitable benefit-sharing is also a multilateral one. In accordance with the basic understanding underlying the Treaty, the Contracting Parties recognize from the outset that facilitated access to PGRFA in the Multilateral System is itself a major benefit of the Multilateral System. But the treaty also provides for other forms of sharing the benefits arising from the commercialization of PGRFA obtained from the Multilateral System. These include the exchange of information regarding PGRFA under the Multilateral System, access to and transfer of technology, capacity building, and the sharing of monetary and other benefits of commercialization.

Before going any further in the description of the Multilateral System, we need to emphasize that all elements of the System are multilateral. Contracting Parties agree to pool PGRFA that is important for food and agriculture and on which countries are interdependent. Conceptually under the Treaty, PGRFA in the Multilateral System are accessed, not from the individual country holding those resources, but from the Multilateral System itself. Equally, the benefits resulting from that access must be pooled and shared among all the Contracting Parties. This is true for the sharing of information and technology as well as for capacity building, none of which is limited to providing benefits only to the country that has provided access to particular resources.

It is also true of the sharing of monetary and other benefits of commercialization. The Treaty provides that all Contracting Parties should take measures to achieve commercial benefit-sharing, through the involvement of the private and public sectors in activities identified in the Treaty. The Treaty also requires that the SMTA should include a requirement that where a recipient of material from the Multilateral System commercializes a product that incorporates material accessed from the Multilateral System, then that recipient should pay an equitable share of the benefits arising from that commercialization into the Multilateral System. Where restrictions are placed on the further availability of the new product for research and breeding, then the payment is mandatory. Where no such restrictions are placed on availability, then the payment is merely to 'be encouraged'. The Governing Body is required to assess within a period of 5 years from the entry into force of the Treaty whether the mandatory payment requirement should be extended to also cover commercial products that are available to others for further research and breeding. At this point it should be pointed out that the benefit-sharing mechanism only applies to new products that fall into the category of PGRFA themselves. It does not, for example, apply to the sale of grain as a commercial commodity. But then again, facilitated access from the Multilateral System is only for the purpose of research, breeding or training. Access for direct commercialization needs to be negotiated with the country of origin on a bilateral basis under the provisions of the CBD.

The Treaty provides that the level, form and manner of payment for commercial benefit-sharing is to be set by the Governing Body and may be reviewed from time to time. In fact the level was set by the Governing Body at its first session at a net rate of 0.77% of the sales of the product (1.1% of gross sales less 30% to take into account trade practices including costs and discounts).

As noted above, the Multilateral System does not include all PGRFA but merely those of a selection of crops and forages listed in Annex 1 of the Treaty and selected on the basis of their importance for food security and the extent to which countries are interdependent on them. In fact the list was negotiated also on the basis of political considerations: developed countries on the whole wanted to see a more complete list and developing countries wanted to start with a smaller list until they were convinced of the advantages that would come from the benefit-sharing arrangements. The list in the annex can be amended, but changes to the annexes as well as to the main body of the Treaty require consensus and will not be easy to achieve.

The Multilateral System does not cover all PGRFA in the territory of a Contracting Party. Only PGRFA that are under the management and control of a Contracting Party and in the public domain[6] are automatically included in the Multilateral System. Other PGRFA can be included in the Multilateral System by the voluntary act of the holder. And the Contracting Parties undertake to take measures to encourage natural and legal persons in their territory holding PGRFA listed in Annex 1 to include them in the Multilateral System. The Governing Body has the power to assess the progress in including such PGRFA in the Multilateral System, and to decide whether or not to continue to facilitate access to those who have not included their material in the Multilateral System.

At the present time the access and benefit-sharing provisions of the Multilateral System are applied mainly to PGRFA contained in government-held or controlled collections. Indeed, as the Technical Advisory Committee on the SMTA pointed out, there is an expectation that all the material contained in the national PGRFA system is included automatically within the Multilateral System. The extent to which PGRFA found in *in situ* conditions that meet the criteria of being under the management and control of a Contracting Party and in the public domain are included automatically in the Multilateral System is still somewhat unclear. Article 12.3.h provides that 'access to PGRFA found in *in situ* conditions will be provided according to national legislation, or in the absence of such legislation, in accordance with such standards as may be set by the Governing Body'. This provision is, however, 'without prejudice to the other provisions under this article', thus indicating that the general criteria for automatic inclusion of PGRFA in the Multilateral System also apply to PGRFA found *in situ*. The import of this paragraph would thus seem to be to confirm the right of the providing country to control the conditions under which collecting may be carried out, rather than to excise PGRFA found in *in situ* conditions from the scope of the Multilateral System altogether, even if the criteria for automatic inclusion are met. That was also the view of the Ad hoc Advisory Technical Committee on the Standard Material Transfer Agreement and the Multilateral System of the Treaty at its first meeting in January 2010. The second meeting of the Committee recognized, however, that the work on Article 12.3h needs to be further developed, in particular with regard to the relationship between Article 12.3h and other provisions of Article 12 and the implications of national legislation for access to *in situ* PGRFA and specifically those resources that are part of the Multilateral System.

Part V of the Treaty provides for a number of supporting components. These include the Global Plan of Action for the Conservation and Sustainable Use of PGRFA, to which reference has already been made, and the need for Contracting Parties to promote its effective implementation. They also include the *ex situ* collections of PGRFA held in trust by the International Agricultural Research Centres of the CGIAR. Because the CGIAR centres have their own legal personality, the provisions of the Treaty cannot apply directly to the centres in the same way as States that are contracting Parties to the Treaty. The importance of the centres' in-trust collections to the Treaty was fully recognized, however, by the Contracting Parties. The Contracting Parties thus called on the centres to sign agreements with the Governing Body placing those collections within the purview of the Treaty. Under the agreements, centres would be required to make Annex 1 PGRFA available in accordance with the provisions set out in Part IV of the Treaty, i.e. under the SMTA and in accordance with its terms. Non-Annex 1 material collected before the entry into force of the Treaty was to be made available in accordance with the MTA currently in use under the 1994 in-trust agreements. However, the Governing Body at its second session approved the suggestion of the centres that non-Annex 1 PGRFA collected before the entry into force of the Treaty should be also made available under the SMTA, with the addition of appropriate explanatory footnotes (FAO, 2007, paragraph 68). Non-Annex 1 material received and conserved by the centres after the entry into force of the Treaty is to be made available on terms consistent with those mutually agreed between the country of origin of the PGRFA or the country that acquired them in accordance with the CBD or other applicable law.

The Agreements between the Governing Body of the Treaty and the 11 CGIAR centres holding in-trust collections were signed on 16 October 2006. Under the agreements the centres recognize the authority of the Governing Body of the Treaty to provide policy guidance relating to *ex situ* collections held by them and subject to the provisions of the Treaty. The total number of accessions placed into the Multilateral System by the centres under these agreements amounts to just less than 700,000.

The supporting components under Part V of the Treaty also include the international plant genetic resources networks, such as the International Network for Genetic Evaluation of Rice (INGER); the Pacific Agricultural Genetic Resources Network (PAPGREN); and the European Cooperative Programme for Plant Genetic Resources (ECPGR). Existing cooperation in such networks will be encouraged and developed on the basis of existing arrangements.

The final supporting component is the Global Information System on Plant Genetic Resources for Food and Agriculture that is to be developed and strengthened through cooperation among Contracting Parties, based on existing information systems. The importance of the development of such a global system cannot be overestimated, both in providing information on the PGRFA available in the Multilateral System and its characteristics, as well as in providing a focal point for the exchange of information on all scientific, technical, environmental and other matters relating to PGRFA, thereby making a substantial contribution to benefit-sharing.

Before leaving this description of the Treaty, we should outline briefly the institutional and financial arrangements provided in the Treaty for its implementation. Policy direction for the Treaty is provided by the Governing Body, on which all Contracting Parties are represented and which meets in regular session at least once every 2 years. Decisions of the Governing Body are normally taken by consensus, except where other methods of decision making are agreed upon by consensus. Consensus is always required for the adoption of amendments to the Treaty itself or to its Annexes, including the list of crops and forages in the Multilateral System. No new financial resources are directly provided for in the Treaty, although the Contracting Parties agree to develop and implement a Funding Strategy for the implementation of the Treaty and to establish funding targets. A benefit-sharing Fund has been established by the Governing Body to receive voluntary contributions as well as the proceeds of the commercial benefit-sharing provided for under the Treaty. A target of US$116 million has been set for the period 2009–2014.

6.3.8 The Global Crop Diversity Trust

Mention has already been made of the obligation of Contracting Parties under the Treaty to cooperate to promote the development of an efficient and sustainable system of *ex situ* conservation. The principle means by which such cooperation has been carried out in practice is through the establishment and operation of the Global Crop Diversity Trust (the Trust). The Trust is an international organization with its own international legal personality set up by an international agreement in 2004 (the Agreement for the Establishment of the Global Crop Diversity Trust), to which some 27 States are now party. The mandate of the Trust, as set out in Article 2 of its Constitution, is to ensure the long-term conservation and availability of PGRFA with a view to achieving global food security and sustainable agriculture. To this end the Trust has set up an endowment fund in order to secure the funding in perpetuity of collections of unique and valuable plant genetic resources held *ex situ*, with priority being given to PGRFA included in Annex 1 of the Treaty. To date the endowment fund stands at some US$135 million. The Trust has also just completed a 3-year project to finance the regeneration of unique accessions of Annex 1 PGRFA that are threatened with extinction. It is now starting on a new 10-year project to collect and conserve wild relatives of Annex 1 crops with the aim of making the genetic resources contained in such wild relatives available to agriculture in order to face the challenges of climate change. The Trust was initially located at FAO Headquarters in Rome but has now moved to new premises in Bonn, Germany.

6.4 How Well is the Treaty being Implemented and What can be Done to Improve it?

A good deal of room has been given in this chapter to a description of the Treaty, because the Treaty and the Multilateral System it creates are of critical importance for food security and the ability of agriculture to respond to the challenges of climate change and the increases in productivity required to feed a growing global population. The Treaty has now been in force for more than eight years. So how well is the Treaty being implemented and how effective has it been so far? To some extent the answers to these questions will depend on the particular aspect from which the Treaty is being viewed.

From the point of view of the International Agricultural Research Centres of the CG system (the CGIAR centres) one can argue that the Treaty is working well. Together the CGIAR centres hold some 693,752 accessions of important Annex 1 materials in their genebanks as part of the Multilateral System. During the period 1 January 2007 through to 31 December 2009, the CGIAR centres distributed a total of 1.15 million samples of PGRFA, of which approximately 84% were sent to developing countries or countries with economies in transition, 9.5% to developed countries and 6.5% among the CGIAR centres. Of these samples, 18% were sent by the centres' genebanks and 82%, mainly improved material, from the centre breeding programmes. During the same period the centres acquired a total of 53,154 new samples. Implementation by the developed countries that are Contracting Parties to the Treaty is similarly encouraging.

Yet despite these encouraging figures, there are signs that the Treaty, and the

Multilateral System in particular, is not yet operating at anything near its full potential and that many of the actors, including both those in developing countries and those in the private sector, are demonstrating reluctance to fully engage (Halewood *et al.*, 2012). Many important countries, such as the USA, Japan, China, Mexico, Russia and Papua New Guinea are not yet party to the Treaty, whereas other countries that have become parties have still not taken action to implement its provisions or are just not responding to requests for material under the Multilateral System.

Of course these are still early days in the life of the Treaty and many countries are still feeling their way in setting up the systems required for its implementation. The Treaty Secretariat and several CGIAR centres such as Bioversity International are providing assistance to developing countries in developing the necessary analyses and implementation systems (FAO, 2011b, paragraphs 96–106). One of the early problems seems to be in reviewing the state of existing laws and determining the need, if any, for new legislation. The Ad hoc Advisory Technical Committee on the Standard Material Transfer Agreement and the Multilateral System of the Treaty set up by the Governing Body at its Third Session in 2009 is providing useful advice on these matters, including legal wording designed to create legal space for the implementation of the Multilateral System in legislation typically designed to implement the CBD, as well as clarifications of the meaning of the criteria for the automatic inclusion of material in the Multilateral System (FAO, 2011b, paragraphs 107–143).

To a large extent, the enthusiasm with which developing countries in particular undertake the work of implementation of the Treaty in general and the Multilateral System in particular, will depend on the extent to which they perceive that real benefits are coming to them from that implementation. In this context it is encouraging to note that the Governing Body at its Third Session in 2009 adopted the target of raising US$116 million in the 5-year period up to 2014. The first tranche of over US$10 million was raised during the first 18 months and the first series of project proposals for this amount have been approved.

Despite these successes, there is still a feeling that the implementation of the Treaty has not been as fast as it should have been. A recent study by Bioversity International (Halewood *et al.*, 2012) has identified some of the underlying causes for this state of affairs and highlighted several weaknesses in the way the Multilateral System is set up. These, according to the authors of the study, include the following:

- Insufficient policy reinforcement for the PGRFA commons created by the Multilateral System. With its lack of clear boundaries, and lack of reciprocity, the authors argue that, in effect, many players can get benefits from the Multilateral System without actually contributing to the system themselves and thus are free riders. The policy of the CGIAR centres in committing themselves to making materials available to non-contracting Parties as well as those that are party to the Treaty is one manifestation of this. So also is the ease with which the same materials as are available under the Multilateral System can be accessed from outside the system without incurring the obligations of the Multilateral System itself.
- The lack of bilateral incentives for countries to place materials into the Multilateral System. Although the CBD stresses the importance of the bilateral sharing of benefits from commercialization, the benefits from the commercialization of PGRFA accessed from the Multilateral System flow to the Multilateral System itself and only indirectly to the individual countries that supplied those resources.

Many of the difficulties currently facing the Treaty and its Multilateral System will undoubtedly fall away with increased participation in the Treaty, particularly by crucial countries such as the USA. Many difficulties could perhaps have been avoided by a simpler construction of the Multilateral

System, such as, for example, by building on a proposal made by the Malaysian delegation during the process of negotiation of the Treaty that all PGRFA of certain crops should be made freely available and payments should be made into a central fund on the basis of the area of land set down to the growing of those crops. Such a system would have avoided some of the complexities of the present contractually based system of benefit-sharing.

That said, the Treaty and its Multilateral System, even in its present form, does represent a great step forward in the conservation and availability of PGRFA, without which agriculture cannot respond effectively to challenges such as those posed by climate change. The Treaty is now in force and cannot be changed easily. There are, however, some recent indications that some countries that were previously opposed to the extension of the list of crops covered by the Multilateral System might now be in favour of seeking an expansion of the list (International Treaty on Plant Genetic Resources for Food and Agriculture, 2012b). The expansion of the list set out in Annex 1 to the Treaty would be a great step forward.

Any radical change to the structure of the Treaty and its Multilateral System of Access and Benefit-sharing would, however, still be difficult to achieve. So we must work with it in its present form and seek ways of improving its effectiveness through the only means available, such as interpretation of the Treaty provisions and policy guidance from the Governing Body of the Treaty.

There are, it is suggested two areas in which that policy guidance may prove effective. The first would be in increasing the flow of funds into the benefit-sharing fund. The second would be in reinforcing the commons set up by the Multilateral System and strengthening reciprocity. The following are more concrete suggestions in each area.

6.5 Increasing the Flow of Funds into the Benefit-sharing Fund

At present the main contributions to the benefit-sharing fund come from governments on a voluntary basis. The actual projected contribution from the sharing of commercial benefits under the SMTA is likely to be very small if not marginal. One way of increasing the flow of money into the benefit-sharing fund could be to encourage a substantial contribution of funds on a voluntary basis from the seed industry, based on the international trade in the crops included in the Multilateral System. If such a contribution could be secured in a deal between the Governing Body and the seed industry, this would have the advantages of increasing the flow of funds into the benefit-sharing mechanism of the Treaty and ensuring that the contribution is available at an earlier date, rather than waiting for mandatory contributions to become due under the SMTA. Some mechanism would of course need to be put in place to ensure that contributors were not penalized financially. For example, such voluntary contributions by the seed industry could be treated as advances on mandatory payments due in the future under the SMTA.

A second way in which the flow of funds into the benefit-sharing mechanism could be increased would be for the Governing Body to exercise its powers under the Treaty to extend mandatory payments under the SMTA to include cases where products incorporating material from the Multilateral System are available for further research and breeding without restriction, whether at the same or a differentiated rate.

6.6 Reinforcing the Commons System and Reciprocity

To a certain extent some of the design flaws noted in the study by Halewood *et al.* (2012) will become less relevant if participation in the Treaty becomes more universal, with the ratification of the Treaty by such countries as the USA, China and Japan. In the meantime, the Governing Body could perhaps improve the situation by adopting recommendations regarding access to germplasm with the Multilateral System by non-parties to the Treaty. Although granting access to non-parties should remain at the discretion of the individual Treaty Parties, a recommendation could be made that any such access should be on terms less favourable than that accorded to Treaty Parties, including the payment of a higher rate of royalties on commercialization of products incorporating material from the Multilateral System. This could include access to materials from the CGIAR centres under the Multilateral System. The Governing Body could also, if it deemed it appropriate, provide policy guidance to the centres to the effect that access to materials obtained from Parties to the Treaty under the Multilateral System should not be given to non-Parties if the Party providing those resources has expressed its wish that such material should not be passed on to non-Parties without its consent.

As pointed out by Halewood *et al.* (2012), countries have less incentive to provide access to materials under the Multilateral System where no share of the benefits of commercial use come directly to the country that supplies the material. Under the Treaty, all such benefits and payments are to be shared multilaterally. Indeed, this is a basic tenet of the Multilateral System: all PGRFA are treated as coming from the Multilateral System and all benefits flow to the Multilateral System. It is, however, a tenet that is hard to explain to many developing countries and that runs counter to the basic ideas of the CBD. Although it is not possible, or necessarily even desirable, to change this concept at this stage, it may be possible for the Governing Body to create some incentive to countries to make Multilateral System material available, by directing that a certain proportion of the benefits accruing to the Multilateral System should be made available to the country supplying the material. This could also apply to materials provided through the medium of the CGIAR centres but not of course to materials coming from the original in-trust collections held by the centres.

Finally, as a means of reinforcing the commons system, the Governing Body may wish to exercise its powers under Article 11.4 of the Treaty to assess the progress in including Annex 1 material held by natural and legal persons in the Multilateral System and to decide on whether access should continue to be facilitated to those that have not included their own resources in the Multilateral System.

6.7 Conclusions

There are several international agreements and mechanisms that together set the legal stage for the conservation and sustainable use of PGRFA. Of these the Treaty is the most important. The provisions on conservation and sustainable use are the most geared to the specific problems of agriculture and food security. So also are the provisions on the Multilateral System that try to meld together the principles of the CBD with the needs that are specific to food and agriculture in general and food security in particular. If the Multilateral System did not exist, the process of negotiating access on a bilateral basis would be prohibitive. And every country and continent is dependent on others for most of its needs with respect to PGRFA, particularly in an era of climate change. The Treaty is not a perfect instrument. But it is the only instrument that we have that would allow for the exchange of PGRFA to continue. And in some aspects, including in particular the flow of PGRFA from the genebanks of the CGIAR centres, the Treaty is in fact working well. Given the difficulties of amending the Treaty and its annexes, there is little prospect that radical changes will be possible in the immediate future. But there may be some ways in which aspects of the Multilateral System could be improved through ways that fall short of formal changes to the text. Some of these have been suggested in this chapter.

Notes

1 The International Rice Research Institute (IRRI) was established in 1960, the Centro Internacional de Mejoramiento de Maíz y Trigo (CIMMYT) in 1966, the International Institute of Tropical Agriculture (IITA) in 1967 and the International Centre for Tropical Agriculture (CIAT) in 1967. A number of new centres were set up in

the 1970s and 1980s: ICRISAT (1972); CIP (1970); ILRAD (1973); IFPRI (1974); ILCA (1974); WARDA (1970); ICARDA (1975); and ISNAR (1980). Other existing institutes were brought within the CG System in the 1990s: ICRAF 1977 (1991); IIMI 1984 (1991); ICLARM 1977 (1992); INIBAP 1984 (1992); and CIFOR 1993 (1993). There have been mergers, renaming and a realignment of centres subsequently.

2 IBPGR subsequently changed its name to the International Plant Genetic Resources Institute (IPGRI), and then to Bioversity International.

3 For example, EMBRAPA in Brazil, the Kenyan Agricultural Research Institute (KARI), CGN in the Netherlands, VIR in Russia, and the Rural Development Administration of Korea.

4 CIAT, ILCA, ICRISAT, CIMMYT, IITA, CIP, IRRI and ICARDA.

5 The use of hybrid varieties also provides a form of 'natural' intellectual property protection in that seeds produced by hybrid varieties do not breed true and thus the farmer has to purchase new seed every season.

6 The meaning of the terms 'under the management and control of a Contracting Party and in the public domain' was considered by the Ad hoc Advisory Technical Committee on the Standard Material Transfer Agreement and the Multilateral System of the Treaty at its first meeting in January 2010. Broadly speaking, the Committee was of the opinion that under the management of a Contracting Party meant that the Contracting Party had the power to undertake acts of conservation and utilization in respect of the material. The term 'control', however, focused on the legal power to dispose of the material and to decide on the treatment to be given to the resources. In the public domain, on the other hand, should be understood as referring to resources or information that is not subject to intellectual property rights.

References

FAO (1983) International Undertaking on Plant Genetic Resources. Resolution 8/83 of the FAO Conference, adopted 23 November 1983. Available at: www.fao.org/docrep/x5563E/X5563e0a.htm#e.%20plant%20genetic%20resources%20(follow%20up%20of%20conference%20resolution%20681) (accessed 6 November 2012).

FAO (2007) Report of the Second Session of the Governing Body of the International Treaty on Plant Genetic Resources for Food and Agriculture, FAO Doc. IT/GB-2/07/Report. Available at: www.planttreaty.org/sites/default/files/gb2re pe.pdf (accessed 6 November 2012).

FAO (2010) The Second Report on the State of the World's Plant Genetic Resources for Food and Agriculture, Rome. Available at: www.fao.org/docrep/013/i1500e/i1500e.pdf (accessed 6 November 2012).

FAO (2011a) Second Global Plan of Action for Plant Genetic Resources for Food and Agriculture. Available at: www.fao.org/agriculture/crops/core-themes/theme/seeds-pgr/gpa/gpa_update/en/ (accessed 6 November 2012).

FAO (2011b) Report on the Implementation of the Multilateral System of Access and Benefit Sharing, IT/GB-4/11/12. Available at: www.planttreaty.org/sites/default/files/gb4w12e.pdf (accessed 6 November 2012).

Halewood, M., López Noriega, I. and Louafi, S. (2012) The global crop commons and access and benefit-sharing laws: examining the limits of policy support for the collective pooling and management of plant genetic resources. In: Halewood, M., López Noriega, I. and Louafi, S. (eds) *Crop Genetic Resources as a Global Commons*. Routledge. London.

International Treaty on Plant Genetic Resources for Food and Agriculture (2012a) What is the SMTA? Available at: www.plant treaty.org/content/what-smta (accessed 6 November 2012).

International Treaty on Plant Genetic Resources for Food and Agriculture (2012b) Report of the Second High-Level Roundtable on the International Treaty on Plant Genetic Resources for Food and Agriculture on the occasion of the United Nations Conference on Sustainable Development ('Rio+20 Summit'), Rio de Janeiro, Brazil, 21 June 2012. Avail-able at: www.planttreaty.org/sites/default/files/HLRT2_Final_Report_1.pdf (accessed 6 November 2012).

United Nations (1992) Convention on Biological Diversity. Available at: www.cbd.int/convention/text/ (accessed 6 November 2012).

UPOV (1991) Act of 1991 – International Convention for the Protection of New Varieties of Plants. Available at: www.upov.int/upovlex/en/upov_convention.html

7 Crop Wild Relatives and Climate Change

Nigel Maxted, Shelagh Kell and Joana Magos Brehm
School of Biosciences, University of Birmingham, Birmingham, UK

7.1 Introduction

Crop wild relatives (CWRs) are taxa closely related to crops (including their progenitors) and are defined by their potential to contribute beneficial traits to crops, such as pest or disease resistance, yield improvement or stability (Ellstrand *et al*., 1999; Maxted *et al*., 2006). The expected increase in the human population to nine billion by 2050 will require an even more significant increase in crop production. Feuillet *et al*. (2008) question the ability of breeders to meet this challenge or to simply sustain current crop yields in the face of the dynamic biotic and abiotic threats attributed to climate change without more systematic utilization of CWR diversity. CWRs have greater intrinsic genetic diversity owing to their adaptation to a wide variety of habitats (Maxted *et al*., 2012) and because they have not gone through the domestication bottleneck (Volbrecht and Sigmon, 2005). They are therefore likely to make a significant and increasing contribution to food security and sustainable agriculture in the face of climate change.

CWR diversity is important because of its intrinsically high level of genetic diversity, which holds potentially beneficial traits that may be transferred to crops using conventional or biotechnological techniques. Breeders' use of CWR taxa has increased year on year during the past 50 years, although they are still far from being systematically exploited. Maxted and Kell (2009) reviewed the use of CWRs in crop improvement programmes for 29 major crops (Fig. 7.1) and noted that:

- For 29 crop species, there were 234 references reporting useful traits in 183 CWR taxa between 1930 and 2008.
- The degree to which breeders use CWR varies between crops. CWR use is particularly prominent in barley, cassava, potato, rice, tomato and wheat, but rice and wheat are the crops in which CWRs have been most widely used, both in terms of number of taxa used and successful attempts to introgress traits from the CWR to the crop.
- The number of published papers detailing use of CWRs in breeding has increased gradually over time, presumably as a result of technological developments for trait transfer, with 2% of citations recorded prior to 1970, 13% in the 1970s, 15% in the 1980s, 32% in the 1990s and 38% after 1999.
- The most widespread CWR use has been and remains in the development of disease and pest resistance, with the references citing disease resistance objectives accounting for 39%, pest resistance 17%, abiotic stress 13%, yield increase 10%, cytoplasmic male sterility and fertility restoration 4%, quality improvement 8% and husbandry improvement 6% of the reported interspecific trait transfers.

The broad conclusion of this analysis was, as originally argued by Tanksley and McCouch (1997), that breeders have not been systematic or comprehensive in their use of CWR diversity and that the vast majority of CWR diversity remains as yet

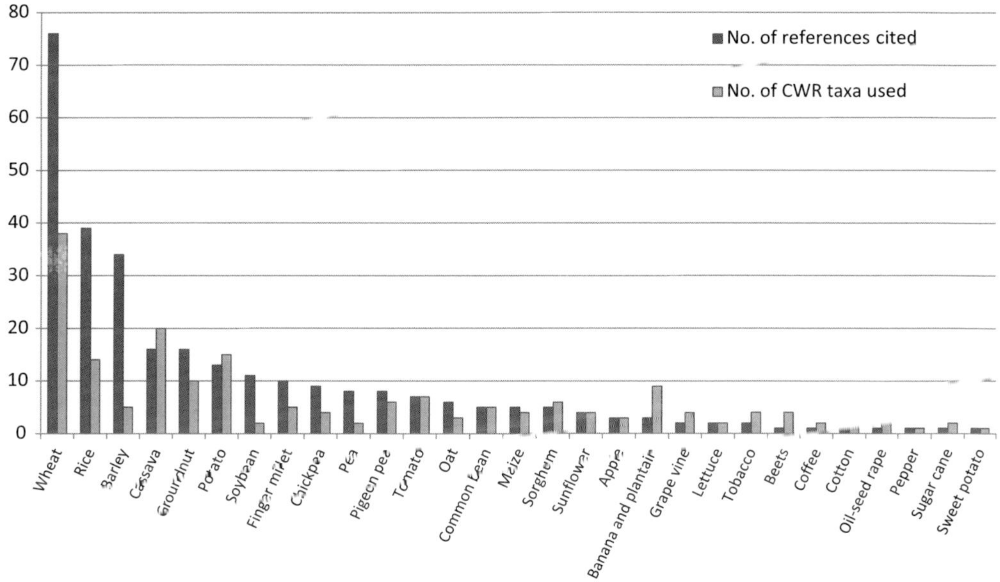

Fig. 7.1. Citation of use of CWRs in breeding and number of CWR taxa used for 29 crop species (Maxted and Kell, 2009).

untapped for utilization and use in climate change mitigation for both major and minor crops. Yet studies of the economic value of the contribution of CWRs to improving food production show it is significant but difficult to quantify precisely. Estimates vary from US$115 billion worldwide annually (Pimentel *et al.*, 1997) to US$10 billion benefit to global wholesale farm values per year (Phillips and Meilleur, 1998). Whatever the precise estimate, it is certain CWRs have a significant and growing value in crop improvement, and combined with advances in breeding technologies and their increased availability, CWRs are a critical natural resource for future food security.

CWRs offer the opportunity for climate change mitigation for our staple crops and so underpin our future food security in the face of climate change. They are, however, also threatened by climate change themselves, leading to calls for their systematic conservation. This is highlighted in international and regional policy instruments, such as the International Treaty on Plant Genetic Resources for Food and Agriculture (FAO, 2001; see Moore and Hawtin, Chapter 6, this volume), the European Strategy for Plant Conservation (Planta Europa, 2008), the Global Strategy for CWR Conservation and Use (Heywood *et al.*, 2008), the FAO Global Plan of Action for the Conservation and Sustainable Utilization of PGRFA (FAO, 2011), and most recently in the CBD Strategic Plan (CBD 2010a), in which Target 13 states that 'by 2020 the status of crop and livestock genetic diversity in agricultural ecosystems and of wild relatives has been improved' ... in addition ... '*in situ* conservation of wild relatives of crop plants could be improved inside and outside protected areas', and Target 9 in the CBD Global Strategy for Plant Conservation 2011–2020 (CBD, 2010b) '70 per cent of the genetic diversity of crops including their wild relatives and other socio-economically valuable plant species conserved'. These include recommendations to establish CWR *in situ* genetic reserves (protected areas where CWR genetic conservation is given priority) and to undertake gap analysis to identify priorities for *ex situ* filling of diversity gaps. CWR conservation and use provides an excellent exemplar of how to address the dual Millennium Development Goals of combining biodiversity conservation with poverty alleviation.

Climate variability is a natural phenomenon that has existed since earliest geological time. But climate change, as defined by United Nations Framework Convention on Climate Change Article 1 is 'a change of climate which is attributed directly or indirectly to human activity that alters the composition of the global atmosphere and which is in addition to natural climate variability observed over comparable time periods' (UNFCCC, 1992). The IPCC (2007), using General Circulation Models, predicts that climate change will affect different areas of the world to different degrees but that for the next two decades a warming of 0.3°C is projected to result in further:

- Melting of glaciers and polar ice, rising sea level and change to oceanic currents and acidity.
- Increased storms, hurricanes, droughts, rain, changes in snow cover and earthquakes.
- Desertification or irrigation and extreme salination.
- Agriculture failures, species extinction and ecosystem disruption.

Therefore, climate change will dramatically impact human food security, well-being and life expectancy within our lifetimes and how we respond now will directly affect future generations. A summary of the likely predicted impacts of climate change on species is provided in Table 7.1.

Specifically for plants this is resulting in species shifting their ranges towards poles (e.g. Norway spruce, Seppä *et al.*, 2009), nival (snow related) species moving upwards in mountainous regions accompanying the receding snow, species germinating and flowering earlier, increased threat from pests and diseases (Gregory *et al.*, 2009; Chakraborty and Newton, 2011) as a result of their arrival earlier and longer time to build up numbers, increased threat from invasive species (Bradley *et al.*, 2010), as well as the obvious associated abiotic changes, such as increased levels of drought and flooding.

Cahill *et al.* (2012) have argued that the most crucial factor affecting survival under climate change is disrupted interaction between species, rather than a simple, physiological intolerance to increased temperatures. They undertook a meta-analysis of 136 published studies that demonstrated a link between extinction and climate change and found only seven studies that identified specific mechanisms by which climate change caused local extinction of populations, none of which involved a straightforward physiological intolerance to higher temperatures. In fact, the majority of the studies suggested that changes in how species interact (e.g. pest or pollinators and host species) is the main cause of extinctions linked to climate change, with declines in food availability being the most common cause. Therefore it is more likely that changes in CWR phenology may perhaps mean that pollinators are not around at the appropriate time for pollination and cause population decline rather than the CWR itself being unable to cope with higher heat levels or less water availability.

The IPCC (2007) estimates that not all species will respond in the same way to similar levels of climatic change. Climate change is likely to impact maize and wheat by reducing yields by as much as 40% at low latitudes and for rice, yields will decrease by up to 30% in China and the Indian subcontinent. Further, looking at the different continental scenarios, excluding Antarctica, the most adverse impact is likely to be in areas of subsistence or marginal agriculture. In Africa, for instance, by 2020, 75–250 million additional people will have crops exposed to increased water stress and rainfed agriculture yield will be reduced by up to 50%, and by 2080 there is predicted to be an increase of 5–8% of arid and semi-arid land – all this in the continent that has the highest levels of malnourishment, the fastest growing population and the most pronounced poverty.

In response to this unprecedented threat, FAO (2008) argues that the existing genetic diversity within plant genetic resources can be used to sustain production systems. Further, the systematic use of genetic

Table 7.1. A summary of some of the predicted aspects of climate change and examples of the effects that these are likely to have on species (adapted from Foden *et al.*, 2009).

Predicted change					
Phenology: Spring arrival Autumn arrival Growing season length	Temperature: Means Extremes Variability Seasonality Sea level rises	Rainfall: Means Extremes Variability Seasonality	Extreme events: Storms Floods Droughts Fires	CO_2 concentrations: Atmospheric Ocean Ocean pH	Effects on species
↳	↳	↳			Desynchronization of dispersal events
↳	↳	↳			Uncoupling of mutualisms (incl. pollinator loss and coral bleaching)
↳	↳	↳			Uncoupling of predator–prey relationships
↳	↳	↳			Uncoupling of parasite–host relationships
↳	↳	↳			Interactions with new pathogens and invasives
	↳	↳	↳		Changes in distribution ranges
	↳	↳	↳		Loss of habitat
	↳	↳	↳	↳	Increased physiological stress causing direct mortality and increased disease susceptibility
	↳	↳	↳		Changes in fecundity leading to changing population structures
	↳	↳			Changes in sex ratios
	↳			↳	Changes in competitive ability
				↳	Inability to form calcareous structures and dissolving of aragonite

diversity will be required if, with a growing human population, future agricultural production targets are to be achieved (Burke *et al.*, 2009; Zimmerer, 2010) and the necessary resilience to sustain small-scale farming systems is met (Hertel *et al.*, 2010). Changes in abiotic stresses at a particular location may be mitigated by adapting crop varieties to permit changes in the timing of sowing or harvesting as well as to increase water use efficiency, heat and salinity tolerance and use of nutrients (see Chapters 12–16, this volume).

There are two distinct approaches to crop improvement in the context of climate change: (i) use of greater CWR diversity within formal plant breeding; and (ii) farmers' use of diversity through the deployment of pre-bred and therefore already adapted material into their own crop management strategies. Both promote increased use of CWR species, and encourage community-based management of a wide portfolio of genetic diversity to facilitate greater adaptive capacity. Further, the impacts of extreme climate events might be mitigated by breeding varieties that delay spring growth to avoid late frosts, the use of genetic diversity of fire-tolerant forest species, and community-based conservation of local varieties adapted to extreme climatic events. Changed biotic stresses at a particular location may also be countered using disease-resistant cultivars or multi-lines to strengthen crop resilience and the promotion of diversification strategies to increase the species and genetic diversity farmed.

There is evidence of the potential use of CWRs in breeding for the traits likely to have climate change mitigation ability through prior use of CWR for trait improvement (Table 7.2) Ultimately, current crop varieties will need replacement or enhancement to enable them to better suit changing agroenvironmental conditions and the trait basis for these 'novel' varieties may draw on as yet untapped genetic variation (e.g., see Jones *et al.*, 2003; Duveiller *et al.*, 2007; FAO, 2008; Deryng *et al.*, 2011; Li *et al.*, 2011; Luck *et al.*, 2011).

7.2 Evidence of Threats to Crop Wild Relative Diversity

Many wild plant species are currently threatened by human mismanagement of the environment, whether it is the species existence *per se* or the genetic diversity within otherwise apparently 'healthy' species. The Sampled Red List Index for Plants project (Brummitt and Backman, 2010) has recently estimated that 20% of all plants are currently threatened with extinction; however, this statistic does not take into account the potential impact of climate change, which is also predicted to have a further and substantial impact on wild plant populations in the future. Further, as pointed out by Maxted *et al.* (1997a), the loss of genetic diversity must always be greater than the loss of species because there will be genetic erosion from the species that remain extant. The taxonomic and genetic diversity of CWRs, like other wild plant species, is currently under significantly greater threat than ever before. A further challenging aspect of climate change is the lack of precision in estimating its potential deleterious impact. For example, Thuiller *et al.* (2005) predicted that by 2080 climate change will cause a 27–42% loss of plant species within Europe, with potential extremes ranging from 2.5–86% loss of current floristic diversity and possible species turnover per 50 × 50 km of 45–63% (17–86%). The lack of precision in the estimates is itself inhibiting mitigation action. There are no comprehensive estimates of the impact of climate change on CWR diversity but the fact that CWRs form a major component of any flora means the impact is likely to be significant.

Using a broad definition of a CWR (i.e. any taxon within the same genus as a crop; see Maxted *et al.*, 2006), Kell *et al.* (2005, 2008) listed in excess of 25,000 food and agriculture, ornamental, medicinal and timber crop and CWR species that occur in the Euro-Mediterranean region, which means that around 80% of the Euro-Mediterranean flora consists of crops and their wild relatives. Maxted and Kell (2009)

Table 7.2. Examples of the transfer of potential climate change mitigation traits from CWR to crops (extracted from Maxted and Kell, 2009).

Crop	CWR	Applications	References
Chickpea (*Cicer arientinum*)	*Cicer reticulatum*	Drought and temperature tolerance	Hajjar and Hodgkin (2007)
Sunflower (*Helianthus annuus*)	*Helianthus paradoxus*	Salt tolerance	Lexer *et al.* (2004)
Barley (*Hordeum vulgare*)	*Hordeum spontaneum*	Drought and temperature tolerance	Hadjichristodoulow (1993) Eglinton *et al.* (2001) Baum *et al.* (2003) Talame *et al.* (2004) Hajjar and Hodgkin (2007) Chen *et al.* (2008)
Tomato (*Lycopersicon esculentum*)	*Lycopersicon cheesmanii*, *Lycopersicon pennellii*, *Lycopersicon peruvianum*	Various forms of resistance, drought and salinity tolerance, soluble solids insect resistance	Stevens and Rick (1986) Rick and Chetelat (1995)
Cassava (*Manihot esculenta*)	*Manihot angustiloba*, *Manihot davisiae*, *Manihot rubricaulis*	Drought tolerance, adaptation to high altitudes and cool temperatures	Jennings (1995)
Banana and plantain (*Musa acuminatae*, *Musa balbisiana*)	*Musa balbisiana*, *Musa itinerans*, *Musa nagensium*	Drought resistance, water-logging resistance	INIBAP/IPGRI (2006)
Rice (*Oryza sativa*)	*Oryza longistaminata*	Drought resistance and yield increase	Brar (2005)
	Oryza rufipogon	Drought resistance	Zhang *et al.* (2006)
Wheat (*Triticum aestivum*)	*Aegilops cylindrica*, *Thinopyrum junceum*	Salt tolerance	Farooq *et al.* (1992) Farooq *et al.* (1995) Wang *et al.* (2003)
	Aegilops spp., *Thinopyrum* spp., *Elytrigia elongata*, *Hordeum* spp.	Salt tolerance	Colmer *et al.* (2006)
	Aegilops speltoides, *T. turgidum*	Salt tolerance	Noori (2005)
	Aegilops tauschii, *Triticum turgidum*	Water-logging tolerance	Villareal *et al.* (2001)
	Agropyron elongatum	Drought tolerance	Goodman *et al.* (1987)
	Thinopyrum bessarabicum	Salt resistance	King *et al.* (1997a,b)

extrapolated from the situation in the Euro-Mediterranean region and estimated that globally there are approximately 50,000 CWR species.

Growing interest in the value of CWRs for crop improvement and the current inadequate measures in place for their conservation has stimulated the commissioning of a global *ex situ* CWR conservation and use project that is being led by the Global Crop Diversity Trust (GCDT, 2011) in partnership with the Royal Botanic Gardens, Kew, working in collaboration with national agricultural research institutes, the Consultative Group on International Agricultural Research and the University of Birmingham. The project involves a range of activities from locating and sampling CWRs from global food crops (including lucerne, apple, Bambara groundnut, banana, barley, bean, carrot, fava bean, lima bean, chickpea, cowpea, aubergine, finger millet, grass pea, lentil, oat, pea, pearl millet, pigeon pea, plantain, potato, rye, rice, sorghum, sunflower, sweet potato, vetch and wheat) to *ex situ* conservation and provision of introgressed material to national crop breeding programmes. The first steps in the project were the production of the first global priority CWR species list (the Harlan and de Wet CWR Inventory) and the collation of the most comprehensive ecogeographic dataset for the taxa included. The Harlan and de Wet CWR Inventory database contains the check-list of priority CWR taxa with key ancillary data (Crop Wild Relatives and Climate Change, 2012) and contains 1667 CWR taxa from 37 families, 109 genera, 1392 species and 299 sub-specific taxa. The inventory was produced using the gene pool, where available, or taxon group concepts (see Maxted *et al.*, 2006) to estimate CWR use potential for 173 food and agricultural crop gene pools (Vincent *et al.*, 2013).

The ecogeographic dataset generated by the Global Crop Diversity Trust CWR project (GCDT, 2011) for the gap analysis of future *ex situ* CWR conservation priorities contains the georeferenced localities of five million CWR populations. It provides the necessary data to estimate the general threat to CWR diversity using the IUCN Red List Categories and Criteria and specific climate change related threat using species distribution modelling. Both threat assessment strategies are currently underway and the results are likely to prove extremely useful to future conservation planning.

7.2.1 IUCN Red List assessment

Whether considering the estimate of all CWR species (Maxted and Kell, 2009) or the more precisely defined Harlan and de Wet CWR Inventory of priority taxa (Vincent *et al.*, 2013) and using the Thuiller *et al.* (2005) predictions, it is likely that a high proportion of CWR species are threatened with extinction and the majority are currently suffering significant genetic erosion. Although precise systematic assessment of threat to global priority CWR diversity is in development, some assessments have already been completed. Hammer and Khoshbakht (2005) were the first to attempt to identify crops and CWRs at risk of extinction by matching the list of crops in Mansfeld's Encyclopaedia of Agricultural and Horticultural Crops (Hanelt and IPK Gatersleben, 2001; IPK Gatersleben, 2003) with the IUCN Red List of Threatened Plants (Walters and Gillett, 1998) and found about 200 threatened crops and 12 threatened CWR taxa. This study did not include novel threat assessment; it simply reviews assessments undertaken for CWR taxa by other assessors.

A more recent study of Bolivian CWRs, however, assessed over 310 CWR species from 36 CWR genera and found that 14.6% (45) of the species are considered as threatened, with 2.3% (7) of them being Critically Endangered, 7.1% (22) Endangered, and 5.2% (16) Vulnerable – a further 6.5% (20) of the species are classified as Near Threatened (Mora *et al.*, 2009). The most comprehensive study to date was the assessment of 572 European CWR species in 25 economically important crop gene pools/groups (Bilz *et al.*, 2011; Kell *et al.*, 2012a). Results of this study showed that at least 8.5% (or 66) of the species are threatened, with 3.3% (19) of them being Critically

Endangered, 4.4% (22) Endangered, and 3.8% (25) Vulnerable. A further 4.5% (or 26) of the species are classified as Near Threatened (19) and one species (*Allium jubatum* J.F. Macbr.) is Regionally Extinct. Although the assessment found that 8.5% of the species are threatened, taken as a proportion of the species that were not assessed because they were Data Deficient, this increases the percentage to around 16%, which is closer to the estimate of Brummitt and Backman (2010) that 20% of all plant species are threatened with extinction.

It is also worth noting that a third of the 54.7% of European CWR species assessed as Least Concern are nationally threatened, which means that there is likely to be some genetic erosion within these species and for those that only have a limited range, this renders them vulnerable to future threats, including climate change. IUCN (2013) reports that of the vascular plant species for which global Red List assessments have been undertaken, 9390 (60%) are threatened, 1321 (8.4%) are Near Threatened, 3443 (22%) are Least Concern and the remainder are Data Deficient or Lower Risk/conservation dependent. The high percentage of threatened species reported is, however, likely to be strongly biased because those species that a priori are considered to be threatened are more likely to be assessed and the assessment is then used to justify conservation action.

Kell *et al.* (2012a) also analysed the factors threatening CWR diversity and reported 31 threats, the most frequent being 'livestock farming and ranching', 'tourism and recreation areas' and 'housing and urban areas' (Fig. 7.2). The review of threats was limited, however, by the IUCN threat descriptors employed and it should be noted that farming *per se* is unlikely to threaten CWR diversity; in fact, farmed areas (including arable land and pasture) are one of the primary habitats of CWR species. Rather it is unsustainable farming practices, such as severe overgrazing, conversion of land to monocultures, and the heavy application of fertilizers, herbicides and pesticides

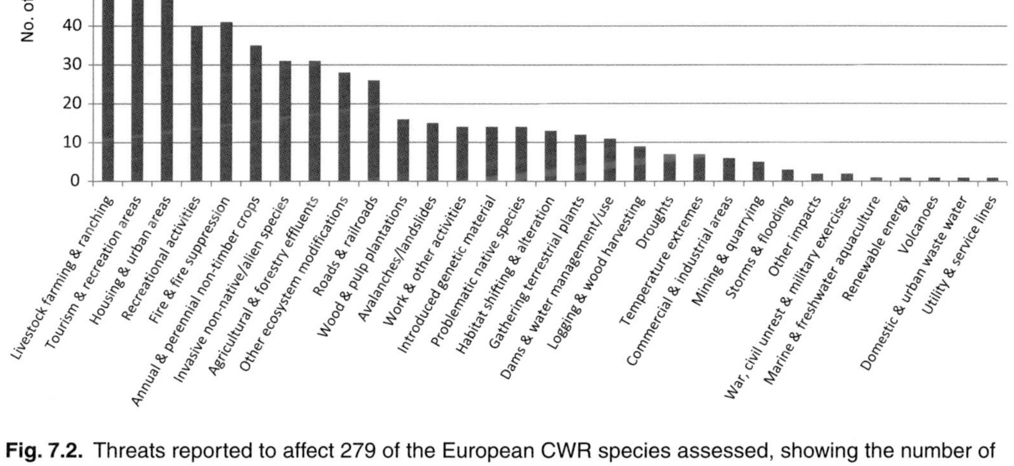

Fig. 7.2. Threats reported to affect 279 of the European CWR species assessed, showing the number of species for which each one was recorded (taken from Kell *et al.*, 2012a).

that are the major threats to CWRs that grow in agricultural areas. IUCN Red List assessments do not directly assess the threat from climate change because the impact is often less direct and so cannot be unequivocally attributed to climate change. What is actually noted is overgrazing, increased fires or competition from alien species, each of which may have at its foundation changes in the biotic or abiotic environment themselves attributable to climate change.

7.2.2 'Risk of extinction' modelling

A more direct means of assessing the impact of climate change on CWR diversity is through modelling. Climate envelope models use environmental data for the locations where a species has been found or where it is absent, to infer its climatic requirements. These inferred requirements can then be used to classify the suitability of any other locations or the same location at a different time under different climate conditions. A number of studies have applied climate-envelope-based species distribution models to the problem of understanding the impacts of climatic change through the use of climatic data for the present, the future (Thomas *et al.*, 2004) and the past (Ruegg *et al.*, 2006). These methods are imperfect and based on the assumption that species cannot demonstrate greater climatic plasticity than currently observed and that they will have insufficient time to evolve with the changing environment. The modelling is further complicated by a lack of specific knowledge concerning the migrational (dispersal) capacities of species and the specific biotic interactions between species (e.g. the species itself may be climatically plastic but its pollinator may not be). There is a lack of sufficiently accurate climatic information needed to build robust distribution models, as well as an absence of data in countries necessitating the use of world databases, which are also insufficiently accurate. Despite these considerable problems, species distribution modelling is now widely used as a means of predicting range shifts and assessing extinction risk in the face of climate change.

Although species distribution modelling has not been comprehensively or systematically used to assess the likely impact of climate change on CWR species, Jarvis *et al.* (2008) used three migrational scenarios to evaluate changes in potential range size and fragmentation of climatically suitable areas for the groundnut (*Arachis* spp.), potato (*Solanum* spp.) and cowpea (*Vigna* spp.) gene pools. The authors generated climate envelopes based on present distributions and then compared present and projected future distributions using predicted climate data for 2055. They predicted that 16–22% of CWR species would go extinct by 2055, that the majority of species showed greater than 50% loss of distributional range and that the ranges that remained were highly fragmented, placing the extant species under greater threat of genetic erosion or extinction. The relative impact of climate change varied depending on the gene pool. While *Vigna* is likely to be least impacted with 0–2 (out of 48) species predicted to go extinct and between 8–41 species predicted to lose more than 50% of their current geographic range, wild *Arachis* species are expected to be far worse affected with 24–31 (out of 50) species predicted to go extinct and 85–94% range decreases for the remaining extant species.

Similar modelling of climate change scenarios in Mexico by Lira *et al.* (2008) highlighted that most of the eight wild cucurbitaceous taxa studied are predicted not to survive under accepted climate change models. A more recent analysis by Davis *et al.* (2012) of indigenous arabica coffee (*Coffea arabica* L.) production found an approximate 65% reduction in the number of pre-existing bioclimatically suitable localities and at worst an almost 100% reduction by 2080. Thus indigenous arabica coffee appears particularly climate sensitive, with populations likely to suffer severe stress and to be at a high risk of extinction.

A potential further threat that is specific to CWR diversity is that many CWRs of major crops are found in disturbed, pre-climax communities (Jain, 1975), which themselves are the habitats that are most likely to be subject to increasing levels of anthropogenic change and destruction as a result of climate change.

7.3 Methods of Crop Wild Relative Threat Assessment

Species may respond to climate change by: (i) adapting to a changed environment (natural evolution); (ii) shifting their range; (iii) being moved or assisted by humans (translocation, assisted migration or population enhancement); or (iv) going extinct (IPCC, 2004; see Ford-Lloyd *et al.*, Chapter 2, this volume). Species vary in their ability to disperse and adapt to new environments so if extinction is to be avoided we need to identify and understand what makes a species vulnerable to climate change and to give conservation priority to these species. For example, we need to know which species have a small population size, where component populations are highly fragmented, have a narrow ecological or geographic range, are poorly adapted to their niche or are poor competitors. Each of these potentially lowers the effective population size for the species below the sustainable minimal viable population number for that species (Foden *et al.*, 2009).

7.3.1 Using the IUCN Red List Categories and Criteria

The most commonly applied means of assessing the threatened status of wild taxa is the application of the IUCN Red List Categories and Criteria (IUCN, 2001), which is a standardized system for the classification of taxa according to their extinction risk (or threatened status). The IUCN Red List assessment does not directly assess the threat from climate change; rather it assesses the indirect impact of climate change using proxy measures (e.g. increased overgrazing, fires or competition from alien species), which themselves could be attributed to climate change. Red List assessment of taxa, most often species, is essentially a two-step process (Kell *et al.*, 2012a):

- Data of seven types are collated and documented, namely: (i) taxonomic; (ii) distribution; (iii) population; (iv) habitat and ecology; (v) use and trade; (vi) threats; and (vii) conservation actions. These data are gathered from a number of sources, including taxon experts, published and grey literature, databases and websites.
- The taxon is evaluated against the IUCN Red List Criteria using the collated data and on the basis of this evaluation the appropriate Red List Category is selected.

There are five main Red List Criteria: (A) population reduction; (B) geographic range; (C) small population size and decline; (D) very small or restricted population; and (E) quantitative analysis indicating the probability of extinction. IUCN Red List Criterion E is based on some form of modelling and quantitative analysis and in theory this could be related to an assessment of climate change vulnerability but the models are still to be developed and Criterion E is still rarely used in Red List assessment.

Each main criterion includes a number of sub-criteria against which the species is evaluated. If the species meets the criteria in at least one of the main classes, it is assigned one of the threatened categories, Critically Endangered (CR), Endangered (EN) or Vulnerable (VU). If the species meets the criteria in more than one main class, it is assigned the highest category of threat but the less threatened category according to the other criterion or criteria is also documented. If the species does not meet any of the criteria A–E needed to evaluate it as threatened, another category is selected; this can be Extinct (EX), Extinct in the Wild (EW), Regionally Extinct (RE), Near Threatened (NT), Least Concern (LC), Data Deficient (DD) or Not Applicable (NA) (for definitions of the categories, see IUCN, 2001). So in each assessment taxon or population data are assessed against established criteria and depending on what levels for the criteria are met an IUCN category is assigned. Figure 7.3 is a

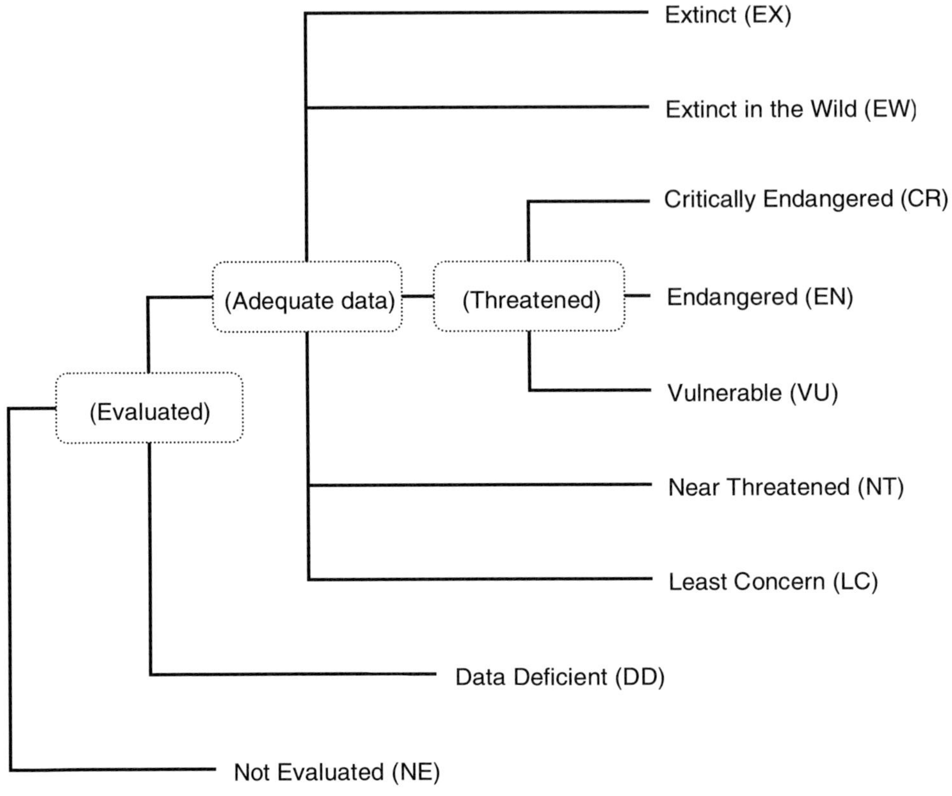

Fig. 7.3. Structure of the IUCN Red List Categories (redrawn from IUCN, 2001).

schematic representation of the IUCN Red List Categories.

Application of IUCN Red List Criteria is so well established as a means of assessing threat to wild plant species that the establishment of 'A preliminary assessment of the conservation status of all known plant species, at national, regional, and international levels' was included as Target two of the 2002 Global Strategy for Plant Conservation (CBD, 2002) to be achieved by 2010, but within the timescale the target was significantly under-achieved. The revised Global Strategy for Plant Conservation 2011–2020 (CBD, 2010a) target of 'An assessment of the conservation status of all known plant species, as far as possible, to guide conservation action' is more easily met, but today still only approximately 15,600 plant species have been assessed out of the estimated 380,000 (IUCN, 2013). Although Red List assessments have been comprehensively completed for birds (BirdLife International, 2008), mammals (Schipper *et al.*, 2008) and amphibians (Stuart *et al.*, 2008), progress for plants has been hampered by the relatively large number of plant species and the lack of readily available data necessary for Red List assessments. Miller *et al.* (2012) propose two streamlined methods using readily available locality data from herbarium specimens – the first is based on the extent of occurrence and the second on the historic timing of collections, relative number of collection localities, absolute number of collections and relative number of recent collections. The methods identified about a quarter of the species assessed as threatened using the IUCN Red List Criteria and so provide a rapid but crude way of identifying plants at risk. However, the full IUCN method is likely to continue to be the standard for assessing threat because it is comprehensive and widely accepted.

7.3.2 Climate change susceptibility assessment

The IUCN Red List Categories and Criteria were developed before the threat of climate change was fully appreciated and IUCN has acknowledged that a more robust method of identifying species at risk due to the impacts of climate change is required to aid conservation planning (Foden *et al.*, 2009). As noted by Foden *et al.* (2009), species respond differently to similar levels of climatic change due to a variety of biological traits, including their life history, ecology, behaviour, physiology and genetic diversity. The authors have therefore proposed that a species' susceptibility to climate change can be estimated on the basis of presence or absence of specific biological traits and species that are both exposed to significant climatic changes and intrinsically susceptible to climate change are at greatest risk of extinction. Identification of susceptibility traits in a species offers foresight into where conservation efforts should be prioritized.

To identify which traits are indicative of climate change susceptibility, Foden and colleagues (2009) conducted a 4-day workshop involving 31 biologists with expertise in a broad range of taxonomic groups and geographic regions. More than 90 susceptibility traits were identified and classified in five trait groups (Foden *et al.*, 2009):

1. Specialized habitat and/or microhabitat requirements.
2. Narrow environmental tolerances or thresholds that are likely to be exceeded due to climate change at any stage in the life cycle.
3. Dependence on specific environmental triggers or cues that are likely to be disrupted by climate change.
4. Dependence on interspecific interactions that are likely to be disrupted.
5. Poor ability to disperse to or to colonize a new or more suitable range.

Although many climate change susceptibility traits were identified, the number that can realistically be used is limited by lack of data availability for many species and by the difficulty of 'defining traits in quantifiable, objective and replicable ways' (Foden *et al.*, 2009, p. 3). To date, Foden and colleagues have assessed susceptibility to climate change of species in three biological groups: birds, amphibians and warm-water reef-building corals. On the basis of a range of information sources and on expert knowledge, the species were assessed against a list of selected traits (17 in total in the study). Any species possessing one or more susceptibility trait was defined as a climate change susceptible species. For these species, they examined the taxonomic and geographical distributions of those that are most susceptible to climate change and compared these to existing Red List assessments published in the 2008 IUCN Red List of Threatened Species (IUCN, 2008). Foden *et al.* (2009) reported that for birds, 11 useful climate change susceptibility traits were identified and 3438 (35%) of the 9856 species assessed are potentially susceptible. Of these, 1288 species possess between two and seven traits. For amphibians, six traits were identified, 3217 of the 6222 species (52%) are potentially susceptible and 962 species possess two to four traits. For warm-water reef-forming coral species, ten useful susceptibility traits were identified, 566 of 799 (71%) species are potentially susceptible, and 253 species possess between two and six traits. For all three taxonomic groups, susceptibility was strongly correlated with taxonomic and geographic groupings.

Results showed that significant percentages of species in each of the three groups (birds, 25%; amphibians, 28%; corals, 51%) are susceptible to climate change but are not threatened according to the 2008 IUCN Red List. This is an important conclusion because it flags the limitations of the current IUCN Red List system of assessing extinction risk. Although these species are currently not considered to be threatened, they are likely to become so in the future as climate change impacts intensify. Further, 10% of birds, 24% of amphibians and 19% of corals are both assessed as threatened in the 2008 Red List and are climate change susceptible. As the authors note, 'a large overlap between threatened and "climate-change-susceptible" species may therefore mean that climate change may cause a sharp rise in both the extinction risk and extinction rate of already threatened species' (Foden *et al.*, 2009, p. 82). They also found a significant overlap between IUCN Red List threatened status and climate change susceptibility, though it is important to identify those susceptible species that, while currently not threatened, are likely to become so in the future, as climate change impacts intensify.

The use of climate change susceptibility traits has proven useful in quantifying climate change vulnerability for corals, amphibians and birds, where 35%, 52% and 71% of the species assessed have traits that render them particularly susceptible to climate change. It is therefore likely to prove useful for other taxonomic groups such as plants and more specifically CWRs, where the methodology could be used as a form of triage to identify the species most likely to be threatened and so help plan more effective conservation actions. There is currently no published application of the climate change susceptibility assessment using biological traits for plants; however, a study of climate change susceptibility of European CWRs is currently underway. Some of the susceptibility traits used for corals, amphibians and birds will prove useful in this and other analyses of plant species (e.g. species with narrow distribution ranges, species restricted to specific habitats or microhabitats, species restricted to specific habitats particularly threatened by climate change, species less able to compete with other species, species already with sub-optimal inter-breeding population sizes), whereas other traits will need to be investigated (e.g. CWRs with obligate interactions such as a restricted range of pollinators or seed dispersers, CWRs with environmental triggers for vernalization, germination, flowering or fruiting, CWRs that are long-lived perennial species or are associated with climax communities). The development of a set of climate change susceptibility traits and their use to assess plants is an obvious short-term priority for CWR conservationists.

7.3.3 Using species distribution modelling

The use of species distribution modelling to assess the impact of climate change on plant genetic resources for food and agriculture has recently been reviewed by Jarvis *et al.* (2010). Although there has been extensive methodological development in recent years in predicting the future geographic distribution of species (Rotenberry *et al.*, 2006), they have yet to be systematically applied to CWR diversity. These methodologies have great potential to use current distributional ranges to construct a statistical model of a species' adaptation range, based on a set of user-defined environmental variables (Guisan and Zimmermann, 2000). The statistical model is then applied over a wider region to locate other areas where the environmental conditions are potentially suitable for the species in question. These species distribution models largely assume that climatic variables are the principal drivers of geographic distribution. Even though other factors such as soils, topography and specific habitat conditions are modelled, they are less commonly used because of the dearth of available datasets and have not been widely used in combination with climatic variables (Jarvis *et al.*, 2010). Another complicating factor is the capacities of species to migrate (Pearson, 2006). Climate change does not create 'new' climates but rather causes geographic shifts in climates, typically to higher altitudes and higher latitudes (Parmesan and Yohe, 2003), so species able to migrate freely are more likely to survive (Menendez *et al.*, 2006). Most models assume that migration is either unlimited or non-existent, yet the reality is likely to be somewhere in between (Pearson, 2006). Obviously, knowing the potential migration rate can strengthen the models, aid species distribution modelling and help assess the impact of climate change on individual species (Jarvis *et al.*, 2010).

7.4 Conservation of Crop Wild Relatives

Given the economic and climate change mitigation potential of CWR diversity, it might be expected that they would be comprehensively and systematically conserved; however, this is not the case – CWR diversity is under-conserved using both *ex situ* and *in situ* techniques.

7.4.1 *Ex situ* CWR conservation

Historically, the most widespread means of conserving CWR diversity has been through population sampling, seed collection and *ex situ* storage in seed genebanks (Maxted *et al.*, 2008a). Yet the Second Report on the State of the World's PGRFA (FAO, 2010) listed only 10% of genebank holdings as wild species. It concludes that 'Interest in collecting and maintaining collections of CWR is growing as land-use systems change, concerns about the effects of climate change grow and techniques for using the material become more powerful and more readily available'. So it appears that *ex situ* collections and interest in conserving CWRs has recently increased. FAO (2010) ultimately concludes, however, that 'For several major crops, such as wheat and rice, a large part of the genetic diversity is now represented in collections. However, for many other crops, especially many neglected and under-utilized species and CWRs, comprehensive collections still do not exist and considerable gaps remain to be filled'. Further, because of the largely *ad hoc* manner in which germplasm has been collected it seems likely that the CWR collections that do exist for major crops may not constitute genetically representative samples.

As an indicator of the coverage of *ex situ* collections, analysis of the data in the European crop genebank portal (EURISCO) revealed that around 9% of total germplasm holdings (accessions) are of CWR species (Dias *et al.*, 2012), which is in line with the global figure. However, only a small proportion of CWR species are included in these collections (1095; O'Regan, 2007) and, surprisingly, the ratio of cultivated to wild species accessions found in *ex situ* collections is 12:1, despite the fact that it is widely recognized that genetic diversity is primarily located in wild species (Maxted *et al.*, 2008a). Kell *et al.* (2012a) reported that for the socio-economically important European CWR species for which Red List assessments were undertaken, germplasm accessions were held by genebanks (as reported by EURISCO) for 273 of the 571 species assessed (nearly 48%) and that these represented 23 of the 25 crop gene pools/groups included for assessment. The authors also reported, however, that most species are represented by very few accessions, are reported by only one genebank and have been collected from only a small part of the species' range. We can therefore reasonably assume that there are large gaps in the *ex situ* conservation of some of the highest priority CWRs in Europe (Kell *et al.*, 2012a).

7.4.2 *In situ* CWR conservation

In situ CWR conservation is even less systematic than *ex situ* conservation. Although CWRs undoubtedly occur in many existing protected areas, they are rarely actively managed and monitored within them. Their conservation status is thus uncertain and populations or entire species could be at risk without the protected area manager being aware of the threat or consequences. Although in recent years considerable attention has been paid to the theory of design, establishment, management and monitoring of CWR diversity in genetic reserves (see Jain, 1975; Hoyt, 1988; Gadgil *et al.*, 1996; Hawkes *et al.*, 1997; Maxted *et al.*, 1997a, b; Safriel *et al.*, 1997; Heywood and Dulloo, 2006; Stolton *et al.*, 2006; Iriondo *et al.*, 2008; Hunter and Heywood, 2011; Kell *et al.*, 2012b), full practical implementation remains limited

(Maxted *et al.*, 1997b; Meilleur and Hodgkin, 2004). There are a few notable examples of activities that have made a significant contribution to the process of conserving CWR *in situ* (see Table 7.3), but even in some of these cases the sites are not managed in the most appropriate manner to conserve CWR genetic diversity.

In practice, *in situ* conservation of CWR is often planned within existing protected areas because: (i) by definition they are protected and managed (although some lack a management plan and conservation may be ineffective for many of the species present), already have an associated long-term conservation ethos and are less prone

Table 7.3. Examples of CWR conserved in protected areas (adapted from Maxted *et al.*, 2010).

CWR	Protected area	Country	References
Teosinte (*Zea diploperennis*)	MAB Sierra de Manantlán Biosphere Reserve	Mexico	Sanchez-Velasquez (1991)
Wild emmer wheat (*Triticum turgidum* subsp. *dicoccoides*)	Ammiad, Galilee	Israel	Anikster *et al.* (1997) Safriel *et al.* (1997)
Wild coffee (*Caffea mauritiana*, *Caffea macrocarpa*, *Caffea myrtifolia*)	Black River Gorges National Park	Mauritius	Dulloo *et al.* (1999)
Wild wheats (*Triticum turgidum* subsp. *dicoccoides*, *Triticum monococcum*, *Aegilops tauschii*, *Aegilops speltoides*)	Ceylanpinar	Turkey	Karagöz (1998)
Chestnut (*Castanea sativa*), wild plum (*Prunus cerasifera* var. *divaricata*)	Kazdağ	Turkey	Küçük *et al.* (1998)
Wild onions (*Allium columbianum*, *Allium geyeri* and *Allium fibrillum*)	Great Basin, Washington State	USA	Hannan and Hellier (1999) Hellier (2000)
Wild grapevine (*Vitis rupestris*, *Vitis shuttleworthii*, *Vitis monticola*)	Witchita Mountains and Ouachita National Forest, Oklahoma, Clifty Creek, Missouri	USA	Pavek *et al.* (2003)
Wild bean populations (*Phaseolus* spp.)	Central valley	Costa Rica	Baudoin *et al.* (2008)
Medicago spp., *Vicia* spp., *Trifolium* spp., *Lathyrus* spp., *Lens* spp., *Triticum* spp., *Avena* spp., *Hordeum* spp., *Aegilops* spp., *Allium* spp., *Amygdalus* spp., *Prunus* spp., *Pyrus* spp., *Pistacia* spp. and *Olea* spp.	Abu Taha, Sale-Rsheida, Ajloon, Wadi Sair	Lebanon Syria Jordan Palestinian Territories	Al-Atawneh *et al.* (2008)
Wild wheats (*Triticum boeoticum*, *Triticum urartu*, *Triticum araraticum*)	Erebuni	Armenia	Avagyan (2008)
Wild beet (*Beta patula*)	Desertas Islands	Portugal	Pinheiro de Carvalho *et al.* (2012)

to short-term management changes; (ii) in some cases, it is relatively easy to negotiate changes in the existing site management plan so as to facilitate genetic conservation of CWR species; and (iii) the creation of new conservation sites can be avoided along with the problems and cost of acquiring new land for conservation (Maxted *et al.*, 2008b). Recently a set of quality standards (Iriondo *et al.*, 2012) for the maintenance of genetic reserves to ensure the effective *in situ* conservation of CWR diversity has been proposed to underpin efficiency and avoid CWR being conserved in name only.

Obviously CWR species are just as likely to be found outside existing protected areas as within them. Various public and private initiatives to provide some degree of protection to species outside protected areas exist, especially in Australia, Brazil, China, Costa Rica, Mexico, South Africa, the USA and several European countries (Maxted *et al.*, 2008c; Heywood, 2009). One approach that can be used to promote *in situ* CWR conservation outside formal protected areas is within the context of Indigenous and Community Conserved Areas (ICCA) (Pathak *et al.*, 2004; Kothari, 2004, 2006, 2008a,b; IUCN 2011) where indigenous people and local communities have for millennia conserved natural environments and species for economic as well as cultural, spiritual and aesthetic reasons independent of more formal conservation sector interventions.

Increasingly, the approach to systematic CWR conservation will involve integrating local, national, regional and global actions (Maxted and Kell, 2009). For conservation strategies to be successful, national governments, the national agencies that oversee protected area conservation and plant genetic resource experts need to work with local people to set the agenda, for it is they who will be responsible for implementation, with international non-governmental organizations (NGOs) and intergovernmental organizations (IGOs) playing a supporting role (Smith *et al.*, 2009). A model for establishing a national CWR strategy has been tested and proven successful in the UK and Portugal (see Maxted *et al.*, 2007; Magos Brehm *et al.*, 2008) and is now being applied in other countries inside and outside of Europe (Maxted *et al.*, 2013). Further, a proposal to establish a worldwide network of *in situ* genetic reserves that is independent of national political borders and focuses on worldwide priority crop gene pools with complementary *ex situ* storage of genetically representative population samples has recently been proposed (Maxted and Kell, 2009; Maxted *et al.*, 2012; FAO, 2013). As mentioned above, a complementary global approach to *ex situ* conservation of 26 crop gene pools is currently being led by the Global Crop Diversity Trust (GCDT, 2011).

It could be argued that the establishment of conservation areas for CWRs in times of a rapidly rising human population, climate change and ecosystem instability, as well as a global economic crisis, is a complex and even risky goal. Sites competing for reserve status would need to be assessed and prioritized for their longer-term sustainability in terms of the predicted impact of climate change on the site and the development plans associated with local communities (Brooks *et al.*, 2006). Some CWR populations conserved in vulnerable locations may be unable to adapt sufficiently quickly to climate change or in doing so may go through a genetic bottleneck reducing the long-term viability of sustaining their diversity *in situ* (see Ford-Lloyd *et al.*, Chapter 2, this volume). On the other hand, basing a strategy on *ex situ* action alone is unlikely to be economically feasible or practical, given: (i) the very large number of species involved (possibly as many as 50,000); (ii) the practicalities and cost of sampling and conserving ecogeographically and genetically diverse populations for each species; and (iii) the fact that few species have immediate known use potential, genebanks may not consider them a priority for collection (Maxted *et al.*, 2012). Therefore, the strategic establishment of a global network of *in situ* CWR genetic reserves to complement targeted *ex situ* collections should be regarded as a priority. Conservation is nearly always a precursor to breeders' use (because breeders have rarely collected CWR material from the wild and prefer to obtain it from genebanks), so unless CWR diversity is conserved, it is unlikely to remain extant. There is therefore an urgent requirement to actively conserve CWR diversity using the full range of *in situ* and *ex situ* techniques.

7.5 Discussion

The importance of CWR diversity as a potentially critical resource for future food security is now widely recognized, particularly in the light of climate change and the rapidly increasing human population. The need to conserve CWR diversity has been placed firmly on the international conservation agenda. CWR diversity through trait transfer offers one means of helping sustain food production under the rapidly changing environmental conditions brought about by climate change. However, like all other wild plant diversity, CWR are under threat from the deleterious impacts of climate change and increases in human population, which in turn presents a tangible threat to global food security. The take-home message is that CWR diversity can be used to mitigate the impacts of climate change but at the same time is severely threatened by climate change. Systematic CWR diversity conservation is the answer to secure its climate change mitigation potential.

The Harlan and de Wet CWR inventory, the first global priority annotated inventory of CWR diversity, accounting for around 0.43% of the world's flora (Vincent *et al.*, 2012), gives us for the first time a clear target for systematic CWR diversity conservation. As suggested by Kappelle *et al.* (1999) for herb, butterfly and bird species, we need to undertake climate change vulnerability assessment, develop and monitor indicators of CWR diversity and focus active conservation on the most threatened species. Hampe and Petit (2005) argue that range analysis should be undertaken and that the conservation of those populations at the 'rear edge' ('low-latitude limit') should be prioritized because they generally have lower genetic diversity and are therefore more susceptible to extinction. We need to identify these CWRs not only because they are vulnerable to extinction but also because they are likely to contain unique diversity not found in the main area of distribution of the species where genetic diversity is likely to be higher overall and therefore these 'rear edge' populations are a conservation priority. Also, regions containing rear edge populations of multiple species ('relict hotspots') are of particularly high conservation interest. In addition, Araújo *et al.* (2004) proposed that we should 'climate proof' the selection of protected areas by selecting sites that are least likely to be impacted and where intrinsic genetic diversity is maximized so species have the best chance of adapting to climate change. Halpin (1997) suggested that protected area management is required that is flexible and easily altered over time to accommodate the changing climate. The active re-distributing of plant species to 'secure' locations has even been suggested (Maunder, 1992).

Ultimately there is unlikely to be one magic bullet for how we maximize CWR diversity conservation (and therefore potential utilization) in the face of climate change. Rather, a combination of many approaches is most likely to succeed. Having a limited number of priority species should, however, make the active conservation a much easier goal to achieve. It is critical that we expand use of CWR diversity in plant breeding if food security is to be maintained, that we have practical methods to promote the effective worldwide *in situ* and *ex situ* conservation strategy, and that we act now before CWR diversity is significantly depleted by climate change.

References

Al-Atawneh, N., Amri, A., Assi, R. and Maxted, N. (2008) Management plans for promoting *in situ* conservation of local agrobiodiversity in the West Asia centre of plant diversity. In: Maxted, N., Ford-Lloyd, B.V., Kell, S.P., Iriondo, J.M., Dulloo, E. *et al.* (eds) *Crop Wild Relative Conservation and Use*. CAB International, Wallingford, pp. 338–361.

Anikster, Y., Feldman, M. and Horovitz, A. (1997) The Ammiad experiment. In: Maxted, N., Ford-

Lloyd, B.V. and Hawkes, J.G. (eds) *Plant Genetic Conservation: the* In Situ *Approach*. Chapman & Hall, London, UK, pp. 239–253.

Araújo, M.B., Cabeza, M., Thuiller, W., Hannah, L. and Williams, P.H. (2004) Would climate change drive species out of reserves? An assessment of existing reserve selection methods. *Global Change Biology* 10, 1618–1626.

Avagyan, A., (2008) Crop wild relatives in Armenia: diversity, legislation and conservation issues. In: Maxted, N., Ford-Lloyd, B.V., Kell, S.P., Iriondo, J.M., Dulloo, E. *et al.* (eds) *Crop Wild Relative Conservation and Use*. CAB International, Wallingford, UK, pp. 58–68.

Baudoin, J.P., Rocha, O.J., Degreef, J., Zoro, Bi, I., Ouédraogo, M. *et al.* (2008) *In situ* conservation strategy for wild Lima bean (*Phaseolus lunatus* L.) populations in the Central Valley of Costa Rica: a case study of short-lived perennial plants with a mixed mating system. In: Maxted, N., Ford-Lloyd, B.V., Kell, S.P., Iriondo, J.M., Dulloo, E. *et al.* (eds) *Crop Wild Relative Conservation and Use*. CAB International, Wallingford, UK, pp. 364–379.

Baum, M., Grando, S., Backes, G., Jahoor, A., Sabbagh, A. *et al.* (2003) QTLs for agronomic traits in the Mediterranean environment identified in recombinant inbred lines of the cross Arta x H. *spontaneum* 41–1. *Theoretical and Applied Genetics* 107, 1215–1225.

Bilz, M., Kell, S.P., Maxted, N. and Lansdown, R.V. (2011) *European Red List of Vascular Plants*. Publications Office of the European Union, Luxembourg.

BirdLife International (2008) *State of the World's Birds: Indicators for our Changing World*. BirdLife International, Cambridge, UK.

Bradley, B.A., Blumenthal, D.M., Wilcove, D.S. and Ziska, L.H. (2010) Predicting plant invasions in an era of global change. *Trends in Ecology and Evolution* 5, 310–318.

Brar, D.S. (2005) Broadening the gene pool of rice through introgression from wild species. In: Toriyama, K., Heong, K.L. and Hardy, B. (eds) *Rice is Life: Scientific Perspectives for the 21st Century*. Proceedings of the World Rice Research Conference held in Tsukuba, Japan, 4–7 November 2004. International Rice Research Institute, Los Baños, Philippines, pp. 157–160.

Brooks, T.M., Mittermeier, R.A., da Fonseca, G.A.B., Gerlach, J., Hoffmann, M. *et al.* (2006) Global biodiversity conservation priorities. *Science* 313, 58–61.

Brummitt, N. and Bachman, S. (2010) *Plants Under Pressure – a Global Assessment*. The first report of the IUCN Sampled Red List Index for Plants. Royal Botanic Gardens Kew, London, UK, the Natural History Museum, London, UK, and the International Union for the Conservation of Nature, Gland, Switzerland.

Burke, M.B., Lobell, D.B. and Guarino, L. (2009) Shifts in African crop climates by 2050, and the implications for crop improvement and genetic resources conservation. *Global Environmental Change* 19, 317–325.

Cahill, A.E., Aiello-Lammens, M.E., Fisher-Reid, M.C., Hua, X., Karanewsky, C.J. *et al.* (2012) How does climate change cause extinction? *Proceedings of the Royal Society B*. doi: 10.1098/rspb.2012.1890.

CBD (2002) *Global Strategy for Plant Conservation*. Secretariat of the Convention on Biological Diversity, Montreal, Quebec, Canada.

CBD (2010a) *Global Strategy for Plant Conservation 2011–2020*. Secretariat of the Convention on Biological Diversity, Montreal, Quebec, Canada.

CBD (2010b) *Strategic Plans for Biodiversity 2011–2020 and the Aichi Targets*. Secretariat of the Convention on Biological Diversity, Montreal, Quebec, Canada. Available at: www.cbd.int/doc/strategic-plan/2011-2020/Aichi-Targets-EN.pdf (accessed 1 January 2013).

Chakraborty, S. and Newton, A.C. (2011) Climate change, plant diseases and food security, an overview. *Plant Pathology* 60, 2–14.

Chen, G., Li, C., Shi, Y. and Nevo, E. (2008) Wild barley, *Hordeum spontaneum*, a genetic resource for crop improvement in cold and arid regions. *Sciences in Cold and Arid Regions* 1, 0115–0124.

Colmer, T.D., Flowers, T.J. and Munns, R. (2006) Use of wild relatives to improve salt tolerance in wheat. *Journal of Experimental Botany* 57, 1059–1078.

Crop Wild Relatives and Climate Change (2012) Online resource. Available at: www.cwrdiversity.org/checklist/ (accessed 29 January 2013).

Davis, A.P., Gole, T.W., Baena, S. and Moat, J. (2012) The impact of climate change on indigenous arabica coffee (*Coffea arabica*): predicting future trends and identifying priorities. *PLoS ONE*, 7, e47981. doi:10.1371/journal.pone.0047981.

Deryng, D., Sacks, W.J., Barford, C.C. and Ramankutty, N. (2011) Simulating the effects of climate and agricultural management practices on global crop yield. *Global Biogeochemical Cycles* 25, GB2006, doi: 10.1029/2009GB 003765.

Dias, S., Dulloo, M.E. and Arnaud, E. (2012) The role of EURISCO in promoting use of agricultural biodiversity. In: Maxted, N., Dulloo, M.E., Ford-Lloyd, B.V., Frese, L., Iriondo, J.M. *et al.* (eds)

Agrobiodiversity Conservation: Securing the Diversity of Crop Wild Relatives and Landraces. CAB International, Wallingford, UK, pp. 270–277.

Dulloo, M.E., Guarino, L., Engelmann, F., Maxted, N., Newbury, H.J.N. *et al.* (1999) Complementary conservation strategies for the genus *Coffea* with special reference to the Mascarene Islands. *Genetic Resources and Crop Evolution* 45, 565–579.

Duveiller, E., Singh, R.P. and Nicol, J.M. (2007) The challenges of maintaining wheat productivity: pests, diseases, and potential epidemics. *Euphytica* 157, 417–430.

Eglinton, J.K., Evans, D.E., Brown, A.H.D., Langridge, P., McDonald, G. *et al.* (2001) The use of wild barley (*Hordeum vulgare* spp. *spontaneum*) in breeding for quality and adaptation. *Proceedings of the 10th Barley Technical Symposium.* Canberra, ACT, Australia, 16–20 September 2001. Available at: www.regional.org.au/au/abts/1999/eglinton.htm (accessed 4 August 2008).

Ellstrand, N.C., Prentice, H.C. and Hancock, J.F. (1999) Gene flow and introgression from domesticated plants into their wild relatives. *Annual Review of Ecology, Evolution and Systematics* 30, 539–563.

FAO (2001) *International Treaty on Plant Genetic Resources for Food and Agriculture.* Food and Agriculture Organization of the United Nations, Rome, Italy. Available at: www.planttreaty.org/texts_en.htm (accessed 1 January 2013).

FAO (2008) *Climate Change and Biodiversity for Food and Agriculture.* Food and Agriculture Organization of the United Nations, Rome, Italy.

FAO (2010) *Second report on the State of the World's Plant Genetic Resources for Food and Agriculture.* Food and Agriculture Organization of the United Nations, Rome, Italy. Available at: www.fao.org/agriculture/seed/sow2/en/ (accessed 1 January 2013).

FAO (2011) *Global Plan of Action for the Conservation and Sustainable Utilization of PGRFA.* Food and Agriculture Organization of the United Nations, Rome, Italy. Available at: www.globalplanofaction.org (accessed 1 January 2013).

FAO (2013) *Towards the Establishment of a Global Network for* In Situ *Conservation and On Farm Management of Plant Genetic Resources for Food and Agriculture.* Technical Workshop Report. Food and Agriculture Organization of the United Nations, Rome, Italy. Available at: www.fao.org/agriculture/seed (accessed 1 March 2013).

Farooq, S., Iqbal, N., Asghar, M. and Shah, T.M. (1992) Intergeneric hybridization for wheat improvement. VI. Production of salt tolerant germplasm through crossing wheat (*Triticum aestivum* L.) with *Aegilops cylindrica* and its significance in practical agriculture. *Journal of Genetics and Breeding* 46, 125–132.

Farooq, S., Asghar, M., Iqbal. N., Asian, E., Arif, M. *et al.* (1995) Production of salt tolerant wheat germplasm through crossing cultivated wheat with *Aegilops cylindrica.* II. Field evaluation of salt tolerant germplasm. *Cereal Research Community* 23, 275–282.

Feuillet, C., Langridge, P. and Waugh, R. (2008) Cereal breeding takes a walk on the wild side. *Trends in Genetics* 24, 24–32.

Foden, W., Mace, G., Vié, J.-C., Angulo, A., Butchart, S. *et al.* (2009) Species susceptibility to climate change impacts. In: Vié, J.-C., Hilton-Taylor, C. and Stuart, S.N. (eds) *Wildlife in a Changing World – An Analysis of the 2008 IUCN Red List of Threatened Species.* IUCN, Gland, Switzerland, pp. 77–88.

Gadgil, M., Niwas Singh, S., Nagendra, H. and Chandran, M.D.S. (1996) In Situ *Conservation of Wild Relatives of Cultivated Plants: Guiding Principles and a Case Study.* Food and Agriculture Organization of the United Nations, Rome, Italy.

GCDT (2011) $50 million pledge for climate change adaptation. Global Crop Diversity Trust, Rome, Italy. Available at: www.croptrust.org/main/index.php?itemid=908 (accessed 1 January 2013).

Goodman, R.M., Hauptli, H., Crossway, A. and Knauf. V.C. (1987) Gene transfer in crop improvement. *Science* 236, 48–54.

Gregory, P.J., Johnson, S.N., Newton, A.C. and Ingram, J.S.I. (2009) Integrating pests and pathogens into the climate change/food security debate. *Journal of Experimental Botany* 60, 2827–2838.

Guisan, A. and Zimmermann, N. (2000) On the use of static distribution models in ecology. *Ecological Modelling* 135, 147–186.

Hadjichristodoulou, A. (1993) The use of wild barley in crosses for grain production under dryland conditions. *Euphytica* 69, 211–218.

Hajjar, R. and Hodgkin, T. (2007) The use of wild relatives in crop improvement: a survey of developments over the last 20 years. *Euphytica* 156, 1–13.

Halpin, P.N. (1997) Global climate change and natural area protection: management responses and research directions. *Ecological Applications* 7, 828–843.

Hammer, K. and Khoshbakht, K. (2005) Towards a 'red list' for crop plant species. *Genetic Resources and Crop Evolution* 52, 249–265.

Hampe, A. and Petit, R.J. (2005) Conserving biodiversity under climate change: the rear edge matters. *Ecology Letters* 8, 461–467.

Hanelt, P. and Institute of Plant Genetics and Crop Plant Research (eds) (2001) *Mansfeld's Encyclopaedia of Agricultural and Horticultural Crops*. Springer, Germany.

Hannan, R. and Hellier, B.C. (1999) In: Pavek, D.S., Lamboy, W.F. and Garvey, E.J. (eds) *Ecogeographic Study of* Vitis *Species: Final Report for Caloosa and Sweet Mountain Grapes*. Report, USDA Pullman, Washington.

Hawkes, J.G., Maxted, N. and Zohary, D. (1997) Reserve design. In: Maxted, N., Ford-Lloyd, B.V. and Hawkes, J.G. (eds) *Plant Genetic Conservation: the* In Situ *Approach*. Chapman & Hall, London, pp. 210–230.

Hellier, B.C. (2000) Genetic, morphologic, and habitat diversity of two species of *Allium* native to the Pacific Northwest, USA and their implications for *in situ* seed collection for the National Plant Germplasm System. MSc. thesis. Washington State University, Pullman, Washington.

Hertel, T.W., Burke, M.B. and Lobell, D.B. (2010) The poverty implications of climate-induced crop yield changes by 2030. *Global Environmental Change* 20, 577–585.

Heywood, V.H. (2009) *The Impacts of Climate Change on Plant Species in Europe*. Convention on the Conservation of European Wildlife and Natural Habitats, Standing Committee, Strasbourg, T-PVS/Inf(2009) 9, 23 November 2009.

Heywood, V.H. and Dulloo, M.E. (2006) In Situ *Conservation of Wild Plant Species – a Critical Global Review of Good Practices*. IPGRI Technical Bulletin No. 8. IPGRI, Rome.

Heywood, V.H., Kell, S.P. and Maxted, N. (2008) Towards a global strategy for the conservation and use of crop wild relatives. In: Maxted, N., Ford-Lloyd, B.V., Kell, S.P., Iriondo, J.M., Dulloo, E. *et al.* (eds) *Crop Wild Relative Conservation and Use*. CAB International, Wallingford, UK, pp. 653–662.

Hoyt, E. (1988) *Conserving the Wild Relatives of Crops*. IBPGR, Rome, Italy; IUCN and WWF, Gland, Switzerland.

Hunter, D. and Heywood, V.H. (2011) *Crop Wild Relatives: A Manual of In Situ Conservation*. Earthscan, UK.

INIBAP/IPGRI (2006) *Global Conservation Strategy for* Musa *(Banana and Plantain)*. INIBAP/IPGRI, Montpellier/Rome, Italy.

IPCC (2007) *Fourth Assessment Report Climate Change 2007: Synthesis Report*. Intergovernmental Panel on Climate Change, Geneva, Switzerland.

IPK Gatersleben (2003) *Mansfeld's Encyclopedia of Agricultural and Horticultural Crops*. Springer, Germany.

Iriondo, J.M., Dulloo, E. and Maxted, N. (eds) (2008) *Conserving Plant Genetic Diversity in Protected Areas: Population Management of Crop Wild Relatives*. CABI Publishing, Wallingford, UK.

Iriondo, J.M., Maxted, N., Kell, S.P., Ford-Lloyd, B.V., Lara-Romero, C. *et al.* (2012) Quality standards for genetic reserve conservation of crop wild relatives. In: Maxted, N., Dulloo, M.E., Ford-Lloyd, B.V., Frese, L., Iriondo, J.M. *et al.* (eds) *Agrobiodiversity Conservation: Securing the Diversity of Crop Wild Relatives and Landraces*. CAB International, Wallingford, UK, pp. 72–77.

IUCN (2001) *IUCN Red List Categories and Criteria: Version 3.1*. IUCN Species Survival Commission. International Union for Conservation of Nature, Gland, Switzerland and Cambridge, UK. Available at: www.iucnredlist.org/static/categories_criteria_3_1 (accessed 1 January 2013).

IUCN (2008) 2008 Red List of Threatened Species. Available at: http://www.iucnredlist.org. (accessed 29 January 2013).

IUCN (2011) Indigenous and community conserved areas. Available at: www.iucn.org/about/union/commissions/ceesp/topics/governance/icca/ (accessed 29 January 2013).

IUCN (2013) *IUCN Red List of Threatened Species. Version 2013*. International Union for Conservation of Nature, Gland, Switzerland and Cambridge, UK. Available at: www.iucnredlist.org/documents/summarystatistics/2012_2_RL_Stats_Table_3b_c.pdf (accessed 1 January 2013).

Jain, S.K. (1975) Genetic reserves. In: Frankel, O.H. and Hawkes, J.G. (eds) *Crop Genetic Resources for Today and Tomorrow*. Cambridge University Press, Cambridge, UK, pp. 379–396.

Jarvis, A., Lane, A. and Hijmans. R.J. (2008) The effect of climate change on crop wild relatives. *Agriculture, Ecosystems and Environment* 126, 13–23.

Jarvis, A., Upadhyaya, H., Gowda, C.L.L., Aggarwal, P.K., Fujisaka, S. *et al.* (2010) *Climate Change and its Effect on Conservation and Use of Plant Genetic Resources for Food and Agriculture and Associated Biodiversity for Food Security*. Background Study Paper. Commission on Genetic Resources for Food and Agriculture, Food and Agriculture Organization of the United Nations, Rome, Italy.

Jennings, D.L. (1995) Cassava, *Manihot esculenta* (Euphorbiaceae). In: Smartt, J. and Simmonds,

N.W. (eds) *Evolution of Crop Plants.* Longman Group, Harlow, Essex, UK, pp. 128–132.

Jones, P.D., Lister, D.H., Jaggard, K.W. and Pidgeon, J.D. (2003) Future climate change impact on the productivity of sugar beet (*Beta vulgaris* L.) in Europe. *Climatic Change* 58, 93–108.

Kappelle, M., Van Vuuren, M.M. and Baas, P. (1999) Effects of climate change on biodiversity: a review and identification of key research issues. *Biodiversity and Conservation* 8, 1383–1397.

Karagöz, A., (1998) *In situ* conservation of plant genetic resources in the Ceylanpinar State Farm. In: Zencirci, N., Kaya, Z., Anikster, Y. and Adams, W.T. (eds) *Proceedings of International Symposium on* In Situ *Conservation of Plant Diversity.* Central Research Institute for Field Crops, Ankara, Turkey, pp. 87–92.

Kell, S.P., Knüpffer, H., Jury, S.L., Maxted, N. and Ford-Lloyd, B.V. (2005) *Catalogue of Crop Wild Relatives for Europe and the Mediterranean.* Available at: www.pgrforum.org/cwris/cwris.asp (accessed 27 January 2013).

Kell, S.P., Knüpffer, H., Jury, S.L., Ford-Lloyd, B.V. and Maxted, N. (2008) Crops and wild relatives of the Euro-Mediterranean region: making and using a conservation catalogue. In: Maxted, N., Ford-Lloyd, B.V., Kell, S.P., Iriondo, J.M., Dulloo, E. *et al.* (eds) *Crop Wild Relative Conservation and Use.* CAB International, Wallingford, UK, pp. 69–109.

Kell, S.P., Maxted, N. and Bilz, M. (2012a) European crop wild relative threat assessment: knowledge gained and lessons learnt. In: Maxted, N., Dulloo, M.E., Ford-Lloyd, B.V., Frese, L., Iriondo, J.M. *et al.* (eds) *Agrobiodiversity Conservation: Securing the Diversity of Crop Wild Relatives and Landraces.* CAB International, Wallingford, UK, pp. 218–242.

Kell, S.P., Maxted, N., Frese, L. and Iriondo, J.M. (2012b) *In situ* conservation of crop wild relatives: a strategy for identifying priority genetic reserve sites. In: Maxted, N., Dulloo, M.E., Ford-Lloyd, B.V., Frese, L., Iriondo, J.M. *et al.* (eds) *Agrobiodiversity Conservation: Securing the Diversity of Crop Wild Relatives and Landraces.* CAB International, Wallingford, UK, pp. 7–19.

King, I.P., Forster, B.P., Law, C.C., Cant, K.A., Orford, S. *et al.* (1997a) Introgression of salt tolerance genes from *Thinopyrum bessarabicum* into wheat. *New Phytologist* 137, 75–81.

King, I.P., Law, C.C., Cant, K.A., Orford, S., Gorham, J. *et al.* (1997b) *Tritipyrum,* a potential new salt-tolerant cereal. *Plant Breeding* 116, 127–132.

Kothari, A., (2004) Protected areas and people: participatory conservation. In: *Biodiversity Issues for Consideration in the Planning, Establishment and Management of Protected Areas Sites and Networks.* CBD Technical Series no. 15. Secretariat of the Convention on Biological Diversity, Montreal, Canada, pp. 94–99.

Kothari, A. (ed.) (2006) Special issue on Community Conserved Areas of the magazine of the IUCN World Commission on Protected Areas. *Parks* 16, 1–79.

Kothari, A. (2008a) Protected areas and people: the future of the past. *Parks* 17, 23–34.

Kothari, A., (2008b) The 4C factor: Community conservation and climate change. *Biodiversity* (Ottawa) 9, 19–23.

Küçük, S.A., Tan, A.Ş., Sabanci, C.O., Cinsoy, A.S., Önul, K. *et al.* (1998) Ecogeographic and floristic differentiation of chestnut gene management zone in Kazdağ. In: Zencirci, N., Kaya, Z., Anikster, Y. and Adams, W.T. (eds) *Proceedings of International Symposium on* In Situ *Conservation of Plant Diversity.* Central Research Institute for Field Crops, Ankara, Turkey, pp. 135–148.

Lexer, C., Lai, Z. and Rieseberg, L.H. (2004) Candidate gene polymorphisms associated with salt tolerance in wild sunflower hybrids: implications for the origin of *Helianthus paradoxus,* a diploid hybrid species. *New Phytologist* 161, 225–233.

Li, X., Takahashi, T., Suzuki, N. and Kaiser, H.M. (2011) The impact of climate change on maize yields in the United States and China. *Agricultural Systems* 104, 348–353.

Lira, R., Téllez, O. and Dávila, P. (2009) The effects of climate change on the geographic distribution of Mexican wild relatives of domesticated Cucurbitaceae. *Genetic Resources and Crop Evolution* 56, 691–703.

Luck, J., Spackmand, M., Freemand, A., Trębicki, P., Griffiths, W. *et al.* (2011) Climate change and diseases of food crops. *Plant Pathology* 60,113–121.

Magos Brehm, J., Maxted, N., Ford-Lloyd, B.V. and Martins-Loução, M.A. (2008) National inventories of crop wild relatives and wild harvested plants: case-study for Portugal. *Genetic Resources and Crop Evolution* 55, 779–796.

Maunder, M. (1992) Plant reintroduction: an overview. *Biodiversity and Conservation* 1, 21–62.

Maxted, N. and Kell, S.P. (2009) *Establishment of a Global Network for the* In Situ *Conservation of Crop Wild Relatives: Status and Needs.* Background Study Paper No. 39. Commission on Genetic Resources for Food and Agriculture, FAO, Rome, Italy. Available at: http://typo3.fao.

org/fileadmin/templates/agphome/documents/PGR/SoW2/CropWildRelatives_2011_final.pdf (accessed 1 January 2013).

Maxted, N., Ford-Lloyd, B.V. and Hawkes, J.G. (1997a) Complementary conservation strategies. In: Maxted, N., Ford-Lloyd, B.V. and Hawkes, J.G. (eds) *Plant Genetic Conservation: the* In Situ *Approach.* Chapman & Hall, London, pp. 20–55.

Maxted, N., Hawkes, J.G., Ford-Lloyd, B.V. and Williams, J.T. (1997b) A practical model for *in situ* genetic conservation. In: Maxted, N., Ford-Lloyd, B.V. and Hawkes, J.G. (eds) *Plant Genetic Conservation: the* In Situ *Approach.* Chapman & Hall, London, pp. 545–592.

Maxted, N., Ford-Lloyd, B.V., Jury, S.L., Kell, S.P. and Scholton, M.A. (2006) Towards a definition of a crop wild relative. *Biodiversity and Conservation* 15(8), 2673–2685.

Maxted, N., Scholten, M.A., Codd, R. and Ford-Lloyd, B.V. (2007) Creation and use of a national inventory of crop wild relatives. *Biological Conservation* 140, 142–159.

Maxted, N., Ford-Lloyd, B.V. and Kell, S.P. (2008a) Crop wild relatives: establishing the context. In: Maxted, N., Ford-Lloyd, B.V., Kell, S.P., Iriondo, J.M., Dulloo, M.E. *et al.* (eds) *Crop Wild Relative Conservation and Use.* CAB International, Wallingford, UK, pp. 3–30.

Maxted, N., Iriondo, J., Dulloo, E. and Lane, A. (2008b) Introduction: The integration of PGR conservation with protected area management. In: Iriondo, J.M., Maxted, N. and Dulloo, E. (eds) *Plant Genetic Population Management.* CAB International, Wallingford, UK, pp. 1–22.

Maxted, N., Iriondo, J., De Hond, L., Dulloo, E., Lefèvre, F. *et al.* (2008c) Genetic reserve management. In: Iriondo, J.M., Maxted, N. and Dulloo, E. (eds) *Plant Genetic Population Management.* CAB International, Wallingford, UK, pp. 65–87.

Maxted, N., Kell, S.P., Toledo, A., Dulloo, M.E., Heywood, V.H. *et al.* (2010) A global approach to crop wild relative conservation: securing the gene pool for food and agriculture. *Kew Bulletin* 65, 561–576.

Maxted, N., Kell, S.P., Ford-Lloyd, B.V., Dulloo, M.E. and Toledo, A. (2012) Toward the systematic conservation of global crop wild relative diversity. *Crop Sciences* 52, 774–785.

Maxted, N., Magos Brehm, J. and Kell, S. (2013) *Resource Book for the Preparation of National Plans for Conservation and Use of Crop Wild Relatives and Landraces.* Commission on Genetic Resources for Food and Agriculture, FAO, Rome.

Meilleur, B.A. and Hodgkin, T. (2004) *In situ* conservation of crop wild relatives. *Biodiversity and Conservation* 13, 663–684.

Menéndez, R., Megias, A.G., Hill, J.K., Braschler, B., Willis, S.G. *et al.* (2006) Species richness changes lag behind climate change. *Proceedings of the Royal Society of London Series B-Biological Sciences* 273, 1465–1470.

Miller, J.S., Porter-Morgan, H.A., Stevens, H., Boom, B., Krupnick, G.A. *et al.* (2012) Addressing target two of the Global Strategy for Plant Conservation by rapidly identifying plants at risk. *Biodiversity and Conservation* 21, 1877–1887.

Mora, A., Zapata Ferrufino, B., Hunter, D., Navarro, G., Galeano, G. *et al.* (2009) *Libro Rojo de Parientes Silvestres de Cultivos de Bolivia.* VMABCC-Bioversity International, Rome.

Noori, S.A.S. (2005) Assessment for salinity tolerance through intergeneric hybridization: *Triticum durum* x *Ae. speltoides. Euphytica* 146, 149–155.

O'Regan, F. (2007) An examination of the European internet search catalogue (EURISCO) with special reference to the quality of passport data and crop wild relatives. MSc thesis. University of Birmingham, Birmingham, UK.

Parmesan, C. and Yohe, G. (2003) A globally coherent fingerprint of climate change impacts across natural systems. *Nature* 421, 37–42.

Pathak, N., Bhatt, S., Balasinorwala, T., Kothari, A. and Borrini-Feyerabend, G. (2004) *Community Conserved Areas: a Bold Frontier for Conservation.* TILCEPA/IUCN, CENESTA, CMWG and WAMIP, Tehran. Available at: www.iucn.org/themes/ceesp/Wkg_grp/TILCEPA/CCA%20Briefing%20Note.pdf (accessed 1 January 2013).

Pavek, D.S., Lamboy, W.F. and Garvey, E.J. (2003) Selecting *in situ* conservation sites for grape genetic resources in the USA. *Genetic Resources and Crop Evolution* 50, 165–173.

Pearson, R.G. (2006) Climate change and the migration capacity of species. *Trends in Ecology and Evolution* 21, 111–113.

Phillips, O.L. and Meilleur, B. (1998) Usefulness and economic potential of the rare plants of the United States: a status survey. *Economic Botany* 52, 57–67.

Pimentel, D., Wilson, C., McCullum, C., Huang, R., Dwen, P. *et al.* (1997) Economic and environmental benefits of biodiversity. *BioScience* 47, 747–757.

Pinheiro de Carvalho, M.Â.A., Nóbrega, H., Freitas, G., Fontinha, S. and Frese, L. (2012) Towards the establishment of a genetic reserve for *Beta patula* Aiton. In: Maxted, N., Dulloo, M.E., Ford-Lloyd, B.V., Frese, L., Iriondo, J.M. *et al.* (eds) *Agrobiodiversity Conservation: Securing the Diversity of Crop Wild Relatives and Landraces*. CAB International, Wallingford, UK, pp. 218–242.

Planta Europa (2008) *A Sustainable Future for Europe; the European Strategy for Plant Conservation 2008–2014*. Plantlife International, Salisbury, UK and the Council of Europe, Strasbourg, France.

Rick, C.M. and Chetelat, R. (1995) Utilization of related wild species for tomato improvement, First International Symposium on Solanaceae for fresh market. *Acta Horticulturae* 412, 21–38.

Rotenberry, J.T., Preston, K.L. and Knick, S.T. (2006) GIS-based niche modeling for mapping species habitat. *Ecology* 87, 1458–1464.

Ruegg, K.C., Hijmans, R.J. and Moritz, C. (2006) Climate change and the origin of migratory pathways in the Swainson's thrush, *Catharus ustulatus*. *Journal of Biogeography* 33, 1172–1182.

Safriel, U.N., Anikster, Y. and Waldman, M. (1997) Management of nature reserves for conservation of wild relatives and the significant of marginal populations. *Bocconea* 7, 233–239.

Sanchez-Velasquez, L.R. (1991) *Zea diploperennis*: Mejoramiento genetico del maiz, ecologia y la conservación de recursos naturales. *Tiempos de ciencia 24 Qulio-septiembre*: 1–8. University of Guadalajara, Jalisco, Mexico.

Schipper, J., Chanson, J.S., Chiozza, F., Cox, N.A., Hoffman, M. *et al.* (2008) The status of the world's land and marine mammals: diversity, threat, and knowledge. *Science*, 322, 225–230.

Seppä, H., Alenius, T., Bradshaw, R.H.W., Giesecke, T., Heikkilä, M. *et al.* (2009) Invasion of Norway spruce (*Picea abies*) and the rise of the boreal ecosystem in Fennoscandia. *Journal of Ecology* 97, 629–640.

Smith, R.J., Verissimo, D., Leader-Williams, N., Cowling, R.M. and Knight, A.T. (2009) Let the locals lead. *Nature* 462, 280–281.

Stevens, M.A., and Rick, C.M. (1986) *The Tomato Crop. A Scientific Basis for Improvement*. Chapman & Hall, New York.

Stolton, S., Maxted, N., Ford-Lloyd, B.V., Kell, S.P. and Dudley, N. (2006) *Food Stores: Using Protected Areas to Secure Crop Genetic Diversity*. WWF Arguments for Protection Series. Gland, Switzerland.

Stuart, S.N., Hoffmann, M., Chanson, J.S., Cox, N.A., Berridge, R.J. *et al.* (2008) *Threatened Amphibians of the World*. Lynx Edicions, Barcelona, Spain, IUCN, Gland, Switzerland, and Conservation International, Arlington, Virginia, USA.

Talame, V., Sanguineti, M.C., Chiapparino, E., Bahri, H., Salem, B.M. *et al.* (2004) Identification of *Hordeum spontaneum* QTL alleles improving field performance of barley grown under rainfed conditions. *Annals of Applied Biology* 144, 309–319.

Tanksley, S.D. and McCouch, S.R. (1997) Seed banks and molecular maps: unlocking genetic potential from the wild. *Science* 277, 1063–1066.

Thomas, C.D., Cameron, A., Green, R.E., Bakkenes, M., Beaumont, L.J. *et al.* (2004) Extinction risk from climate change. *Nature* 427, 145–148.

Thuiller, T., Lavorel, S., Araújo, M.B., Sykes M.T. and Prentice, I.C. (2005) Climate change threats to plant diversity in Europe. *Proceedings of the National Academy of Sciences USA* 102, 8245–8250.

UNFCCC (1992) *United Nations Framework Convention on Climate Change*. Secretariat of the United Nations Framework Convention on Climate Change, Bonn, Germany.

Villareal, R.L., Sayre, K., Banuelos, O. and Mujeeb-Kazi, A., (2001) Registration of four synthetic hexaploid wheat (*Triticum turgidum/Aegilops tauschii*) germplasm lines tolerant to waterlogging. *Crop Science* 41, 274.

Vincent, H., Wiersema, J., Kell, S.P., Dobbie, S., Fielder, H. *et al.* (2013) A prioritised crop wild relative inventory as a first step to help underpin global food security. *Biological Conservation* (in press).

Volbrecht, E. and Sigmon, B. (2005) Amazing grass: Developmental genetics of maize domestication. *Biochemical Society Transactions* 33, 1502–1506.

Walters, K.S. and Gillett, H.J. (1998) *1997 IUCN Red List of Threatened Plants*. Compiled by the World Conservation Monitoring Centre. IUCN – The World Conservation Union, Gland, Switzerland and Cambridge, UK.

Wang, W., Vinocur, B. and Altaian, A. (2003) Plant responses to drought, salinity and extreme temperatures: towards genetic engineering for stress tolerance. *Planta* 218, 1–14.

Zhang, X., Zhou, S., Fu, Y., Su, Z., Wang, X. *et al.* (2006) Identification of a drought tolerant introgression line derived from Dongxiang common wild rice (*Oryza rufipogon* Griff.). *Plant Molecular Biology* 62, 247–259.

Zimmerer, K.S. (2010) Biological diversity in agriculture and global change. *Annual Review of Environment and Resources* 35, 137–166.

8 Climate Change and On-farm Conservation of Crop Landraces in Centres of Diversity

Mauricio R. Bellon[1] and Jacob van Etten[2]

[1]*Bioversity International, Rome, Italy and* [2]*Bioversity International, Regional Office for the Americas, Recta Cali, Colombia*

8.1 Introduction

Global climate change will have important effects on the ways in which humanity feeds itself (Fischer *et al.*, 2005; Lobell *et al.*, 2008). Increasing temperatures, changing rainfall patterns and increases in frequency of extreme events, such as droughts, storms, floods and weather extremes, will present important challenges to agricultural and food systems (Ortiz, 2011; Vermeulen *et al.*, 2012). Agricultural biodiversity in general, and plant genetic resources for food and agriculture in particular, will play a fundamental role in the capacity of agricultural and food systems to adapt and respond to climate change (Jarvis *et al.*, 2010; Ortiz, 2011). The fundamental role of these resources still remains, however, relatively unacknowledged (Ortiz, 2011). This is changing, however, and there is increasing attention to the complex relationships between climate change and the use and conservation of agricultural biodiversity, of which this volume is a good example. The conservation of crop genetic diversity has been a worldwide concern for many decades, particularly owing to the worry that much of this diversity would disappear with agricultural and economic development, i.e. genetic erosion (Hawkes, 1983; Harlan, 1992). Although this has happened for certain crops and certain areas, considerable diversity is still grown in developing countries by smallholder farmers, particularly in centres of crop domestication and diversity (Brush, 2004; Jarvis *et al.*, 2008). Efforts to conserve crop diversity by the scientific community have led to the collection and conservation of seeds in *ex situ* genebanks (Plucknett *et al.*, 1987). In the past two decades, there has been a growing interest in on-farm conservation of landraces in centres of crop diversity (Bellon *et al.*, 1997; Bretting and Duvick, 1997; Brush, 2004; Gepts, 2006). It refers to the management of landraces in farmers' fields and orchards where they originated, and aims at maintaining the processes of crop evolution (Brush, 1995). Landraces have been defined as dynamic populations of a cultivated plant with a historical origin, distinct identity, often genetically diverse and locally adapted, and associated with a set of farmers' practices of seed selection and field management as well as with a knowledge base (Camacho Villa *et al.*, 2005). To function, on-farm conservation depends then on farmers' preferences, knowledge, management, practices and social organization. It contrasts with *ex situ* conservation, where the aim is to maintain the genes and genotypes contained in samples of seeds or planting material representative of the diversity of a particular crop without change over a long time. These two strategies are aimed at conserving different things. On-farm conservation is about maintaining processes, whereas *ex situ* conservation is about maintaining specific results of these processes (specific genes and genotypes sampled at a particular point in time). Both

 Plant Genetic Resources and Climate Change (eds M. Jackson, B. Ford-Lloyd and M. Parry)

types of conservation can be treated as complementary (Bretting and Duvick, 1997; Brush, 2004).

Climate change is predicted to have major impacts on small-scale farmers in the developing world but these impacts are likely be complex, locally specific and hard to predict (Morton, 2007). Therefore, the climatic suitability for many crops will change with particularly negative consequences for crops in tropical areas of the world (Lane and Jarvis, 2007). Because many centres of crop diversity are located in those areas and the small-scale farmers there are key players in on-farm conservation, it is clear that climate change will have an important impact on the viability of on-farm conservation (Jarvis *et al.*, 2010). The future of crop genetic resources maintained on-farm will depend on the responses of landraces – and the farmers who grow them – to climate change, particularly on the tolerance and adaptive capacity of landraces to the novel conditions climate change will bring about (Mercer and Perales, 2010). The sensitivity of landraces to climate variability and their capacity to evolve and adapt to these changes is central to assess both their fate and value under a changing climate. Changes may be of such magnitude and speed that the evolutionary potential of crop populations may be limited (Gepts, 2006) and novel climatic conditions may adversely affect currently adapted landraces. This will reduce their performance and increase the vulnerability of the associated farming systems, probably requiring the adoption of new crops or varieties, leading to further genetic erosion (Jarvis *et al.*, 2010). Ecologists have, however, already documented rapid rates of evolution in response to climate change in several species (Davis *et al.*, 2005; Reusch and Wood, 2007), a pattern observed also in crops (Vigouroux *et al.*, 2011), which suggests that climate change can lead to the adaptation of landraces (Mercer and Perales, 2010) and provide options to farmers. These dynamics will depend not only on the inherent genetic and biological properties of the specific crop landraces in particular locations but also on the knowledge and management practices of the farmers who maintain them. In this chapter we review the potential impacts of climate change on on-farm conservation of crop diversity and on the contribution that on-farm conservation can make to agricultural adaptation to climate change, and ask two questions:

1. How might climate change affect the viability of on-farm conservation of landraces in centres of crop diversity?
2. How can on-farm conservation contribute to the adaptation of small-scale farmers to climate change?

8.2 What is On-farm Conservation?

On-farm conservation involves farmers' continued cultivation and management of a diverse set of crop populations in the agroecosystem where the crop evolved or in secondary centres of diversity. It depends on farmers' active participation based on their reasons and incentives for maintaining diversity (Bellon *et al.*, 1997). Crop genetic diversity is unequally distributed around the world and is concentrated in centres of diversity that often coincide with centres of crop domestication (Gepts, 2006). In these locations, besides the broad genetic and phenotypic diversity present, there is a long history of co-evolution between humans and the crops (Brush, 2004; Zimmerer, 2010), reflected in:

- The cultivation of a diverse set of landraces with an associated knowledge base.
- The existence of multiple uses and preparations, usually linked with particular cultural preferences.
- Specific management practices such as intercropping or rotations, as well as seed selection and sharing.
- Matching specific landraces to particular environmental niches for optimizing production and managing risk.
- Social norms and organization that underpin all of these aspects.

The structure and dynamics of these landraces are the result of both natural and human selection. Even natural selection is influenced by human actions such as moving and planting crop species in particular locations and environments, thereby exposing the crops to different natural selection pressures such as climatic conditions, specific soils, pests and diseases. As indicated by Gepts (2006): 'Through farming practices (time of planting, thinning, and seed selection), farmers are able to keep landraces adapted to their growing conditions and socio-cultural preferences'.

The objective of on-farm conservation is therefore to maintain crop evolution in farmers' fields, farms and landscapes. The reason to maintain evolutionary processes in crops is 'to generate new potentially useful genetic variation, which in turn contributes to maintain the capacity of agricultural and food systems to adapt to change, particularly if it is unpredictable' (Bellon, 2009). Clearly this capacity involves many other dimensions but, given that crops are central to these systems, the crop dimension is of great importance. The outcome of on-farm conservation can be conceptualized as an 'evolutionary service' to agricultural and food systems, and, although it is implicit in the definitions and rationale for on-farm conservation, it has not been conceptualized explicitly as such. The idea of ecosystem services has gained recognition and value as a way to conceptualize how nature contributes to human well-being (Millennium Ecosystem Assessment, 2005), but it is only recently that the idea of evolutionary services is beginning to be explicitly recognized as a category in its own right. Evolutionary services have been defined as 'all of the uses or services to humans that are produced from the evolutionary process' (Faith *et al.*, 2010) and include novel uses from known and from not-yet-known elements of biodiversity. These unknown uses are part of what economists call option values, which are to do with the idea that maintaining diversity keeps our options open to benefit from unanticipated future uses of biodiversity (Faith *et al.*, 2010).

A crucial aspect of on-farm conservation is the seed systems that are associated with the maintenance and management of landraces in centres of crop diversity (Pautasso *et al.*, 2012). A seed system refers to the interrelated set of participants, rules, interactions and infrastructure by which farmers obtain seed or planting material through time and space. Historically seed systems have been in the hands of farmers and communities, and are usually referred to as local, informal or traditional seed systems. In them, farmers rely mainly on themselves to obtain seed and these systems are still common in the developing world for a large number of crops (Pautasso *et al.*, 2012), particularly in centres of crop diversity, such as maize in Mexico and Guatemala (van Etten and de Bruin, 2007; Dyer and Taylor, 2008; Bellon *et al.*, 2011), potatoes in the Andes (Thiele, 1999; Zimmerer, 2003), durum wheat (Tsegaye and Berg, 2007) and sorghum (McGuire, 2008) in Ethiopia, and millet in India (Nagarajan and Smale, 2007). In these systems farmers typically save seed from one season to the next and may share seed with other farmers inside or outside their communities. Seed sourcing is embedded in well-structured traditional systems with rules and expectations based on family and local social networks and regulated by ideas of fairness, and of respect for the seed (Badstue *et al.*, 2007), though farmers may also obtain seed – including commercial varieties – from strangers, in village markets and from the formal seed system through purchase or as aid (Lipper *et al.*, 2010).

Traditional seed systems are not closed or static but open and dynamic with seed coming in and out of the systems, and with farmers experimenting and incorporating new seeds, while keeping and discarding others. They can be quite decentralized because farmers and communities make different and independent decisions in multiple locations, environments and situations. These systems, however, tend to be quite local with a limited spatial scope (Bellon *et al.*, 2011), although they can include long distance seed flows (Van Etten and de Bruin, 2007). This contrasts with formal seed systems that are in the hands of plant breeders and seed companies that are driven by profits, specialization and economies of scale, mediated by commercial transactions of homogenized products, and can be global in their scope.

Seed systems influence which alleles and genotypes pass from one generation to the next, shaping the traits that are under human selection and, by influencing the movement of and location where a crop is planted, the specific natural selection pressures to which it is exposed. Depending on the reproductive system of the crop, sharing of seed and planting in close proximity can foster gene flow. Therefore, understanding the structure and functioning of seed systems is fundamental to maintain and influence the generation of new and useful genetic variation in agricultural systems and highlight the landscape nature of on-farm conservation. On-farm conservation does not depend then on just a few farmers managing diverse crop populations in one village or even a group of farmers in a few villages in a particular area. It is tied to broader social and ecological landscapes of many farmers and villages interlinked to various degrees, and distributed across different types of environments facing diverse selection pressures (both natural and human) leading to the generation of a broad array of genetic variation. Because conditions and environments change and some of those that are rare today may become common tomorrow – and vice versa – having a diversity of 'winning' (adaptive) combinations of genes and traits that are constantly being updated in response to changing situations and new knowledge should allow us to cope and adapt better to change. This is precisely the idea of the option value of the evolutionary services that on-farm conservation delivers (Bellon, 2009).

An important conceptual issue is that the socio-biological systems that maintain landraces in centres of crop diversity are producing two distinct types of benefits: private and public (Smale and Bellon, 1999). The private benefits refer to those that accrue and are captured directly by farmers who maintain these systems, such as the food and products they consume or sell (and associated income), the insurance they gain and the cultural values they fulfil. The public benefits refer to those that accrue to others besides the farmers themselves and that can happen at different scales. Pest control properties of planting varieties with different resistance genes by different farmers (occurring at the local or regional level; Rebaudo and Dangles, 2011) or the new potentially useful genetic variation generated by their management of landraces (the evolutionary services described above) and that can be available through seeds locally, regionally or globally to other farmers for direct use or to plant breeders for further crop improvement are good examples. The problem is that the resources needed to generate these benefits, such as land, labour, capital and knowledge, are limited, whereas public and private benefits can often diverge, creating trade-offs for individuals and society (Heal *et al.*, 2004; Smale and Bellon, 1999). For example, the conventional explanation for crop genetic erosion is that farmers increasingly specialize and replace their diverse sets of landraces with a few scientifically bred varieties that provide them with higher yields and more income. Although farmers pursue their legitimate private interest (higher incomes), crop genetic diversity that may be central to ensure the adaptation of other farmers to changing conditions or the needs of future generations (public benefits) may be lost. Farmers as individuals may tend to underinvest in the conservation of landraces and associated genetic diversity relative to what might be considered optimal for society at large (Heal *et al.*, 2004; Smale and Bellon, 1999).

This explanation underpins two important and related criticisms levelled at on-farm conservation. The first is that replacement of landraces by scientifically bred varieties is an inexorable process that sooner or later will take place everywhere, including centres of crop diversity. Hence on-farm conservation is not a viable and sustainable strategy in the long run. The second criticism is that on-farm conservation perpetuates poverty among farmers who maintain diverse landraces by promoting them at the expense of more productive scientifically bred varieties that would provide farmers with higher incomes and improved welfare (Brush, 1995; Bellon, 2009).

Research among small-scale farmers in centres of crop diversity over the past 20 years has shown that these two criticisms are not necessarily valid and involve more complex issues than appear at first sight (Brush, 2004). Many small-scale farmers still have strong private incentives to maintain a diversity of landraces. There are many species of regional and local importance where no breeding has taken place, and hence local landraces are still the mainstay for farmers who grow them (Gruere *et al.*, 2009; Padulosi *et al.*, 2011). Even for major crops, scientifically bred varieties are often inadequate for farmers' circumstances, seed may be unavailable, or they complement rather than replace landraces and hence both are grown together. Because it is common among small-scale farmers that production and consumption decisions are linked, profit maximization is not a main production objective, and therefore consumption preferences continue to influence their decisions. This can lead to very high willingness to pay for landraces in some systems, and shows that market prices only capture a fraction of the private value that farmers attach to their landraces (Smale and Bellon, 1999; Arslan and Taylor, 2009). This in turn means that cultural preferences play a role in their decision making, even when farmers can be quite commercially oriented (Bellon and Hellin, 2011). Landraces can perform well under improved management and can provide important commercial opportunities (e.g. maize; Perales *et al.*, 1998), particularly as new uses are discovered and products developed as knowledge progresses and new markets are created.

This does not mean that there are no challenges to on-farm conservation. Besides the substitution of a diverse set of landraces by a few scientifically bred varieties, other mechanisms that can lead to the loss of crop biological and genetic diversity in farming systems include: the replacement of a native crop with great diversity by another crop with little local diversity (e.g. sorghum by maize in some parts of sub-Saharan Africa); farmers abandoning agriculture altogether and shifting to other activities; or even migration. Maintaining crop diversity on-farm can entail important costs to farmers, but as long as they have private incentives to maintain landraces, there are opportunities for on-farm conservation to continue. Although on-farm conservation may continue on its own, as has already been documented extensively, in certain circumstances it needs to be supported by outside interventions (Bellon, 2004). The methods to target and prioritize systematically the specific areas and systems where these interventions should take place are still poorly developed, and this is an area for further research. Nevertheless, on the basis of the arguments presented above, these methods should be based on evolutionary and landscape approaches with a strong socioeconomic perspective, given that the processes and the socio-biological systems that generated them are what should be maintained.

In any case, in terms of outside interventions to support on-farm conservation there is already a wealth of knowledge at the farmer and community levels. In the past two decades there have been many projects worldwide to support on-farm conservation implemented by many different types of institutions such as national and international non-governmental organizations (NGOs), farmers' organizations, universities, international research organizations (Jarvis *et al.*, 2011). However, even though placed in centres of crop diversity, these projects usually tend to be ad hoc and opportunistic. Projects usually implement a series of interventions aimed at changing the way crop diversity is accessed, managed, used, consumed and/or marketed. Interventions can either influence the demand for crop diversity by aiming to increase the value of crop diversity for farmers or decreasing the opportunity costs of maintaining it, or its supply by aiming to decrease the costs of access to diversity (Bellon, 2004). A recent and extensive review by Jarvis *et al.* (2011) identified 59 different types of interventions for supporting on-farm conservation, which can also be conceptualized as influencing either the demand or supply of crop diversity. Unfortunately, there is still scant rigorous evidence that such interventions actually make a difference, an area in much need of further research and evidence. The challenge of any on-farm conservation project then is to identify, design and implement interventions that make the conservation of crop diversity on-farm compatible with improved livelihoods and well-being among the farmers who conserve it and to demonstrate that interventions work (Bellon, 2004). The basic principle is not to keep farmers poor, but to enable them to capture more benefits from the diversity they maintain, including the public benefits resulting from that conservation.

8.3 Changes in Climatic Suitability due to Climate Change for Selected Crops in their Centres of Diversity

The impact of climate change on crop productivity and land suitability depends not only on global climate trends but also on a range of local factors, such as soil characteristics, crop management, as well as specific adaptation measures taken by farmers. Also, it may be that extreme weather events are equally or even more important than shifts in average values (Trnka *et al.*, 2011). Therefore, it is difficult to estimate climate change impact in a quantitative way. Simple, heuristic models work well to make broad comparisons and make it easier to assess shortcomings.

In order to assess how climate change can affect the land suitability for crops in their areas of origin and/or centre of diversity, the focus for on-farm conservation, we determined the future trends in land suitability for a small number of important crops in relevant areas (Table 8.1). We used a model that determines suitability on the basis of monthly averages in temperature and precipitation (Hijmans *et al.*, 2012). The model has been shown to function reasonably well for a range of crops (Jarvis *et al.*, 2012; Ramirez *et al.*, 2013). Depending on crop-specific parameters, we determined the suitability based on current climate data and 19 general circulation models (GCMs). For each region, we took the 'majority vote' of the models, and in Table 8.1 we indicate whether ten or more models projected an increase (+) or decrease (–) in area. This analysis does not take into account inter-annual variability, even though this may increase in the future, with negative effects for land suitability.

The results are mixed. Rice, sorghum, pearl millet, cassava and *Musa* crops show a

Table 8.1. Climate change impact on land suitability for selected crops.

Crop	Geographical origin / centre of diversity	Trend in area suitable for this crop in this region, current to 2030s[a]
Wheat	West Asia	–
Maize	Mesoamerica	–
Asian rice	China	+
Barley	West Asia	–
Sorghum	North-east tropical Africa	+
Sorghum	Indian subcontinent	–
Pearl millet	Sahel	+
Potatoes	Andes	–
Cassava	Amazon Basin	+
Banana and plantains	Papuasia	+

[a]Change in arable land suitable for the crop under current climate conditions and land suitable in the 2030s. EcoCrop results of 'majority vote' of 19 general circulation models (down-scaled using delta-method, Ramirez and Jarvis, 2008), scenario A2, assuming rainfed conditions.

net expansion in the broad region around their centre of diversity. In broad terms, these crops may play an important role in climate change adaptation. The results for rice may not be fully representative because rainfed conditions were assumed (as for the other crops) but rice is predominantly an irrigated crop in its area of origin. Also, rice may suffer much from weather variability, given its sensitivity to high night temperatures during the flowering period (Jagadish *et al.*, 2007; see Djanaguiraman and Prasad, Chapter 12, this volume). Interestingly, the suitability of sorghum will decrease on the Indian subcontinent, a secondary area of diversity for this crop. Wheat, maize, barley and potatoes tend to decrease their area of suitability in their centres of origin. Interestingly, these are crops that are widely grown in temperate as well as tropical climates. Potatoes are a highland crop in the tropics and suffer under higher temperatures. Although some high altitude environments will become more suitable for potato, their overall area is projected to decrease in the Andean region. Maize is an interesting case owing to its wide area of adaptation and large genetic diversity. Maize landraces in Mexico – a centre of origin and diversity for this crop – show remarkable diversity and climatic adaptability growing from arid to humid environments and from temperate to very hot environments (Ruiz Corral *et al.*, 2008). Under climate change, the area of adaptation of some maize populations may expand, whereas others such as highland maize contract (Bellon *et al.*, 2011; Ureta *et al.*, 2011), but the great diversity present suggests that there is already enough to allow the crop to adapt to new conditions fostered by climate change (Mercer *et al.*, 2011; Ureta *et al.*, 2011). Nevertheless, some particular races may be threatened as well, such as those from the highlands (Mercer *et al.*, 2008; Bellon *et al.*, 2011), and because these are not well represented in genebanks, they should be a priority for further *ex situ* conservation efforts.

Clearly the implications of these results are variable. They depend on the crop and the environments in which it grows. There is no widespread threat for crops in their centres of diversity from climate change but there are some specific ones. Hence on-farm conservation may continue to be a viable strategy for some crops but not necessarily for others. This will depend on the capacity of these crops to evolve under climate change conditions. In particular for crops with a wide distribution, such as maize, the threat may be for specific populations. Obviously, this a simplistic analysis but it is useful as a first approximation, and to identify further areas of research such as additional analysis at the infraspecific level related to the ranges of adaptation of the crops in their centres of diversity and to prioritize additional *ex situ* conservation efforts.

8.4 On-farm Conservation as an Evolutionary Service in the Context of Climate Change

Predictions of the impact of climate change on the distribution and productivity of crops rely on models for identifying novel climates and to assess the potential response of crops to them. Although it is extremely useful to assess some of the challenges that agricultural and food systems may face with climate change, they provide only a partial picture because they do not take into account the potential that crops have for change and evolution, which can vary widely within and among populations of a species and depend on the presence of genetic variation in populations (Sgro *et al.*, 2011). Due to genetic variation, climate change will affect populations differently throughout the species range and populations will vary in the rate of adaptation (Davis *et al.*, 2005). Hence an evolutionary perspective is needed to assess how crops may respond to climate change and the extent that these responses can be useful for farmers to adapt (Mercer and Perales, 2010). In ecology, there is an increasing recognition of the importance of an evolutionary framework to assess species responses to climate change and for biodiversity conservation (Davis *et al.*, 2005; Sgro *et al.*, 2011). The assumption that species tolerance limits to climate remain stable over time is increasingly being challenged and there is evidence that selection under climate change conditions can lead to adaptive changes regardless of whether ranges shift or not (Davis *et al.*, 2005). Although plant species, and hence crops, may evolve and adapt to new conditions, there are limits to this capacity as well. The capacity of plant species to evolve and adapt to climate change will depend not only on the genetic variation present in populations, but also on the magnitude, direction and speed of change that climate change will bring about.

Mercer and Perales (2010) have already used an evolutionary framework to explore the potential impacts of climate change on landraces in centres of crop domestication. These authors note that the future of crop genetic resources maintained on-farm will depend on the responses of landraces and the farmers who grow them, particularly on the tolerance and adaptive capacity of landraces to novel climate conditions. The issue is complex and they identify four potential responses of landraces to climate change: plasticity; evolution; gene flow; and extinction. Although the first three factors offer opportunities for landraces to adapt to the novel conditions induced by climate change, there are also limits on how far these can go and hence extinction is always a possibility. Particularly, the multiplicity of biophysical and socio-economic factors that can affect negatively the performance of landraces (stressors) associated with climate change, and the existence of negative correlations among traits that are adaptive to these stressors individually but which act antagonistically when stressors occur simultaneously or in close succession, can limit the capacity of landraces to adjust. These antagonistic correlations can be due to pleiotropy, when the effect of an allele on one trait enhances fitness but its effect on another reduces fitness, or to linkage disequilibrium, which is the association of alleles at different loci (Davis *et al.*, 2005). Although local adaptation is seen as a positive feature of landraces and a reason for the great diversity observed in centres of crop diversity, it could cause difficulties for farmers who grow them under climate change by placing limits on their future plastic responses. Though the use of theory to explore these issues is very useful, as these authors also note, there is a need for empirical evidence on how landraces may respond to novel conditions associated with climate change. Although there is increasing understanding on how improved germplasm responds to changing conditions associated with climate change from multi-locational trials (Lobell *et al.*, 2011), there is still limited empirical knowledge on how landraces may respond.

Two examples of the type of work needed to address this gap are reported by Mercer *et al.* (2008) and Vigouroux *et al.* (2011). Mercer *et al.* (2008) established two experimental gardens (referred to by the authors as 'common gardens') at two altitudes (1500 and 2150 m above sea level), planting a set of 21 maize landraces from three altitudinal ranges (lowland, mid-elevation and highland) collected in the Mexican state of Chiapas, and recorded their performance in terms of the likelihood of producing good quality seed and the total mass of good quality seed per plant. The results show that landraces are well adapted to the altitude where they were collected. When planted at different altitudes, however, they showed asymmetric local adaptation. Although mid-elevation and lowland landraces did not produce as well as highland ones in the highland site, they still produced about 80% of seeds per plant compared with the highland types. This was not the case with the highland landraces when planted in the mid-elevation site, which only produced 33% of the seeds per plant compared with what the mid-altitude types produced there; it appears that highland landraces do not express the necessary plasticity to sustain productivity under warmer conditions. Highland maize environments in Mexico are likely to show the most dramatic shifts in climate under climate change (Bellon *et al.*, 2011), which suggests that highland Mexican races may be the most threatened by climate change due to their strong local adaptation, and therefore merit special attention from a genetic resources conservation perspective, particularly because of the high genetic diversity among New World maize races (Vigouroux *et al.*, 2008).

Vigouroux *et al.* (2011) made agronomic and genetic comparisons of 136 paired varieties of pearl millet in Niger, a very dry country in West Africa, from 79 different communities distributed throughout the country collected in 1976 and 2003. Their results show no evidence of genetic erosion between the two periods, but significant changes in flowering time (shorter), plant height (smaller) and spike size (smaller) among the sampled pearl millet populations, showed a correlation between time to flowering and annual precipitation. They were also able to show that these changes were linked to changes in the locus PHYC associated with flowering time. In the context of climate change, a shorter flowering cycle allows flowering and seed production under drier conditions.

Another crucial aspect regarding on-farm conservation is the impact of climate change on farmer seed systems (Bellon *et al.*, 2011). As indicated above, farmer seed systems are the backbone of the socio-biological systems that underpin on-farm conservation of landraces. Although historically these systems have worked very well, climate change may disrupt their functioning, not only as providers of actual seed but also as providers of valuable information about seeds, the traits they contain and the adaptation they show. These systems may not be able to provide small-scale farmers with adapted local varieties in the face of climate change because they may be 'too local' relative to the spatial scope of environmental shifts expected with climate change; climate change may render local information and knowledge about the performance of varieties less useful because it makes past and current crop performance an unreliable indicator for future one, making farmers' decisions more difficult and riskier. Climate change destroys valuable agricultural information, especially informal knowledge of particular climates and their interaction with crops that farmers have acquired through experience over a long period of time (Quiggin and Horowitz, 2003).

An evolutionary perspective of on-farm conservation actually coincides with current developments in the conservation of biodiversity. For example, Sgro *et al.* (2011) argue that evolution should be taken into consideration in the management and planning for biodiversity conservation, in order to develop resilient landscapes where evolutionary potential of species and populations can be maintained. They introduced the concept of evolutionary resilience, defined as the ability of populations to persist in their current state and to undergo evolutionary adaptation in response to changing environmental conditions. This approach places an explicit emphasis on maintaining genetic diversity and the processes that support ongoing evolutionary processes (notice the parallel with on-farm conservation). There starts to be a convergence in the thinking about the conservation of wild and domesticated biodiversity as processes that maintain evolution *in situ* and that are particularly relevant under climate change.

8.5 Responses to Climate Change – Supporting On-farm Conservation

Supporting on-farm conservation of landraces as a means for farmers and society to adapt to climate change will require several interventions. The status quo is not enough. From the outset there is a need for a global information system that monitors changes in adaptation and evolution processes in selected landscapes or across environmental gradients, and enables scientists and farmers to identify new genes and genotypes that can be used in different places as needed. Such a global information system will require the development of a new set of tools and methods to monitor evolution and adaptation (and not only genetic erosion). This is one means by which on-farm conservation can generate global option values. This system will be challenged by issues related to farmers' rights, economic incentives and cross-border access to germplasm.

At the community and farmer level there will be a need to:

- Strengthen the capacity of farmers to store seeds of multiple varieties given that increases in extreme climatic events may increase the risk of field losses, reducing the capacity to save seed (as well as share it).
- Increase capacity to save seed for replanting from plants that survive under extreme conditions, and hence may have adaptive traits particularly by avoiding consumption or sale of seeds.
- Link farmers' seed systems to seed systems from areas that have the adapted types required under novel climatic conditions.

This will probably require the establishment of new relationships within farmer seed networks that go beyond their traditional spatial scopes so as to connect communities in current and future analogue climates. For example, in the case of maize in Mexico, this would entail farmers in the highland agroclimate environment linking with communities in the dry mid-altitude environment. In practice, broadening the geographical reach of farmers' seed networks could be achieved through: exchange visits; linking farmer groups in different locations; fostering the exchange of germplasm, knowledge and practices among them; and encouraging cross-community experimentation with local and introduced crop varieties.

Evolutionary plant breeding is beginning to be reconsidered as an option to deal with current and predicted threats to agriculture, even in countries with highly industrialized agriculture (Goldringer *et al.*, 2001; Döring *et al.*, 2011). Evolutionary plant breeding involves subjecting crop populations with high levels of genetic diversity to the forces of natural selection, leading to evolving populations that are in constant change responding to the strength and direction of environmental variables (Döring *et al.*, 2011; see Ceccarelli, Chapter 13, this volume). It is not a new idea but goes back more than half a century (Suneson, 1956), with well documented research supporting it (Allard, 1988). The process has parallels with the management of landraces by farmers in developing countries (and hence is relevant to our discussion of on-farm conservation), but evolutionary plant breeding is usually based on purposefully created crop populations either through crosses or varietal mixtures from the outset. Although there may be trade-offs between overall performance, diversity and stability in an evolving population, a crucial advantage is that, under large environmental variance, evolving populations will maintain diversity, providing a buffer against environmental fluctuations through compensatory effects (Döring *et al.*, 2011). This approach has limitations, however, particularly related to varietal and seed legislation and incentives to make it attractive to the private sector. For our purposes it illustrates and stresses again the importance of genetic diversity and evolution to cope with unpredictable change in agricultural systems, and hence the link between climate change and on-farm conservation. One can imagine some type of strategic evolutionary landrace breeding by which a diversity of landrace populations are purposefully located under different environments (targeting analogue climates) and are allowed to evolve, using the concepts presented by Döring *et al.*, (2011), as well as replicating the experimental garden approach used by Mercer and Perales (2010) in a wider and strategic manner. This strategy may need an outside intervention because farmers themselves may not wish to apply it because of the risks involved. A similar approach has already being pioneered by French scientists for research purposes, indicating its potential viability (Goldringer *et al.*, 2001, 2006).

Climate change may also have indirect effects on on-farm conservation, by prompting policies to support adaptation measures that have implications for crop diversity and on the incentives for on-farm conservation. For example, to manage climate risk, different forms of financial risk transfer, and specifically index-based insurance, have often been proposed and implemented to reduce income instability of smallholder farmers (Hansen *et al.*, 2007). The need for risk management should become more acute as climate variability increases (Hansen *et*

1

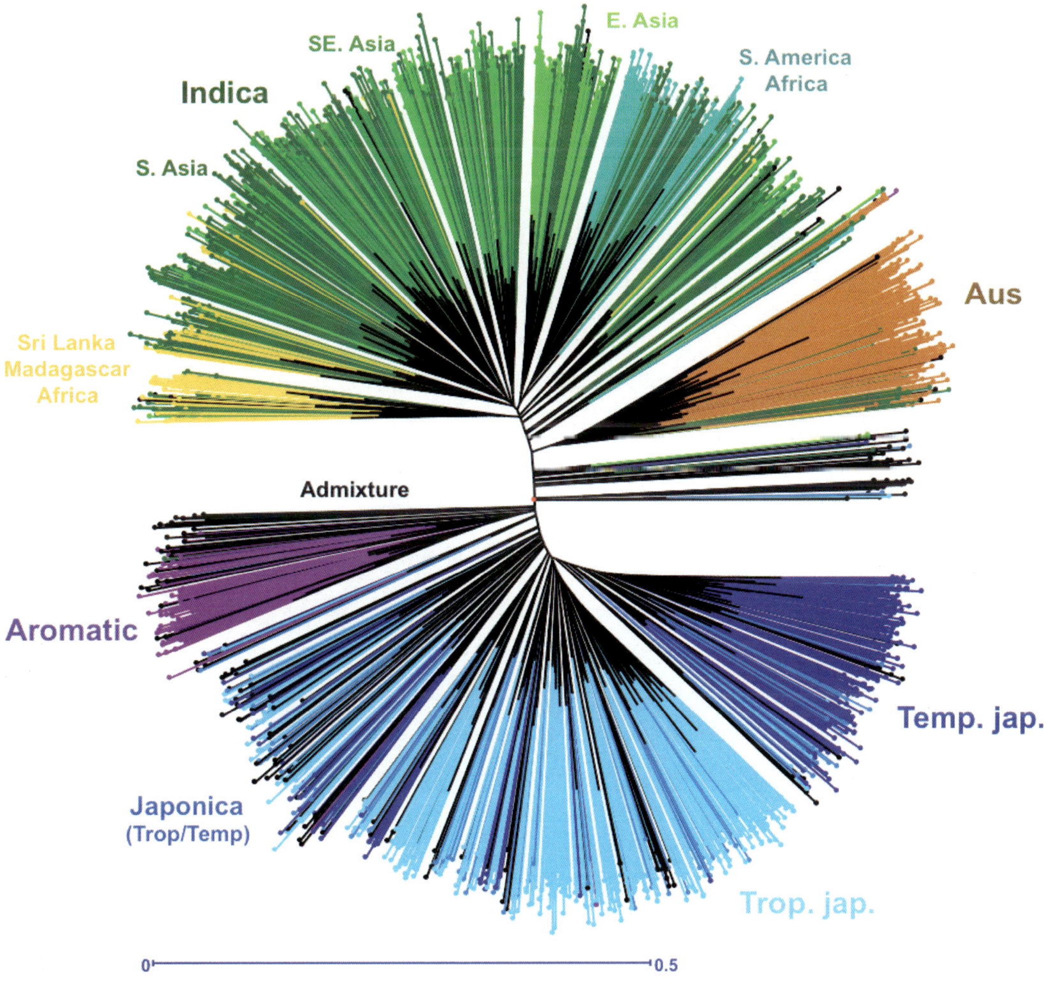

Plate 1. Population structure for *Oryza sativa* represented by a neighbour joining dendrogram showing the five variety groups (indica, aus, aromatic, tropical japonica and temperate japonica) and admixed types for 2252 rice accessions analysed using 45 SSR loci (built by IRRI's T.T. Chang, Genetic Resources Center, unpublished; based on Garris *et al.*, 2005).

2

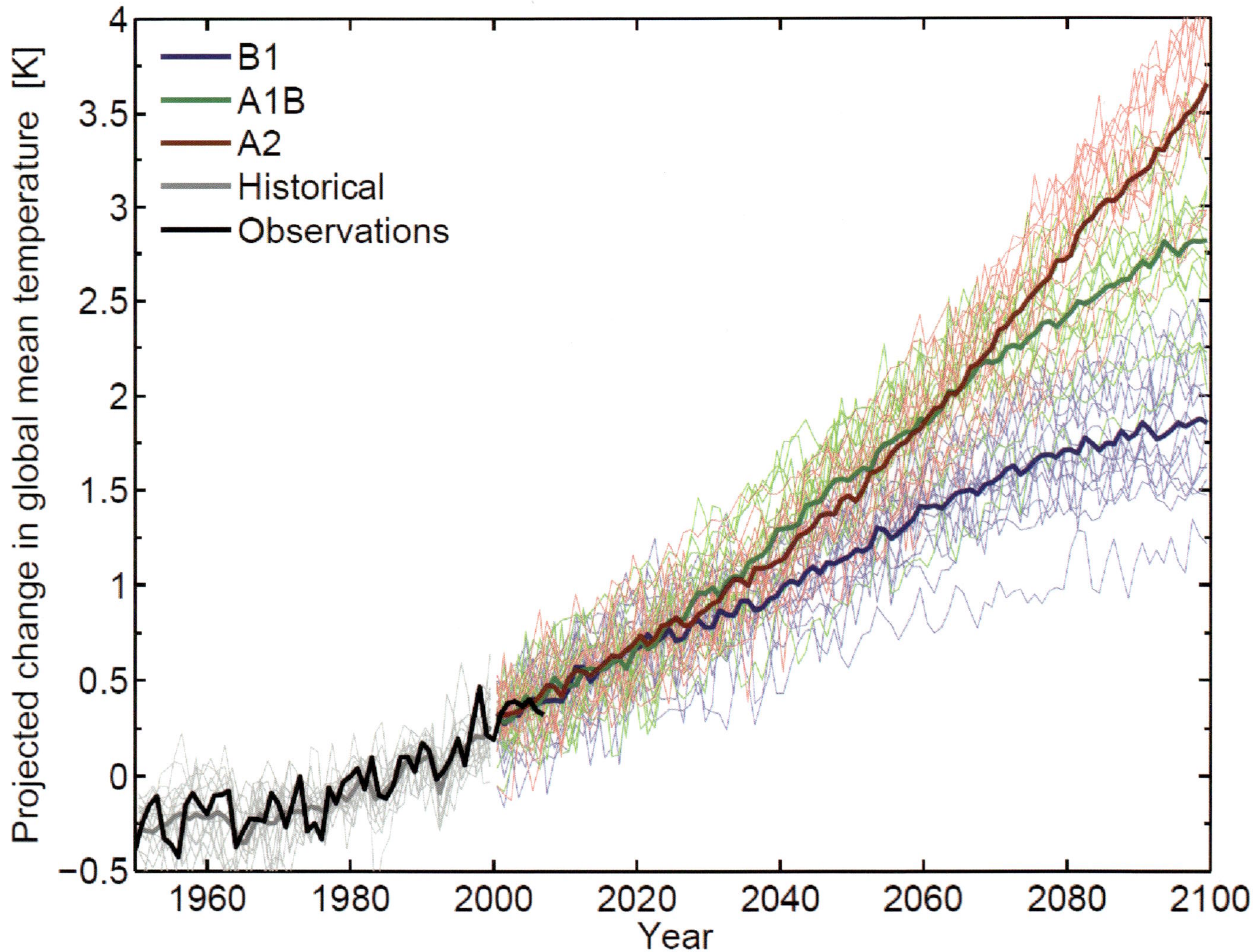

Plate 2. GCM projections of global mean temperature change for the A2, A1B and B1 scenarios from the 3rd Coupled Model Intercomparison Project (CMIP3) multi-model ensemble. Reproduced from Hawkins and Sutton (2009).

3

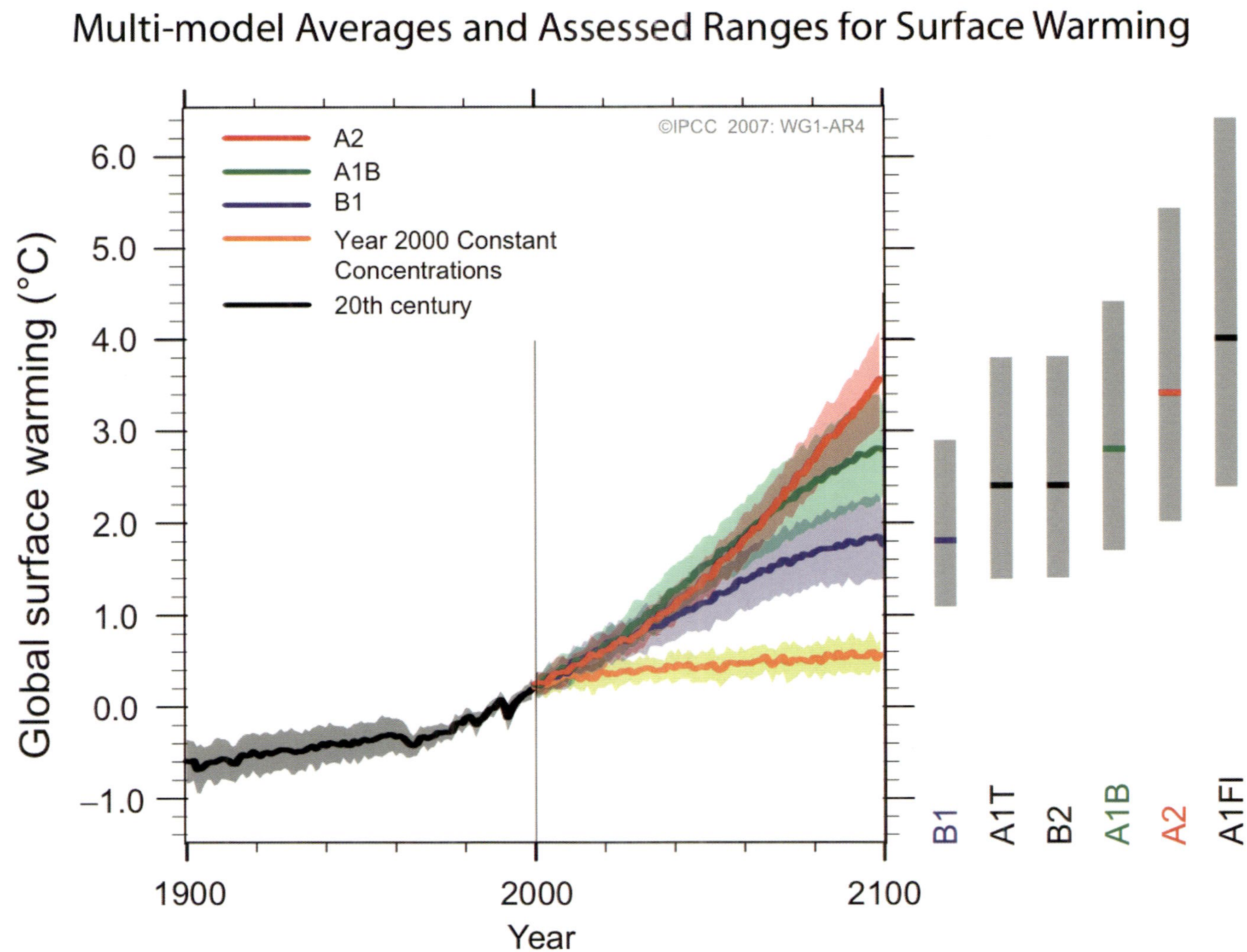

Plate 3. Projections of global mean temperature change for the A2, A1B and B1 scenarios from the CMIP3 multi-model ensemble (coloured curves) and uncertainty range for these and the A1FI, A1T and B2 scenarios from simple climate models (grey bars). Reproduced from IPCC (2007), copyright IPCC (2007).

4

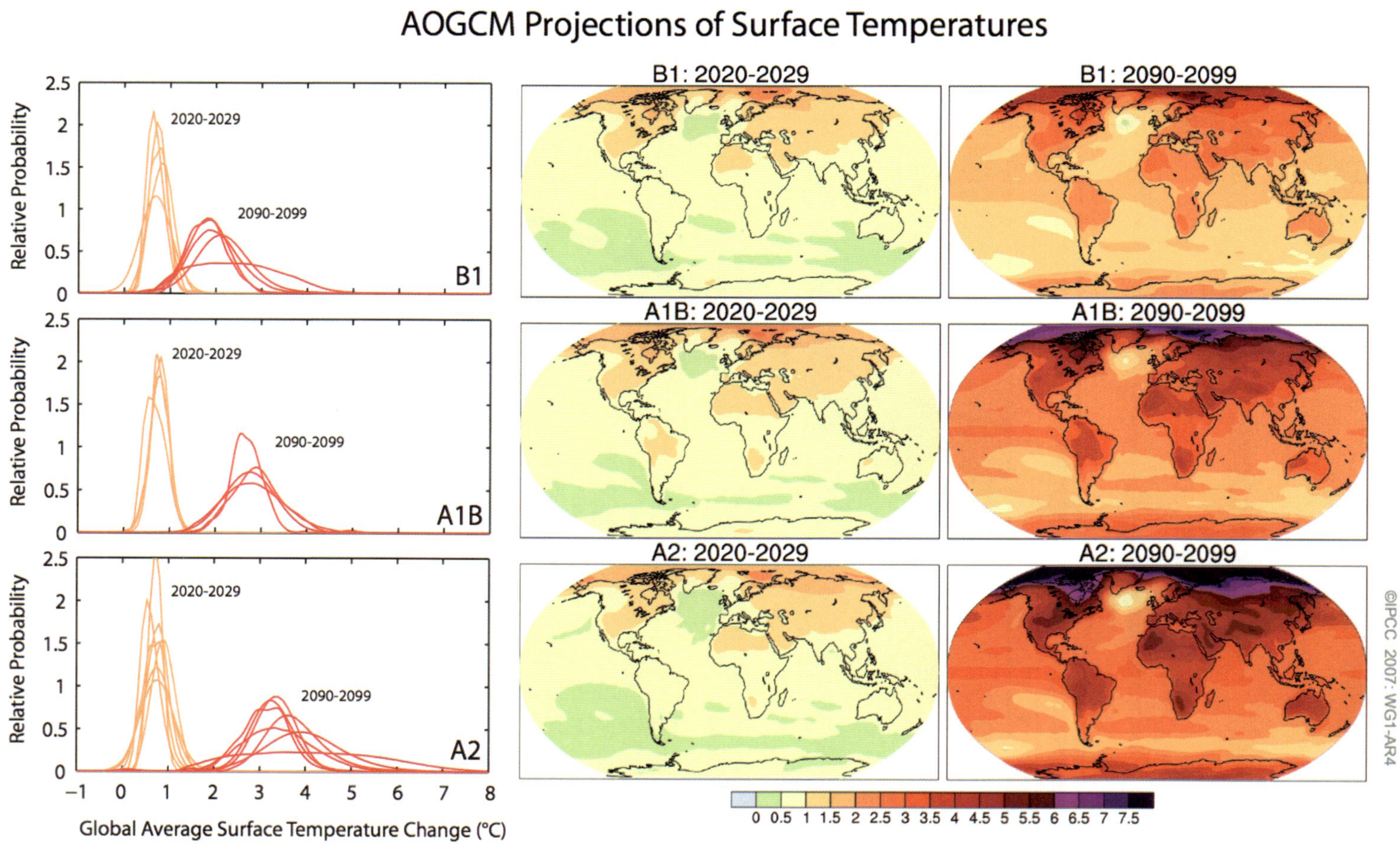

Plate 4. Annual mean temperature change projected by the CMIP3 multi-model ensemble for 2020–2029 and 2090–2099 relative to 1980–1999 under the A1B scenario; mean across all ensemble members.

5

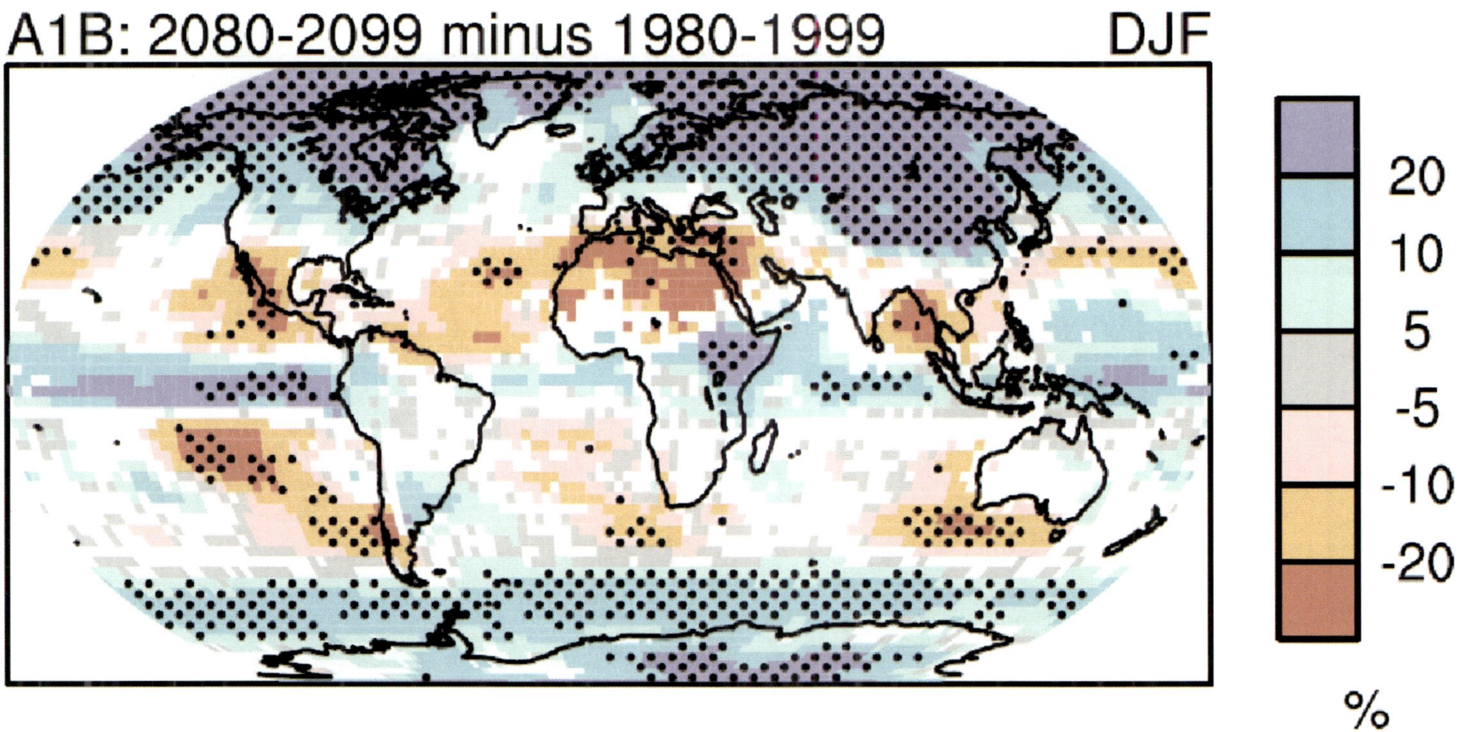

Plate 5. Mean percentage change in precipitation projected by the CMIP3 multi-model ensemble by 2080–2099 relative to 1980–1999 for the A1B scenario, for December–January–February. Colours are shown where 66% of the models agree on the sign of the change, black stipples where 90% of the models agree, as in standard IPCC AR4 method. Reproduced from Tebaldi *et al.* (2011).

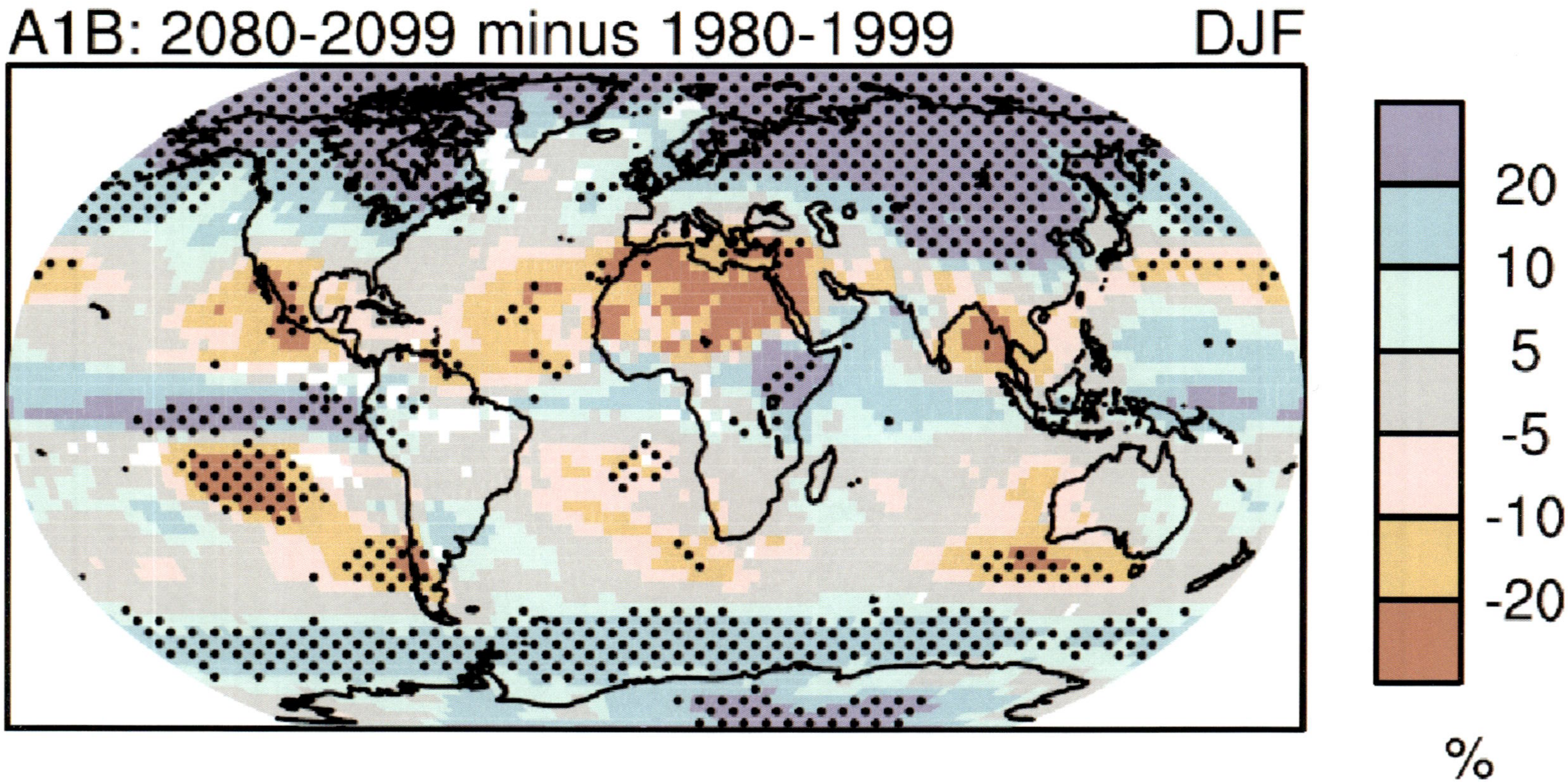

Plate 6. Mean percentage significant change in precipitation projected by the CMIP3 multi-model ensemble by 2080–2099 relative to 1980–1999 for the A1B scenario, for December–January–February. Colours within stipples indicate where less than 50% of models show significant change, white shows where more than 50% show significant change but less than 80% agree on sign of change, stipples show where more than 80% agree on sign of change. This allows areas of low signal to be distinguished from disagreements between models. Reproduced from Tebaldi *et al.* (2011).

7

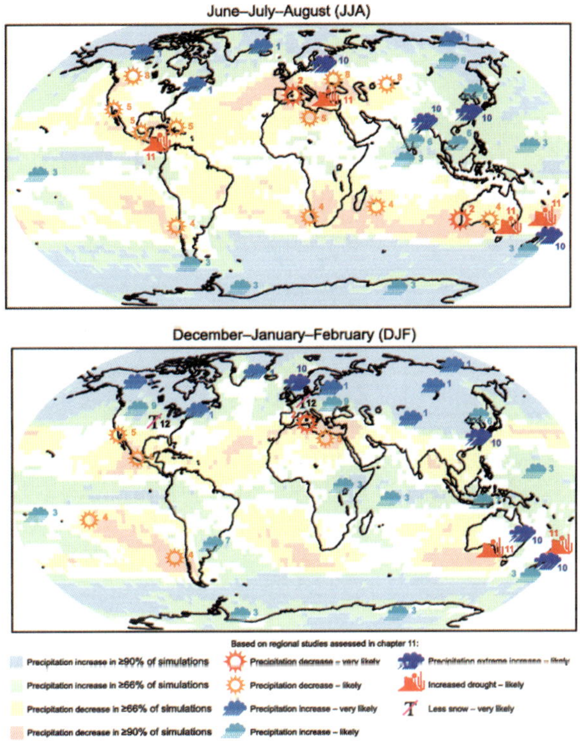

8

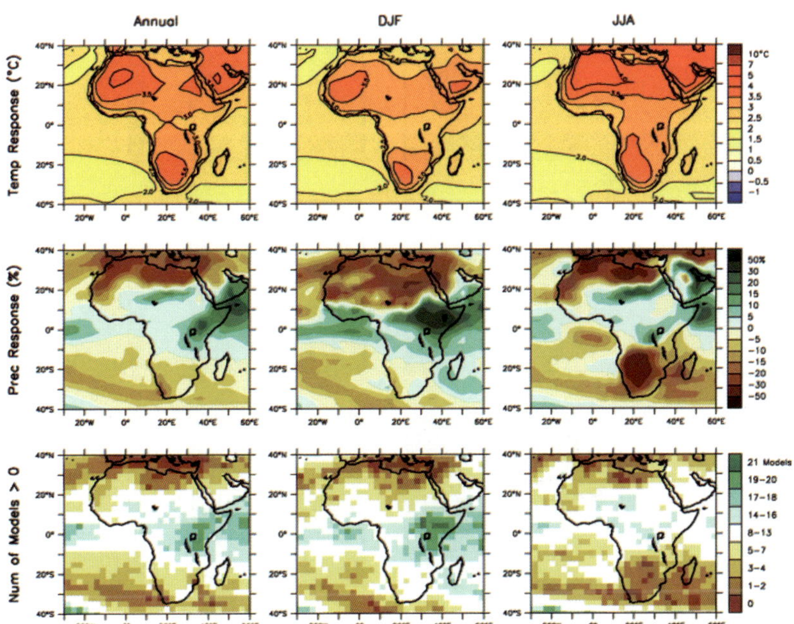

Plate 7. Changes in seasonal precipitation by 2080–2099 relative to 1980–1999, projected by the CMIP3 multi-model ensemble for the A1B scenario. Top: June–July–August; Bottom: December–January–February. Symbols indicate changes assessed as robust on the basis of inter-model agreement and physical understanding. Further details, including an explanation of the numbering associated with the symbols, are in the original publication. Reproduced from Christensen *et al.* (2007), copyright IPCC (2007).

Plate 8. Changes in precipitation projected for Africa by IPCC AR4 multi-model ensemble driven by the A1B scenario, by 2080–2099. Left column: annual mean. Centre column: December–January–February. Right column: June–July–August. Top row: multi-model mean percentage precipitation change. Bottom row: number of models projecting an increase in precipitation. Reproduced from Christensen *et al.* (2007), copyright IPCC (2007).

9

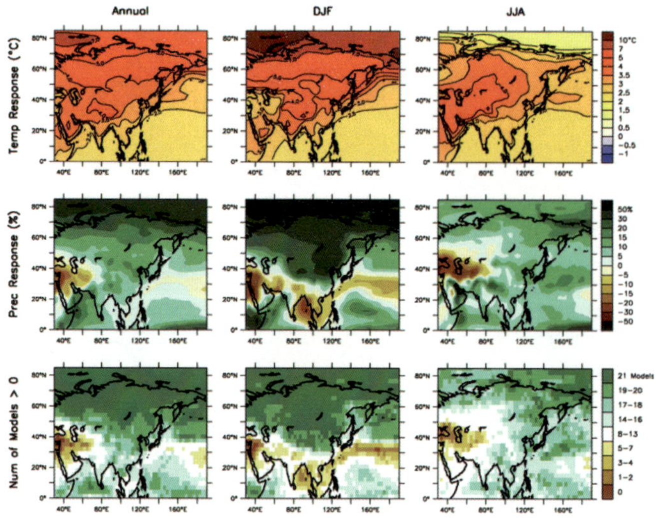

10

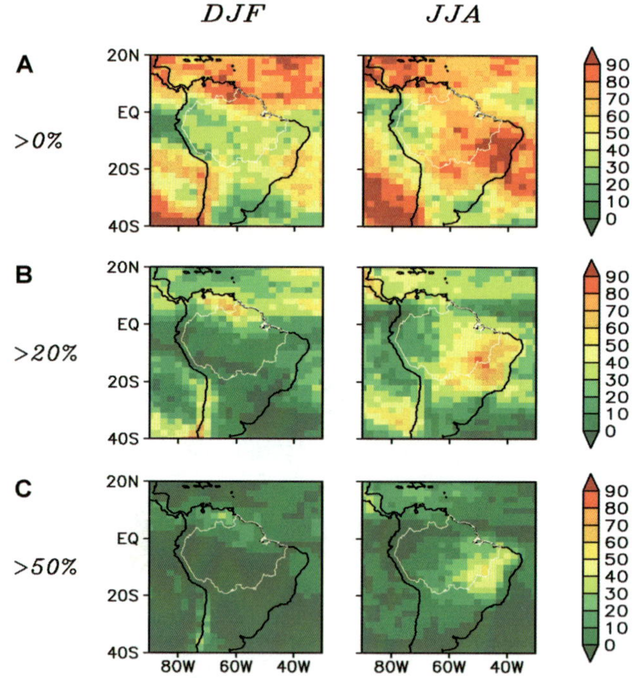

Plate 9. Changes in precipitation projected for Asia by IPCC AR4 multi-model ensemble driven by the A1B scenario, by 2080–2099. Left column: annual mean. Centre column: December–January–February. Right column: June–July–August. Top row: multi-model mean percentage precipitation change. Bottom row: number of models projecting an increase in precipitation. Reproduced from Christensen *et al.* (2007), copyright IPCC (2007).

Plate 10. Precipitation changes simulated by the IPCC AR4 multi-model ensemble driven by the A1B scenario by 2080–2099. Left column: December–January–February. Right column: June–July–August. Top row: percentage of models simulating any level of precipitation reduction. Middle row: percentage of models simulating at least a 20% precipitation reduction. Bottom row: percentage of models simulating at least a 50% precipitation reduction. Reproduced from Malhi *et al.* (2008).

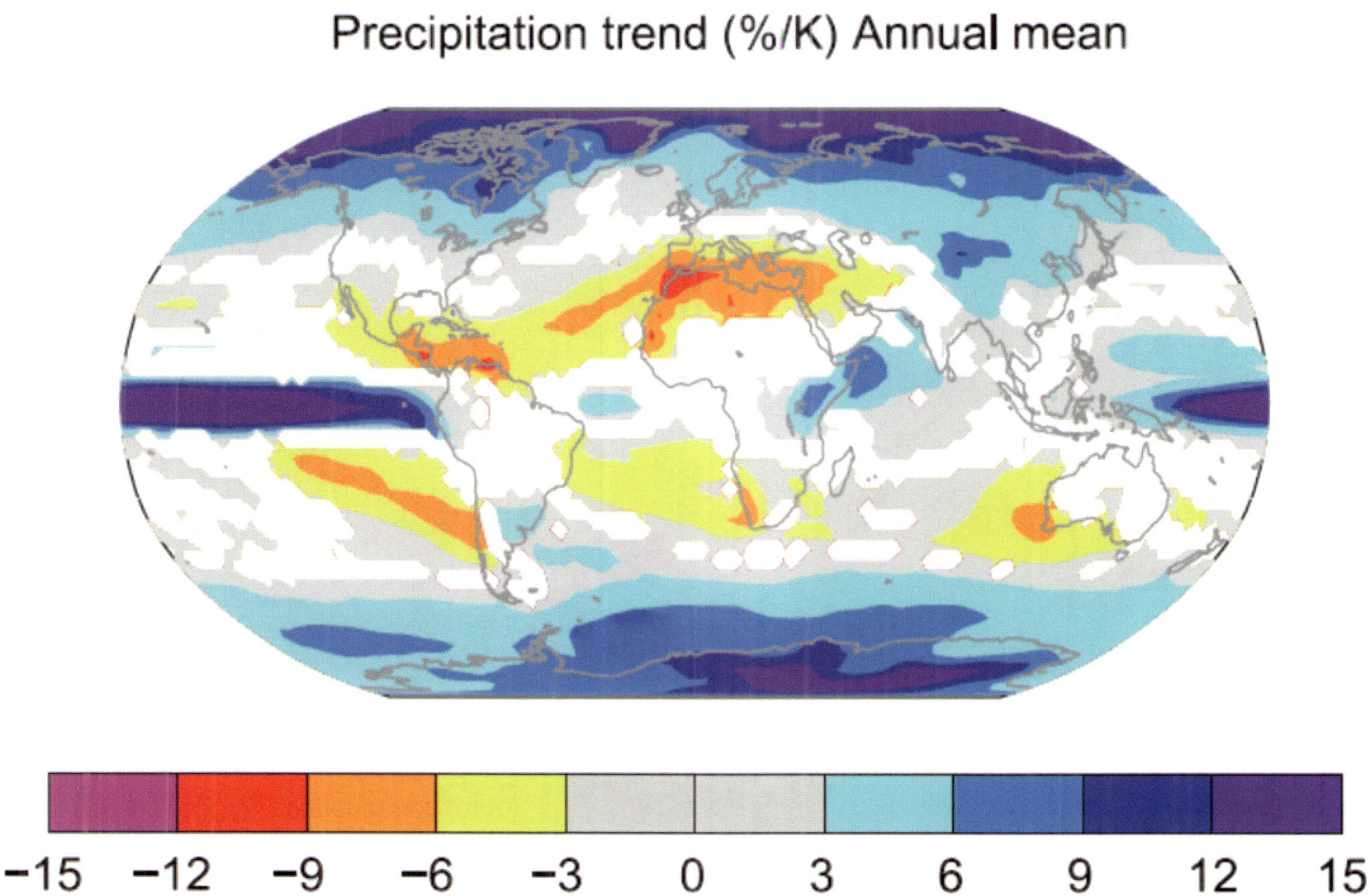

Plate 11. Percentage change in annual mean precipitation per K global warming; mean across CMIP3 multi-model ensemble projections for the A1B scenario. Reproduced from Knutti *et al.* (2010).

12a

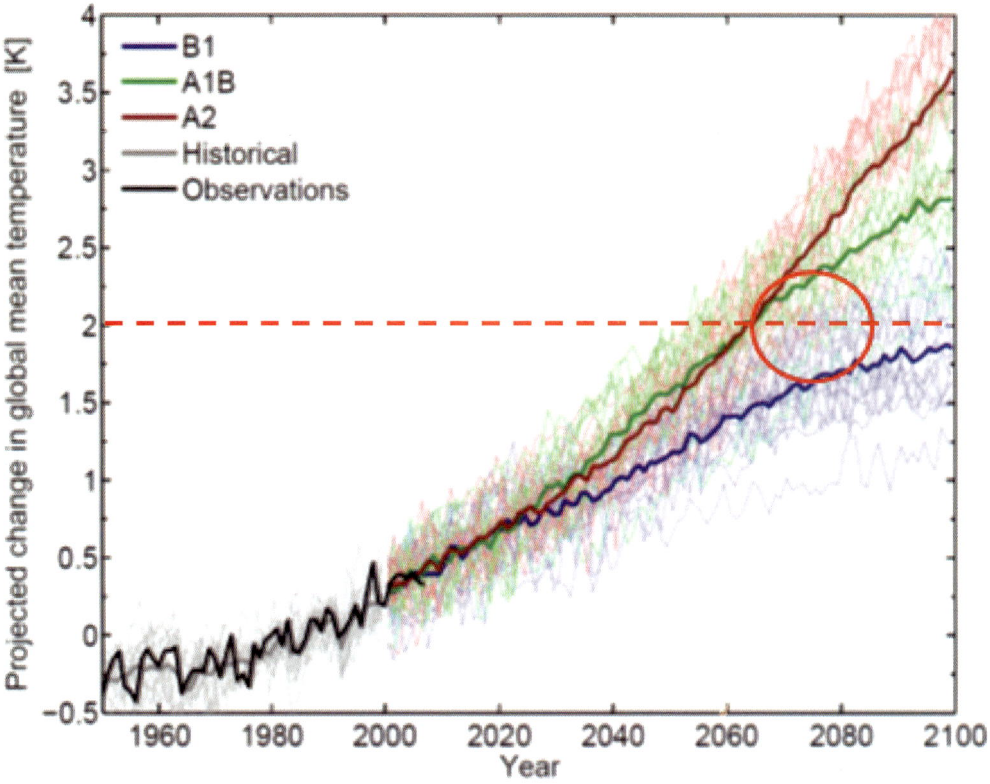

12b

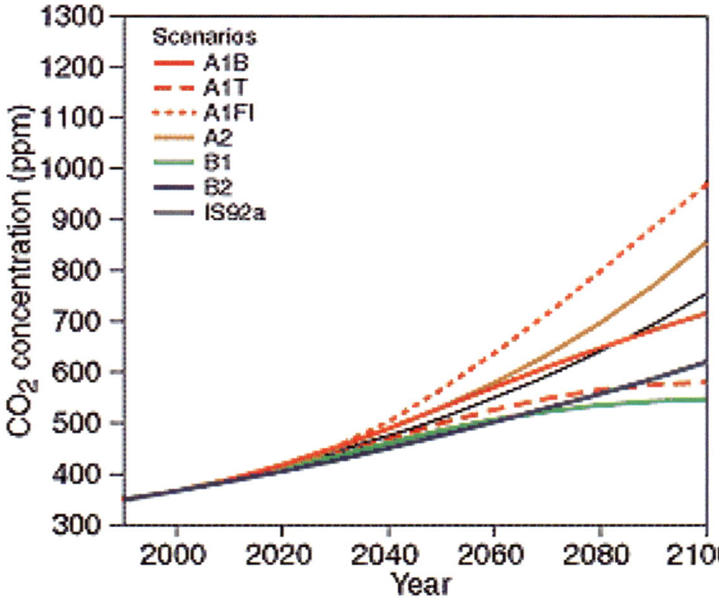

Plate 12. (a) Multi-model GCM projections of global mean temperature change relative to 1971–2000 for the A2, A1B and B1 scenarios, showing overlap of the ranges of the A2 and B1 projections at 2°C warming (red dashed line) in the 2070s (red circle). (b) CO_2 concentrations for all SRES marker scenarios, including the concentrations applied as input to the GCM used for the A2, A1B and B1 projections in (a). (a) Reproduced from Hawkins and Sutton, 2009. (b) Reproduced from IPCC (2001), copyright IPCC (2001).

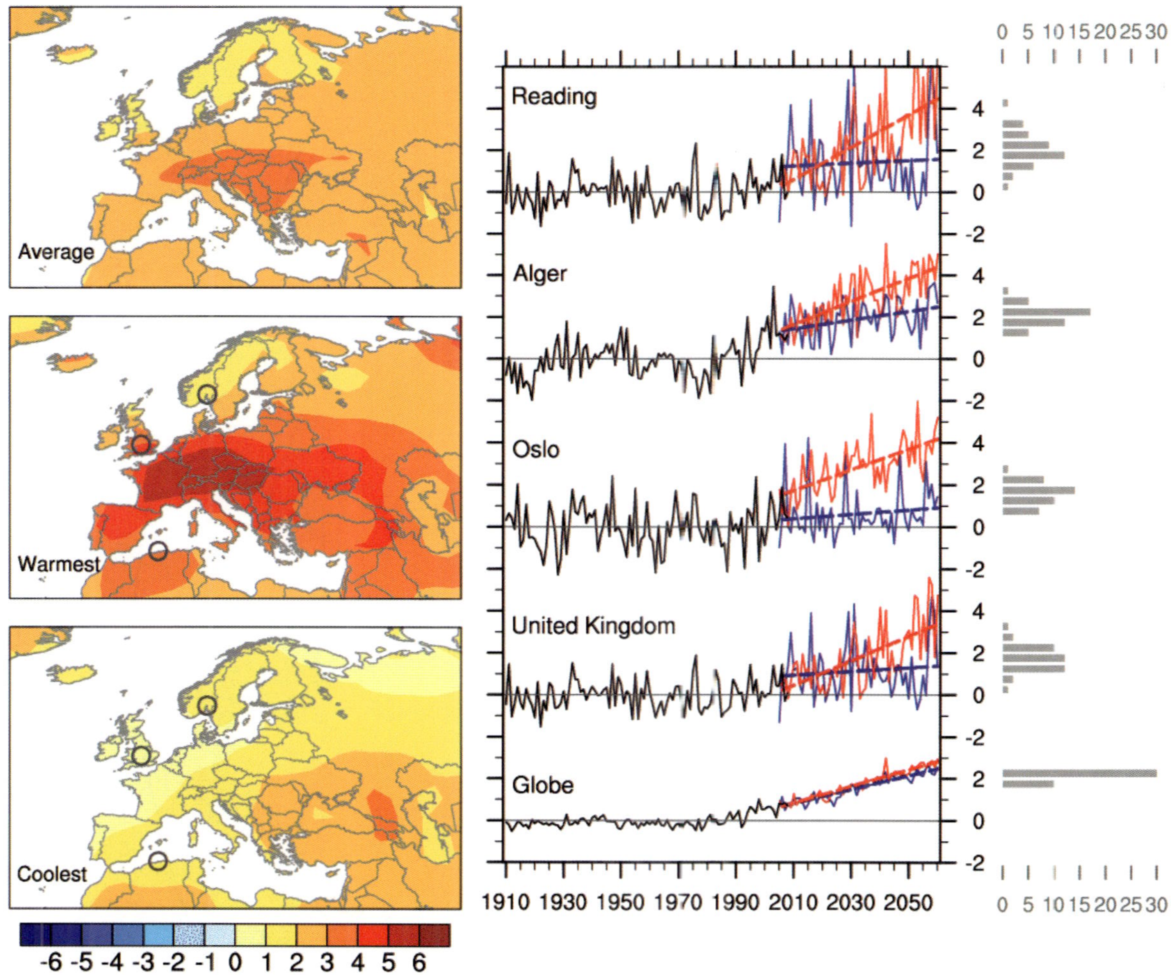

Plate 13. Change in June–July–August mean temperature projected for the A1B scenario by the CCSM3 GCM, with different projections starting from different initial conditions. The top map shows the average warming by 2060 across 40 simulations, the middle map shows the results from the simulation with the greatest warming over the UK, and the bottom shows the simulation with least UK warming. The time series on the right show observations (black) and the warmest (red) and coolest (blue) projections for each location or region, all under the same external forcing. Figure provided by C. Deser.

14

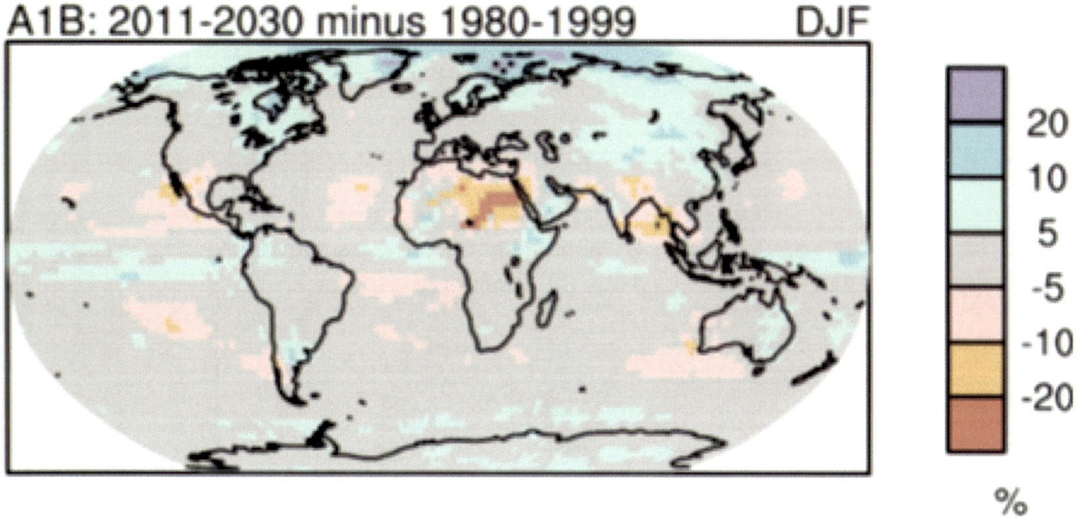

15

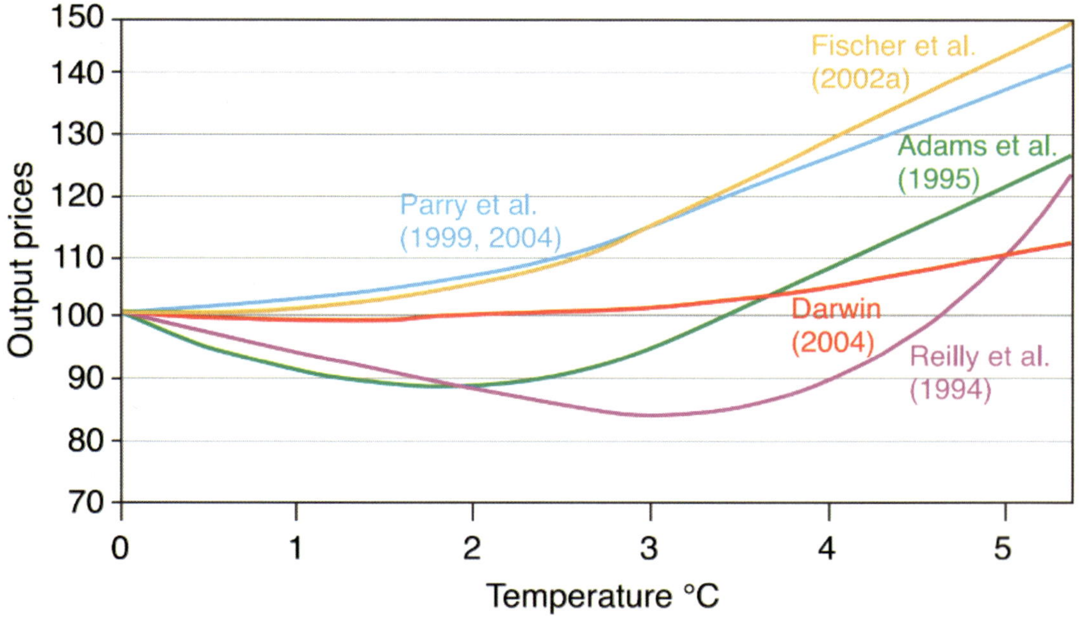

Plate 14. Projected changes in precipitation for December–January–February over 2011–2030 relative to 1980–1999, including consideration of signal-to-noise ratios (Tebaldi *et al.*, 2011).

Plate 15. Cereal prices (percentage of baseline) versus global mean temperature change for some modelling studies. Prices interpolated from point estimates of temperature effects (from Easterling and Aggarwal, 2007).

16

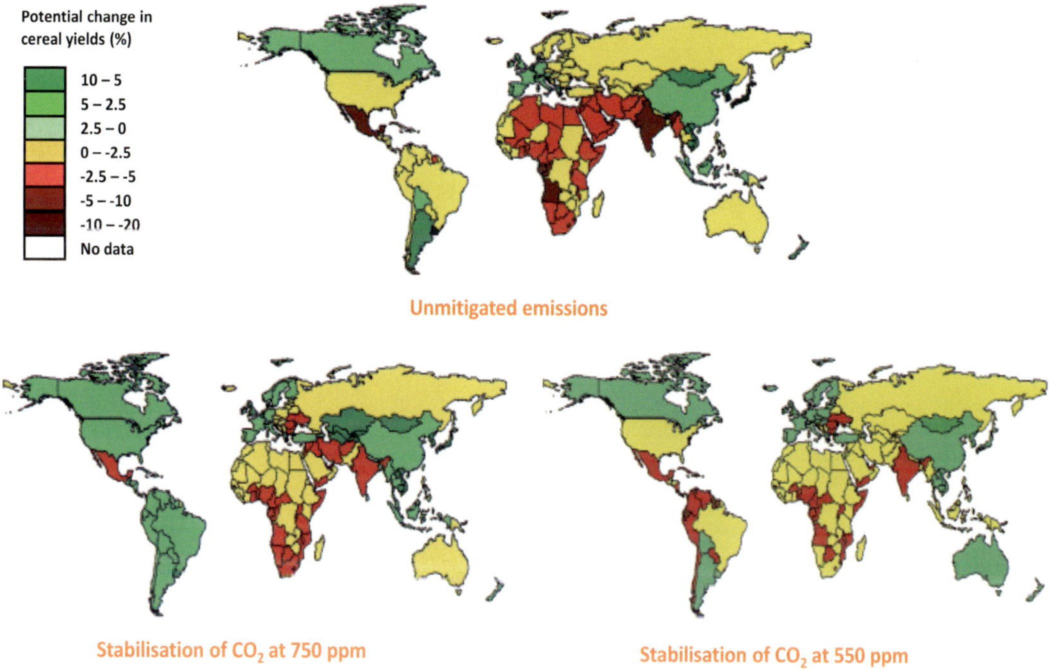

17

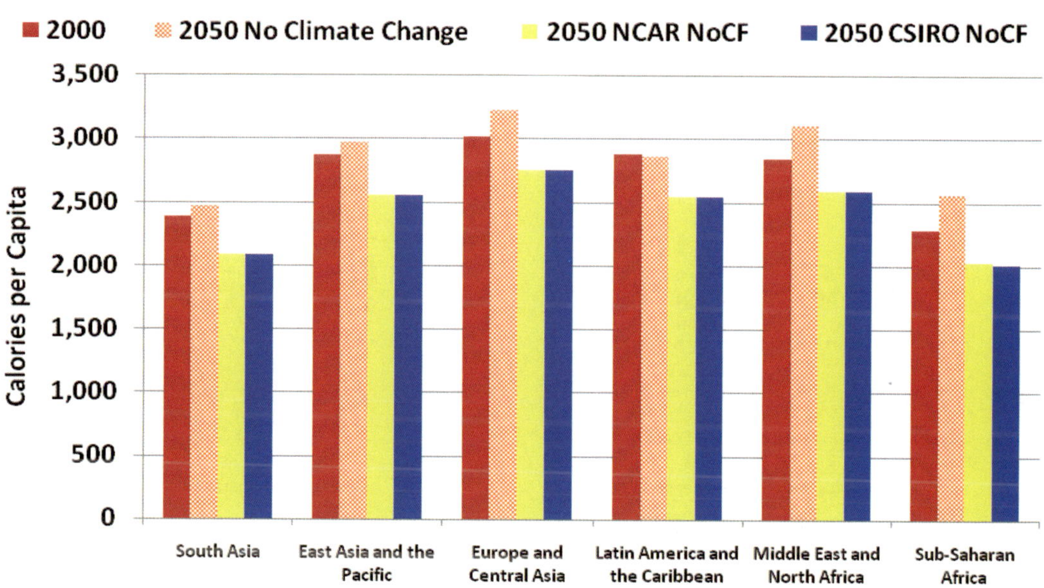

Plate 16. Changes in national cereal crop yields by the 2080s under three different emissions scenarios: unmitigated (IS92a: top map), S750 (bottom left map) and S550 (bottom right map) (from Arnell *et al.*, 2001).
Plate 17. Daily per capita calorie availability with and without climate change. No CF, no carbon fertilization (from Nelson *et al.*, 2009).

18

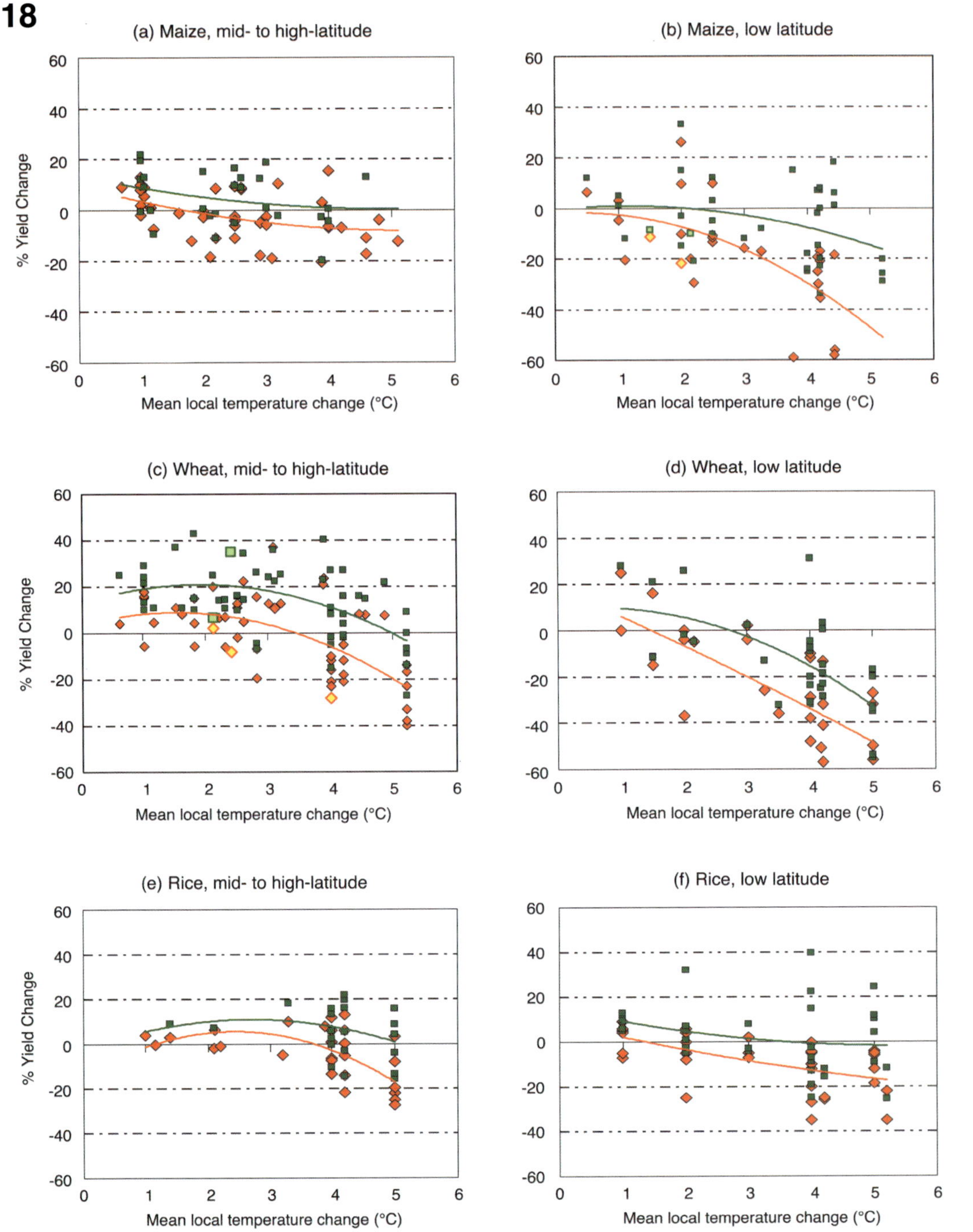

Plate 18. Sensitivity of cereal yield to climate change for maize (a,b), wheat (c,d) and rice (e,f), as derived from the results of 69 published studies at multiple simulation sites, against mean local temperature change used as a proxy to indicate magnitude of climate change in each study. Responses include cases without adaptation (red shapes) and with adaptation (dark green shapes). Adaptations+ represented in these studies include changes in planting, changes in cultivar, and shifts from rainfed to irrigated conditions. Lines are best-fit polynomials and are used here as a way to summarize results across studies rather than as a predictive tool. The studies span a range of precipitation changes and CO_2 concentrations, and vary in how they represent future changes in climate variability. For instance, lighter-coloured shapes in (b) and (c) represent responses of rainfed crops under climate scenarios with decreased precipitation. From Easterling *et al.* (2007).

19a b c d e f

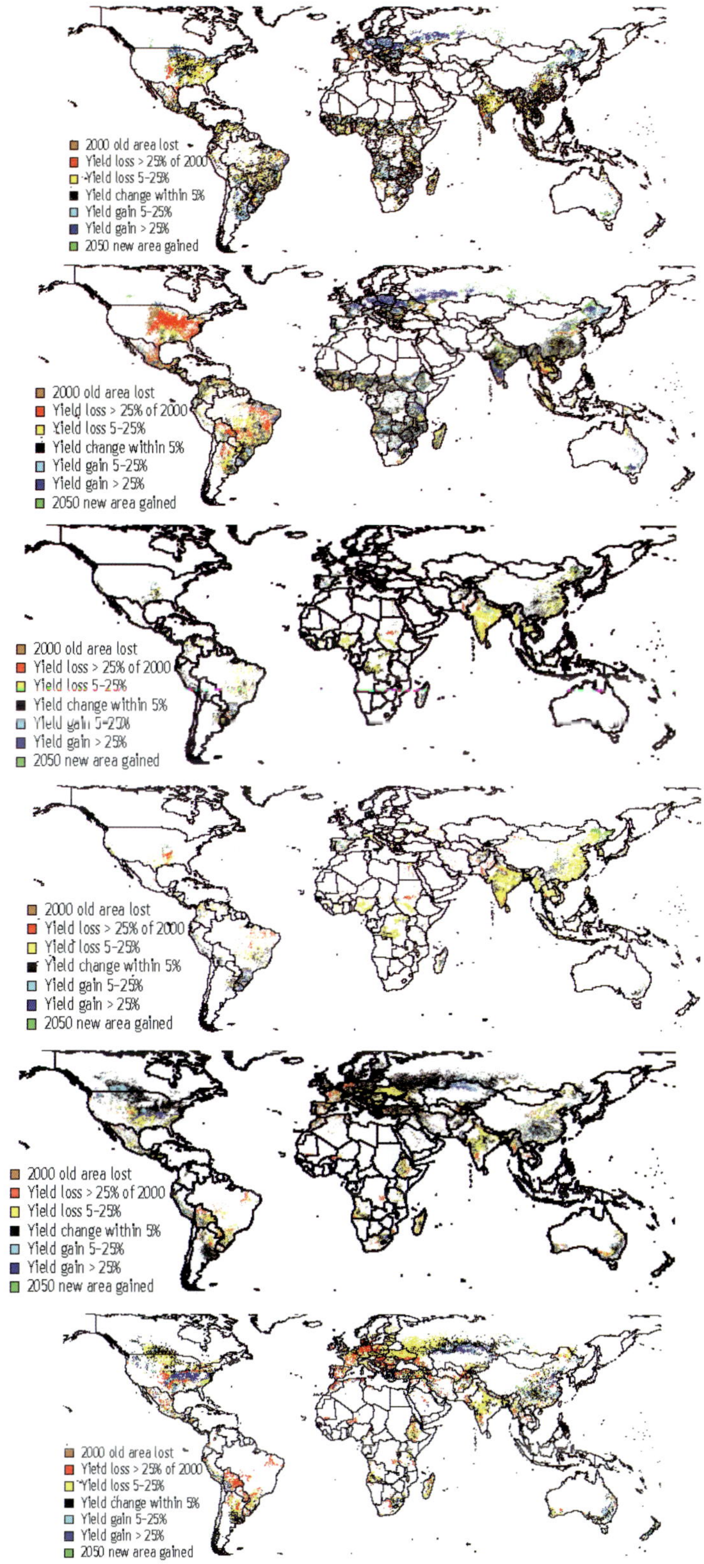

Plate 19. Modelled potential effect of climate change on yields of (a,b) rainfed maize, (c,d) irrigated rice and (e,f) rainfed wheat in a 2050 climate (from Nelson *et al.*, 2010, with permission from IFPRI).

20

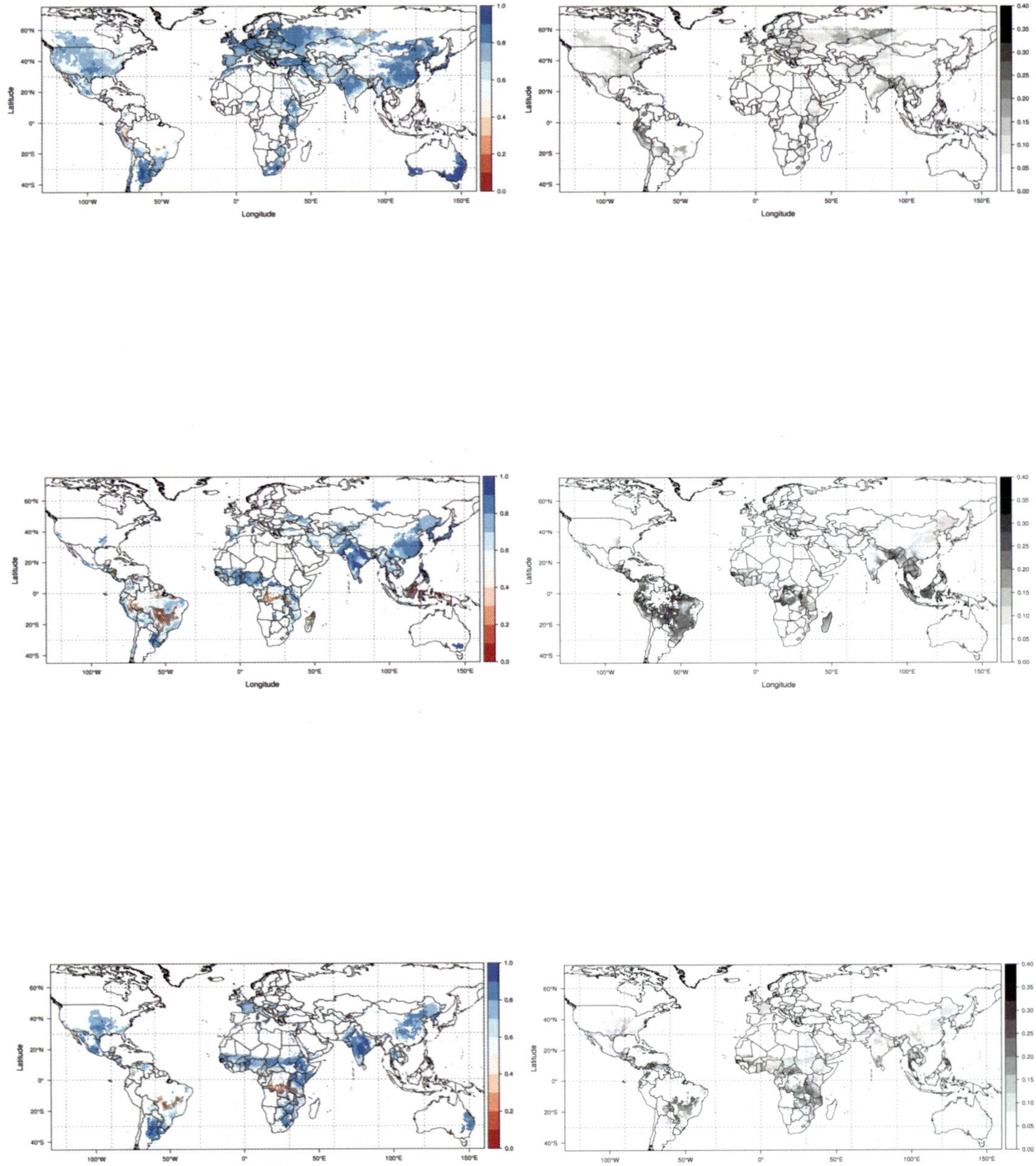

Plate 20. Self-overlap between historical (1966–2005) and future (1956–2095) climates as average (left) of ten individual GCMs for wheat (top), rice (middle) and sorghum (bottom). Uncertainties (right) are expressed as standard deviations of the ten GCMs. Hatched areas in the left panels indicate locations where self-overlap (as average of ten GCM) is below 50%, but where at least 75% of the novel climate exists in neighbouring areas (i.e. 250 km neighbourhood, country).

al., 2012). Improved access to financial risk management is, however, expected to lead to a reduced role for crop diversity as a natural insurance against risk (Baumgärtner and Quaas, 2009), reducing the incentives that farmers have to maintain crop diversity and thus to engage in on-farm conservation. For farmers growing crops in their centres of diversity, badly designed, over-subsidized insurance policies could therefore have a negative effect not only on farmers' welfare by allowing them to take more risk that would make sense economically (this will be true also for non centres of crop diversity), but also on public benefits such as the evolutionary services derived from on-farm conservation. It is worth noting that there is already some anecdotal evidence of this type of effect (though in a developed country and not a centre of diversity) where a representative of a private seed company commented, in the context of the 2012 drought in the USA, that the market demand for drought tolerance traits was fairly low, possibly due to the effect of crop insurance (Keim, 2012).

A key aspect in tackling climate change is global interdependence among nations with respect to plant genetic resources. This interdependence has always existed but it will be even more evident with climate change. For example, Burke *et al.* (2009) have studied the distribution of maize in current and future climates (in 2050) in sub-Saharan Africa, and included Mexico as a centre of diversity for this crop. The results show three types of situations:

1. Countries with a low overlap between current and future climates, but with many similar climates in other countries, hence the former may obtain genetic resources from the latter.
2. Countries with low overlap between current and future climates, but with few similar future climates in other countries, which puts them in a very difficult situation and a high vulnerability condition.
3. Countries where current climates are similar to those in other countries and therefore future climates are a source of crop varieties.

In this respect, the role of Mexico should be noted because its current climates are analogous to many future climates in Africa, thus underlining the global and future value of centres of crop diversity for climate change adaptation. Although this interdependence is increasingly being recognized, there are also increasing restrictions on farmer and breeder access to seeds and germplasm, locally and globally. Local constraints on access reflects national seed policies that favour the recognition of only scientifically bred varieties that are distinct, uniform and stable, discouraging the use of more heterogeneous, variable landraces. Global constraints on access result from asserting the sovereignty of countries over the plant genetic resources found within their national boundaries, coupled with the belief that there are major monetary benefits to be gained by restricting the access of other countries to these resources. Unfortunately, this largely mistaken perception has contributed to restrictions in the global flow of plant genetic resources (Falcon and Fowler, 2002). The problem has been identified but is only partially addressed for some crops by the International Treaty for Plant Genetic Resources for Food and Agriculture (see Moore and Hawtin, Chapter 6, this volume). This is a crucial area to address for the value of on-farm conservation to respond to climate change to be realized.

8.6 Conclusions

The impact of climate change on the viability of on-farm conservation of crops in their centres of origin and/or diversity and the potential contribution of on-farm conservation to climate adaptation strategies are complex. These impacts will vary by crop, and the environments and conditions present in its centre of diversity. Climate change can affect the viability of on-farm conservation of landraces by reducing the range of adaptation of a crop – changing the environmental conditions so much that growing a crop in its centre of diversity becomes non-viable – leading to its 'extinction' in particular regions or agroecosystems. At the same time, on-farm conservation can remain viable, depending on the sensitivity of landraces to climate variability and their capacity to evolve and adapt to these changes. The capacities depend not only on the genetic and biological characteristics of the crop, but also on the management, preferences and incentives of farmers that grow them. Depending on the evolutionary capacity of landraces, on-farm conservation can provide options to farmers and society to adapt to climate change. Although this still requires further analysis and action, recognizing this need and potential is fundamental to advance further.

References

Allard, R.W. (1988) Genetic changes associated with the evolution of adaptedness in cultivated plants and their wild progenitors. *Journal of Heredity* 79, 225–238.

Arslan, A. and Taylor, J.E. (2009) Farmers' subjective valuation of subsistence crops: the case of traditional maize in Mexico. *American Journal of Agricultural Economics* 91, 956–972.

Badstue, L.B., Bellon, M.R., Berthaud, J., Ramirez, A., Flores, D. *et al.* (2007) The dynamics of seed flow among small-scale maize farmers in the Central Valleys of Oaxaca, Mexico. *World Development* 35, 1579–1593.

Baumgärtner, S. and Quaas, M.F. (2009) Agrobiodiversity as natural insurance and the development of financial insurance markets. In: Kontoleon, A., Pascual, U. and Smale, M. (eds) *Agrobiodiversity and Economic Development.* Routledge, London and New York, pp. 293–317.

Bellon, M.R. (2004) Conceptualizing interventions to support on-farm genetic resource conservation. *World Development* 32, 159–172.

Bellon, M.R. (2009) Do we need crop landraces for the future? Realizing the global option value of *in situ* conservation. In: Kontoleon, A., Pascual, U. and Smale, M. (eds) *Agrobiodiversity and Economic Development.* Routledge, London and New York, pp. 51–61.

Bellon, M.R. and Hellin, J. (2011) Planting hybrids, keeping landraces: agricultural modernization and tradition among small-scale farmers in Chiapas, Mexico. *World Development* 39, 1434–1443.

Bellon, M.R., Pham, J.L. and Jackson, M.T. (1997) Genetic conservation: a role for rice farmers. In: Maxted, N., Ford-Lloyd, B.V. and Hawkes, J.G. (eds) *Plant Conservation: the In Situ Approach.* Chapman and Hall, London, pp. 263–289.

Bellon, M.R., Hodson, D. and Hellin, J. (2011) Assessing the vulnerability of traditional maize seed systems in Mexico to climate change. *Proceedings of the National Academy of Sciences USA* 108, 13432–13437.

Bretting, P.K. and Duvick, D.N. (1997) Dynamic conservation of plant genetic resources. *Advances in Agronomy* 61, 1–51.

Brush, S.B. (1995) *In situ* conservation of landraces in centers of crop diversity. *Crop Science* 35, 346–354.

Brush, S.B. (2004) *Farmers' Bounty. Locating Crop Diversity in the Contemporary World.* Yale University Press, New Haven, Connecticut.

Burke, M.B., Lobell, D.B. and Guarino, L. (2009) Shifts in African crop climates by 2050, and the implications for crop improvements and genetic resources conservation. *Global Environmental Change* 19, 317–325.

Camacho-Villa, T., Maxted, N., Scholten, M. and Ford-Lloyd, B. (2005) Defining and identifying crop landraces. *Plant Genetic Resources: Characterization and Utilization* 3, 373–384.

Davis, M., Shaw, R.G. and Etterson, J.R. (2005) Evolutionary responses to changing climate. *Ecology* 86, 1704–1714.

Döring, T., Knapp, S., Kovacs, G., Murphy, K. and Wolfe, M.S. (2011) Evolutionary plant breeding in cereals – into a new era. *Sustainability* 3, 1944–1971.

Dyer, G.A. and Taylor, J.E. (2008) A crop population perspective on maize seed systems in Mexico. *Proceedings of the National Academy of Sciences USA* 105, 470–475.

Faith, D.P., Magallon, S, Hendry, A.P., Conti, E., Yahara, T. *et al.* (2010) Ecosystem services: an evolutionary perspective on the links between biodiversity and human well-being. *Current Opinion in Environmental Sustainability* 2, 1–9.

Falcon, W.P. and Fowler, C. (2002) Carving up the commons – emergence of a new international regime for germplasm development and transfer. *Food Policy* 27, 197–222.

Fischer, G., Mahendra, S., Tubiello, F.N. and van Velhuizen, J. (2005) Socio-economic and climate change impacts on agriculture: an integrated assessment, 1990–2080. *Philosophical Transactions of the Royal Society B* 360, 2067–2083.

Gepts, P. (2006) Plant genetic resources conservation and utilization: the accomplishments and future of a societal insurance policy. *Crop Science* 46, 2278–2292.

Goldringer, I., Enjalbert, J., David, J., Paillard, S., Pham, J.L. *et al.* (2001) Dynamic management of genetics resources: a 13-year experiment on wheat. In: Cooper, H.D., Spillane, C. and Hodgkin, T. (eds) *Broadening the Genetic Base of Crop Production.* IPGRI/FAO, Rome, Italy, pp. 245–260.

Goldringer, I., Prouin, C., Rousset, M., Galic, N. and Bonnin, I. (2006) Rapid differentiation of experimental populations of wheat for heading time in response to local climatic conditions. *Annals of Botany* 98, 805–817.

Gruere, G., Guiliani, A. and Smale, M. (2009) Marketing underutilized plant species for the poor: a conceptual framework. In: Kontoleon, A., Pascual, U. and Smale, M. (eds) *Agrobiodiversity and Economic Development.* Routledge, London and New York, pp. 62–81.

Hansen, J.W., Baethgen, W., Osgood, D., Ceccato, P. and Ngugi, R.K. (2007) Innovations in climate risk management: protecting and building rural livelihoods in a variable and changing climate. *SAT eJournal* 4, 1–38.

Hansen, J., Sato, M. and Ruedy, R. (2012) Perception of climate change. *Proceedings of the National Academy of Sciences USA.* doi:10.1073/pnas.1205276109.

Harlan, J.R. (1992) *Crops and Man* (2nd edn). American Society of Agronomy, Madison, Wisconsin.

Hawkes, J.R. (1983) *The Diversity of Crop Plants.* Harvard University Press, Cambridge, Massachusetts.

Heal, G., Walker, B., Levin, S., Arrow, K., Dasgupta, P. *et al.* (2004) Genetic diversity and interdependent crop choices in agriculture. *Resources and Energy Economics* 26, 175–184.

Hijmans, R.J., Guarino, L. and Mathur, P. (2012) DIVA-GIS 7.5 Manual. Available at: www.diva-gis.org/docs/DIVA-GIS_manual_7.pdf (accessed 3 January 2013).

Jagadish, S.K., Craufurd, P.Q. and Wheeler, T.R. (2007) High temperature stress and spikelet fertility in rice (*Oryza sativa* L.) *Journal of Experimental Botany* 58, 1627–1635.

Jarvis, A., Upadhaya, H., Gowda, C.L.L., Aggarwal, P.K., Fujisaka, S. and Anderson, B. (2010) Climate change and its effect on conservation and use of plant genetic resources for food and agriculture and associated biodiversity for food security. Thematic Background Study. In: *The Second Report on the State of the World's Plant Genetic Resources for Food and Agriculture.* Food and Agriculture Organization of the United Nations, Rome.

Jarvis, A., Ramirez-Villegas, J., Herrera Campo, B.V. and Navarro-Racines, C. (2012) Is cassava the answer to African climate change adaptation? *Tropical Plant Biology* 5, 9–29.

Jarvis, D.I., Brown, A.H.D., Cuong, P.H., Collado-Panduro, L., Latournerie-Moreno, L. *et al.* (2008) A global perspective of the richness and evenness of traditional crop variety diversity maintained by farming communities. *Proceedings of the National Academy of Sciences USA,* 105, 5326–5331.

Jarvis, D.I., Hodgkin, T., Sthapit, B.R., Fadda, C., and Lopez-Noriega, I. (2011) Identifiying multiple ways of supporting the conservation and use of traditional crop varieties within the agricultural production system. *Critical Reviews in Plant Sciences* 30, 125–176.

Keim, B. (2012) Why King Corn wasn't ready for the drought. Available at: www.wired.com/wiredscience/2012/08/drought-resistant-corn/ (accessed 29 November 2012).

Lane, A. and Jarvis, A. (2007) Changes in climate will modify the geography of crop suitability: agricultural biodiversity can help with adaptation. *SAT eJournal* 4, 1–12.

Lipper, L., Anderson, C.L. and Dalton, T.J. (eds) (2010) *Seed Trade in Rural Markets: Implications for Crop Diversity and Agricultural Development.* Food and Agriculture Organization of the United Nations and Earthscan, London.

Lobell, D., Burke, M.B., Tebaldi, C., Mastrandrea, M.D., Falcon, W.P. *et al.* (2008) Prioritizing climate change adaptation needs for food security in 2030. *Science* 319, 607–610.

Lobell, D., Banziger, M., Magorokosho, C. and Bindiganavile, V. (2011) Nonlinear heat effects on African maize as evidenced by historical yield trials. *Nature Climate Change* 1, 42–45.

McGuire, S.J. (2008) Securing access to seed: social relations and sorghum seed exchange in eastern Ethiopia. *Human Ecology* 36, 217–229.

Mercer, K.L. and Perales, H.R. (2010) Evolutionary response of landraces to climate change in centers of crop diversity. *Evolutionary Applications* 3, 480–493.

Mercer, K.L., Martínez-Vásquez, A. and Perales, H.R. (2008) Asymmetrical local adaptation of maize landraces along an altitudinal gradient. *Evolutionary Applications* 1, 489–500.

Mercer, K., Perales, H.R. and Wainwright, J.D. (2011) Climate change and the transgenic adaptation strategy: smallholder livelihoods, climate justice and maize landraces in Mexico. *Global Environmental Change* 22, 495–504.

Millennium Ecosystem Assessment (2005) *Ecosystems and Human Well-being: Synthesis.* Island Press, Washington, DC.

Morton, J. F. (2007) The impact of climate change on smallholder and subsistence agriculture. *Proceedings of the National Academy of Sciences USA* 104, 19680–19685.

Nagarajan, L. and Smale, M. (2007) Village seed systems and the biological diversity of millet crops in marginal environments of India. *Euphytica* 155, 167–182.

Ortiz, R. (2011) Agrobiodiversity management and climate change. In: Lenné, J.M. and Wood, D. (eds) *Agrobiodiversity Management for Food Security: a Critical Review.* CAB International, Wallingford, UK, pp. 189–211.

Padulosi, S., Heywood, V., Hunter, D. and Jarvis, A. (2011) Underutilized species and climate change: current status and outlook. In: Yadav, S.S., Redden, R.J., Hatfield, J.L., Lotze-Campen, H. and Hall, A.E. (eds) *Crop Adaptation to Climate Change.* Wiley-Blackwell, Oxford, UK, pp. 507–521.

Pautasso, M., Aistara, G., Barnaud, A., Caillon, S., Coomes, O. *et al.* (2012) Seed exchange networks for agrobiodiversity conservation. A review. *Agronomy for Sustainable Development.* doi:10.1007/s13593-012-0089-6.

Perales, H.R, Brush, S.B. and Qualset, C.O. (1998) Agronomic and economic competitiveness of maize landraces and *in situ* conservation in Mexico. In: Smale, M. (ed.) *Farmers, Gene Banks and Crop Breeding.* Kluwer Academic Publishing, Boston, Massachusetts, pp 109–125.

Plucknett, D.L., Smith, N.H.J., Williams, J.T. and Anishetty, N.M. (1987) *Gene Banks and the World's Food.* Princeton University Press, Princeton, New Jersey.

Quiggin, J.C. and Horowitz, J.K. (2003) Costs of adjustment to climate change. *Australian Journal of Agricultural and Resource Economics* 47, 429–446.

Ramirez, J. and Jarvis, A. (2008) *High Resolution Statistically Downscaled Future Climate Surfaces.* CGIAR Research Program on Climate Change, Agriculture and Food Security (CCAFS). CIAT, Cali, Colombia.

Ramirez-Villegas, J., Jarvis, A. and Läderach, P. (2013) Empirical approaches for assessing impacts of climate change on agriculture: The EcoCrop model and a case study with grain sorghum. *Agricultural and Forest Meteorology* 170, 67–78.

Rebaudo, F. and Dangles, O. (2011) Coupled information diffusion – pest dynamics models predict delayed benefits of farmer cooperation in pest management programs. *PLoS* 7, 1–10.

Reusch, T.B. and Wood, T. (2007) Molecular ecology of global change. *Molecular Ecology* 16, 3973–3992.

Ruiz Corral, J.A., Duran Puga, N., Sanchez Gonzalez, J.J., Ron Parra, J., Gonzalez Eguiarte, D.R. *et al.* (2008) Climatic adaptation and ecological descriptors of 42 Mexican maize races. *Crop Science* 48, 1502–1512.

Sgro, C.M., Lowe, A.J., and Hoffmann, A.A. (2011) Building evolutionary resilience for conserving biodiversity under climate change. *Evolutionary Applications* 4, 326–337.

Smale, M. and Bellon, M.R. (1999) A conceptual framework for valuing on-farm genetic resources. In: Wood, D. and Lenné, J.M. (eds) *Agrobiodiversity: Characterization, Utilization and Management.* CAB International, Wallingford, UK, pp. 387–408.

Suneson, C.A. (1956) An evolutionary plant breeding method. *Agronomy Journal* 48, 188–191.

Thiele, G. (1999) Informal potato seed systems in the Andes: why are they important and what should we do with them? *World Development* 27, 83–99.

Trnka, M., Olesen, J.E., Kersebaum, K.C., Skjelvag, A.O., Eitzinger, J. *et al.* (2011) Agroclimatic conditions in Europe under climate change. *Global Change Biology* 17, 2298–2318.

Tsegaye, B. and Berg, T. (2007) Utilization of durum wheat landraces in eastern Shewa, central Ethiopia: are home uses an incentive for on-farm conservation? *Agriculture and Human Values* 24, 219–230.

Ureta, C., Martinez-Meyer, E., Perales, H.R. and Alvarez-Buyllas, E.R. (2011) Projecting the effects of climate change on the distribution of maize races and their wild relatives in Mexico. *Global Change Biology* 18, 1073–1082.

Van Etten, J. and de Bruin, S. (2007) Regional and local maize seed exchange and replacement in the western highlands of Guatemala. *Plant Genetic Resources: Characterization and Utilization* 5, 57–70.

Vermeulen, S., Aggarwal, P., Ainslie, A., Angelone, C., Campbell, B. *et al.* (2012) Options for support to agriculture and food security under climate. *Environmental Science & Policy* 15, 136–144.

Vigouroux, Y., Glaubitz, J.C., Matsuoka, Y., Goodman, M.M., Sanchez, G. *et al.* (2008) Population structure and genetic diversity of New World maize races assessed by DNA microsatellites. *American Journal of Botany* 95, 1240–1253.

Vigouroux, Y., Cedric, M., De Mita, S., Pham, J.L., Gerard, B. *et al.* (2011) Selection for earlier flowering crop associated with climatic variations in the Sahel. *PLoS ONE* 6, 1–9.

Zimmerer, K.S. (2003) Geographies of seed networks for food plants (potato, ulluco) and approaches to agrobiodiversity conservation in the Andean countries. *Society and Natural Resources* 16, 583–601.

Zimmerer, K.S. (2010) Biological diversity in agriculture and global change. *Annual Review of Environment and Resources* 35, 137–166.

9 Germplasm Databases and Informatics

Helen Ougham and Ian D. Thomas
Institute of Biological, Environmental and Rural Sciences, Aberystwyth University, UK

9.1 Introduction

In order to develop crop varieties better able to cope with the demands of changing environmental conditions, many breeders need access to a wider range of genetic variation than is present within their existing germplasm resources or breeding materials. Although collecting trips may in some cases provide this novel variation, they are expensive to carry out and cannot be guaranteed to deliver the required genepool. It may often be more effective and economical to acquire new germplasm from genebanks. However, in order to identify appropriate material, breeders and researchers need access to information about the accessions held in those genebanks, in their own countries, and worldwide. Fortunately, computational developments within the past two decades have greatly facilitated the sharing of such information.

It was probably in the 1960s that organizations maintaining collections of plant germplasm began to capitalize on developments in computer technology in order to digitize records of the accessions they held. Initially, information would have been held as simple listings; but as software became more sophisticated, genebanks moved on to use database management systems. The early databases were designed for internal use within an institute, genebank or other organization, and if their contents were shared with other groups it was generally in the form of printouts. By the end of the 1980s, however, increasing access to the developing internet allowed organizations to begin to disseminate germplasm information electronically. With the advent of the World Wide Web in the 1990s, it became a realistic possibility for databases of germplasm information to be shared freely with any individual or group possessing a computer with internet connectivity.

Germplasm databases form an important component of the wider field of plant biodiversity informatics, which encompasses analysis and visualization tools as well as databases for taxonomy, species lists and ecological datasets, among others. A recent paper by Agrawal *et al.* (2012) reviews biodiversity informatics and includes a list of relevant websites. This chapter focuses solely on germplasm databases, specifically those that are publicly accessible through the World Wide Web and that are hosted and curated by well-established organizations, although there are, or have been, numerous other such databases. Doubtless many more are maintained privately by for-profit organizations, some of which may be willing to make data and germplasm available on the basis of licensing or reciprocal exchange. These are, however, also outside the scope of the present discussion. An additional criterion for choosing the databases introduced in this chapter is that they catalogue germplasm – generally conserved *ex situ* in genebanks – much of which is available free of charge (such as the collections held in the centres of the Consultative Group on International Agricultural Research – CGIAR) or may possibly be for purchase at a modest price. Access to germplasm is normally subject to the recipient agreeing to the terms of a Material Transfer Agreement (MTA), which defines the responsibilities of both donor and recipient of the germplasm (see Moore and Hawtin, Chapter 6, this volume, for a more detailed discussion about policies and access to germplasm).

In searching for useful material in the databases listed here, users should be aware that not all germplasm will be available in a form suitable for introduction into a crop improvement programme. For example, some material may be held as cryopreserved specimens for reference only, whereas in other cases seed stocks may be extremely limited.

9.2 Descriptors for Plant Genetic Resources

The concept of standardized descriptors for germplasm resources pre-dates the development of databases. It was recognized by the 1960s, if not earlier, that effective curation and exchange of germplasm would be greatly facilitated by the use of standard terminology to describe the accessions held by all institutes. Of course, descriptions of plant morphology and traits must of necessity be, at least in part, species specific. One set of terms has, however, become established as the international standard used (and in some cases extended) by a large number of international databases: these are the Multi-Crop Passport Descriptors (MCPDs).

9.2.1 Passport descriptors and the MCPDs

Passport descriptors are designed to provide essential information about the organization holding the germplasm, as well as the source and nature of the material. In the 1980s, when it was apparent that a core set of descriptors, independent of species, was required, the MCPDs were developed jointly by the International Plant Genetic Resources Institute (IPGRI, now Bioversity International; Gotor *et al.*, 2008) and the Food and Agriculture Organization (FAO). The original list was published in 1987 and revised in 2001 (Alercia *et al.*, 2001). A summary of the current version (version 2) is given in Table 9.1. Each descriptor has a defined format such as dates of acquisition

Table 9.1. The FAO/Bioversity Multi-Crop Passport Descriptors (version 2) used to describe accessions in plant germplasm collections, summarized from FAO/Bioversity (2012).

	Descriptor	Code	Description
1	Institute code	INSTCODE	Code of the institute where the accession is maintained
2	Accession number	ACCENUMB	A unique identifier for an accession in a genebank collection
3	Collecting number	COLLNUMB	Original number assigned by the collector(s) of the sample; essential for identifying duplicates held in different collections
4	Collecting institute code	COLLCODE	Code of the Institute collecting the sample. If the holding institute has collected the material, COLLCODE should be the same as INSTCODE
5	Genus	GENUS	Genus name for taxon, e.g. 'Triticum'
6	Species	SPECIES	Specific portion of the binomial scientific name, e.g. 'aestivum'
7	Species authority	SPAUTHOR	The authority for the species name
8	Subtaxa	SUBTAXA	Subtaxa can be used to store any additional taxonomic identifier, e.g. 'subsp.' (for subspecies) or 'var.' (for variety)
9	Subtaxa authority	SUBTAUTHOR	The subtaxa authority
10	Common crop name	CROPNAME	Name of the crop in colloquial language, preferably English (e.g. 'malting barley', 'cauliflower')

	Descriptor	Code	Description
11	Accession name	ACCENAME	Registered or other formal designation(s) given to the accession (e.g. Rheinische Vorgebirgstrauben, Emma or Avalon)
12	Acquisition date (YYYYMMDD)	ACQDATE	Date on which the accession entered the collection where YYYY is the year, MM is the month and DD is the day
13	Country of origin	ORIGCTY	The 3-letter ISO 3166-1 extended country code of the country in which the sample was originally collected
14	Location of collecting site	COLLSITE	Information that describes where, within a country, the accession was collected; this might include the distance in kilometres and direction from the nearest town, village or map grid reference point, (e.g. 7 km south of Curitiba in the state of Paraná, Brazil)
15	Latitude of collecting site1	LATITUDE	Degree (2 digits), minutes (2 digits), and seconds (2 digits) followed by N (north) or S (south) (e.g. 103020S)
16	Longitude of collecting site1	LONGITUDE	Degree (3 digits), minutes (2 digits), and seconds (2 digits) followed by E (east) or W (west) (e.g. 0762510W)
17	Elevation of collecting site (masl)	ELEVATION	Elevation (altitudo) of collecting site expressed in metres above sea level. Negative values are allowed
18	Collecting date of sample (YYYYMMDD)	COLLDATE	Collecting date of the sample where YYYY is the year, MM is the month and DD is the day
19	Breeding institute code	BREDCODE	Institute code of the institute that bred the material; if the holding institute bred the material, BREDCODE should be the same as the INSTCODE
20	Biological status of accession	SAMPSTAT	The coding scheme can be used at different levels of detail, by using either general codes such as 100 (wild), 200 (weedy), 300 (traditional cultivar/landrace) or 400 (breeding/research material), or more specific codes such as 414 (inbred line; parent of hybrid cultivar) or 420 (mutant/genetic stock)
21	Ancestral data	ANCEST	Information about pedigree or other description of ancestral information (i.e. parent variety in case of mutant or selection)
22	Collecting/acquisition source	COLLSRC	The coding scheme can be used at two different levels of detail, by using either general codes such as 10 (wild habitat), 20 (farm or cultivated habitat) etc., or the more specific codes such as 11 (forest/woodland), 12 (shrubland) etc.
23	Donor institute code	DONORCODE	Code for the donor institute
24	Donor accession number	DONORNUMB	Number assigned to an accession by the donor
25	Other identification (numbers) associated with the accession	OTHERNUMB	Any other identification (numbers) known to exist in other collections for this accession
26	Location of safety duplicates	DUPLSITE	Code of institute where a safety duplicate of the accession is maintained
27	Type of germplasm storage	STORAGE	If germplasm is maintained under different types of storage, multiple choices are allowed
28	Remarks	REMARKS	The remarks field is used to add notes or to elaborate on certain descriptors

and collection that are expressed in the form YYYYMMDD, where YYYY is the year, MM is the month and DD is the day. The codes used to identify holding, collecting and breeding institutes are those assigned by FAO and are listed on the FAO website (FAO, 2011) and new codes can also be obtained through this site.

For a given accession in a genebank, not all descriptors will be used. Where an accession is a breeder's line, descriptors relating to collection site are not relevant; safety duplicates are not stored in all cases; and very frequently the holding institute is the same as the breeding or collecting institute. However, certain descriptors are essential for all accessions: the holding institute, the unique accession number by which the accession is identified at that institute and the plant genus. Figure 9.1 shows examples of the descriptors used for two contrasting accessions of the grass species *Lolium multiflorum* held in Aberystwyth, UK and listed in the relevant European Central Crop Database, discussed later; one is an accession collected from the wild, the other is a breeder's line.

Although the MCPDs form a good basis for cataloguing and sharing germplasm, in some cases they allow for insufficient detail, so extended descriptor lists have been developed to annotate, for example, site types of particular relevance to a given crop. Thus in the case of the *L. multiflorum* database, the descriptors for habitat type have been extended to include details of grassland management regime, such as 'cut only', 'grazed only', 'grazed and cut' and 'sports', whereas the European *Allium* database includes descriptors for mode of reproduction (vegetative, seed or both), daylength requirement and availability of material (available, not available or reference collection only).

9.2.2 Characterization and evaluation descriptors

In addition to passport data, many databases contain, for some accessions at least, information about plant form and performance. Data of this type are catalogued using characterization and evaluation descriptors. The difference between the two types is that characterization data are constant for a given accession: they include features such as seed or flower colour, chromosome number and leaf shape. Evaluation data, on the other hand, depend on the conditions under which a plant is grown, including temperature regime, nutrient availability and presence of other, symbiotic or pathogenic, organisms. Examples of evaluation data types are yield of grain or fruit, resistance to disease-causing organisms, and number of days to maturity.

Unlike passport descriptors, many characterization and evaluation descriptors are necessarily crop specific. Like passport data, however, there are considerable advantages to using a universal language that is comprehensible to anyone working on a given crop or group of crops. Accordingly, lists of appropriate descriptors have been, or are being, developed by acknowledged experts on many crops, and are used in published reports as well as in germplasm databases. Bioversity International recently published a paper (Alercia, 2011) containing a progress report on the characterization and evaluation descriptors for 22 major food crops. Table 9.2 shows as an example the list of key plant characterization and evaluation descriptors recommended for lentil (*Lens* Miller), illustrating the range of descriptor types used. They include measured numerical variables such as weight and seed number, human-assigned scores on a 1–9 scale, and descriptive terms.

9.3 International Germplasm Databases

In this section, some of the major international germplasm databases are introduced. The European central crop databases, EURISCO, and GENESYS collate data about germplasm accessions collected from, and held in, multiple countries, making it easier to find information without the need to carry out extensive online searches. The nature of the data accessible directly through

(a)

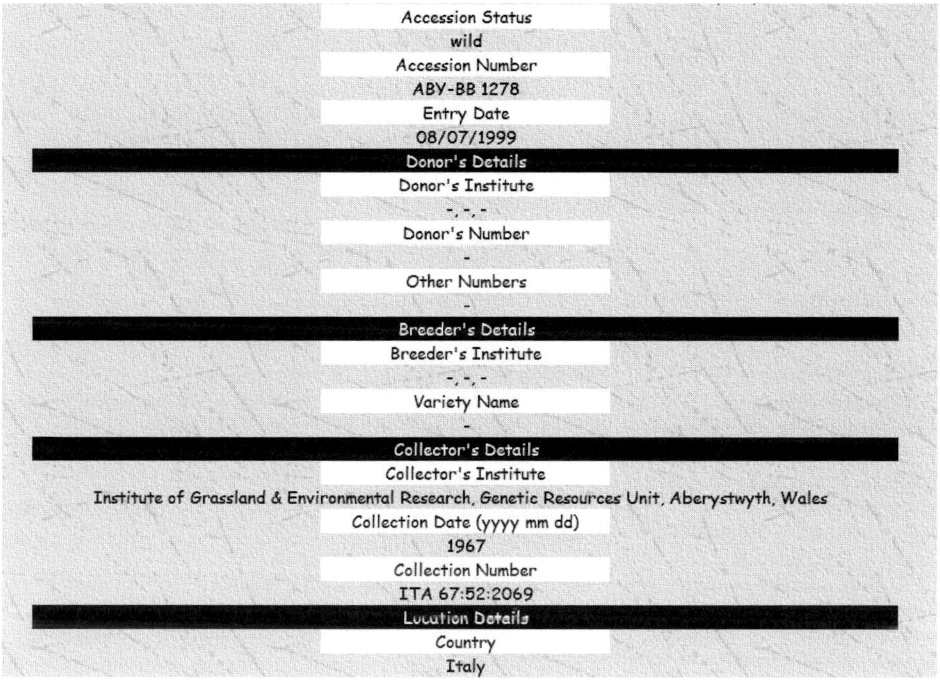

Accession Status
wild
Accession Number
ABY-BB 1278
Entry Date
08/07/1999
Donor's Details
Donor's Institute
-, -, -
Donor's Number
-
Other Numbers
-
Breeder's Details
Breeder's Institute
-, -, -
Variety Name
-
Collector's Details
Collector's Institute
Institute of Grassland & Environmental Research, Genetic Resources Unit, Aberystwyth, Wales
Collection Date (yyyy mm dd)
1967
Collection Number
ITA 67:52:2069
Location Details
Country
Italy

(b)

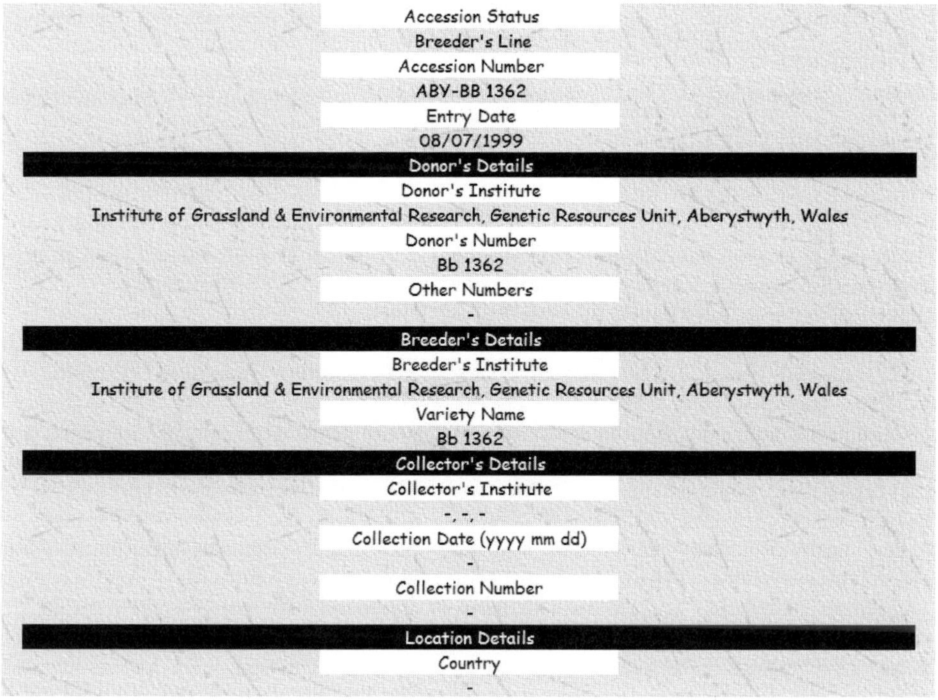

Accession Status
Breeder's Line
Accession Number
ABY-BB 1362
Entry Date
08/07/1999
Donor's Details
Donor's Institute
Institute of Grassland & Environmental Research, Genetic Resources Unit, Aberystwyth, Wales
Donor's Number
Bb 1362
Other Numbers
-
Breeder's Details
Breeder's Institute
Institute of Grassland & Environmental Research, Genetic Resources Unit, Aberystwyth, Wales
Variety Name
Bb 1362
Collector's Details
Collector's Institute
-, -, -
Collection Date (yyyy mm dd)
-
Collection Number
-
Location Details
Country
-

Fig. 9.1. Examples of descriptors for two accessions of *Lolium multiflorum* catalogued in the European Central Crop Database for *Lolium* species. (a) Accession collected from the wild; (b) Breeder's line (available from Genetic Resources Unit – IGER, 2002).

Table 9.2. List of key plant characterization and evaluation descriptors for lentil, summarized from Alercia (2011).

Descriptor	Description
Plant height (cm)	Height of plant measured from the ground to the tip of the extended foliage, at maturity; average height of 10 plants
Plant growth habit	Observed after flowering: prostrate, semi-prostrate, intermediate, upright, erect or other
Days to 50% flowering	Number of days from sowing until 50% of the plants are in flower. However, in dry land areas when planting in dry soils, it is counted from the first day of rainfall or irrigation, which is sufficient for germination
Days to physiological maturity	Number of days from sowing until 90% of the pods are golden brown
Number of seeds per pod	Average number of seeds of 10 dry pods
100-seed weight (g)	Average weight of two samples of 100 randomly chosen seeds
Ground colour of seed testa	To be observed on seeds less than three months old: green, grey, brown, black or pink
Pattern of seed testa	Absent, dotted, spotted, marbled or complex
Cotyledon colour	To be observed on seeds less than three months old: yellow, orange-red or olive-green
Lodging susceptibility	Scored at maturity on a scale 1–9
Biological yield per plant (g)	Yield of dried mature plants after pulling
Harvest index (%)	Weight of seeds relative to total biomass
Number of pods per peduncle	Maximum number of pods per peduncle on 10 representative plants
Height of lowest pod (cm)	Estimate of the average height above ground of the lowest pod on unlodged plants at harvest
Pod shedding	Scored after or during harvesting one week after maturity on a scale 1–9
Pod dehiscence	Scored one week after maturity on a scale 1–9
Number of pods per plant	Average number of pods; recorded from randomly selected plants at physiological maturity
Seed yield per plant (g/plant)	Yield of seed after drying

these portals varies, and there is no doubt that while any information about an accession is potentially valuable, availability of searchable characterization and/or evaluation data adds greatly to the practical utility of that accession. So, for example, the European central crop database for wheat allows users to search characterization data for important traits such as 100-kernel weight or protein content. The GENESYS portal provides powerful crop-specific search tools to assist in the identification of accessions resistant to abiotic stresses, insect pests and fungal pathogens, among many other traits. Many other databases provide only passport data, but even in these cases the holding or donating institute may have additional information available about certain accessions.

9.3.1 The European central crop databases

The European Cooperative Programme for Plant Genetic Resources (ECPGR), working with individual institutes, has for many years supported an initiative to establish databases for passport data and, in some cases, characterization and primary evaluation data for the major collections of crops important in Europe, and their wild relatives. These European central crop databases are classified into six groups: Cereals; Forages; Oil and Protein Crops; Fruit; Sugar, Starch and Fibre crops; and Vegetables (ECPGR, 2013a). Figure 9.2 shows the databases in the Sugar, Starch and Fibre grouping, together with the names, institutions and countries of the database managers. The

Database	Database Manager	Institute	Country	Email
ECPGR Cultivated Potato Database	Stuart Cernegie	Potato Section Science and Advice for Scottish Agriculture (SASA), Science and Advice for Scottish Agriculture (SASA)	United Kingdom	⌧
ECPGR Flax Database	Martin Pavelek	AGRITEC Research, Breeding and Services, Ltd.	Czech Republic	⌧
ECPGR Hemp Database	Gianpaola Grassi	CRA-CIN Centro di Ricerca per le Colture Industriali	Italy	⌧
ECPGR Wild Potato Database	Roel Hoekstra	Centre for Genetic Resources, the Netherlands (CGN), Wageningen University and Research Centre	Netherlands	⌧
International Database for *Beta*	Lothar Frese	Julius Kühn-Institute (JKI), Institute for Breeding Research on Agiricultural Crops	Germany	⌧
International Database for *Beta*	Christopher Germeier	Julius Kühn-Institute (JKI), Institute for Breeding Research on Agiricultural Crops	Germany	⌧

Fig.9.2. European Central Crop Databases in the Sugar, Starch and Fibre species grouping (from ECPGR, 2013a).

germplasm documented in each database is generally maintained at a number of different institutes throughout Europe, and to request material catalogued in a database, the user must contact the institute that maintains the germplasm accessions, rather than the database manager.

Because each database has its own manager, the mode of access differs from one database to another. In many cases the entire contents of the databases are available for download, sometimes in a choice of formats (for example, Microsoft Excel spreadsheet or Access database; text-only listing). Often, but not always, there is a web search interface, created by the team managing the database. It is appropriate to compare the interface available for the database of *Phaseolus* species (Fig. 9.3a) with that for *Avena* species (Fig. 9.3b) to gain an idea of the types of interface that may be encountered.

9.3.2 EURISCO and the European National Inventories of plant genetic resources

The EURISCO web catalogue (EURISCO, 2013a) brings together data on germplasm held in over 40 European countries. Each of these countries has nominated a National Focal Point (NFP), an individual who is responsible for collecting information about the material present in that country's genebanks and other facilities (EURISCO, 2013b), and making it available to EURISCO. The catalogue was first made public in 2003, at the end of a project funded by the European Union in order to improve access to germplasm resources. It is currently hosted by Bioversity International, on behalf of the ECPGR Secretariat, and as of September 2012 it contained over 1.1 million entries, representing more than 36,000 plant species (EURISCO, 2013a). Data are not restricted to those species represented in the European Central Crop Databases, discussed earlier, but accessions for the latter do tend to predominate in EURISCO.

EURISCO currently contains only passport data, although in many cases the holding institutes have additional characterization, evaluation or other information that they can make available. The EURISCO descriptors are based on the standard FAO/IPGRI MCPDs, but there are six additional EURISCO-specific descriptors. The first of these identifies the National Inventory from which a data entry originated, another

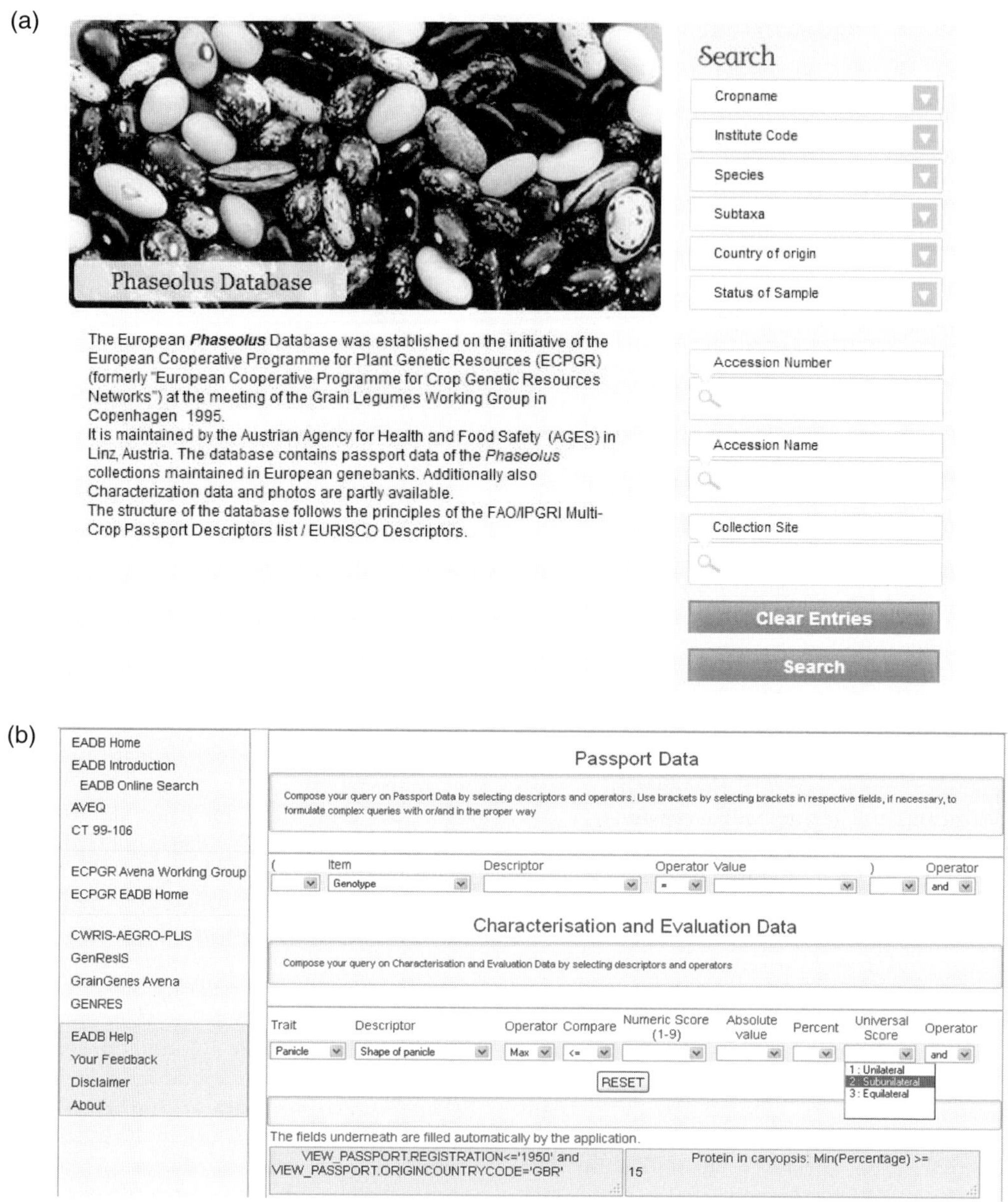

Fig. 9.3. Examples of two contrasting interfaces for searching European Central Crop Databases. (a) *Phaseolus* species. (b) *Avena* species (see AGES, 2013 and Julius Kühn-Institut, 2012).

allows inclusion of a URL linking to additional information about the accession (for example, in another germplasm database), and the remaining four are required only when the organization collecting, breeding or holding the material, or maintaining a safety duplicate, has not been assigned an FAO code. As in the case of the standard multi-crop descriptors, the use of most of these is optional; only the National Institute code, the FAO code of the institute maintaining the accession, the accession number and the plant genus are mandatory in EURISCO.

The individual National Inventories vary greatly in size and nature. Their contents range from a little over 200 accessions, up to more than 200,000 in the case of the Russian Federation; these accessions are all held at the N.I. Vavilov All-Russian Scientific Research Institute of Plant Industry. For some countries, developing a national inventory was straightforward, because of an existing national strategy of maintaining central records (and, in some cases, a single national germplasm bank); in other cases, creating the inventory required bringing together data from a large number of genebanks, some of which had not previously been able to make their data web-accessible.

In addition to dissemination through EURISCO, some national focal points make their national inventories independently accessible; most frequently this is done through a more comprehensive national germplasm information system of the type described later. For example, Germany hosts its national inventory via its online database, GENRES, of genetic resources in Germany (IBV, 2013), whereas the five Nordic countries (Denmark, Finland, Iceland, Norway and Sweden) have a joint genebank documentation system, SESTO (NordGen, 2013), through which the individual National Inventories can be accessed. Figure 9.4 shows an example of characterization data for a barley (*Hordeum vulgare*) accession catalogued in the Austrian National Inventory and available from the Austrian portal (AGES, 2013); the EURISCO entry for the same accession contains only the passport data.

9.3.3 GENESYS

In 2011, the portal GENESYS (2011) was launched, with the aim of providing a 'one-stop' source of information about accessions held in germplasm collections worldwide. It was developed through a collaboration between Bioversity International, the Global Crop Diversity Trust and the Secretariat of the International Treaty on Plant Genetic Resources for Food and Agriculture (Arnaud *et al.*, 2010). At the time of writing, GENESYS gives access to data from three major sources:

- EURISCO.
- The National Germplasm System of the US Department of Agriculture.
- SINGER, the System-wide Information Network for Genetic Resources.

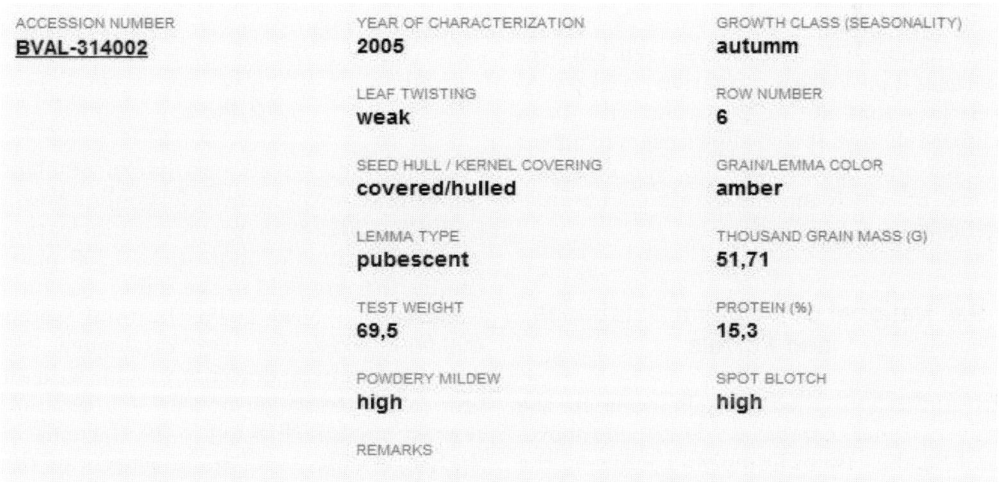
ACCESSION NUMBER
BVAL-314002
YEAR OF CHARACTERIZATION
2005
GROWTH CLASS (SEASONALITY)
autumm
LEAF TWISTING
weak
ROW NUMBER
6
SEED HULL / KERNEL COVERING
covered/hulled
GRAIN/LEMMA COLOR
amber
LEMMA TYPE
pubescent
THOUSAND GRAIN MASS (G)
51,71
TEST WEIGHT
69,5
PROTEIN (%)
15,3
POWDERY MILDEW
high
SPOT BLOTCH
high
REMARKS

Fig. 9.4. Characterization data for a *Hordeum vulgare* accession catalogued in the Austrian National Inventory of plant genetic resources conserved *ex situ*.

SINGER, the germplasm information exchange network formerly maintained by the CGIAR and its partners, documents more than three-quarters of a million accessions of crop, forage and tree germplasm held in the collections of the 12 SINGER members: Africa Rice Center, AVRDC – The World Vegetable Center, Bioversity International, CIAT, CIMMYT, CIP, ICARDA, ICRAF, ICRISAT, IITA, ILRI and IRRI. Following the launch of GENESYS, the separate SINGER website was decommissioned.

GENESYS currently provides access to well over two million accessions of crop and crop wild relative germplasm, a number estimated to represent about a third of all the germplasm accessions held worldwide; the intention is to increase coverage over time. In many cases, GENESYS accession records include links to the corresponding records in the source database, and these original records may include characterization and evaluation as well as passport data, which are not obtainable directly from GENESYS.

9.3.4 The Crop Wild Relatives Global Portal

Although its main function is not that of making germplasm databases accessible, the Crop Wild Relatives Global Portal (Bioversity International, 2013) deserves mention here because it provides information on crop wild relatives that are at particular risk from climate change. It also allows users to access 12 national portals for crop wild relatives, some of which include access to national genebanks.

9.4 National Germplasm Databases and Portals

In addition to their contributions to the international databases described above, and (in the case of European countries) the National Inventories of germplasm conserved *ex situ* that together make up EURISCO, many countries maintain their own databases of plant genetic resources. Often these are made available through web portals that provide access to a range of data useful in plant genetic improvement and germplasm research, including, for example, information on *in situ* conservation efforts and extensive trait data. A reasonably comprehensive list of such national provision is provided by ECPGR (2013b) that gives links to national multi-crop databases and portals in over 20 countries (Fig. 9.5). Here we give three specific examples.

9.4.1 Japan

Staff members of the Plant Section of Japan's National Institute of Agrobiological Sciences (NIAS) Genebank (2013) have responsibility for collecting, evaluating and characterizing, as well as conserving and documenting, germplasm of plants in 12 important agricultural groups. Searches can be made on the basis of either passport descriptors or (where available) evaluation data. Figure 9.6 shows some of the options available for locating soybean (*Glycine max*) accessions in the NIAS Genebank. Germplasm, generally in the form of seeds, is made available free of charge to many organizations meeting certain criteria (for example, being not-for-profit); otherwise, the recipient is required to pay a small fee (a few tens of dollars).

9.4.2 India

In 2012, India's National Bureau of Plant Genetic Resources launched its online portal (NBPGR, 2012). It provides information about accessions of more than 1800 plant species that are held in the Indian National Genebank in New Delhi. In total, about 400,000 accessions are catalogued. At present the portal does not provide a direct link for ordering germplasm – a separate request must be directed to the Germplasm Exchange Unit – and some aspects of the search and display tools are still being optimized, particularly with respect to characterization and evaluation data. There

National Multi-crop Databases

- Australian Plant Genetic Resource Information Service (AusPGRIS)
- Austria – National Inventory of Austria
- Belgium – National Botanic Garden of Belgium – The Database of Living Plant Collections (LIVCOLL)
- Bulgaria – National Seed Genebank
- Canada – Plant Gene Resources of Canada (GRIN-CA)
- China – Chinese Crop Germplasm Information System (CGRIS)
- Czech Republic – Information System on Plant Genetic Resources (EVIGEZ)
- France – BRG – collections de ressources génétiques végétales
- Germany – BIG-Flora, Zentralstelle für Agrardokumentation und -information (ZADI)
- Germany – Federal Research Centre for Cultivated Plants – Julius Kuhn Institute
- Germany – Leibniz Institute of Plant Genetics and Crop Plant Research (IPK)
- India – National Bureau of Plant Genetic Resources
- Israel – Israel Plant Gene Bank
- Israel – The Harold and Adele Lieberman Germplasm Bank
  Institute for Cereal Crops Improvement (ICCI)
  Tel Aviv University, The George S. Wise Faculty of Life Sciences
- Italy – CRA – Consiglio per la Ricerca e Sperimentazione in Agricoltura
- Italy – IGV – Germplasm database
- Japan – NIAS genebank
- New Zealand – Arable crop genebank and online database, New Zealand Institute for Crop & Food Research
- Romania – Suceava Genebank Databases
- Russian Federation – N.I. Vavilov All-Russian Scientific Research Institute of Plant Industry (VIR)
- Spain – INIA – Centro de Recursos Fitogenéticos – Genebank
- Sweden – Stored material at the Nordic Genebank
- Switzerland – Conservation of PGRFA – Swiss National Database
- The Netherlands – Centre for Genetic Resources (CGN)
- United Kingdom – Millennium Seed Bank Seed List
- United States of America – National Plant Germplasm System

Fig. 9.5. National multi-crop databases and portals listed by ECPGR, as of October 2012.

is no doubt, however, that this portal will prove to be extremely valuable for those interested in germplasm held in India or adapted to Indian conditions.

9.4.3 USA

The National Plant Germplasm System of the Unites States Department of Agriculture (USDA-ARS, 2012) is part of the broader Genetic Resources Information Network (GRIN) in the USA. It offers access to US-held germplasm through a range of search tools from the simple to the advanced, and includes characterization and evaluation data as well as passport information for many of the more than 500,000 accessions that it catalogues. Users can download data, selected by genus and country of origin, in MCPD format. There is also an online germplasm ordering service for researchers and breeders. Much, but not all, of the information in GRIN is now accessible through the GENESYS portal.

9.5 Germplasm Data in Genome Databases

From the early 1990s, the flood of data that became available as molecular marker systems were developed for plant species, together with the results of DNA sequencing studies, provided the impetus for establishing genome databases for many major crops, as well as for the model organism *Arabidopsis thaliana*. A more detailed discussion of these

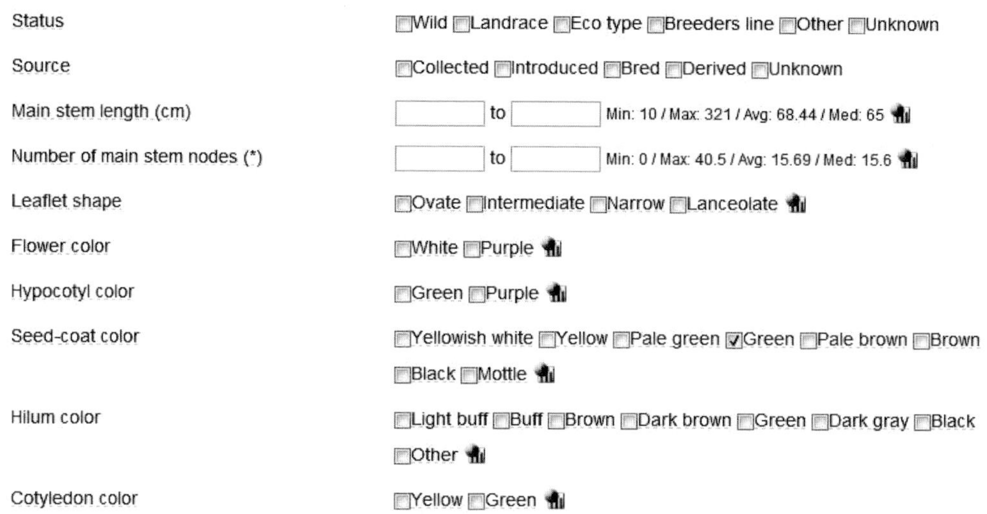

Fig. 9.6. Options for identifying *Glycine max* accessions in Japan's NIAS Genebank, according to characteristics required.

databases (and informatics in general) is provided in another chapter (see McNally, Chapter 10, this volume). A list of the major ones is, however, available on the Arabidopsis Information Resources website (TAIR, 2013). Some genome database websites also provide direct or indirect access to germplasm data. They may prove particularly useful for those wishing to obtain mutant stocks and other genetic variants for use in research aimed at identifying the genetic basis of traits, or developing new markers, based on DNA sequence variants, for marker-assisted selection programmes. Frequently such germplasm is not listed in national or international databases of plant genetic resources, so genome databases can provide an additional route to access the relevant data. The following section, which is not intended to be comprehensive, describes three genome databases that contain extensive germplasm data.

9.5.1 Some genome databases with significant germplasm data

GrainGenes (USDA-ARS, 2013), which focuses on wheat, barley and other temperate cereals, is not itself a portal for direct access to germplasm but it maintains extensive links to sites worldwide through which users may obtain genetic resources for cereal improvement, together with links to evaluation and characterization data.

The Maize Genetics and Genomics Database, MaizeGDB (2013), provides powerful tools for searching the contents of the Maize Genetic Stock Center, with options for specifying phenotypic traits of interest (Fig. 9.7), and for placing orders with the Stock Center.

The Breeder Toolbox component of the CottonDB database (USDA-ARS, 2006) allows many types of search for cotton germplasm, on the basis of holding institute, pedigree or evaluation data (Fig. 9.8). Links to the corresponding genebanks allow users to order stocks, where they are available.

9.6 Future Developments

In the next few years, it is likely that the major effort in the field of germplasm databases and informatics will be to increase the number of genebank accessions held worldwide that are catalogued in web-accessible databases (or at least in a form

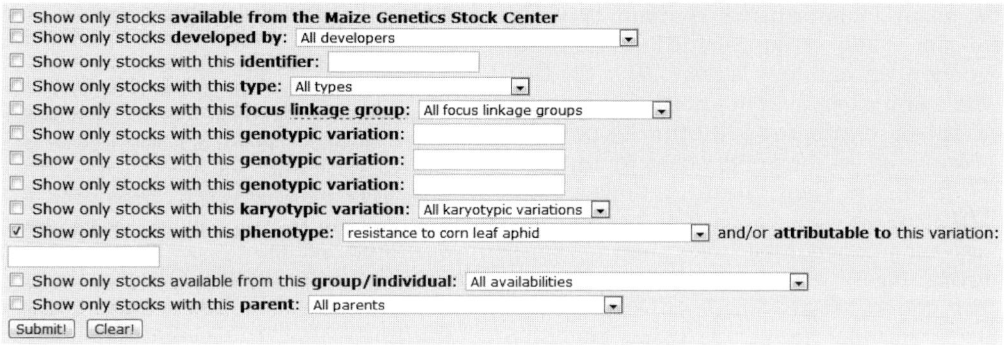

Fig. 9.7. Interface for searching the Maize Genome Database, MaizeGDB, to identify germplasm with specific traits of interest.

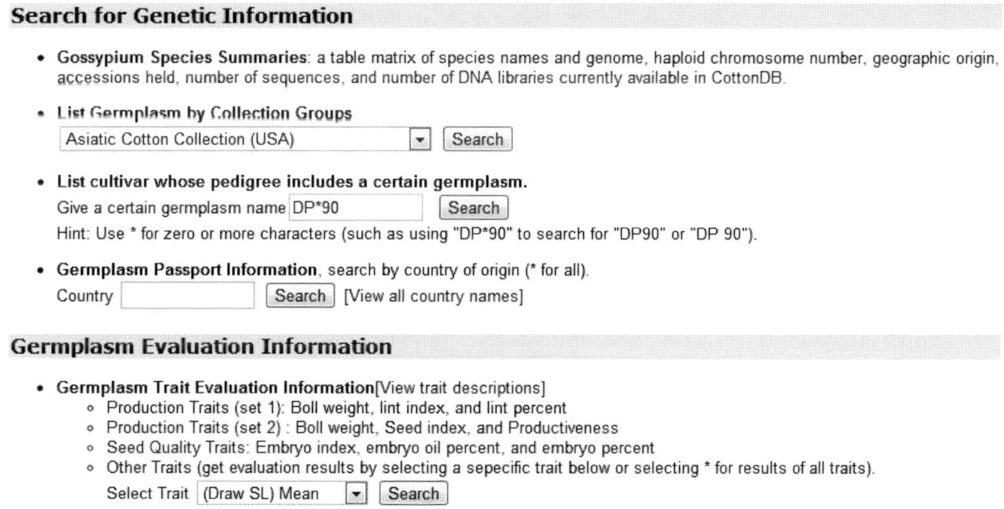

Fig. 9.8. The search interface of the Breeder Toolbox component of CottonDB.

downloadable from the Web); to add to the characterization, evaluation and other data types available for accessions; and to improve the user-friendliness of database interfaces and associated portals. In the latter context, visualization and other geographic information system tools allowing users to identify the origin of an accession on a map, and overlay, for example, meteorological, soil type or epidemiological data, are attractive possibilities, particularly for those needing to predict responses of crops to climate change and to breed for improved tolerance to environmental variables. Many of the necessary computational tools are already available to do this; the primary requirement is for an investment of manpower (and therefore, of course, funding) in order to deploy the tools in this context.

Closer integration between germplasm informatics systems and other areas of plant bioinformatics also offers opportunities. Conceptually, characterization and evaluation descriptors bear some similarity to the ontology terms that are being developed by the plant bioinformatics community with the aim of annotating genetic loci and gene function. An ontology seeks to describe the whole of a given field of knowledge, in this case plant science, and in the process provide a controlled vocabulary of terms to describe the field. The Plant Ontology Consortium (2012) has developed a controlled vocabulary for plant anatomy, morphology and developmental stages – analogous to characterization descriptors – whereas the Trait Ontology for plants includes terms for many environmentally dependent characters such as plant height and disease resistance, in the same way that a set of evaluation descriptors may do. Unlike descriptors, ontology terms are not designed to provide quantitative assessments such as trait scores but only to associate a gene (or other entity such as a protein, genetic locus or pathway) with a character. Nevertheless, some groups are attempting to introduce the use of ontologies into the area of plant genetic resources conservation (Dias, 2012), and there is little doubt that the use of common standards will eventually make it easier for plant breeders and genebank curators to take advantage of the wealth of genome and marker data available for crop species and their relatives, in order to breed plants better equipped to feed the world of the future.

References

AGES (2013) *Phaseolus* database. Available at: www.genbank.at/en/ecpgr-phaseolus.html (accessed 6 February 2013).

Agrawal, R.C., Archak, S. and Tyagi, R.K. (2012) An overview of biodiversity informatics with special reference to plant genetic resources. *Computers and Electronics in Agriculture* 84, 92–99.

Alercia, A. (2011) *Key Characterization and Evaluation Descriptors: Methodologies for the Assessment of 22 Crops*. Bioversity International, Rome. Available at: www.bioversityinternational.org/nc/publications/publication/issue/key_characterization_and_evaluation_descriptors.html (accessed 23 January 2013).

Alercia, A., Diulgheroff, S. and Metz, T. (2001) List of Multi-Crop Passport Descriptors. FAO/IPGRI, Rome. Available at: www.bioversityinternational.org/fileadmin/bioversity/publications/pdfs/124_FAOIPGRI_Multi-crop_passport_descriptors.pdf (accessed 23 January 2013).

Arnaud, E., Dias, S. and Mackay, M. (2010) A global portal enabling worldwide access to information on conservation and use of biodiversity for food and agriculture. In: Maurer, L. (ed.) *ICT for Agriculture and Biodiversity Conservation*. ICT Ensure, Graz University of Technology, Graz, Austria, pp. 168–180.

Bioversity International (2013) *Crop Wild Relatives Global Portal*. Available at: www.cropwildrelatives.org/ (accessed 6 February 2013).

Dias, S. (2012) Pieces of the puzzle – Trait Information Portal. *Crop Wild Relatives* 8, 28–30.

ECPGR (2013a) Database available at: www.ecpgr.cgiar.org/germplasm_databases/central_crop_databases.html (accessed 6 February 2013).

ECPGR (2013b) Database available at: www.ecpgr.cgiar.org/germplasm_databases/national_multicrop_databases.html (accessed 6 February 2013).

EURISCO (2013a) EURISCO Catalogue. Available at: http://eurisco.ecpgr.org/ (accessed 24 September 2012).

EURISCO (2013b) National focal points database. Available at: http://eurisco.ecpgr.org/about/the_network/national_focal_points.html (accessed 6 February 2013).

FAO (2011) *World Information and Early Warning System on PGRFA*. Available at: http://apps3.fao.org/wiews/wiewspage.jsp?show=DownloadinstEN.jsp (accessed 6 February 2013).

FAO/Bioversity International (2012) FAO/Bioversity Multi-Crop Passport Descriptors V.2. Available at: http://eurisco.ecpgr.org/fileadmin/www.eurisco.org/documents/MCPD_V2_2012_Final_PDFversion.pdf (accessed 23 January 2013).

Genetic Resources Unit-IGER (2002) Database available at: www.igergru.ibers.aber.ac.uk/Welcome/ECCDB/eccdb.htm (accessed 6 February 2013).

GENESYS (2011) Database available at: www.genesys-pgr.org/ (accessed 6 February 2013).

Gotor, E., Alercia, A., Ramanatha Rao, V., Watts, J. and Caracciolo, F. (2008) The scientific information activity of Bioversity International: the descriptor lists. *Genetic Resources and Crop Evolution* 55, 757–772.

IBV – Information and Coordination Centre for Biological Diversity (2013) *GENRES – Information System Genetic Resources*. Available at: www.genres.de/nc/en/titletext/ (accessed 6 February 2013).

Julius Kühn-Institut (2012) *EADB – European* Avena *Database*. Available at: http://eadb.jki.

bund.de/eadb/index.php (accessed 6 February 2013).
MaizeGDB (2013) *Maize Genetics and Genomics Database*. Available at: www.maizegdb.org/ (accessed 6 February 2013).
NBPGR (2012) *PGR Portal*. Available at: www.nbpgr.ernet.in/PGRPortal/ (accessed 6 February 2013).
NIAS Genebank (2013) Available at: www.gene.affrc.go.jp/index_en.php (accessed 6 February 2013).
NordGen (2013) *SESTO Gene Bank Documentation System*. Available at: www.nordgen.org/index.php/en/content/view/full/344 (accessed 6 February 2013).
Plant Ontology Consortium (2012) *Plant Ontology Database*. Available at: www.plantontology.org/ (accessed 6 February 2013).
TAIR (2013) *Other Genome Databases and Sequencing Efforts.* Available at: www.arabidopsis.org/portals/genAnnotation/other_genomes/index.jsp (accessed 6 February 2013).
USDA-ARS (2006) *CottonDB – Cotton Genome Database*. Available at: www.cottondb.org/wwwroot/cdbhome.php (accessed 6 February 2013).
USDA-ARS (2012) *National Plant Germplasm System*. Available at: www.ars-grin.gov/npgs/searchgrin.html (accessed 6 February 2013).
USDA-ARS (2013) *GrainGenes: a Database for Triticeae and* Avena. Available at: http://wheat.pw.usda.gov/GG2/index.shtml (accessed 6 February 2013).

10 Exploring 'Omics' of Genetic Resources to Mitigate the Effects of Climate Change

Kenneth L. McNally

International Rice Research Institute (IRRI), Los Baños, Philippines

10.1 Introduction

One of the biggest challenges society has to face is the relentless change in climate brought about by our activities coupled with burgeoning population growth. These changes include drought, flooding, lowered daylight intensity and higher greenhouse gas levels. Besides effects on the environment, changes in disease and pest pressure are also expected. All of these will be experienced alongside the increasing demand for more food. What tools can we use to identify plant genes that will enable us to meet these challenges?

Molecular research into the genetic diversity across a range of germplasm at a deep level provides a portfolio of technologies for identifying the genomic, allelic, transcriptomic, proteomic, metabolomic, epigenomic and systems interactions of crop species and the organisms that live in close association with them. Analysing and integrating this information is key to understanding the links between genotypes, phenotypes and environment and their inter-control. Advanced systems for annotating, curating and visualizing these datasets are key to the effective use and application of 'omics' data. Once these are in place, the stage is then set for targeted breeding and use of plant genetic resources to mitigate the effects of climate change while improving yield and human nutrition.

The purpose of this chapter is to provide a glimpse into the molecular technologies that could be used to address and mitigate the effects that climate change will impose on food production. In particular, the chapter provides an overview of these molecular technologies and their current state. Because informatics analyses are integral to the effective exploration and use of high-density sequence and other 'omics' datasets, informatic tools and approaches for analyses are also explored. Some examples of their application in relevant crop species are discussed. Game-changing technologies that may significantly accelerate our understanding are highlighted and ways that certain information can be used to fine-tune genotypic selection for targeted phenotyping and breeding is explored. Approaches to improve data mining and its use in breeding are addressed.

Because other chapters in this book have discussed climate change in depth, only a brief introduction is given here. The genetic resources that are the foundation for data generation and subsequent application are briefly discussed. This chapter will not, however, discuss crop-specific perspectives on the future of genebanks because other recent reviews have introduced the topic (McCouch *et al.*, 2012; Kilian and Graner, 2012). A simple overview of different 'omics' technologies is given. Informatics that can drive the change from just large datasets to useful information and some specific current and envisioned tools for data mining are explored. In conclusion, the vision of a transformed system of improved technologies for investigating and accessing genetic resources is presented together with its implication for crop improvement in the face of climate change.

10.2 The Challenge of Climate Change for Agriculture

The consumption of fossil fuels is the single strongest driver of climate change. Without converting to alternate sources of energy, demand for fossil fuels will remain prominent. Some offsets can be achieved through the use of biofuels but this can be at the expense of food production (FAO, 2009). One solution for an alternative renewable fuel is to use the residual biomass from crops such as maize or sorghum and return the waste minerals (micronutrients) to the soil. Another avenue is to identify plants for biofuel that can grow on marginal unused lands. Switchgrass is one such example (Chuck *et al.*, 2011). However, attempts to domesticate these wild species for use as crops may be difficult, and may require new breeding programmes and greater understanding of their biology. New domestication as a response to salinization is discussed later in this book (see Erskine *et al.*, Chapter 14, this volume).

Climate change will severely impact rainfall patterns and increases in sea level that will dramatically affect coastal regions. Water-related effects are expected to include drought, flooding, submergence and salinity. Areas affected by drought will shift with some highly productive areas being the most affected. Timing and intensity of droughts will exacerbate problems with food production. A recent example was presented by the drought of 2012 in the North American corn and soybean belt (USDA-ERS, 2013). Flooding in deltas and low-lying regions will intensify, necessitating improved tolerance to submergence under short- and long-term durations (see Ismail and Mackill, Chapter 15, this volume). Such flooding will also be accompanied by increase in salinity exposure (Negrão *et al.*, 2011).

Temperature effects will be extreme with more severe storms as a result. Changes in diurnal patterns of temperature, particularly high temperatures in the morning, combined with humidity levels will affect pollen and embryo fertility (Wassmann *et al.*, 2009). Other regions will see the need for improved cold tolerance because timing of frosts will be less predictable. Hence, improved heat and cold tolerance will be needed for these affected environments (Wahid *et al.*, 2007; Sanghera *et al.*, 2011).

Disease pressures may intensify owing to population shifts and pest adaptation (Yáñez-López *et al.*, 2012). Pathogen and pest population dynamics will change as a result of the climate. Adaptation to the new conditions may alter the dynamics of resistance/tolerance particularly because the same pathways for stress response are involved. Windfall and dispersion patterns are also expected to change, necessitating that varieties targeted for specific regions also include the appropriate sources of resistance or tolerance. Hence, maintenance breeding alone will not suffice and new sources of horizontal and durable resistance alongside thorough epidemiology will be needed (Evans *et al.*, 2008; Pautasso *et al.*, 2012; see Chapters 12 to 16, this volume, for examples of responses to these threats).

10.3 The Opportunities for Germplasm Use

The single most important resource available to mitigate the effects of climate change on crop production is the wealth of germplasm across the cultivated species and crop wild relatives. Fortunately, foresight of this need was at hand early enough to preclude loss of much of the genetic diversity, although not entirely (especially in regions not accessible for collection). The end result has been the conservation of over 1.5 million accessions of primary cereal crop species including rice, maize, wheat and barley.

For rice alone, global holdings total about 600,000 accessions largely held by the Consortium of International Agricultural Research Centers (CGIAR) genebanks at the International Rice Research Institute (IRRI) and AfricaRice (Africa Rice Center in Benin,

West Africa) and national genebanks in India, China (2), Japan, Korea, the USA and Thailand (IRRI, 2010). For example, even considering duplication of holdings, this represents, on average, one unique variety type per several small villages over the rice-growing regions of the world. The history and ecogeography of rice has led to a diversification of types across the regions of primary diversity. Estimates of the number of domestication events range from one to two or more independent events followed by migration, drift, introgression and selection (Vitte *et al.*, 2004; Londo *et al.* 2006; He *et al.*, 2011; Molina *et al.*, 2011; Huang *et al.*, 2012). Interpretation of the number of domestication events has been dependent on the dataset and analyses performed. Regardless of when and where rice arose as a crop, five main variety groups of cultivated Asian rice, *Oryza sativa*, exist with ecogeographical partitioning (see Plate 1 from Zeigler, Chapter 1, this volume). These are the *indica*, *aus* (and *boro*), aromatic, *tropical* and *temperate japonica* types. Taxonomic and molecular evidence support the division of rice into these clades. Origins are from south and east Asia (*indica*), the Ganges delta (*aus*), the Himalayan foothills (*aromatic*), Southeast Asia (*tropical japonica*) and east Asia (*temperate japonica*). The two minor variety groups identified by isozyme analysis (Glaszmann *et al.*, 1995), the deep water types 3 and 4 for Aswina and Rayada, when analysed by other molecular data fall into either the *indica* or the *aus* groups, respectively. Another species cultigen from Africa, *Oryza glaberrima* also exists. There are six other related species with the AA genome and a range of related species with other genome types in the genus (Lu *et al.*, 2001; Vaughan *et al.*, 2003; see Zeigler, Chapter 1, this volume).

Although much data exist regarding passport and basic morphological and agronomic characterization (see Ougham and Thomas, Chapter 9, this volume), the latter were often not evaluated under experimental design and many traits of interest were not observed. Further, sequencing and molecular studies need purified materials for success, and these purified accessions require multiplication under quality control conditions (McCouch *et al.*, 2012). The maintenance of these stocks as independent entities with clear histories must be assured. Targeted or complete panels depending on the trait then need to undergo phenotypic characterization to provide the necessary information for association studies. The phenotypic data need curation through the same strategies as 'omics' data and using the same ontologies.

Choice of entries for initial genotyping and sequencing is usually based on representative sampling using molecular or, at the least, morphological data and geographical partitioning for selection. On purification and multiplication, these materials are subjected to genotyping and phenotyping, of which 'omics' is one part, laying the foundation for a general diversity platform. Figure 10.1 presents a view of the process. This figure shows genetic diversity as the foundation with application of 'omics' bridging the genotype to phenotype divide. Identified alleles and germplasm affiliated with the desired traits can then be targeted for breeding and selection programmes. The intersection of 'omics' data provides one level of confidence in allele choice. Specific alleles can also be validated through gene-based pipelines with or without transgenic support. Kilian and Graner (2012) discuss the impact that next generation sequencing will have on genebanks.

10.4 The Application of Molecular Technologies

Biology has seen the rapid proliferation and rise of molecular technologies over the past quarter of a century, often driven by human genetics studies. The number of sequences in public databases continues to increase exponentially while concomitant datasets such as those from ENCODE are also growing. Recently, 31 main concurrent and 14 additional publications appeared from the ENCODE project (Nature ENCODE, 2013), leading to a much expanded interpretation of active sequences in the human genome. The publications have been marked

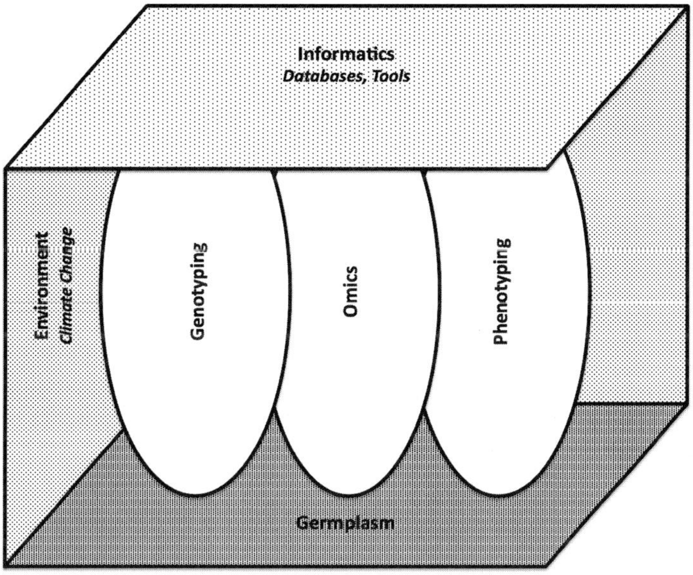

Fig. 10.1. The foundation of genetic resources characterized by genotyping, 'omics' and phenotyping technologies and explored through the application of informatics to address the challenges imposed by climate change.

and linked allowing view of content in a 'threaded' manner.

Even though this is now cliché, the rate of genomic sequence data production has significantly outpaced that of the increases in computer chip speed and storage density of hard drives (National Human Genome Research Institute, 2013). Hence, more and more processors and drives are needed to handle the data. This alone presents one of the biggest challenges that we face. It also begs for alternative strategies for data encoding, reduction, and optimization of algorithms for large and often sparse datasets.

The rapid rise of large datasets being the foundation of molecular science has pushed informatics to the forefront and has even been reported in the popular press (Pollack, 2011). Hence, no discussion of technology would be complete without addressing aspects of the new informatics. This topic is at the forefront of current thinking as discussed by a panel of experts (Mak, 2012). It is not, however, the purpose of that section to deal with the current state of genome assembly and annotation pipelines unless the new technology serves to improve their efficiency.

Although not a plant species used for food, thale cress (*Arabidopsis thaliana*) is a model organism that has greatly increased our understanding of and been used in studies at the leading edge of plant 'omics'. Hence, many of the relevant examples in the following sections made use of this organism.

10.5 Genomics

Broadly speaking genomics refers to DNA sequence(s) of the genome or genes with the concomitant annotation and organization of the information into a coherent body that enables use and exploration. Besides the experimental methods, there are the affiliated tools for analysis and interpretation. A tour of databases such as Gramene (2013) gives a flavour of the current state of genomic analysis and visualization. Nucleic Acids Research (2013) publishes a special issue in January of each year with database and tool updates.

Sequence data from individual genes to complete genomes have relied on various sequencing technologies whose history, development, and applications relevant to genetic resources are discussed below.

10.5.1 Sanger sequencing

Initially, the backbone of gene and genome sequencing was Sanger dideoxy sequencing. Sanger sequencing involves the incorporation of chain terminating nucleotides into a growing DNA strand that are typically labelled with fluorescent dyes (Sanger *et al.*, 1977). Separation of chain lengths is then performed by capillary electrophoresis on an automated system. Application en masse led to quite a number of high-quality references by physical-map assembly for rice, *Arabidopsis*, human and other genomes (*Arabidopsis* Genome Initiative, 2000; IHGSC, 2001, IRGSP, 2005). Whole genome shotgun approaches using libraries of Sanger reads have been used to produce draft genomes (Venter *et al.*, 2001; Yu *et al.*, 2002).

During the height of Sanger sequencing, over 900 gigabases of DNA were archived in public databases. To put this into perspective, in 1989 the size of Genbank was only 13 megabases. In conjunction, there was a rapid rise in tool development for improving assemblies, annotation and predictions. Furthermore, databases have seen much improvement in their structure, organization and presentation.

10.5.2 Next generation sequencing (NGS)

Several technologies based on sequencing by synthesis have supplanted Sanger sequencing as the method of choice. These are the so-called 'next generation sequencing' (NGS) approaches and have been recently reviewed by Pareek *et al.* (2011). The Illumina sequencing system (based on PCR) is the sequencing by synthesis (SBS) approach where DNA strands are elongated by fluorescently labelled nucleotides at physically isolated clusters in a flow-cell (Bentley *et al.*, 2008). This approach entails the need for long runs. Extensive image analysis is required prior to obtaining nucleotide calls and quality scores. Billions of short reads are obtained per run – alignment and assembly is a significant bioinformatics task with many challenges to improve efficiency and accuracy. Although the pyrosequencing based Roche 454 delivers longer reads (Margulies *et al.*, 2005), it is at a significantly higher cost and lower throughput. Short runs on the 454 have proven useful for linking and resolving scaffolds. Other platforms like Complete Genomics are proprietary (Drmanac *et al.*, 2010). Still others such as ABI-SOLID have a low market profile (Valouev *et al.*, 2008). Improvements in existing chemistries extending read lengths to 10 kb are in the pipeline for Illumina.

The advent of NGS has put the identification of new sequences into overdrive. Sequence data generation by NGS will reach a prospective output of more than 1 Pbase (10^{15} bases) per day within the year largely driven by increased lengths on established platforms. Now, the pace of new sequence data production is double that of microchip power and data storage density increases. This can be seen as the most significant obstacle for distillation of the sequences into significant chunks of useful data. The section after 'omics' technologies explores the informatics for dynamic curated analyses.

NGS continues to be the driver for change on the genomic landscape with tens of thousands of complex genomes (including genotypes of species) and millions of simpler genomes in the pipeline. In 2011, Feuillet *et al.* estimated that there are complete or ongoing projects for 29 separate crop species. For example, 473 *Arabidopsis* genomes are now available (Cao *et al.*, 2011; Schneeberger *et al.*, 2011; 1001 Genomes Project, 2012) and 50 moderate (Xu *et al.*, 2011) and 1400 low coverage rice genomes (Huang *et al.*, 2010, 2012). Many of these projects are being driven by the BGI (originally called the Beijing Genomics Institute; based in Shenzhen, China and elsewhere) or have BGI as a key scientific collaborator. The rice 3k genome sequencing project is one such project with over 17

TBases of reads now generated (SciDev Net, 2011).

Although moderate- to high-depth sequencing is needed for establishing *de novo* genome builds or identifying low frequency alleles, low coverage NGS followed by reference-guided single-nucleotide polymorphism (SNP) calling can be used to identify sufficient SNPs for diversity, mapping or association studies. Multiple samples can be pooled prior to sequencing, resulting in reduced assay costs. The depth of this coverage can be from a few tenths of a percent or higher. Because coverage is quite low, however, error rates including false discovery rates may be high. Genotyping by sequencing (GBS) is performed by using a complexity reduction step (typically some combination of restriction cleavage) followed by sequencing of tagged sites (Elshire *et al.*, 2011). Typically, millions of tags are generated per run and the cost per sample (with multiplexing) can be relatively low. For both techniques, however, few loci in common among all genotypes might be found, possibly limiting the utility of the datasets or requiring that they be grouped into common haplotypes prior to analysis. GBS is being used by 'The Seeds of Discovery project' at the International Maize and Wheat Improvement Center (CIMMYT) to survey tens of thousands of wheat and maize accessions (CIMMYT, 2013). Poland *et al.* (2102) applied GBS to create high-density linkage maps for wheat and barley that lack reference genomes and can be used to facilitate anchoring of physical maps during sequencing of their genomes. Morris *et al.* (2013) used GBS to identify about ~265,000 SNPs from 971 sorghum accessions that were applied to genome-wide association studies. Byrne *et al.* (2013) used GBS to create genome-wide allele fingerprints from populations of eight ryegrass cultivars.

10.5.3 Third generation sequencing technology

Although NGS produces short reads for many fragments, it still works on a macro scale necessitating the need for relatively large amounts of sample DNA. New methods that enable single or very few molecules of DNA to be sequenced comprise the 3rd generation of sequencing technologies. They have entailed refinement of methods for labelling, synthesis and capture and improved instrumentation with highly improved sensitivities of detection. Because read lengths may be in the order of the length of Sanger reads – 1 kb – to many kilobases, assembly of the reads into builds is far easier. These datasets can provide adjunct data for improving NGS assemblies.

The Pacific Biosciences system sequences DNA strands in nanometer-scale wells, using tethered DNA polymerase (Eid *et al.*, 2009). This single-molecule real-time or SMRT technology has several modes of operation. One mode involves adaptation of primers with single strand loops to create double strand templates that can be circularized to single strand templates. Sequencing proceeds in tandem for both strands multiple times improving the accuracy and depth of calls. Melters *et al.* (2013) utilized the SMRT technology to sequence centromeric regions from four plant species (*Zea mays*, *Zea luxurians*, *Panicum virgatum* and *Panicum capillare*). SMRT technology is also being used for *de novo* genome builds of plant genomes at Cold Spring Harbor for two types of rice – the *indica* variety IR 64 and *aus* variety DJ 123 (R.W. McCombie, New York, 2013, personal communication).

Another 3rd generation technology is the semiconductor-based system by Ion Torrent (Life Technologies) where the incorporation of different bases into nascent DNA strands is detected by discriminating their hydrogen ion effluxes (Rothberg *et al.*, 2011). Ion current traces are then analysed to assign nucleotide calls. Turnaround time is very short and throughput rates with the newly released 416 (1 GB) and upcoming 418 (10 GB) chips enable deep re-sequencing. Monsanto's genome analysis centre has used this platform for survey sequencing of the facultative crassulacean acid metabolism plant *Sedum album* (Genome Web, 2012). Lee *et al.* (2013) have used the Ion Torrent system to study the transcriptome of *Arabidopsis* mutants with defective chloroplast development.

10.6 Other 'omics'

The use of the term 'omics' has quickly accelerated into anything that generates its own large dataset(s) being termed an 'omic'. Even, topics outside of the typical realm of molecular biology are getting branded as 'omics'. Phenotyping and the emergence of 'phenomics' is one of these. In the following discussion, the technologies of epigenomics, transcriptomics, proteomics, metabolomics, ionomics, systems biology with one of its primary tools, interactomics and metagenomics are described. In fact, the plethora of studies called 'omics' has led Eisen (2012) to question ad hoc use of the term.

These 'omics' technologies all deal with aspects of the organism that might be influenced by the environment. Hence, proper experimental design and controls for treatment and sampling need to be followed in order to extract the most useful information. The environmental cues to which molecules respond add significant cost, time and complexity to the analyses. Sufficient replications and samples need to be run in order to have sufficient statistical power for detecting the signal from the noise. Nicotra *et al.* (2010) provide a framework for better understanding of the phenotypic plasticity of plants in response to climate change wherein 'omics' technologies will be crucial for an integrated, multidisciplinary approach that bridges ecology, evolution, physiology and molecular biology perspectives.

10.6.1 Epigenome

The investigation of the fraction of the genome that is effectively silenced requires novel sequencing approaches and merits its own discussion. Both DNA and histone methylation play a role in this and are subject to environmental control (Harris *et al.*, 2010) and hence could be important when considering the effects of climate change. For example, a recent review by Chinnusamy and Zhu (2009) summarized epigenomic responses across a range of species to stress including drought, salinity, submergence, cold and aluminium toxicity, as well as hormonal and other responses. Furthermore, the methylation status is heritable across generations. Thus, the conditions to which a grandparent was exposed in some cases may still affect the genetic regulation and gene expression in their grandchildren.

DNA methylation plays a key role in regulating transcription by limiting access to regions containing genes that are not methylated. Histone methylation and modification also play a role in determining what regions are available. Capture and cleavage for removing sequences bound to histone can be performed, and the exposed regions subjected to conventional or bisulfite sequencing. Bisulfite sequencing allows DNA methylation sites to be identified by the specific conversion of non-methylated cytosines to uracils. Cross-referencing those to the genome sequence allows methylation patterns to be discerned. Full exploration of the epigenome will require much effort and expense due to tissue, developmental and environmental conditioning. Hence, studies should be well designed for impact. Church (2013) has proposed a grand challenge to study the epigenome along with the complexity of environmental and other interactions.

Although relatively few studies of the effect of the epigenome among diverse accessions and their responses have been accomplished to date, this area of research deserves further attention and will be important in understanding environmental and regulatory aspects in plant genetic resources. Li *et al.* (2012) compared the methylation status of cultivars of *indica* and *japonica* types of rice with their wild relatives, *Oryza rufipogon* and *Oryza nivara*, and identified significant differences in degree of methylation across genic regions and transposable elements between the domesticates and the wild progenitors. Their sample size was small, however, and they conclude by stating that a clearer understanding of the impact of methylation on gene expression across the primary gene pool will only be evident on the analysis of a much larger sample. Schmitz *et al.* (2013) compared the epigenomes of 152 genotypically distinct, wild *Arabidopsis* accessions. In their study, clear differences among the methylation status of the 152 accessions were observed in gene families such as the defence response NBS-LRRs, f-box genes involved in protein degradation and MADS-box transcription factors. These gene families are involved in responses to biotic and abiotic stresses and reproductive development. The same genes were shown to have extensive differences at the gene level, indicating that they are hypermutable.

10.6.2 Transcriptome

The genetic complement of an organism that is being transcribed under various conditions is its transcriptome. It provides a snapshot into the dynamic interplay of environment and trait at the level of the DNA and RNA. Previously, transcriptomics was performed using spotted or printed arrays at considerable cost per sample with the limitation that only those genes with probes could be analysed and the abundance of a signal needed to be above a certain threshold (Clarke and Zhu, 2006). This threshold frequently meant that many potentially interesting genes could not be detected. Nevertheless this technology has identified thousands of responding genes with patterns of expression and has allowed key genes and mechanisms to be elucidated that may be relevant to climate change stresses (Brady *et al.*, 2006). Mining the data still can be challenging, especially if gene annotation is lacking and the expressed genes do not have sequence/function similarity to known genes. A recent example of the use of spotted microarrays for gene expression in diverse accessions of *A. thaliana* is the study by Rasmussen *et al.* (2013). They studied the effects on the transcriptome of heat, cold, salt, flagellin and high light, plus double combinations of these stresses and found that the majority of responses to single stresses were not predictive of the responses to double stresses.

Now, RNA-seq is starting to supplant use of arrays (Wang *et al.*, 2009; Wang *et al.*, 2010), where cDNAs are made from a tissue sample and are then subjected to deep NGS. Counts of read numbers per gene provide a mechanism to quantify expression as long as other controls are in place. For complex genomes, RNA-seq can provide a window into the gene space by greatly reducing the fraction needing to be sequenced compared to whole genome sequencing.

Some recent applications of RNA-seq to lesser studied crops and plant species include survey of:

- Transcripts from five cultivars of rye at five developmental stages from plants subjected to cold, dehydration shock or nutrient starvation (Haseneyer *et al.*, 2011).
- Two commercial hybrids of *Eucalyptus* under water stress (Villar *et al.*, 2011).
- *De novo* transcriptome profiling of lignin and cell wall formation genes in two species of *Acacia* (Wong *et al.*, 2011).
- Transcript profiling of leaf and root tissue from wild oat, *Avena barbata*, under low and high rainfall conditions (Swarbreck *et al.*, 2011).

Perhaps the most extensive gene expression profiling done in a crop plant is the work on transcriptomics in rice under uncontrolled environmental conditions in the field (Jaeger *et al.*, 2012). In this study, conducted by the team at the National Institute of Agrobiological Sciences in Japan, 1249 separate microarray experiments from field-grown temperate *japonica* cultivars with temporal sampling of various growth stages and tissues were performed (Nagano *et al.*, 2012). Analyses of the transcriptome along with meteorological data allowed them to develop statistical models that explain both endogenous and external factors influencing gene expression; these models also allow prediction of the effects of increased temperatures on gene expression.

10.6.3 Proteome

The protein and enzyme constituents can be explored in a high-throughput manner by applying proteomics. Proteins are isolated using various methodologies and subjected to separation into individual components (Thelen and Peck, 2007). Component proteins are treated as fractions for assignment to spectral signals such as those in an NMR or mass spectrometry database. Post-translational modifications and control of expression can be investigated. The classical approach was to perform 2D gel electrophoresis followed by isolation of spots that were subjected to mass spectrometry. Coupling of this to automatic spot capture and analysis has further enhanced the throughput. Much remains to be done, however, for improved sensitivity of signal detection and more detailed libraries of signals (Bindschedler and Cramer, 2011; Heazlewood, 2011; Wan *et al.*, 2012). This area has the potential to be the next frontier, particularly with respect to high-throughput structure determination (Hemmer, 2013). Two very recent publications demonstrate the feasibility of nanoscale magnetic resonance imaging (MRI) measurements on effectively single proteins *in situ* (Mamin *et al.*, 2013; Staudacher *et al.*, 2013).

A number of reviews on different aspects of plant proteomics have recently appeared. Malcevschi and Marmiroli (2012) discussed various proteomic technologies applied to plants and different, general studies. Hashiguchi *et al.* (2010) highlighted plant proteomics of stresses in the context of climate change. Alvarez and Hicks (2011) examined the status of root proteomics; Kosová *et al.* (2011) gave a view of proteomics of plant responses to abiotic stresses. Rampitsch and Bykova (2012) summarized proteomics in relation to plant disease. Alexandersson *et al.* (2013) reviewed plant secretory proteomics.

Here, some examples of recent proteomic studies involving stresses relevant to climate change are discussed. The response of grain proteins in wheat to drought by comparing a tolerant to a sensitive cultivar showed responses in storage, carbohydrate metabolism and detoxification proteins (Jiang *et al.*, 2012). A study of three wheat cultivars with different yields under drought identified protein changes associated with increased reactive oxygen species scavenging (Ford *et al.*, 2011). Galant *et al.* (2012) studied proteomic responses to ozone stress in soybean observing significant changes in a broad range of protein families. Riaño-Pachón *et al.* (2010) looked at the complement of *A. thaliana* proteins with phosphorylation sites and showed that presence/absence of these sites was highly variable across diverse accessions for signal transduction and receptor proteins important in stress and other responses.

10.6.4 Metabolome

The extent of primary, secondary and tertiary compounds existing in an organism comprises its metabolome. Typically, DNA, RNA, proteins and ions are excluded because they have their own specific domains of study. Compounds are isolated wholesale by water or organic phase solvents, separated by gas chromatography, liquid chromatography or high-performance liquid chromatography on various column types with or without derivatization, and then subjected to mass spectrometry after desorption or ionization. Hence, many combinations of isolation, separation method and type of detection contribute to the power. They also provide a drawback in that several extraction types and separation methods will need to be used to obtain a more complete picture of any given metabolome under a specific environment.

Regardless, typical studies identify hundreds of peaks for both known and unknown compounds. The areas under the peaks correlate to concentrations. The signatures of the spectra are compared with libraries of known compounds. The abundance of compounds under different conditions allows enzymatic reactions to be characterized. When quantification is combined with time course analyses, kinetics of the reaction systems can be explored. This methodology shows great promise in identifying the key steps to focus on general response mechanisms that are sometimes hidden in background expression.

Obata and Fernie (2012) and Arbona *et al.* (2013) provide examples of metabolomics for understanding abiotic stress responses, many of which are relevant to climate change. Carreno-Quintero *et al.* (2013) reviewed metabolomics in the framework of interactions with phenotypes for guiding application of information from model species to crops. A study of comparing salt tolerant and sensitive *Populus* species indicated that the tolerant species constitutively activated osmotic protection, reactive oxygen scavenging, and ion compartmentalization (Janz *et al.*, 2010). These results were confirmed through pathway analysis by concurrent transcriptome surveys. In the Mediterranean shrub *Erica multiflora* the stoichiometry between key elements (C, N, P and K) was highly correlated with changes in the metabolome under moderate drought and increased temperatures in field conditions (Rivas-Ubach *et al.*, 2012). Chan *et al.* (2010) have studied complex interactions of metabolite profiles with >200,000 SNPs through association analysis in *Arabidopsis*. Their results indicated that interactions between metabolites and genotypes were sensitive to the environment and that association studies will need to account for this.

10.6.5 Ionome

The elemental composition of an organism comprises the ionome. Salt *et al.* (2008) recently reviewed the status of ionome studies in plants. Ionomics provides a view on free ions and elemental cofactors necessary for metabolic function plus environmental pollutants such as arsenic. Besides quantification of the elemental composition, ascertaining the dynamics of reduction/oxidation states becomes possible.

In a study of the ionome of 12 accessions of *A. thaliana*, Buescher *et al.* (2010) demonstrated significant differences in elemental composition among the accessions. In their study, they also analysed three recombinant inbred line (RIL) populations and identified more than a hundred quantitative trait loci (QTLs) exhibiting strong environmental effects on element accumulation and correlations between elements. Sanchez *et al.* (2011) compared the ionomes, transcriptomes and metabolomes of six *Lotus* species under salinity stress and found significant differences in ion concentrations among genotypes and under stress conditions. These two examples indicate that analyses of the ionomes of diverse plant genetic resources under verifying environmental conditions will be a necessary area of research for understanding adaptation to climate change.

10.6.6 Systems biology and the interactome

Systems biology implies integration across diverse datasets and their interpretation at different stages of organismal development under defined conditions and treatments (Snoep and Westerhoff, 2005). This is largely an arena where informatics across multiple variables comes into play with mapping and graph theory applied extensively. Analysis and presentation of the resulting networks of physical and regulatory interactions between genes and proteins is referred to as the interactome. Machine learning and pattern recognition approaches are key to understanding the biology as a holistic 'system'. Adaptive methods that model multiple interactions in a non-linear fashion need to be explored (Soon *et al.*, 2013). Sufficiently evolved database and mining tools are needed to make headway into modelling the organism or a fraction of it (Church, 2013).

A number of experimental methods allow the binding interactions between proteins and between proteins and nucleotide sequences to be ascertained and their networks explored. Braun *et al.* (2013) have recently reviewed methods and results from various approaches. Prominently featured is the yeast two-hybrid assay. Chip-Chip is the pull down of one protein in solution by another that is tagged (Aparicio *et al.*, 2004). This allows complexes of proteins that bind to one another to be determined. Then, the interactions between the proteins can be modelled as a graph. Likewise, Chip-Seq is a method that enables the capture of specific protein sequences by a nucleotide probe (Jothi *et al.*, 2008). Binding of transcription and other factors to DNA sites can then be validated. Graphs of these interactions can also be presented.

The graphical presentations of these interactions are referred to as the interactome. Braun (2012) recently reviewed interactome mapping. Other approaches use gene annotation and pathway information to classify sequences into groups and model their interactions, as, for example, in the predicted rice stress interactome (Seo *et al.*, 2011) and protein interactome (Ho *et al.*, 2012), from microarray data (Nitsch *et al.*, 2011) and from the literature (Stark *et al.*, 2011). This non-hierarchical view encompasses classes and domains of functionality and forms. Interactome graphs are increasingly used to study the inter-relationships and connectedness of processes (Blinov and Moraru, 2012). These can be visualized as dynamic 3D representations where science approaches art and the subtlety of interactions can be explored. A caveat is, however, that computing interaction models assume complete data that is truly never met (Blivnov and Moraru, 2012).

Although there has been much hype about systems biology, much work to integrate data is needed (Colmsee *et al.*, 2012). Perhaps, the most promising success story might be that of engineering C_4 and Krantz anatomy into a C_3 plant – rice (von Caemmerer *et al.*, 2012). Using integrative comparative genomics, candidate genes are being identified with involvement in anatomical development and control. Because the biochemistry is well understood and an engineered approach for multigene transformation and targeted expression is underway, a C_4 or C_4-like rice with much higher yields could be on the horizon.

10.6.7 Metagenome

It is now very clear that an organism consists of a composite between the recognized entity and a plethora of symbionts and commensals from known but mostly unknown bacteria, fungi and protists. This asssembly of organisms residing together for mostly mutual benefit may best be described as a 'collective'. Sequencing the genome of an organism, hence, leads to a sequence of the collective genome or metagenome (Knight *et al.*, 2012). Placing these sequences into context when no reference exists could mean that they are either discarded or are placed incorrectly into an assembly (Thomas *et al.*, 2012). More development needs to happen before these organisms can be reliably identified and, at the most basic, classified. Assembly tools for metagenomes use assembly into clusters then segregation to types based on genetic distances (Treangen *et al.*, 2013). These assembly tools may also prove useful as a means to create 'type' specific references that are consensus builds across multiple genomes that share a population type.

A number of studies have been recently published that investigate the metagenomes of microbial and viral populations associated with plants. Lundberg *et al.* (2012) sampled 600 *A. thaliana* plants of different genotypes and at different developmental stages for their rhizospheric and endophytic microbial populations. Significant quantitative variations for some types of bacteria occurred among different accessions and development stages. Community types varied according to bulk soil type, and endophytic bacteria populations were of lower complexity. Endophytes from rice roots grown under field conditions were subjected to metagenomic analysis by Sessitch *et al.* (2012). Genes identified in their study included ones involved in degradation of plant cell walls, protein secretion, quorum sensing and reactive oxygen detoxification. They also detected sequences involved in N_2 fixation, nitrification and denitrification. These results indicate that the endophytes may be involved in promoting plant growth and adaptation to stress. Dunbar *et al.* (2012)

monitored changes in microbial populations under elevated CO_2 levels in six different ecosystems over a span of 10 years. Common effects on community composition and bacterial richness across ecosystems were not observed but dramatic shifts in bacterial populations occurred within each ecosystem. This study indicates that soil populations affiliated with plants may undergo major shifts as atmospheric CO_2 levels continue to rise. Metagenomics has also been applied extensively to studies of plant viruses, as recently reviewed by Roossinck (2012).

10.7 Informatics

An overview of some of the databases, tools and approaches for the curation, visualization and analysis of plant 'omics' data has been given by Mochida and Shinozaki (2011). Although many of the tools they review are organism centric, their application to diverse genotypes is feasible. A thorough treatment of the plethora of current databases and bioinformatics methods is beyond the scope of this chapter. Instead, I will describe some of the recent advancements that are impacting how the deluge of new 'omics' data is being handled. This is a rapidly moving discipline and we can expect further developments to alter the informatics landscape.

The current era is defined by 'BigData' (Weiss and Indurkha, 1998), data on the scale encompassed by the whole of the internet and World Wide Web on the one extreme and the smaller if not immense datasets generated by modern high-throughput molecular sciences (Howe *et al.*, 2008). Data and analyses generated by a single experiment can now reach the petabyte scale necessitating new approaches for archive, query and analysis. Fortunately, the whole of informatics is grappling with these issues with many developments and much progress. Derrick Harris has referred to 'data as the fourth paradigm of science because the sheer amount of data available and the new technologies and techniques for working with it are fundamentally changing how scientists go about their research' (Harris, 2012). Baker (2012) has touched on some of the approaches that are now being applied to biological BigData to render its curation adaptive to changes over time.

Cloud computing is one proposed mechanism for addressing the challenges of data and sharing (Stein, 2010). Projects such as 'Fourth generation sequencing data – Infrastructure and architectures for the analysis of third and fourth generation sequencing data' (KU Leuven, 2011) propose to address some of the issues, such as data transfer, storage and distributed analyses across clusters.

At the most basic, ways to improve the storage and indexing of data need to be addressed. For example, Cochrane *et al.* (2012) have proposed standards for lossy compression (a data encoding method whereby some content is lost) of sequence data that retain a level of information content in the sequence files graded by the quality and ease of repeating the experiment. Other approaches to improving storage are the use of HDF5 or hierarchical data format v5 (HDF Group, 2011a) and FastBit, an efficient compressed bitmap index technology (Rübel *et al.*, 2010).

BioHDF5 is an algorithmic approach to store biological data at a low level with metadata indices allowing entry into the address space (HDF Group, 2011b). This data format is very suitable for discrete data. Hoffman *et al.* (2010) have adapted HDF5 for storing large-scale functional genomics datasets where the functional data are numeric tracks associated with a genome. Millard *et al.* (2011) developed an approach using BioHDF5 and extensible markup language (XML) for adaptive curation of high-throughput 'omics' data. Sun and colleagues at Cornell are applying BioHDF5 for storing large-scale SNP datasets from diverse germplasm (Q. Sun, New York, 2013, personal communication). For quantitative data, FastBit indexing (Berkeley Lab, 2013) may be better. FastBit is being applied for improving the speed and throughput of searching mass-spectrometry databases for proteomics (Yu *et al.*, 2011). Erik Lyons at Arizona State University is implementing a FastBit solution for storing data for comparative genomics along with visualization tools as part of the CoGe toolkit (CoGE, 2013; E. Lyons, Arizona, 2013, personal communication).

Interface tools to query data stores using either BioHDF5 or FastBit comprise types of NoSQL type database – meaning a not only SQL interface to the data (NoSQL, 2013). Other examples of NoSQL databases that are being applied to bioinformatics are Hadoop (2012; Taylor, 2010); implementations of mongoDB (2013) such as that by Borozan *et al.* (2012) and CouchDB (2012; Manyam *et al.*, 2012). Although certain aspects of this are rigidly controlled, their simple structures allow very large datasets to be integrated and efficiently queried. The message here is that these database systems are likely to be very important in the future for curating and manipulating the genomics and other 'omics' datasets derived from plant genetic resources.

10.7.1 Controlled vocabularies and ontologies

Controlled vocabularies are a means to strengthen the clarity of definitions and avoid redundancy of terms that are used to describe genes, morphology, anatomy, traits, chemicals, environments, etc. Ontologies impose structures of relationships to terms defined using the controlled vocabularies. The structure is a directed acyclic graph because one-to-many but not many-to-one relationships are allowed (i.e. no cyclic behaviour in the graph). Ontologies for genetic and other information are really important because they serve as the source of terms to annotate database entities, giving them the dimension to be queried from external sources.

The most widely used ontology is the gene ontology (Ashburner *et al.*, 2000). This ontology alone has been cited 5 million times in the literature (Google Scholar search conducted in January 2013). Of particular relevance to crops and other plant genetic resources are the plant, trait and environment ontologies and various crop specific user ontologies (Walls *et al.*, 2012). These are being re-curated to become reference ontologies that are built by cross-referencing non-redundant existing terms that define measurements, structures, genes, compounds, etc. (Arnaud *et al.*, 2012). These terms are centrally archived and retrievable using web services and semantic web technologies (Berners-Lee *et al.*, 2001; Stephens *et al.*, 2006).

10.7.2 Databases and tools

Web portals serve as the entry point to systems of independently curated databases that can be integrated through the use of Semantic Web (a collaborative movement promoting common data formats), Web 2.0 and ontologies. For the long-term curation of *Arabidopsis* information beyond the *Arabidopsis* Information Resource or TAIR (2013), the International *Arabidopsis* Informatics Consortium (2010, 2012, 2013) is undertaking the development of an *Arabidopsis* Information Portal. Integration across databases results in a federated, distributed system of cross-queryable data sources.

Adherence to the new web standards of data curation imparts the ability to compute using the data in ways not previously possible. The curated data along with integrated tools, in its entirety might reach a critical boundary state – a state of convergence, a singularity – where data mining tools such as those used to find patterns in communication may identify their own new testable hypotheses. When such a system is coupled with tools to visualize, export and analyse subsets, the user may achieve a 'eureka' moment imparted by viewing data in a novel way. Eureqa (2013) is a tool that uses algorithms that refine themselves over time and allows a scientist to submit a dataset that is analysed to derive equations that best fit the problem (Schmidt and Lipson, 2009). Schmidt *et al.* (2011) have applied the Eureqa system to a priori discover equations that fit those in a known metabolic network. One can imagine its extension to pattern recognition and relationships in other 'omics' and systems biology data. Other approaches are described in section 10.9 on novel tools.

Apart from the demands of users, the long-term management and sustainability of a database needs to be ensured. This is increasingly a requirement of funding organisations to ensure that years of investment are not squandered. Public–private partnerships for database licensing and time-bound access are becoming more and more prevalent. These models have various modes of public and private access to data sources. For example, Kyoto Encyclopedia of Genes and Genomes (KEGG, 2012) uses a paid subscription model for its human curated pathway information, whereas TransFac is free online for not-for-profit organizations and by subscription for off-line and other users (Gene-regulation.com, 2011). Updating curation and tools, and ensuring long-term sustainability for end-users is a current topic under much discussion in the crop communities, as evident from sessions and working group meetings held at the Plant & Animal Genome XXI conference in January 2013 (PAG XXI, 2013).

In summary, a database system can be created that enables thousands of genome sequences and associated information for an organism to be accessed. Merging genomes in the same population group to a composite reference genome such that every allele is represented by either an insertion or SNP (even if rare) will reduce the complexity of mapping because the composite genome has a single coordinate space. Other 'omics' data and annotations could then be curated onto the composite genome as numerical variants hierarchically arranged by class. Such a strategy might rely on using binary or indexed trees at each annotation position to create a new data structure of the genome and its 'omics' data. Embedding analytical tools into the system might create a hybrid 'kernel' optimized for modelling the organism's biology. This would be a departure from the computer as a universal machine to the machine becoming an *in silico* avatar of the organism being modelled.

Crucial to any database system integrating 'omics' information for plant genetic resources will be tracking at the level of sample so that identity of a sequenced or otherwise studied individual can be linked to the sample from which it was derived (McCouch *et al.*, 2012). Without this, there is a serious risk that invalid conclusions would be reached. For example, 'omics' data from a purified genetic stock should not be linked to the possibly heterogeneous mixture of the ancestral accession from which it was derived.

10.8 Bridging the Genotype to Phenotype Divide

Genome-wide association studies or GWAS and certain studies on artificial populations serve to connect phenotyping information to genomics data (Klein *et al.*, 2005). Highly significant association tests are the key for mining the data and building testable hypotheses (Hall *et al.*, 2010; Rafalski, 2010). When genome-wide analyses are placed in an environmental context, they can elucidate the topography of the genotype to phenotype landscape (Brachi *et al.*, 2011). Ward and Kellis (2012) examine the current status of GWAS tools and cross-interpretation to other 'omics' data. Statistical tests for association identify strongly correlated trait–loci pairs using a mixed linear model with kinship as a covariate (Zhang *et al.*, 2010). On physical mapping clusters significant SNPs appear as sharply defined peaks above the background threshold. Underlying gene models are then candidates for validation. Genes underlying peaks that have support from other sources of information such as QTL co-localization and functional annotation may be deemed testable without validation. Huang *et al.* (2010) examined agro-morphological traits in rice, while Pasam *et al.* (2012) have recently published GWAS for agronomic traits in barley and Morris *et al.* (2013) studied agroclimactic traits in sorghum.

Candidate genes under testing can undergo gene knockout and gene activation tagging for validating function assignment. This requires an efficient targeted transformation, e.g. TALE (Transcription Activator-Like type III Effector) or CRISPR

(Clustered Regularly Interspaced Short Palindromic Repeats) technology (Church, 2013; Mali *et al.*, 2013) or access to a library of genome-wide mutants that have been sequenced. A drawback is the scarcity of easily identifiable phenotypes. Yet, although a discernable phenotype may be absent, the effect on expression can still be ascertained and used as a trait.

The large phenotyping datasets such as those combined across studies, with missing data, or unbalanced designs may require Bayesian refinement to remove confounding effects prior to testing. This process reduces variance in the dataset by iterative cycles of Markov chain Monte-Carlo beginning with a t-test (A. Greenberg and S.R. McCouch, New York, 2013, personal communication). The genotyping datasets can also be improved by applying k-means nearest neighbour imputation or, preferably, by linkage disequilibrium (LD)-based imputation to reduce the number of states that are missing (Pei *et al.*, 2008). Besides being used as threshold values, quality scores and their levels can be used as covariates to refine imputation of calls on locally structured data, i.e. haplotype blocks whether or not they are introgressed from one genotypic background to another. Application of these methods improves the power to detect significant associations between phenotypes and SNPs.

10.9 Exploratory Tools

Tools that allow the diversity space of sequence and annotated data to be explored are critical for effective use of 'omics' resources. The depth and breadth of the data allow analyses to be undertaken that, while they may require significant computation, will improve our knowledge immensely. Some tools include the modelling of allele flow in space and time, the coalescent simulation of likely ancestral states, the creation of allele interaction models, targeted selection of donors based on haplotype association to beneficial alleles, focused identification of germplasm stocks, modelling gene to landscape environmental interactions, and ascertaining recombination breakpoints through genetic algorithms and likelihoods. A brief exploration of these prospective tools follows.

10.9.1 Allele flow

Human migration and other means of dispersal led to the historic spread of crops over a wide geographic range. During these events, favoured alleles were naturally introgressed across types and subjected to purifying selection. Signatures of this effect can be calculated from the allelic diversity and richness along the genome. Regions appearing identical by descent can be pinpointed. Following trends of the values allows historical allele flow and drift to be estimated. These can be correlated to a molecular clock and geographic distance. Until recently, very large datasets were difficult to handle as a result of the computational complexity of modelling using coalescent simulation, but now access to high-performance cluster computing with hundreds or thousands of CPUs (scalar processor units) and/or GPUs (graphical processing units using vector computations) have made such calculations more feasible (Liang *et al.*, 2007).

10.9.2 Coalescent modelling

Bayesian modelling allows the testing of a genetic model that is developed by improvement from a random starting state. The coalescent estimate for descent can be determined by Bayesian analysis of the likelihood of a neighbourhood of states (most commonly by descent) leading to the observed state (Donnelly and Tavaré, 1995). These estimates also allow constraints such as testing of isolation with founder effects from admixture without isolation models. The main constraint is that the problem is polynomially complex and many years of processor time can be consumed for a single analysis to determine the best model. For example, a study of isolation-with-migration in rice populations from Madagascar required 168,000 core hours of computation (Mather *et al.*, 2010). Hence, high degrees of parallelization are needed and algorithms need to be optimized for effective use of system resources. The area of coalescent modelling is very active with novel approaches under development (Zeng, 2012). Application of coalescent simulation to plant genetic resources will, for example, allow better understanding of the selection imposed by man during domestication of crops and creation of the diversity of landraces over the centuries.

10.9.3 Allele interactions

The status from moment to moment in an organism is governed by allele (or allele product) interactions. Embedding interactome and pathway information into the database of a crop will enable specialized queries that trace interactions across networks and identify key regulatory points with higher degrees of connectivity to other nodes (genes, proteins or regulatory elements). The entities at these 'hubs' could then be targets for use in crop improvement because they represent likely points of failure across the network (Braun, 2012). For example, Seo *et al.* (2011) verified that 'hub' proteins function in rice stress tolerance and are candidates for use and selection in crop improvement.

10.9.4 Targeted selection

Breeders and geneticists need an interface to the crop genome and 'omics' databases that enables them to effectively choose convenient and, hopefully, high value genotypes from a collection. This targeted selection scheme needs to factor in the most reliable convergent data from the database based on user parameters, such as duration, trait and quality under stress, because they are associated with specific sets of genotypes (Kilian and Graner, 2012). The tools should also provide a snapshot of the supporting information for user assessment. These tools are under development as part of a 'breeders toolkit' suite of database applications such as the Integrated Breeding Platform of the CGIAR Generation Challenge Programme (2013), the *Hordeum* toolkit (USDA-NIFA, 2009), and the BreeDB tools being developed at Wageningen UR (2013) such as Marker2sequence (Chibon *et al.*, 2012). Although the underlying data are complex, the tools should, however, be as simple as possible to use and have appealing user interfaces.

10.9.5 Focused identification

Focused identification of germplasm stocks or FIGS is an approach developed to sample germplasm over ecogeographic or climatic clines factoring in genetic information and possible adaptation to these micro-environments (Endresen, 2011). This approach requires detailed collection information that is frequently lacking especially for older accessions where collection history and geographic detail were ignored. Hence, use of FIGS could be constrained to samples from limited regions (these are usually recent when collection standards significantly improved and a geographic information system (GIS) was available). However, if sufficient samples exist from other locations with similar clines and modes of adaptation, then inference might be possible. It may also be feasible to combine FIGS with simulations of allele dispersal and drift to identify new sources of genes responsive to climate. The FIGS approach has been successfully applied to identify new sources of resistance to wheat kernel smut races (Bhullar *et al.*, 2009; see Pritchard *et al.*, Chapter 16, this volume).

10.9.6 Genotype x environment interactions

Bridging the genotype × phenotype divide requires a thorough understanding of the genotype by environment interaction space. This is potentially one of the greatest challenges at our feet. The landscape embodied by the interactions can be explored by taking 'walks' across multidimensional surfaces in the phase space (Martens *et al.*, 2009). Models of the interactions that factor in physical processes as well as genetic mechanisms are considered as the best choices for fitting experimental and environmental data into the multi-dimensional parameter space.

10.10 Recombination Breakpoints

Elucidating the composite nature of genome patterns exhibited across diverse germplasm can be achieved by defining the fine-scale structure of recombination breakpoints between haplotype blocks. Several approaches can be applied for this. Standard LD approaches can be used but do not define clear margins. Phylogenetic trees focusing on specific segments of the genome can be constructed followed by the assignment of the genetic samples to particular groups (Strimmer *et al.*, 2001). Alternatively, breakpoints can be modelled by Bayesian sampling or by fitting to samples selected by a genetic algorithm for pattern recognition (Posada and Crandall, 2001). The mosaics uncovered by breakpoint analysis can be used to control local kinship across the genome while weighting for introgression effect relative to background.

10.11 Conclusions

Although a veritable deluge of data from molecular technologies has washed over us, we have tools at our disposal to cope. Several times I have alluded to the convergence of computation and learning. I hope this process will move us beyond sheer accumulation of data to an emergent system wherein the whole plant is modelled and the effects of new allelic combinations can be estimated prior to breeding.

Creating new crop databases and integrative portals are needed to impose order on pools of information. Modelling data from diverse germplasm leads to a digital proxy for the plant until the plant is used in highly optimized breeding schemes. Hence, the seed genebank goes *in silico* and meets the sequence genebank head on. I have overviewed molecular experimental procedures and explored some of the challenges faced by informatics. Yet, the future developments promise to address the needs and deliver the best possible solutions to end users.

Although this has not been an exhaustive review, perspectives have been given on creating flexible database structures that will allow all curated traits to be explored. Prioritization of choice then comes from the user and the genebank database manager does not control extent of choices available beyond those imposed by the datasets. Hence, all traits where prior information has been collected and applied for analysis can be used to identify new sources of tolerance to those traits. On a final note, an International Climate-Resilient Crop Genomics Consortium (ICRCGC, 2012) was formed at the Plant and Animal Genome XX conference in January 2012 that intends to apply genomics and 'omics' tools to crop species and their wild relatives so that the challenges imposed by climate change will not constrain our ability to produce sufficient food for humanity.

References

1001 Genomes Project (2012) A catalog of *Arabidopsis thaliana* genetic variation. Available at: www.1001genomes.org/ (accessed 15 March 2013).

Alexandersson, E., Ali, A., Svante, R. and Andreasson, E. (2013) Plant secretome proteomics. *Frontiers in Plant Science* 4, 9.

Alvarez, S. and Hicks, L.M. (2011) Survey in plant root proteomics: to know the unknown. In: Benkeblia, N. (ed.) *Sustainable Agriculture and New Biotechnologies*. CRC Press, Boca Raton, Florida, pp. 215–256.

Aparicio, O., Geisberg, J.V. and Struhl, K. (2004) Chromatin immunoprecipitation for determining the association of proteins with specific genomic sequences in vivo. *Current Protocols in Cell Biology* Chapter 17:Unit 17.7 doi:10.1002/0471143030.cb1707s23.

Arabidopsis Genome Initiative (2000) Analysis of the genome sequence of the flowering plant *Arabidopsis thaliana*. *Nature* 408, 796–815.

Arbona, V., Manzi, M., de Ollas, C. and Gómez-Cadenas, A. (2013) Metabolomics as a tool to investigate abiotic stress tolerance in plants. *International Journal of Molecular Sciences* 14, 4885–4911.

Arnaud, E., Cooper, L., Shrestha, R., Menda, N., Nelson, R.T. *et al.* (2012) Towards a reference plant trait ontology for modeling knowledge of plant traits and phenotypes. In: *Proceedings of the 4th Conference on Knowledge Engineering and Ontology Development*, 4–7 October 2012, Spain.

Ashburner, M., Ball, C.A., Blake, J.A., Botstein, D., Butler, H. *et al.* (2000) Gene ontology: tool for the unification of biology. *Nature Genetics* 25, 25–29.

Baker, M. (2012) Quantitative data: learning to share. *Nature Methods* 9, 39–41.

Bentley, D.R., Balasubramanian, S., Swerdlow, H.P., Smith, G.P., Milton, J. *et al.* (2008) Accurate whole human genome sequencing using reversible terminator chemistry. *Nature* 456, 53–59.

Berkeley Lab – Scientific Data Management Research Group (2013) FastBit: an efficient compressed bitmap index technology. Available at: https://sdm.lbl.gov/fastbit/ (accessed 23 March 2013).

Berners-Lee, T., Hendler, B. and Lassila, O. (2001) The semantic web. *Scientific American Magazine.* http://www.scientificamerican.com/article.cfm?id=the-semantic-web (retrieved 26 March 2008).

Bhullar, N.K., Street, K., Mackay, M., Yahiaoui, N. and Keller, B. (2009) Unlocking wheat genetic resources for the molecular identification of previously undescribed functional alleles at the Pm3 resistance locus. *Proceedings of the National Academy of Sciences USA* 106, 9519–9524.

Bindschedler, L.V. and Cramer, R. (2011) Quantitative plant proteomics. *Proteomics* 11, 756–775.

Blinov, M.L. and Moraru, I.I. (2012) Logic modeling and the ridiculome under the rug. *BMC Biology* 10, 92.

Borozan, I., Wilson, S., Blanchette, P., Laflamme, P., Watt, S.N. *et al.* (2012) CaPSID: a bioinformatics platform for computational pathogen sequence identification in human genomes and transcriptomes. *BMC Bioinformatics* 13, 206.

Brachi, B., Morris, G.P. and Borevitz, J.O. (2011) Genome-wide association studies in plants: the missing heritability is in the field. *Genome Biology* 12, 232.

Brady, S.M., Long, T.A. and Benfey, P.N. (2006) Unraveling the dynamic transcriptome. *Plant Cell* 18, 2101–2111.

Braun, P. (2012) Interactome mapping for analysis of complex phenotypes: insights from benchmarking binary interaction assays. *Proteomics* 12, 1499–1518.

Braun, P., Aubourg, S., Van Leene, J., De Jaeger, G. and Lurin, C. (2013) Plant protein interactomes. *Annual Reviews in Plant Biology* 64, 6.1–6.27.

Buescher, E., Achberger, T., Amusan, I., Giannini, A., Ochsenfeld, C. *et al.* (2010) Natural genetic variation in selected populations of *Arabidopsis thaliana* is associated with ionomic differences. *PLoS ONE* 5, e11081.

Byrne, S., Czaban, A., Studer, B., Panitz, F., Bendixen, C. *et al.* (2013) Genome wide allele frequency fingerprints (GWAFFs) of populations via genotyping by sequencing. *PLoS ONE* 8, e57438.

von Caemmerer, S., Quick, W.P. and Furbank, R.T. (2012) The development of C4 rice: Current progress and future challenges. *Science* 336, 1671–1672.

Cao, J., Schneeberger, K., Ossowski, S., Günther, T., Bender, S. *et al.* (2011) Whole-genome sequencing of multiple *Arabidopsis thaliana* populations. *Nature Genetics* 43, 956–963.

Carreno-Quintero, N., Bouwmeester, H.J. and Keurentjes, J.J.B. (2013) Genetic analysis of metabolome–phenotype interactions: from model to crop species. *Trends in Genetics* 29, 41–50.

Chan, E.K.F., Rowe, H.C., Hansen, B.G. and Kliebenstein, D.J. (2010) The complex genetic architecture of the metabolome. *PLoS Genetics* 6, e1001198.

Chibon, P.-Y., Schoof, H., Visser, R.G.F. and Finkers, R. (2012) Marker2sequence, mine your QTL regions for candidate genes. *Bioinformatics* 28, 1921–1922.

Chinnusamy, V. and Zhu, J.K. (2009) Epigenetic regulation of stress responses in plants. *Current Opinion in Plant Biology* 12, 133–139.

Chuck, G.S., Tobias, C., Sun, L., Kraemer, F., Li, C. *et al.* (2011) Overexpression of the maize Corngrass1 microRNA prevents flowering, improves digestibility, and increases starch content of switchgrass. *Proceedings of the National Academy of Sciences USA* 108, 17550–17555.

Church, G.M. (2013) Reading and writing omes. *Molecular Systems Biology* 9, 642.

CIMMYT (2013) *Seeds of Discovery.* Available at: www.cimmyt.org/en/where-we-work/global-activities/seeds-of-discovery (accessed 23 March 2013).

Clarke, J.D. and Zhu, T. (2006) Microarray analysis of the transcriptome as a stepping stone. *Plant Journal* 45, 630–650.

Cochrane, G., Cook, C.E. and Birney, E. (2012) The future of DNA sequence archiving. *GigaScience Journal* 1, 2.

CoGe (2013) Accelerating comparative genomics. Available at: http://genomevolution.org/CoGe/ (accessed 23 March 2013).

Colmsee, C., Mascher, M., Czauderna, T., Hartmann, A., Schlüter, U. *et al.* (2012) OPTIMAS-DW: A comprehensive transcriptomics, metabolomics, ionomics, proteomics and phenomics data resource for maize. *BMC Plant Biology* 12, 245.

CouchDB (2012) Available at: http://couchdb.apache.org/ (accessed 23 March 2013).

Donnelly, P. and Tavaré, S. (1995) Coalescents and genealogical structure under neutrality. *Annual Review of Genetics* 29, 401–421.

Drmanac, R., Sparks, A.B., Callow, M.J., Halpern, A.L., Burns, N.L. *et al.* (2010) Human genome sequencing using unchained base reads in self-assembling DNA nanoarrays. *Science* 327, 78–81.

Dunbar, J., Eichorst, S.A., Gallegos-Graves, L.V., Silva, S., Xie, G. *et al.* (2012) Common bacterial responses in six ecosystems exposed to 10 years of elevated atmospheric carbon dioxide. *Environmental Microbiology* 14, 1145–1158.

Eid, J., Fehr, A., Gray, J., Luong, K., Lyle, J. *et al.* (2009) Real-time DNA sequencing from single polymerase molecules. *Science* 323, 133–138.

Eisen, J.A. (2012) Badomics words and the power and peril of the ome-meme. *GigaScience Journal* 1, 6.

Elshire, R.J., Glaubitz, J.C., Sun, Q., Poland, J.A., Kawamoto, K. *et al.* (2011) A robust, simple genotyping-by-sequencing (GBS) approach for high diversity species. *PLoS One* 6, e19379.

Endresen, D.T.F. (2011) Utilization of plant genetic resources: a lifeboat to the gene pool. PhD thesis. Copenhagen University, Copenhagen.

Eureqa (2013) Available at: http://creativemachines.cornell.edu/eureqa (accessed 23 March 2013).

Evans, N., Baierl, A., Semenov, M.A., Gladders, P. and Fitt, B.D.L. (2008) Range and severity of a plant disease increased by global warming. *Journal of the Royal Society Interface* 5, 525–531.

FAO (2009) *Report on the High Level Expert Forum – How to Feed the World in 2050.* 23–24 October 2009, Rome, Italy.

Feuillet, C., Leach, J.E., Rogers, J., Schnable, P.S. and Eversole, K. (2011) Crop genome sequencing: lessons and rationales. *Trends in Plant Science* 16, 77–88.

Ford, K.L., Cassin, A. and Bacic, A. (2011) Quantitative proteomic analysis of wheat cultivars with differing drought stress tolerance. *Frontiers in Plant Science* 2, 44.

Galant, A., Koester, R.P., Ainsworth, E.A., Hicks, L.M., and Jez, J.M. (2012) From climate change to molecular response: redox proteomics of ozone-induced responses in soybean. *New Phytolologist* 194, 220–229.

Gene-regulation.com (2011) Public Databases for Academic and Non-profit Organizations. Available at: www.gene-regulation.com/pub/databases.html (accessed 23 March 2013).

Generation Challenge Programme (2013) Integrated Breeding Platform. Available at: www.generationcp.org/platforms/integrated-breeding-platform (accessed 23 March 2013).

Genome Web (2012) Succulent plant sequenced with ion torrent PGM, researcher reports at PAG. Available at: www.genomeweb.com/sequencing/succulent-plant-sequenced-ion-torrent-pgm-researcher-reports-pag (accessed 20 March 2013).

Glaszmann, J.-C., Mew, T., Hibino, H., Kim, C.K., Mew, T.I. *et al.* (1995) Molecular variation as a diverse source of disease resistance in cultivated rice. In: Khush, G., Hettel, G. and Rola, T. (eds) *Rice Genetics III.* IRRI, Los Baños, Philippines, pp. 460–466.

Gramene (2013) Available at: www.gramene.org/ (accessed 15 March 2013).

Hadoop (2012) Available at: http://hadoop.apache.org/ (accessed 23 March 2013).

Hall, D., Tegstrom, C. and Ingvarsson, P.K. (2010) Using association mapping to dissect the genetic basis of complex traits in plants. *Briefings in Functional Genomics* 9, 157–165.

Harris, D. (2012) Why cloud could make crowdsourcing the norm for scientists. Available at: http://gigaom.com/2012/01/03/why-cloud-could-make-crowdsourcing-the-norm-for-scientists/ (accessed 20 March 2013).

Harris, R.A., Wang, T., Coarfa, C., Nagarajan, R.P., Hong, C. *et al.* (2010) Comparison of sequencing-based methods to profile DNA methylation and identification of monoallelic epigenetic modifications. *Nature Biotechnology* 28, 1097–1105.

Haseneyer, G., Schmutzer, T., Seidel, M., Zhou, R., Mascher, M. *et al.* (2011) From RNA-seq to large-scale genotyping – genomics resources for rye (*Secale cereale* L.). *BMC Plant Biology* 11, 131.

Hashiguchi, A., Ahsan, N. and Komatsu, S. (2010) Proteomics application of crops in the context of climatic changes. *Food Research International* 43, 1803–1813.

HDF Group (2011a) What is HDF5? Available at: www.hdfgroup.org/HDF5/whatishdf5.html (accessed 23 March 2013).

HDF Group (2011b) About BioHDF. Avail-able at: www.hdfgroup.org/projects/biohdf/ (accessed 23 March 2013).

He, Z., Zhai, W., Wen, H., Tang, T., Wang, Y. *et al.* (2011) Two evolutionary histories in the genome of rice: the roles of domestication genes. *PLoS Genetics* 7, e1002100.

Heazlewood, J.L. (2011) The Green proteome: challenges in plant proteomics. *Frontiers in Plant Sciences* 2, 1–3.

Hemmer, P. (2013) Toward molecular-scale MRI. *Science* 339, 529–530.

Ho, C.-L., Wu, Y., Shen, H.-B., Provart, N.J. and Geisler, M. (2012) A predicted protein interactome for rice. *Rice* 5, 15.

Hoffman, M.M., Buske, O.J. and Noble, W.S. (2010) The Genomedata format for storing large-scale functional genomics data. *Bioinformatics*, 26, 1458–1459.

Howe, D., Costanzo, M., Fey, P., Gojobori, T., Hannick, L. *et al.* (2008) Big data: the future of biocuration. *Nature* 455, 47–50.

Huang, X., Wei, X., Sang, T., Zhao, Q., Feng, Q. *et al.* (2010) Genome-wide association studies of 14 agronomic traits in rice landraces. *Nature Genetics* 42, 961–967.

Huang, X., Kurata, N., Wei, X., Wang, Z., Wang, A. *et al.* (2012) A map of rice genome variation reveals the origin of cultivated rice. *Nature* 490, 497–501.

ICRCGC (International Climate-Resilient Crop Genomics Consortium) (2012) Climate change genomics. Available at: www.climatechange genomics.org/ (accessed 23 March 2013).

IHGSC (International Human Genome Sequencing Consortium) (2001) Initial sequencing and analysis of the human genome. *Nature* 409, 860–921.

International *Arabidopsis* Informatics Consortium (2010) An International bioinformatics infrastructure to underpin the *Arabidopsis* community. *Plant Cell* 22, 2530–2536.

International *Arabidopsis* Informatics Consortium (2012) Taking the next step: building an *Arabidopsis* information portal. *Plant Cell* 24, 2248–2256.

International *Arabidopsis* Informatics Consortium (2013) Available at: www.arabidopsisinformatics.org/ (accessed 23 March 2013).

IRGSPC (International Rice Genome Sequencing Project) (2005) The map-based sequence of the rice genome. *Nature* 436, 793–800.

IRRI (2010) Global strategy for the *ex situ* conservation of rice genetic resources. Available at: www.croptrust.org/documents/cropstrategies/Rice%20Strategy.pdf (accessed 20 March 2013).

Jaeger, P., Doherty, C. and Ideker, T. (2012) Modeling transcriptome dynamics in a complex world. *Cell* 151, 1161–1162.

Janz, D., Behnke, K., Schnitzler, J.-P., Kanawati, B., Schmitt-Kopplin, P. *et al.* (2010) Pathway analysis of the transcriptome and metabolome of salt sensitive and tolerant poplar species reveals evolutionary adaption of stress tolerance mechanisms. *BMC Plant Biology* 10, 150.

Jiang, S.-S., Liang, X.-N., Li, X., Wang, S.-L., Lv, D.-W. *et al.* (2012) Wheat drought-responsive grain proteome analysis by linear and nonlinear 2-DE and MALDI-TOF mass spectrometry. *International Journal of Molecular Sciences* 13, 16065–16083.

Jothi, R., Cuddapah, S., Barski, A., Cui, K. and Zhao, K. (2008) Genome-wide identification of in vivo protein–DNA binding sites from ChIP-seq data. *Nucleic Acids Research* 36, 5221–5231.

KEGG (2012) *Kyoto Encyclopedia of Genes and Genomes*. Available at: www.genome.jp/kegg/ (accessed 23 March 2013).

Kilian, B. and Graner, A. (2012) NGS technologies for analyzing germplasm diversity in genebanks. *Briefings in Functional Genomics* 11, 38–50.

Klein, R.J., Zeiss, C., Chew, E.Y., Tsai, J.Y., Sackler, R.S. *et al.* (2005) Complement factor H polymorphism in age-related macular degeneration. *Science* 308, 385–389.

Knight, R., Jansson, J., Field, D., Fierer, N., Desai, N. *et al.* (2012) Unlocking the potential of metagenomics through replicated experimental design. *Nature Biotechnology* 30, 513–520.

Kosová, K., Vítámvás, P., Prášil, I.T. and Renaut, J. (2011) Plant proteome changes under abiotic stress – contribution of proteomics studies to understanding plant stress response. *Journal of Proteomics* 74, 1301–1322.

KU Leuven (2011) Fourth generation sequencing data – Infrastructure and architectures for the analysis of third and fourth generation sequencing data. Available at: www.esat.kuleuven.be/scd/project.php?prid=622 (accessed 23 March 2013).

Lee, K.H., Park, J., Williams, D.S., Xiong, Y., Hwang, I. *et al.* (2013) Defective chloroplast development inhibits maintenance of normal levels of abscisic acid in a mutant of the Arabidopsis RH3 DEAD-box protein during early post-germination growth. *Plant Journal* 73, 720–732.

Li, P., Li, X., Zhu, J., Hu, F., Ge, S., Ye, M. *et al.* (2012) Single-base resolution maps of cultivated and wild rice methylomes and regulatory roles of DNA methylation in plant gene expression. *BMC Genomics* 13, 300.

Liang, L., Zöllner, S. and Abecasis, G.R. (2007) GENOME: a rapid coalescent-based whole genome simulator. *Bioinformatics* 23, 1565–1567.

Londo, J.P., Chiang, Y., Hung, K., Chiang, T. and Schaal, B.A. (2006) Phylogeography of Asian wild rice, *Oryza rufipogon*, reveals multiple independent domestications of cultivated rice, *Oryza sativa*. *Proceedings of the National Academy of Sciences USA* 103, 9578–9583.

Lu, B.-R., Ge, S., Sang, T., Chen, J.-K. and Hong, D.-Y. (2001) The current taxonomy and perplexity of the genus *Oryza* (Poaceae). *Acta Phytotaxonomica Sinica* 39, 373–388.

Lundberg, D.S., Lebeis, S.L., Paredes, S.H., Yourstone, S., Gehring, J. *et al.* (2012) Defining the core *Arabidopsis thaliana* root microbiome. *Nature* 488, 86–90.

Mak, H.C. (2012) Genome interpretation and assembly – recent progress and next steps. *Nature Biotechnology* 30, 1081–1083.

Malcevschi, A. and Marmiroli, N. (2012) Plant protein analysis. In: Heazlewood, J. (ed.) *Proteomic Applications in Biology*. InTech, Manhattan, New York.

Mali, P., Yang, L., Esvelt, K.M., Aach, J., Guell, M. *et al.* (2013) RNA-guided human genome engineering via Cas9. *Science* 339, 823–826.

Mamin, H.J., Kim, M., Sherwood, M.H., Rettner, C.T., Ohno, K. *et al.* (2013) Nanoscale nuclear magnetic resonance with a nitrogen-vacancy spin sensor. *Science* 339, 557–560.

Manyam, G., Payton, M.A., Roth, J.A., Abruzzo, L.V. and Coombes, K.R. (2012) Relax with CouchDB – into the non-relational DBMS era of bioinformatics. *Genomics* 100, 1–7.

Margulies, M., Egholm, M., Altman, W.E., Attiya, S., Bader, J.S. *et al.* (2005) Genome sequencing in microfabricated high-density picolitre reactors. *Nature* 437, 376–380.

Martens, H., Veflingstad, S.R., Plahte, E., Martens, M., Bertrand, D. *et al.* (2009) The genotype–phenotype relationship in multicellular pattern-generating models – the neglected role of pattern descriptors. *BMC Systems Biology* 3, 87.

Mather, K.A., Rubinstein, S., Molina, J., Flowers, J.M., Rauh, B.L. *et al.* (2010) Migration, isolation and hybridization in island crop populations: the case of Madagascar rice. *Molecular Ecology* 19, 4892–4905.

McCouch, S., McNally, K., Wang, W. and Hamilton, R.S. (2012) Genomics of gene banks: a case study in rice. *American Journal of Botany* 99, 407–423.

Melters, D.P., Bradnam, K.R., Young, H.A., Telis, N., May, M.R. *et al.* (2013) Comparative analysis of tandem repeats from hundreds of species reveals unique insights into centromere evolution. *Genome Biology* 14, R10.

Millard, B.L., Niepel, M., Menden, M.P., Muhlich, J.L. and Sorger, P.K. (2011) Adaptive informatics for multifactorial and high-content biological data. *Nature Methods* 8, 487–492.

Mochida, K. and Shinozaki, K. (2011) Advances in omics and bioinformatics tools for systems analyses of plant functions. *Plant Cell Physiology* 52, 2017–2038.

Molina, J., Sikora, M., Garud, N., Flowers, J., Rubinstein, S. *et al.* (2011) Molecular evidence for a single evolutionary origin of domesticated rice. *Proceedings of the National Academy of Sciences USA* 108, 8351–8356.

MongoDB (2013) Available at: www.mongodb.org/ (accessed 23 March 2013).

Morris, G.P., Ramu, P., Deshpande, S.P., Hash, C.T., Shah, T. *et al.* (2013) Population genomic and genome-wide association studies of agroclimatic traits in sorghum. *Proceedings of the National Academy of Sciences USA* 110, 453–458.

Nagano, A.J., Sato, Y., Mihara, M., Antonio, B.A., Motoyama, R. *et al.* (2012) Deciphering and prediction of transcriptome dynamics under fluctuating field conditions. *Cell* 151, 1358–1369.

National Human Genome Research Institute (2013) DNA sequencing costs. Available at: www.genome.gov/sequencingcosts/ (accessed 12 March 2013).

Nature ENCODE (2013) Available at: www.nature.com/encode/#/threads (accessed 15 March 2013).

Negrão, S., Courtois, B., Ahmadi, N., Abreu, I., Saibo, N. *et al.* (2011) Recent updates on salinity stress in rice: from physiological to molecular responses. *Critical Reviews in Plant Sciences* 30, 329–377.

Nicotra, A.B., Atkin, O.K., Bonser, S.P., Davidson, A.M., Finnegan, E.J. *et al.* (2010) Plant phenotypic plasticity in a changing climate. *Trends in Plant Sciences* 15, 684–692.

Nitsch, D., Tranchevent, L.C., Gonçalves, J., Vogt, J.K., Madeira, S.C. *et al.* (2011) PINTA – A web server for network-based gene prioritization from expression data. *Nucleic Acids Research, Web Server Issue* 39 (suppl. 2), 334–338.

NoSQL (2013) List of NoSQL databases. Available at: http://nosql-database.org/ (accessed 23 March 2013).

Nucleic Acids Research (2013) Database issue – Volume 41 Issue D1, 1 January 2013. Available at: http://nar.oxfordjournals.org/content/41/D1.toc (accessed 20 March 2013).

Obata, T. and Fernie, A.R. (2012) The use of metabolomics to dissect plant responses to abiotic stresses. *Cellular and Molecular Life Sciences* 69, 3225–3243.

PAG XXI (2013) Available at: www.intlpag.org/ (accessed 23 March 2013).

Pareek, C.S., Smoczynski, R. and Tretyn, A. (2011) Sequencing technologies and genome sequencing. *Journal of Applied Genetics* 52, 413–435.

Pasam, R.K., Sharma, R., Malosetti, M., van Eeuwijk, F.A., Haseneyer, G. *et al.* (2012) Genome-wide association studies for agronomical traits in a world wide spring barley collection. *BMC Plant Biology* 12, 16.

Pautasso, M., Döring, T.F., Garbelotto, M., Pellis, L. and Jeger, M.J. (2012) Impacts of climate change on plant diseases – opinions and trends. *European Journal of Plant Pathology* 133, 295–313.

Pei, Y.-F., Li, J., Zhang, L., Papasian, C.J. and Deng, H.-W. (2008) Analyses and comparison of accuracy of different genotype imputation methods. *PLoS ONE* 3, e3551.

Poland, J.A., Brown, P.J., Sorrells, M.E. and Jannink, J.-L. (2012) Development of high-density genetic maps for barley and wheat using a novel two-enzyme genotyping-by-sequencing approach. *PLoS One* 7, e32253.

Pollack, A. (2011) DNA sequencing caught in deluge of data. *New York Times*, 30 November 2011.

Posada, D and Crandall, K.A. (2001) Evaluation of methods for detecting recombination from DNA sequences: Computer simulations. *Proceedings of the National Academy of Sciences USA* 98, 13757–13762.

Rafalski, J.A. (2010) Association genetics in crop improvement. *Current Opinion in Plant Biology* 13, 174–180.

Rampitsch, C. and Bykova, N.V. (2012) Proteomics and plant disease: Advances in combating a major threat to the global food supply. *Proteomics* 12, 673–690.

Rasmussen, S., Barah, P., Suarez-Rodriguez, M.C., Bressendorff, S., Friis, P. *et al.* (2013) Transcriptome responses to combinations of stresses in *Arabidopsis thaliana*. *Plant Physiology* 161, 1783–1794. doi:dx.doi.org/10.1104/pp.112.210773.

Riaño-Pachón, D.M., Kleessen, S., Neigenfind, J., Durek, P., Weber, E. *et al.* (2010) Proteome-wide survey of phosphorylation patterns affected by nuclear DNA polymorphisms in *Arabidopsis thaliana*. *BMC Genomics* 11, 411.

Rivas-Ubach, A., Sardans, J., Pérez-Trujillo, M., Estiarte, M. and Peñuelas, J. (2012) Strong relationship between elemental stoichiometry and metabolome in plants. *Proceedings of the National Academy of Sciences USA* 109, 4181–4186.

Roossinck, M.J. (2012) Plant virus metagenomics: biodiversity and ecology. *Annual Review of Genetics* 46, 359–369.

Rothberg, J.M., Hinz, W., Rearick, T.M., Schultz, J., Mileski, W. *et al.* (2011) An integrated semiconductor device enabling non-optical genome sequencing. *Nature* 475, 348–352.

Rübel, O., Ahern, S., Bethel, E.W., Biggin, M.D., Childs, H. *et al.* (2010) Coupling visualization and data analysis for knowledge discovery from multi-dimensional scientific data. *Procedia Computer Science* 1, 1757–1764.

Salt, D.E., Baxter, I. and Lahner, B. (2008) Ionomics and the study of the plant ionome. *Annual Review of Plant Biology* 59, 709–733.

Sanchez, D.H., Pieckenstain, F.L., Szymanski, J., Erban, A., Bromke, M. *et al.* (2011) Comparative functional genomics of salt stress in related model and cultivated plants identifies and overcomes limitations to translational genomics. *PLoS ONE* 6, e17094.

Sanger, F., Nicklen, S. and Coulson, A.R. (1977) DNA sequencing with chain-terminating inhibitors. *Proceedings of the National Academy of Sciences USA* 74, 5463–5467.

Sanghera, G.S., Wani, S.H., Hussain, W. and Singh, N.B. (2011) Engineering cold stress tolerance in crop plants. *Current Genomics* 12, 30–43.

Schmidt, M. and Lipson, H. (2009) Distilling free-form natural laws from experimental data. *Science* 324, 81–85.

Schmidt, M.D., Vallabhajosyula, R.R., Jenkins, J.W., Hood, J.E., Soni, A.S. *et al.* (2011) Automated refinement and inference of analytical models for metabolic networks. *Physical Biology* 8, 055011.

Schmitz, R.J., Schultz, M.D., Urich, M.A., Nery, J.R., Pelizzola, M. *et al.* (2013) Patterns of population epigenomic diversity. *Nature* 495, 193–198.

Schneeberger, K., Ossowski, S., Ott, F., Klein, J. D., Wang, X. *et al.* (2011) Reference-guided assembly of four diverse *Arabidopsis thaliana* genomes. *Proceedings of the National Academy of Sciences USA* 108, 10249–10254.

SciDev Net (2011) Scientists ramp up sequencing of rice varieties. Available at: www.scidev.net/en/agriculture-and-environment/news/scientists-ramp-up-sequencing-of-rice-varieties.html (accessed 21 March 2013).

Seo, Y.-S., Chern, M., Bartley, L.E., Han, M., Jung K.-H. *et al.* (2011) Towards establishment of a rice stress response interactome. *PLoS Genetics* 7, e1002020.

Sessitsch, A., Hardoim, P., Döring, J., Weilharter, A., Krause, A. *et al.* (2012) Functional characteristics of an endophyte community colonizing rice roots as revealed by metagenomic analysis. *Molecular Plant Microbe Interactions* 25, 28–36.

Snoep, J.L. and Westerhoff, H.V. (2005) From isolation to integration, a systems biology approach for building the silicon cell. In: Alberghina, L. and Westerhoff, H.V. (eds) *Systems Biology: Definitions and Perspectives. Topics in Current Genetics.* Springer-Verlag, Berlin, Germany, pp. 13–30.

Soon, W.W., Hariharan, M. and Snyder, M.P. (2013) High-throughput sequencing for biology and medicine. *Molecular Systems Biology* 9, 640.

Stark, C., Breitkreutz, B.-J., Chatr-aryamontri, A., Boucher, L., Oughtred, R. *et al.* (2011) The BioGRID interaction database: 2011 update. *Nucleic Acids Research* 39 (suppl 1), D698–D704.

Staudacher, T., Shi, F., Pezzagna, S., Meijer, J., Du, J. *et al.* (2013) Nuclear magnetic resonance spectroscopy on a (5-Nanometer)^3 sample volume. *Science* 339, 561–563.

Stein, L.D. (2010) The case for cloud computing in genome informatics. *Genome Biology* 11, 207.

Stephens, S., LaVigna, D., DiLascio, M. and Luciano, J. (2006) Aggregation of bioinformatics data using semantic web technology. *Web Semantics: Science, Services and Agents on the World Wide Web* 4, 216–221.

Strimmer, K., Wiuf, C. and Moulton, V. (2001) Recombination analysis using directed graphical models. *Molecular Biology and Evolution* 18, 97–99.

Swarbreck, S.M., Lindquist, E.A., Ackerly, D.D. and Andersen, G.L. (2011) Analysis of leaf and root transcriptomes of soil-grown *Avena barbata* plants. *Plant Cell Physiology* 52, 317–332.

TAIR (2013) The *Arabidopsis* Information Resource. Available at: www.arabidopsis.org/ (accessed 23 March 2013).

Taylor, R.C. (2010) An overview of the Hadoop/MapReduce/HBase framework and its current applications in bioinformatics. *BMC Bioinformatics* 11, S1.

Thelen, J.J. and Peck, S.C. (2007) Quantitative proteomics in plants: Choices in abundance. *Plant Cell* 19, 3339–3346.

Thomas, T., Gilbert, J. and Meyer, F. (2012) Metagenomics - a guide from sampling to data analysis. *Microbial Informatics and Experimentation* 2, 3.

Treangen, T.J., Koren, S., Sommer, D.D., Liu, B., Astrovskaya, I. *et al.* (2013) MetAMOS: a modular and open source metagenomic assembly and analysis pipeline, *Genome Biology* 14, R2.

USDA-ERS (2013) US drought 2012: farm and food impacts. Available at: www.ers.usda.gov/topics/in-the-news/us-drought-2012-farm-and-food-impacts.aspx (accessed 5 March 2013).

USDA-NIFA (2009) Barley CAP. Available at: www.barleycap.org/ (accessed 23 March 2013).

Valouev, A., Ichikawa, J., Tonthat, T., Stuart, J., Ranade, S. *et al.* (2008) A high-resolution, nucleosome position map of *C. elegans* reveals a lack of universal sequence-dictated positioning. *Genome Research* 18, 1051–1063.

Vaughan, D.A., Morishima, H. and Kadowaki, K. (2003) Diversity in the *Oryza* genus. *Current Opinion in Plant Biology* 6, 139–146.

Venter, J.C., Adams, M.D., Myers, E.W., Li, P.W., Mural, R.J. *et al.* (2001) The sequence of the human genome. *Science* 291, 1304–1351.

Villar, E., Klopp, C., Noirot, C., Novaes, E., Kirst, M. *et al.* (2011) RNA-Seq reveals genotype-specific molecular responses to water deficit in eucalyptus. *BMC Genomics* 12, 538.

Vitte, C., Ishii, T., Lamy, F., Brar, D. and Panaud, O. (2004) Genomic paleontology provides evidence for two distinct origins of Asian rice (*Oryza sativa* L.). *Molecular Genetics and Genomics* 272, 504–511.

Wageningen UR (2013) BreeDB. Available at: www.plantbreeding.wur.nl/UK/software_breedb.html (accessed 23 March 2013).

Wahid, A., Gelania S., Ashrafa, M. and Fooladb, M.R. (2007) Heat tolerance in plants: An overview. *Environmental and Experimental Botany* 61, 199–223.

Walls, R.L., Athreya, B., Cooper, L., Elser, J., Gandolfo, M.A. *et al.* (2012) Ontologies as integrative tools for plant science. *American Journal of Botany* 99, 1263–1275.

Wan, C., Liu, J., Fong, V., Lugowski, A., Stoilova, S. *et al.* (2012) ComplexQuant: High-throughput computational pipeline for the global quantitative analysis of endogenous soluble protein complexes using high resolution protein HPLC and precision label-free LC/MS/MS. *Journal of Proteomics* 81, 102–111.

Wang, L., Li, P. and Brutnell, T.P. (2010) Exploring plant transcriptomes using ultra high-throughput sequencing. *Briefings in Functional Genomics* 9, 118–112.

Wang, Z., Gerstein, M. and Snyder, M. (2009) RNA-Seq: a revolutionary tool for transcriptomics. *Nature Reviews in Genetics* 10, 57–63.

Ward, L.D. and Kellis, M. (2012) Interpreting noncoding genetic variation in complex traits and human disease. *Nature Biotechnology* 30, 1095–1106.

Wassmann, R., Jagadish, S.V.K., Sumfleth, K., Pathak, H., Howell, G. *et al.* (2009) Regional vulnerability of climate change impacts on Asian rice production and scope for adaptation. *Advances in Agronomy* 101, 59–122.

Weiss, S.M. and Indurkha, N. (1998) *Predictive Data Mining: A Practical Guide.* Morgan Kaufmann Publishers, Inc., San Francisco, California.

Wong, M.M., Cannon, C.H. and Wickneswari, R. (2011) Identification of lignin genes and regulatory sequences involved in secondary cell wall formation in *Acacia auriculiformis* and *Acacia mangium* via de novo transcriptome sequencing. *BMC Genomics* 12, 342.

Xu, X., Liu, X., Ge, S., Jensen, J.D., Hu, F.Y. *et al.* (2011) Resequencing 50 accessions of cultivated and wild rice yields markers for identifying agronomically important genes. *Nature Biotechnology* 30,105–111.

Yáñez-López, R., Torres-Pacheco, I., Guevara-González, R.G., Hernández-Zul, M.I., Quijano-Carranza, J.A. *et al.* (2012) The effect of climate change on plant diseases. *African Journal of Biotechnology* 11, 2417–2428.

Yu, J., Hu, S., Wang, J., Wong, G.K., Li, S. *et al.* (2002) A draft sequence of the rice genome (*Oryza sativa* L. ssp *indica*). *Science* 296, 79–92.

Yu, W., Wu, K.W., Ku, W.-S., Xu, C. and Gao, J. (2011) BMF: Bitmapped Mass Fingerprinting for Fast Protein Identification. In: *CLUSTER '11 Proceedings of the 2011 IEEE International Conference on Cluster Computing.* IEEE Computer Society, Washington, DC, pp. 17–25.

Zeng, K. (2012) A coalescent model of background selection with recombination, demography and variation in selection coefficients. *Heredity* 110, 363–371.

Zhang, Z., Ersoz, E., Lai, C.Q., Todhunter, R.J., Tiwari, H.K. *et al.* (2010) Mixed linear model approach adapted for genome-wide association studies. *Nature Genetics* 42, 355–360.

11 Harnessing Meiotic Recombination for Improved Crop Varieties

Susan J. Armstrong
School of Biosciences, University of Birmingham, Birmingham, UK

11.1 Introduction

Food security will become increasingly important in the near future owing to the challenges of the global demand for human and animal food sources as well as for bioenergy. The world's population is projected to increase to more than nine billion, requiring up to a 70% increase in agricultural production (Feed the Future, 2012). The impact of climate changes and the dependence on inorganic fertilizers and biocides means that we rapidly need to improve generation of novel varieties that are suitable not only for producing high-yielding varieties but that will also grow in suboptimal environments. We thus need to maximize genetic variability to breed new crop varieties suitable to sustain the growing world population, both now and in the future. To meet this challenge it will be necessary to develop crops that are matched to the requirements of human populations, soils and climates around the world.

Plant breeders are constantly striving to produce novel varieties but it takes at least 8–10 years before a plant with improved agronomic qualities can be released to the farmers for production (L. Ramsay, James Hutton Institute, 2013, personal communication). In an attempt to generate crop plants with desirable characteristics, a worldwide plant breeding programme supported by the United Nations and the International Atomic Energy Agency (IAEA, 2012) was set up around 60 years ago. The project relies on the induction of random mutations in genomes of established crop plants with subsequent selection of desired traits and is as such a 'blunt tool'. In addition to generating desirable traits this strategy may also result in considerable collateral genomic damage and chromosome instabilities that will have to be bred out through multiple rounds of backcrossing. Furthermore, only a limited number of genes that account for a certain trait may be successfully altered by a mutational approach.

Marker assisted selection (MAS) and, increasingly, genome whole association studies (GWAS) have contributed to speeding up introgression and making this more efficient (see McNally, Chapter 10, this volume). Currently, however, the way of introducing single or few novel alleles is by using genetic modification (GM) but this is unlikely to be accepted in many countries (Royal Society, 2009). In Europe, the general population is suspicious of the safeguards in place and is concerned by the comments in popular press referring to them as 'Frankenstein foods' (*The Telegraph*, 2011). In recognition of this problem, in the UK and Europe the Biotechnology and Biological Sciences Research Council (BBSRC) and European Framework 7 (EU FP7), respectively, have identified food security as a research priority area.

Genetic recombination is the natural process used as a tool in which genetic variation and varietal resources are incorporated into breeding germplasm. Much improvement has been through heterotic improvements in F_1 hybrids, for example in maize. Traditionally, plant breeders have relied on repeated backcrossing of new lines to remove undesirable characters from introgressed material, an expensive and time-consuming procedure because the breeder may need to disrupt linkage drag in order to break up linkage between favourable and unfavourable traits. This can be a problem even in conventional germplasm where it is clear that levels of recombination are greatly reduced in some areas of the genome and of course is true for many crop species. This effect is clearly visible from the marked clustering of randomly selected markers on the genetic map.

At Birmingham we are investigating how to translate our knowledge of meiotic recombination in *Arabidopsis thaliana* into an ability to alter recombination frequency and distribution in the diploid crop barley and in the allopolyploid, canola or oilseed rape (OSR), *Brassica napus*.

11.1.1 Barley

Barley belongs to the *Triticeae* subfamily of the Poaceae, which represents all important small grain cereal species including wheat, rice and rye. Its diploid genome structure (2n = 14), a strongly inbreeding-based mating system and a long tradition in breeding have made barley the model crop species for the *Triticeae*. Barley ranks number four in world crop production (harvested area) with the EU producing over 60% of the world harvest (FAOSTAT, 2012). As well as being the world's largest producer, Europe is also the leading exporter of barley. Thus, any investment in this crop species will significantly benefit European agriculture. Barley is important for industrial fodder as well as for malt production (beer, regional whisky and bioethanol production). Of a total cereal production of 215 Mt, barley accounts for 22% and wheat for 49% in 15 EU member states (FAOSTAT, 2012), respectively. Germany, the UK and Denmark produce 21.1 Mt of barley.

11.1.2 Brassicas

Brassicas are central to EU agriculture. For example the allopolyploid OSR (2n = 38; *Brassica napus AACC*) has emerged as an important agricultural plant, both as a food and increasingly as a biofuel. It is now the second largest oilseed crop with an annual worldwide production of 38 million tons of oil (FAOSTAT, 2012). The EU aims to promote a stepwise substitution of conventional fuel by crop-derived biofuels. This will significantly increase the pressure for an efficient breeding programme, with the potential for increasing harvest index of a major oil crop. Importantly, improved yields of biofuel crops will mitigate the competition for land use between food and non-food crop uses.

11.1.3 Current state of crop improvement programmes for barley and OSR

Yields of barley in the UK have increased by just less than 1% per year during the past 40 years, largely as a result of conventional genetic improvement, and OSR yields are increasing owing to the introduction of new varieties. The harvest index (harvestable yield as a proportion of total biomass) of OSR crops is low compared with other arable crops such as cereals.

There is considerable scope for improvement for both of these crops, by reducing nitrogen use and increasing the harvest index. But, although yields are constantly increasing, they are not increasing as fast as the human population, which is also showing a capacity to transfer their diet to eating meat – an inherently inefficient use of grain resources – giving a significant decrease in the per capita availability.

If we are to feed the growing world population then it is imperative to increase yields. Although yield per hectare of these crops varies tenfold between different countries, reflecting different soils and rainfall, as well as levels of inputs and the efficiency of cultivation, this highlights the scope for crop improvement. Yield will continue to be the primary focus of many crop improvement programmes and the relatively low harvest index of brassica oil crops compared with other arable crops thus provides a considerable impetus for improvement. Modern OSR is a good example for genetics research and development as applied to crop improvement. The crop did not exist on any substantial scale until identification and introgression of two traits (low seed glucosinolate and low erucic acid = double low) from exotic germplasm. The rate of adoption of the crop has been phenomenal since these introductions in 1970s; on the other hand, linkage drag from this still slows down improvement of OSR (Graham King, Southern Cross University, 2012, personal communication).

Plant breeders have been aware of the importance and impact of meiotic recombination to their breeding programmes. They have been constrained in generating varieties with optimal combinations of alleles due to the fundamental limitations imposed by the frequency and distribution of meiotic recombination, and, until recently there was insufficient mechanistic understanding of meiosis in plants directly to benefit their activities. This chapter explains why meiotic recombination lies at the heart of generating novel varieties.

11.2 Meiosis

Meiosis is a highly conserved process that occurs in all sexually reproducing organisms resulting in the production of four haploid gametes. During the meiotic pathway, a single round of DNA replication is followed by two rounds of chromosome segregation, which results in a halving of the chromosome number (Roeder, 1997). The diploid state is then restored during sexual reproduction by fusion of male and female gametes. Meiosis ensures that the correct number of chromosomes is passed on from generation to generation. It also recombines or mixes the genes inherited from the previous generation.

11.2.1 Outline of the meiotic pathway

A unique set of programmed events is initiated at meiotic prophase I. This consists of five cytologically recognizable stages (leptotene, zygotene, pachytene, diplotene and diakinesis) and prophase I the longest duration of stages in the meiotic pathway. For example prophase I of *A. thaliana* meiosis occupies 30 hours of a total duration of approximately 33 hours (Armstrong *et al.*, 2003). The establishment of a meiotic time-course has provided a framework for determining the relative timing and durations of key molecular events in meiosis in *A. thaliana* in relation to cytologically identified landmarks. In contrast to mitosis, specific meiotic chromatin components must be synthesized and reorganized to allow for chromosome pairing, recombination and synapsis.

Homologous recombination is initiated during leptotene by the formation of double strand breaks (DSBs). The homologous paired telomeres are clustered on the nuclear envelope. Subsequently homologous chromosomes begin to pair (align) along their length, a process most likely commencing from the paired telomeres becoming connected by a proteinaceous structure called the synaptonemal complex (SC). Formation of the SC is initiated during zygotene, often from the telomeric ends and progresses through to pachytene where homologues can be observed to be fully synapsed (Armstrong *et al.*, 2001; Armstrong and Jones, 2003; Jones *et al.*, 2003; Lopez *et al.*, 2008). The SC begins to break down during diplotene and the products of recombination can be observed cytologically at diakinesis. At this stage homologues are held together at sites of chiasmata, reciprocal crossovers (COs) between non-sister chromatids that involve the exchange of genetic material between homologues (Jones, 1984). These connections, together with attachment of the spindle apparatus, are important for ensuring the faithful segregation of homologues at anaphase I. The segregation of homologues at anaphase I and sister chromatids at anaphase II is permitted by progressive loss of meiotic cohesion proteins (Ishiguro and Watanabe, 2007). This second division resembles haploid mitosis, with sister chromatids separating to each pole. Thus homologous recombination generates genetic variation through the production of genetically distinct gametes.

Many highly conserved genes are involved in the progression of the meiotic pathway: following on from the generation of DSBs by SPO11 in leptotene, the MRE11–RAD50–XRS2 complex coats the ends and the breaks are resected from the 5' end to give 3' single strand (ss) overhangs. Strand invasion of both ends (DMC1, RAD51) forms a D loop and, during zygotene, DNA synthesis from the invading ends fills in regions lost by end resection. Ligation of extended strands then forms two Holliday junctions. The majority of these recombination events results in non-COs via a synthesis-dependent strand annealing pathway and around only 10% of initial breaks are converted to mature COs during pachytene. Crossovers are mediated by MSH4 and MSH5 and subsequently by MLH1and MLH3 (Fig. 11.1; Osman *et al.*, 2011).

Research in the past 15 years into the genes and proteins involved in meiotic recombination in the model plant species *A. thaliana (*thale cress*)* is beginning to identify the factors involved in the control of this process. Novel approaches in genomics and systems biology have been developed and research has provided important insights into the control of meiotic recombination in plants. Indeed, 80 plant meiotic genes/proteins have now been studied and a central database has been recently been established for the compilation and exchange of data in *A. thaliana*, brassica and barley (MEIOsys, 2009). These studies have demonstrated that the controlled formation of COs is

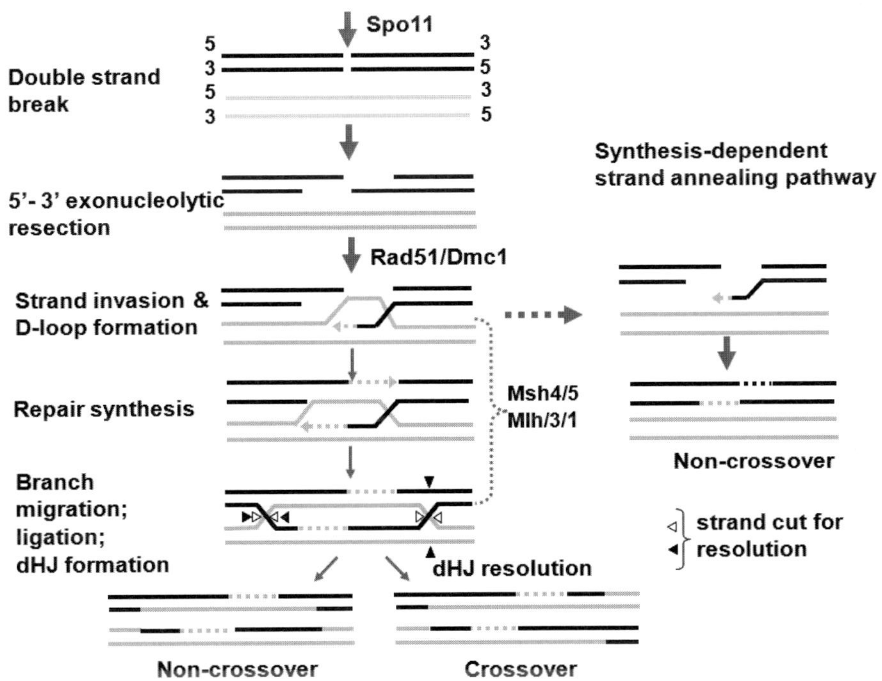

Fig 11.1. A current model for meiotic homologous recombination during yeast meiosis. The recombination pathway is shown but with only a few of the recombination proteins indicated (see Osman *et al.*, 2011 for a more detailed survey of current recombination proteins). dHJ, double Holliday junction.

dependent on the relationships between the proteins that catalyse recombination and those that regulate the extensive remodelling of chromosomes during prophase I of meiosis (Osman *et al.*, 2011).

Our approach is providing us a detailed knowledge of the poorly understood factors that control meiotic recombination. We are now in a position to use this information to speed up novel methods for introducing genetic variation into crop species.

11.2.2 Control of the frequency and distribution of crossovers

The question of how the frequency and distribution of meiotic COs is controlled has been a focus of research for more than a century and is currently a highly active area of basic research. CO control is highly coordinated in both diploid and established polyploid backgrounds. It appears that there is a tendency of COs in the majority of all species to be non-random, reflecting several layers of regulation; for example, in *A. thaliana*, Drouaud *et al.* (2007) have published a detailed genetic map of chromosome 4 along which they demonstrated that CO rates were highly variable. On a finer scale, COs tend to be clustered in small regions only a few kilobases in size, as has been clearly demonstrated in many eukaryotes (de Massy, 2003). These meiotic 'hot spots of recombination' are now known to be formed as a result of preference for meiotic DSB formation and from which meiotic recombination is initiated, although the factors that define these 'hotspots' remain poorly understood.

11.2.3 Factors controlling hotspots of recombination

Studies by Jones (1984) suggested that pairing and CO control operates at the chromosomal autonomous level. Subsequently Kleckner *et al.* (2004) have proposed that there is a link between chromosome organization and meiotic progression, and we now believe that meiotic recombination occurs within the context of dramatic changes in the organization of meiotic chromosomes. A growing body of evidence indicates that chromatin organization and meiotic progression are indeed functionally coupled (Higgins *et al.*, 2004; Jackson *et al.*, 2006). This has led to the proposal of an intra-prophase I surveillance mechanism that couples recombination with dynamic changes in chromosome/chromatin organization to allow regulated meiotic progression.

A link between histone acetylation and CO distribution has been confirmed in *A. thaliana*. Perrella *et al.* (2010) used an activation tagging approach to isolate a dominant mutation, *MCC1* (meiosis chromatin condensation), that leads to the over-expression of a gene encoding a GCN-5-related N-acetyltransferase. Male meiocytes progressing in meiosis showed an alteration in CO distribution and pairing failure at a low frequency (8%) in one chromosome pair. Several studies have also shown that meiotic CO frequency is susceptible to epigenetic remodelling in *Arabidopsis* DNA mutant lines (reviewed by Pecinka and Mittelsten Scheid, 2012). Mirouze *et al.*, (2012) have shown that in hypo methylated *Arabidopsis* plants the loss of DNA methylation affects the recombination landscape. Studies in budding yeast and mammals have provided evidence that recombination patterns or 'hot-spots' along chromosomes are also influenced by chromatin organization. In mammals, it appears that the histone modification H3 lysine 4 tri-methyl (H3K4me3) in conjunction with the histone H3 methyltransferase PRDM9 defines preferred recombination hotspots (Baudat *et al.*, 2010). Thus, overall it seems highly likely that the chromatin landscape is an important factor in defining patterns of recombination in plants.

11.2.4 Distribution and frequency of CO formation in barley and the cereals

Understanding the factors that control meiotic recombination is of great significance for the improvement of crop breeding and requires the development and application of new techniques. In brassicas and most strikingly in the cereals, recombination is skewed to the distal region of the chromosomes and restricts the ability for the breeder to produce improved varieties involves changes in genes in the remaining parts of the chromosomes. Research in cereals indicates that a non-uniform distribution of recombination events is a characteristic feature of these species (Künzel *et al.*, 2000) and knowledge of the control of these COs would be particularly useful for the improvement of temperate grass species (e.g. wheat, barley, oats and forage grasses such as *Lolium* and *Festuca*). These plants are all members of the grass family Poaceae, with large genomes and very close syntenic relationships (Moore *et al.*, 1995), and all exhibit a non-random pattern of recombination relative to the gene distribution in their genomes. A clear suppression of recombination has been shown in pericentromeric regions, with chiasmata/COs predominately occurring in distal regions of the chromosomes. On the other hand, gene density studies reveal that a substantial proportion of the gene complement is within these regions (Sandhu and Gill, 2002). For example, analyses in wheat and barley indicate that >30% of the genes are located in recombination-cold chromosomal regions (Künzel and Waugh, 2002). Moreover, the level of recombination within the gene-rich regions can vary many-fold. This combination with gene distribution and pattern of recombination has several consequences. First, it reduces the genetic variation that is available to breeders and second it limits the potential for map-based cloning of genes. A further consequence of this CO distribution can be linkage drag when attempting to introgress new genetic traits. This means that large areas of the chromosome around the centromeric region rarely recombine, even though these regions represent substantial proportions of the physical maps of the chromosomes (Künzel *et al.*, 2000; Künzel and Waugh, 2002). Because the genes in this area are inherited together as a large linkage block, the generation of novel gene combinations is prevented and useful variation that could be exploited in breeding programmes remains cryptic.

We have investigated factors underlying the pattern of chiasma formation in barley and show that the chiasma distribution reflects polarization in the spatiotemporal function of recombination, chromosome pairing and synapsis. As a consequence meiotic progression in distal chromosomal regions occurs in coordination with the chromatin cycles that are a conserved feature of the meiotic programme. Recombination initiation in interstitial and proximal regions occurs later than distal events, is not coordinated with the cycles, and rarely if ever progresses to form chiasmata (Higgins *et al.*, 2012). Small changes in CO frequency and distribution, particularly changes that promote re-combination in centromeric regions, could therefore have a significant effect on the efficiency of breeding in these crops by breaking up some of the extensive linkage blocks. Such an objective is not unrealistic because partial loss of control of chiasmata distribution can occur in certain inbred lines of normally outbreeding species such as rye and *Lolium* (Jones, 1967; Karp and Jones, 1983). Barley offers a critical natural advantage for studies in the control of recombination in that it is an inbreeding diploid unlike the other crop species, and also benefits from the wide range of genetic and genomic resources that have been developed for this species.

11.2.5 Distribution and frequency of CO in polyploids

CO control is also highly coordinated in established polyploid backgrounds and this requires an additional level of meiotic regulation that we need to enable meiotic stabilization of newly synthesized and established polyploids. A newly synthesized F_1 comprising one set of chromosomes is often likely to fail, e.g. *Allium cepa* × *Allium fistulosum* (Stevenson *et al.*, 1998), owing to failure in pairing and COs between the homoeologues (closely related but non-homologous chromosomes). Analyses in newly established *A. thaliana* autopolyploids and subsequent generations (Santos *et al.*, 2003) show that pairing behaviour that disfavours homoeologous interactions is developed by the seventh generation, when pairing is strictly 'diploidized'; in earlier generations quadrivalent formation is likely to be accompanied by loss of fertility.

This is of particular importance in plant allopolyploid species, where they comprise at least two genomes from hybridization events between related diploid species, and this has been recognized to have a role in the evolution of new species. In order to avoid the undesirable effects on fertility and stability in both types of polyploidy, efficient homologous chromosome pairing is restricted to pairwise association. Differential control of chromosome pairing during meiosis I is a key factor in the stabilization of allopolyploid species. This mechanism enables homologues to pair but restricts pairing between homoeologues. Many crop species are allopolyploids, including wheat, cotton and oilseed rape. Genetic diversity in allopolyploids is often lower than in closely related diploid species and plant breeders frequently wish to introduce useful genes for traits – such as disease resistance from diploids into the allopolyploid. In the monocotyledon wheat, *Triticum aestivum* (AABBDD), pairing between homologues is strictly controlled and, although this gives stability to the species, it restricts the ability of breeders to introgress traits. When *Ph1*, the major pairing locus, is deleted, however, pairing and recombination can occur between homoeologues, aiding wheat breeding.

Recently it has been shown that *Ph1* is associated with a multigene family of *Cdk-like* genes (Griffiths *et al.*, 2006; Al-Kaff *et al.*, 2008). It appears that deletion of the *Ph1* locus compromises meiotic chromatin condensation and thus allows homoeologous pairing. Apart from in wheat, information on the mechanisms responsible for the meiotic diploidization of allopolyploid crop species is limited (Jenczewski and Alix, 2004). Natural established stable-pairing OSR (AACC) is likely to have arisen relatively recently, in the past 1000–2000 years (Cheung *et al.*, 2009) compared with ~10,000 years for hexaploid wheat (Moore *et al.*, 1995). The linkage groups of the *B. napus* A and C genomes are collinear with those of the extant diploid species *Brassica rapa* (A genome, n = 10) and *Brassica oleracea* (C genome, n = 9). Pairing and recombination can occur between the *B. rapa* and *B. napus* A genomes and the *B. oleracea* and *B. napus* C genomes, suggesting that major changes in chromosome structure have not occurred in *B. napus*. Recombination generally occurs between homologues within *B. napus*, but there is a low incidence of recombination between A and C genome homoeologues within cultivars resources (Gaeta *et al.*, 2007). In resynthesized *B. napus* this occurs more frequently (Parkin *et al.*, 1995), leading to chromosomal rearrangements and genetically unbalanced gametes. It appears that the restriction of pairing to homologues rather than homoeologues in *B. napus* is under genetic control (Jenczewski *et al.*, 2003; Liu *et al.*, 2006).

11.3 Generating Novel Recombinant Crop Species

Mapping loci that control homoeologous recombination in the dicotyledon *B. napus* is important for breeding improved cultivars. This amphidiploid appears to have arisen more than once from chance hybridization events between domesticated *B. rapa* and *B. oleracea*, probably within the last 2000 years (Cheung *et al.*, 2009; Allender and King, 2010). With recent advances in genome sequencing and molecular marker techniques aiding conventional breeding (see McNally, Chapter 10, this volume), plant breeders are trying to introduce useful genes from the *B. rapa* and *B. oleracea* genepools into OSR. One way that this can be achieved is by producing 're-synthesized' *B. napus* by crossing appropriate lines. Correct pairing is crucial to the production of gametes with a balanced number of chromosomes. In the established OSR, each chromosome pairs with its partner within a genome but in re-synthesized lines, the pairing control is more 'relaxed' and a chromosome can often pair with a closely related chromosome in the other genome (i.e. a homoeologous chromosome from the *B. oleracea* genome can pair with a chromosome from the *B. rapa* genome). Jenczewski *et al.* (2003) have used cytogenetic techniques to assess pairing and recombination directly during meiosis in pollen mother cells. They used a population consisting of haploid plants (AC n = 19) rather than allopolyploid OSR plants. The mean number of univalents at metaphase I was calculated from ~20 pollen mother cells for each plant. A major locus controlling homoeologous pairing in haploids, *PrBn*, was found on C9 (DY15). Only 24% of the variance was explained by this locus and other additive and epistatic quantitative trait loci were detected (Jenczewski *et al.*, 2003; Liu *et al.*, 2006).

At Birmingham, we have used a cytogenetic approach to investigate pairing in natural OSR (Howell *et al.*, 2008). We are currently attempting to define the gene(s) involved in the control of pairing between partners and to determine the mechanism which ensures that chromosome pairing is restricted to partners. This knowledge will enable plant breeders to develop modern varieties of OSR using re-synthesized lines without the significant problems of genome instability.

11.3.1 Using chemicals to alter chromatin organization

One potential route to modifying the CO distribution and frequency is through the use of chemicals that may alter chromatin organization. Recently, we showed that an *Arabidopsis* line MCC1, which over-expresses a GCN5-related histone acetylase, exhibited an alteration in CO distribution relative to wild-type plants (Perrella *et al.*, 2010). This effect could be phenocopied with the histone deacetylase inhibitor trichostatin A. This suggested a potential route to modify recombination in barley and we are investigating a range of chemicals that have previously been implicated in modification of chromatin (Amritpal Sandu – doctoral researcher, University of Birmingham, 2013, personal communication).

11.3.2 Changing environmental conditions

We have investigated the effect of temperature changes in barley (Higgins *et al.*, 2012). We have found that a modest temperature shift (22°C to 30°C) during meiosis is sufficient to alter meiotic progression in relation to the chromosome cycles. Here the spatiotemporal polarization of the meiotic processes is reduced and this is accompanied by a significant shift in chiasma distribution with an increase in interstitial and proximal chiasmata. This observation suggests a potential route for breeders to modify recombination in cereals (Higgins *et al.*, 2012).

11.3.3 Targeting gene expression

Although altering gene expression may be achieved through non-specific DNA lesion causing mutagens such as X-rays and ethyl methane sulphonate (EMS), specific genes are not targeted. RNA interference (RNAi) where a double-stranded RNA molecule is generated to specifically recognize and degrade an endogenous transcript and gene targeting where an endogenous gene may be replaced by an exogenous gene by homologous recombination. We have recently commenced a programme to introduce RNAi to specific meiotic genes and are currently evaluating this approach.

11.4 Knowledge Transfer to Crop Species

One key to transferring novel genes to crops is likely to rely on effective exploitation of natural genetic diversity. This is frequently considered to be a question relating to biodiverse wild or unadapted germplasm where considerable amounts of genetic variability remain untapped, even in elite germplasm genepools, due to factors that limit the distribution of meiotic COs along chromosomes discussed earlier (Mézard *et al.*, 2007). We do not intend to modify crops with foreign genes that are transferred across species barriers. Our approach can be seen as a highly advanced strategy to accelerate traditional breeding, which has been applied for hundreds of years, and will thus address societal objectives within the EU. We suggest utilizing the wealth of natural variation, by enhancing the efficiency of meiotic recombination at a desired locus. The transgenes that we will use to modify recombination will only be present transiently in the crop genome and will be eliminated after successful recombination has been achieved. The desired alleles, combined in a common genetic background, will not have been altered and will be represented in their natural state. The entire strategy of 'accelerated breeding' has an identical outcome to conventional breeding, with the advantage of saving time and being cost effective. Further, many breeding programmes rely on doubled haploid approaches that involve limited scope for recombination. There may be circumstances where a reduction in COs would be valuable, to create, for example, less variation and greater uniformity in progeny. Stability will be increased in those lines with favourable genes in areas where COs are low. Likewise, an increase in COs may allow linkage to be broken between desired and undesired traits, such that the latter can be efficiently selected against.

Our knowledge from *Arabidopsis* will be implemented and assessed by experiments aimed at altering meiotic recombination in crop species, initially using the closely related brassicas and the cereal crop barley, with the overall aim of providing the plant breeder with the knowledge to modify recombination and potentially develop new varieties. With the ongoing requirements for the production of cultivars resistant to plant pathogens and to develop cultivars with enhanced yields in order to sustain a growing population, the plant breeder will need to develop cultivars that will cope with the new environmental challenges that we are likely to encounter during this century. In times of environmental change the ability to use natural diversity and variation in order to breed favourable cultivars more rapidly would be a considerable advantage to meet the needs of society.

Plants are the basis of many European industries with an annual turnover of more than €1 trillion. Intensive plant breeding has led to improvements in yield, quality and tolerance to biotic and abiotic stress. However, the development of new approaches to breeding, on the basis of the systematic analysis of key biological processes, will be necessary to ensure the world's future nutritional requirements. Our goal of modulating recombination distribution and frequency will not only speed up traditional routes for breeders, but will also give us the potential to generate a new combination of genes. New varieties are urgently required if we are to meet the increased pressure from the combined challenges of population growth and climate change.

11.5 Acknowledgements

We acknowledge funding in the S.J.A. laboratory:

- Barley research is funded by BBSRC LOLA award to F.C.H. Franklin /S.J.A. as well as a case studentship with Limagrain UK.
- The brassica project is also funded by BBSRC.
- Our European work is funded from the European Community's Seventh Framework Programme FP7/2007-2013 under grant agreement number KBBE-2009-222883.

References

Al-Kaff, N., Knight, E., Bertin, I., Foote, T., Hart, N. *et al.* (2008) Detailed dissection of the chromosomal region containing the *Ph1* locus in wheat *Triticum aestivum*: with deletion mutants and expression profiling. *Annals of Botany* 101, 863–872.

Allender, C.J. and King, G.J. (2010) Origins of the amphiploid species *Brassica napus* L. investigated by chloroplast and nuclear markers. *BMC Plant Biology* 10, 54.

Armstrong, S.J. and Jones, G.H. (2003) Meiotic cytology and chromosome behaviour in wild-type *Arabidopsis thaliana*. *Journal of Experimental Botany* 54, 1–10.

Armstrong, S.J., Franklin, F.C.H. and Jones, G.H. (2001) Nucleolus-associated telomere clustering precedes meiotic chromosome synapsis in *Arabidopsis thaliana*. *Journal of Cell Science* 114, 4207–4217.

Armstrong, S.J., Franklin, F.C.H. and Jones G.H. (2003) A meiotic time course for *Arabidopsis thaliana*. *Sexual Plant Reproduction* 16, 141–149.

Baudat, F., Buard, J., Grey, C., Fledel-Alon, A., Ober, C. *et al.* (2010) PRDM9 is a major determinant of meiotic recombination hotspots in humans and mice. *Science* 327 836–840.

Cheung, F., Trick, M., Drou, N., Lim, P.Y., Park, J.-Y. *et al.* (2009) Comparative analysis between homoeologous genome segments of *Brassica napus* and its progenitor species reveal extensive sequence-level divergence. *Plant Cell* 21, 1912–1928.

Drouaud, J., Mercier, R., Chelysheva, L., Bérard, A., Falque, M. *et al.* (2007) Sex-specific crossover distributions and variations in interference level along *Arabidopsis thaliana* chromosome 4. *PLoS Genetics* 3, e106.

FAOSTAT (2012) Available at: http://faostat.fao.org/ (accessed 20 December 2012).

Feed the Future (2012) Available at: www.feedthefuture.gov/ (accessed 1 February 2012).

Gaeta, R.T., Pires, J.C., Iniguez-Luy, F., Leon, E. and Osborn, T.C. (2007) Genomic changes in resynthesized *Brassica napus* and their effect on gene expression and phenotype. *Plant Cell* 19, 3403–3417.

Griffiths, S., Sharp, R., Foote, T.N., Bertin, I., Wanous, M. *et al.* (2006) Molecular characterisation of *PH1* as a major chromosome pairing locus in polyploidy wheat. *Nature* 439, 749–752.

Higgins, J.D., Armstrong, S.J., Franklin, F.C.H. and Jones, G.H. (2004) The *Arabidopsis MutS* homolog *AtMSH4* functions at an early step in recombination: evidence for two classes of recombination in *Arabidopsis*. *Genes & Development* 18, 2557–2570.

Higgins, J.D., Perry, R.M., Barakate, A., Ramsay, L., Waugh, R. *et al.* (2012) Spatiotemporal asymmetry of the meiotic program underlies the predominantly distal distribution of meiotic crossovers in barley. *Plant Cell* 24, 4096–4109.

Howell, E.C., Kearsey, M.J., Jones, G.H., King, G.J. and Armstrong, S.J. (2008) A and C genome and chromosome identification in *Brassica napus* by sequential FISH and GISH. *Genetics* 180, 1849–1857.

IAEA (2012) *Plant Breeding and Genetics*. Joint FAO/IAEA Programme – Nuclear Techniques in Food and Agriculture. Available at: www-naweb.iaea.org/nafa/pbg/index.html (accessed 15 January 2013).

Ishiguro, K.I. and Watanabe, I. (2007) Chromosome cohesion in mitosis and meiosis. *Journal of Cell Science* 120, 367–369.

Jackson, N., Sanchez-Moran, E., Buckling, E., Armstrong, S.J., Jones, G.H. *et al.* (2006) Reduced meiotic crossovers and delayed prophase I progression in AtMLH3-deficient *Arabidopsis. Embo Journal* 25, 1315–1323.

Jenczewski, E. and Alix, K. (2004) From diploids to allopolyploids: the emergence of pairing control genes. *Critical Reviews in Plant Sciences* 23, 21–45.

Jenczewski, E., Eber, F., Grimaud, A., Huet, S., Lucas, M.O. *et al.* (2003) PrBn, a major gene controlling homoeologous pairing in oilseed rape (*Brassica napus*) haploids. *Genetics* 164, 645–654.

Jones, G.H. (1967) The control of chiasma distribution in rye. *Chromosoma* 22, 69–90.

Jones, G.H. (1984) The control of chiasma distribution. In: Evans, C.W. and Dickinson, H.G. (eds) *Symposia of the Society for Experimental Biology* 38, 293–320.

Jones, G.H., Armstrong, S.J., Caryl, A.P. and Franklin, F.C.H. (2003) Meiotic chromosome synapsis and recombination in *Arabidopsis thaliana;* an integration of cytological and molecular approaches. *Chromosome Research* 11, 205–215.

Karp, A. and Jones, R.N. (1983) Cytogenetics of *Lolium perenne. Theoretical and Applied Genetics* 6492, 137–146.

Kleckner, N., Zickler, D., Jones, G.H., Decker, J., Padmore, R. *et al.* (2004) A mechanical basis for chromosome function. *Proceedings of the National Academy of Sciences USA* 101, 12592–12597.

Künzel, G. and Waugh, R. (2002) Integration of microsatellite markers into the translocation-based physical RFLP map of barley chromosome 3H. *Theoretical and Applied Genetics* 105, 660–665.

Künzel, G., Korzun, L. and Meister, A. (2000) Cytologically integrated physical restriction fragment length polymorphism maps for the barley genome based on translocation breakpoints. *Genetics* 154, 397–412.

Liu, Z., Adamczyk, K., Manzanares-Dauleux, M., Eber, F., Lucas M.O. *et al.* (2006) Mapping PrBn and other quantitative trait loci responsible for the control of homeologous chromosome pairing in oilseed rape (*Brassica napus* L.) haploids. *Genetics* 174, 1583–1596.

Lopez, E., Pradillo, M., Romero, C., Santos, J.L. and Cunado, N. (2008) Pairing and synapsis in wild-type *Arabidopsis thaliana*. *Chromosome Research* 16, 701–708.

de Massy, B. (2003) Distribution of meiotic recombination sites. *Trends in Genetics* 19, 513–522.

MEIOsys (2009) *EU Collaborative Project*. Available at: www.meiosys.org (accessed 20 December 2012).

Mézard, C., Vignard, J., Drouaud, J. and Mercier, R. (2007) The road to crossovers: plants have their say. *Trends in Genetics* 23, 91–99.

Mirouze, M., Liberman-Laraovich, M., Aversano, R., Bucher, E., Nicolet, J., *et al.* (2012) Loss of DNA methylation affects the recombination landscape in *Arabidopsis. Proceedings of the National Academy of Sciences USA* 109, 5880–5885.

Moore, G., Devos, K.M., Wang, Z. and Gale, M.D. (1995) Cereal genome evolution: grasses, line up and form a circle. *Current Biology* 5, 737–739.

Osman, K., Higgins, J.D., Sanchez-Moran, E., Armstrong, S.J. and Franklin, F.C.H. (2011) Pathways to meiotic recombination in *Arabidopsis thaliana. New Phytologist* 190, 523–544.

Parkin, I.A.P., Sharpe, A.G., Keith, D.G., and Lydiate, D.J. (1995) Identification of the A and C genomes of the amphidiploid *Brasscia napus* (oilseed rape). *Genome* 38, 1122–1131.

Pecinka, A. and Mittelsten Scheid, O. (2012) Stress-induced chromatin changes: a critical review on their heritability. *Plant & Cell Physiology* 53, 801–808.

Perrella, G., Consiglio, M.F., Aiese-Cigliano, R., Cremona, G., Sanchez-Moran, E. *et al.* (2010) Histone hyperacetylation affects meiotic recombination and chromosome segregation in Arabidopsis. *Plant Journal* 62, 796–806.

Roeder, S. (1997) Meiotic chromosomes: it takes two to tango. *Genes & Development* 11, 2600–2621.

Royal Society (2009) Reaping the Benefits – Science and the Sustainable Intensification of Global Agriculture. Available at: http://royalsociety.org/uploadedFiles/Royal_Society_Content/policy/publications/2009/4294967719.pdf (accessed 20 March 2013).

Sandhu, D. and Gill, K.S. (2002) Gene-containing regions of wheat and the other grass genomes. *Plant Physiology* 128, 803–811.

Santos, J.L., Afara, D., Sanchez-Moran, E., Armstrong, S.J., Franklin, F.C.H. *et al.* (2003) Chromosome-dependent partial diploidisation of meiosis in established autotetraploid lines of *Arabidopsis thaliana*. *Genetics* 165, 1533–1540.

Stevenson, M., Armstrong, S.J., Ford-Lloyd, B.V. and Jones, G.H. (1998) Comparative analysis of crossover exchanges and chiasmata in *Allium cepa* x *fistulosum* after genomic *in situ* hybridization (GISH). *Chromosome Research* 6, 567–574.

The Telegraph (2011) Britons must swallow fear of GM crops to feed world. Available at: www.telegraph.co.uk/foodanddrink/8277851/Britons-must-swallow-fear-of-GM-crops-to-feed-world.html (accessed 24 January 2011).

12 High Temperature Stress*

Maduraimuthu Djanaguiraman and P.V. Vara Prasad

Department of Agronomy, Kansas State University, Manhattan, Kansas, USA

12.1 Introduction

Global climate change and increasing human population challenge global food production systems. New arable land is limited so increased productivity has to come from existing land, which necessitates minimizing losses resulting from abiotic and biotic stresses along with efficient harvesting and storage techniques. Global surface temperatures increased by about 0.6°C during the last century. Climate models project global mean temperatures to increase (above values in 1990) in the range of 0.4 to 1.1°C by 2025, 0.8 to 2.6°C by 2050 and 1.4 to 5.8°C by 2100. Projections also indicate increased frequency of short episodes of hot, cold and frost days. Adaptation strategies to combat the negative effects of climate change and climate variability are needed. One promising strategy is to improve existing genotypes and develop new genotypes that can tolerate abiotic stresses, including high temperature stress.

Targeted and controlled breeding programmes generally result in narrow genetic diversity. In most crops, present tolerance to environmental stresses is limited, particularly in the case of high temperature. In the process of domestication or evolution, crops go through a genetic bottleneck that leads to more uniformity, which can make them more vulnerable to abiotic and biotic stresses. Genotype development needs sources of new genes and tolerance traits. Exploration of wild relatives of cultivated crop species, landraces and other genetic resources can provide that opportunity and play a critical role in developing climate-smart genotypes. Wild species have served as gene donors to modern cultivated crops and represent a vital source of untapped genetic diversity that needs to be preserved, expanded and utilized. Wild relatives often contain characteristics useful for tolerance to high temperature, drought, salinity tolerance and disease resistance, which can permit the adaptation of crops to a far wider range of environments and stresses.

12.2 Impact of High Temperature Stress on Crops and their Wild Relatives

Climatic conditions, particularly temperature and photoperiod, determine the distribution of a plant species by acting directly through physiological constraints (growth and reproduction) or indirectly through ecological factors (competition for resources). Changes in climatic conditions, including temperature in the past century, have had significant effects on the distribution, abundance, phenology and physiology of a wide range of crop species. Modelling studies indicate that climate change may lead to large-scale shift in geographical location and survival of species. Species capable of migrating at high rates and higher fitness are more likely to survive and dominate, and in some cases may gain geographic range (Thomas *et al.*, 2004; Menendez *et al.*, 2006).

Jarvis *et al.* (2008) reported that climate change significantly decreased the suitable area of cultivation of 80 to 100% of species, and the size of suitable area of some crop species such as groundnut, potato and cowpea (*Arachis, Solanum* and *Vigna*) decreased by 98 to 100%, 80 to 100% and 63 to 100%, respectively. The study also indicated that 16 to 22% of all species modelled were predicted to become extinct. Fischer *et al.* (2002) predicted the loss of suitable area for crop production in sub-Saharan Africa, which already has harsh environments; however, some regions will experience considerable expansion of suitable arable land (higher altitudes) and have the potential to increase crop production area. Similarly, Lane and Jarvis (2007) also reported loss in suitable area in sub-Saharan Africa, the Caribbean, India and northern Australia, and gains in the northern USA, Canada and most of Europe.

12.3 Impact of High Temperature Stress on Species Fitness

Species fitness described here refers to the germinability and vigour of seeds produced under different environmental conditions. The offspring (seed) must survive to contribute to the following generation, so fitness has a generational value. Limited information is available on the effects of climate change, particularly the effects of high temperature on fitness. Daytime maximum and night-time minimum temperatures of 35/30°C, 38/33°C and 38/27°C during seed filling stages decreased seed germination of several cultivars of soybean (Gibson and Mullen, 1996; Spears *et al.*, 1997; Egli *et al.*, 2005). Zanakis *et al.* (1994) observed that germination and vigour decreased by 28% and 38%, respectively, when the plants were exposed to high temperature stress (35–43°C) during seed development and maturation periods of soybean. The degree of decrease in germination potential due to high temperature stress was influenced by the duration of temperature and the phase of reproductive growth when the stress occurred. Greater decreases were observed with longer duration of exposure to high temperatures, especially during the seed filling period and maturation. Thomas *et al.* (2009) observed that growth temperature influenced percentage seedling emergence and early seedling vigour in dry bean. Growth temperatures of 34/24°C decreased total percentage seedling emergence compared with 28/18°C. Similarly, developmental time to the second trifoliate stage was longer, and the rate of development was slower in seeds that formed and matured at growth temperature of 34/24°C compared with 28/18°C. Ellis (2011) reported that seed quality in rice may be less sensitive to high temperatures during later stages of seed development and maturation than during early seed development stages.

Baskin and Baskin (1998) observed that an increase in temperatures during seed growth and development has a pre-conditioning effect that decreases the seed dormancy in some plant species. Fenner (1991) listed 15 cases in which high temperatures during maturation resulted in lower dormancy and three plant species in which they resulted in higher dormancy.

12.4 Role of Crop Wild Relatives and Landraces for High Temperature Stress Tolerance

Limited information is available on the role of wild relatives in improving high temperature tolerance in crop species. Within wild species, wide genetic variation for high temperature tolerance exists among accessions originating from a diverse range of environments (Nautiyal *et al.*, 2008). Crops evolved from hot environments have superior alleles for thermostable proteins encoding important enzymes such as Rubisco activase and superoxide dismutase (Salvucci and Crafts-Brandner, 2004; Khanna-Chopra and Sabarinath, 2004). Hajjar and Hodgkin (2007) reported that a chickpea cultivar with enhanced thermotolerance was derived from wild species (*Cicer reticulatum*); similarly, the wild potato *Solanum gandarillasii* exhibited higher membrane stability than cultivated potato (Coleman, 2008). Researchers also have explored wild relatives and landraces for wheat and rice.

Hede *et al.* (1999) evaluated 2255 Mexican wheat landraces for traits associated with high temperature tolerance (leaf chlorophyll content, canopy temperature depression and kernel weight) in two Mexican environments. Three landrace accessions (CWI 59788, CWI 60155 and CWI 60391) were identified as having superior performance in both environments. In wheat, high temperature tolerance during the vegetative stage was higher in wild relatives of *Aegilops speltoides* and *Aegilops tauschii* than in *Triticum urartu* and *Triticum boeoticum* (Waines, 1994). A high level of tolerance was also observed in the line *Triticum dicoccoides* 600808 from Jordan (Rekika *et al.*, 1997).

Pradhan *et al.* (2012) evaluated 52 accessions belonging to five *Aegilops* species at optimum temperature (25/19°C day/night) and high temperature (36/30°C). Stress was imposed at anthesis and continued for 16 days. Across species, high temperature decreased chlorophyll, grain number spike^{-1}, individual grain weight and grain yield plant^{-1} by 38%, 40%, 56% and 70%, respectively. Among the five species, *A. speltoides* and *Aegilops geniculata* had greater tolerance to high temperature for yield (58–61% decline from optimum temperature), and *Aegilops longissima* was highly susceptible (84% decline). Tolerance was associated with greater grain number spike^{-1} and/or individual grain weight. Within *A. speltoides*, accession TA 2348 was highly tolerant to high temperature, with a 13.5% decline in grain yield and a high temperature susceptibility index (HSI) of 0.23. The highly susceptible accessions were TA 1787 and TA 2097, with >82% yield decline and HSI >1.4. Among *A. geniculata* accessions, two were moderately high temperature tolerant, TA 2899 and TA 1819, with an HSI of 0.80. These results suggest that genetic variability among *Aegilops* species can be utilized for improving high temperature tolerance in wheat during the reproductive stages of crop development (Pradhan *et al.*, 2012).

High-temperature-tolerant traits also can be derived from interspecific or intraspecific hybrids such as synthetic wheat (Van Ginkel and Ogbonnaya, 2007). Some primary synthetic wheats have higher chlorophyll content under high temperature growth conditions and more tolerance to high-temperature-induced acceleration of grain filling duration, kernel weight and yield than wheat cultivars (Yang *et al.*, 2002).

Rice species differences in spikelet fertility and harvest index under high temperature stress showed that N-22, an *indica* type, was the most tolerant, and L-204, a *japonica* type, was most susceptible (Prasad *et al.*, 2006b). It is not possible, however, to generalize that *indica* types are more tolerant to high temperatures than *japonica* types. Liao *et al.* (2011) developed backcross recombinant lines from the cross of Xieqingzao B × N22// Xieqingzao B for high temperature tolerance at the milky stage. For F_1, N22 was used as the male parent and Xieqingzao B as the female, and the F_1 was backcrossed with Xiegingzao to obtain BC_1F_1 and the generations were forwarded until BC_1F_8. The crossing yielded 12 (combination of six pairs of tolerant and susceptible lines) rice lines with similar phenotypic characters but significantly greater tolerance to high temperature. Finally, two backcross introgression rice lines, 703T (heat tolerant) and 407S (heat sensitive), with similar genetic background (genomic polymorphism 1.8%) but that significantly differed in heat tolerance at milking stage, were finally selected and renamed as XN0437T (heat-tolerant) and XN0437S (heat-sensitive), respectively.

In rice, time of day of flowering can be used as a mechanism to avoid high temperatures during anthesis, pollen germination and pollen tube growth. Genotypes are known to differ in time of day of flowering (Prasad *et al.*, 2006b; Jagadish *et al.*, 2010). Ishimaru *et al.* (2010) and Jagadish *et al.* (2010) reported that more than 4000 *Oryza sativa* ssp. *indica* accessions were field-tested at the International Rice Research Institute (IRRI) in the Philippines to evaluate the use of the early morning flowering (EMF) trait as a viable strategy to overcome the adverse effect of high temperature stress during the later hours of the morning. They identified few *O. sativa* accessions that possessed this trait. Similarly, screening of wild rice accessions at IRRI has reconfirmed the EMF trait in *Oryza minuta* and *Oryza officinalis*. Ishimaru *et al.* (2010) successfully introgressed the EMF trait from a wild rice *O. officinalis* into the modern variety Koshihikari (*O. sativa*), and the EMF-introgressed line started and completed flowering a few hours before Koshihikari. This shift in flowering time toward the cooler early morning resulted in significantly lower high-temperature-induced spikelet sterility than in the wild type. Scafaro *et al.* (2010) reported that *Oryza meridionalis*, a wild relative of *O. sativa*, was found to be high temperature tolerant compared with *O. sativa* ssp. *japonica* cv. Amaroo. Heat tolerance in *O. meridionalis* was established by comparing leaf elongation, photosynthetic rates and accumulation of heat-induced protein, namely Rubisco activase, transketolase, HSP70, Cpn60, GDC-P and the putative thiamine biosynthesis protein at high temperature. The available wild accessions that have the genes for high temperature stress tolerance in some selected plants are listed in Table 12.1.

12.5 Impact of High Temperature Stress on Physiological, Growth and Yield Processes

High temperature stress is known to influence plant growth and development and various physiological and yield processes. Some critical responses are discussed below and presented in Table 12.2.

12.5.1 Physiological processes

High temperature stress decreases leaf chlorophyll content (Prasad *et al.*, 2008a,b; Djanaguiraman *et al.*, 2011b). The loss of chlorophyll during high temperature stress resulted in change in the chlorophyll *a:b* ratio due to premature leaf senescence (Al-Khatib and Paulsen, 1984). The degradation of chlorophyll molecules may be associated with production of reactive oxygen species under high temperature stress (Camejo *et al.*, 2006; Guo *et al.*, 2006). Xu *et al.* (2000) identified three quantitative trait loci (QTLs) for chlorophyll content (Chl1, Chl2 and Chl3) that coincided with three stay-green QTL regions (Stg1, Stg2 and Stg3) in sorghum. The Stg1 and Stg2 regions also contain the genes for key photosynthetic enzymes and heat shock proteins.

Table 12.1. Examples of available wild accessions sources of tolerance to high temperature stress.

Crop	Wild accessions/relatives	Trait	Reference
Rice	*Oryza glaberrima* *Oryza minuta* *Oryza officinalis* *Oryza meridionalis*	Early morning flowering Leaf elongation rate and photosynthetic rates	Ishimaru *et al.* (2010); Scafaro *et al.* (2010)
Wheat	*Aegilops speltoides* *Aegilops geniculata*	Stay-green trait Thylakoid membrane damage Chlorophyll content and yield and yield components	Reynolds *et al.* (1994); Rekika *et al.* (1997) Pradhan *et al.* (2012)
Chickpea	*Cicer reticulatum*	High temperature resistance visual score	Canci and Toker (2008)
Potato	*Solanum gandarillasii*	Membrane stability	Coleman (2008)

Table 12.2. General effects of high temperature stress on various growth and yield parameters of most plant species; exceptions occur at non-optimal growth conditions (any abiotic or biotic stress).

Trait	High temperature stress
Photosynthesis	–
Photorespiration	=
Dark respiration	+
Stomatal conductance	–
Transpiration	+
Water use	+
Water use efficiency	–
Phenology	– +
Senescence	+
Plant height	–
Branching	–
Total leaf area	–
Leaf growth and expansion	–
Leaf thickness	–
Root growth	–
Nodulation and N fixation	–
Dry mater production	–
Flowering	–
Percentage fruit-set	–
Pollen viability	–
Pollen production	–
Fruit (grain) number	–
Seed, grain or fruit size	–
Harvest index	–
Quality (Seed N)	–
Fitness	–

+, increases; –, decreases; =, no change.

High temperature stress increases thylakoid membrane damage (Prasad *et al.*, 2008a,b; Ristic *et al.*, 2008b; Prasad and Djanaguiraman, 2011). Photosystem (PS) II is considered to be highly temperature sensitive and its activity decreases at high temperatures (Camejo *et al.*, 2006) due to the changes in the properties of thylakoid membranes where PS II is located (Ristic *et al.*, 2008b). High temperature stress causes dissociation of the oxygen-evolving complex, resulting in an imbalance in electron flow from the complex toward the acceptor side of the PS II reaction centre (De Ronde *et al.*, 2004). The PS I system is usually more conserved under high temperature stress than the PS II (Heckathorn *et al.*, 1998).

High temperature stress can influence photosynthesis either through regulation by stomatal closure and decreasing flow of CO_2 into mesophyll tissue or by directly impairing metabolic activities. High temperature stress causes impaired photosynthetic electron transport, ribulose-1,5-bisphosphate (RuBP) regeneration capacity (Wise *et al.*, 2004), and ribulose 1,5-bisphosphate carboxylase/oxygenase (Rubisco) protein content and decreased Rubisco activity (Sharkey, 2005). In general, during the initial stages of a stress period, decreased conductance through stomata is the primary cause of decline in photosynthesis. Stomatal conductance and net photosynthesis are inhibited by high temperature stress owing to decreased Rubisco activase enzyme (Crafts-Brander and Salvucci, 2002; Morales *et al.*, 2003). However, high temperature increases rubisco catalytic activity, and low affinity between enzyme and CO_2 and enhanced oxygenase-type activity decreases the net photosynthetic rate. Respiration rates are generally known to increase with increasing temperatures but the regulation of respiration under high temperature stress conditions is less well understood. Understanding these responses is important because photosynthesis is temporally (only during daytime) and spatially (only in green tissues) restricted, whereas respiration occurs throughout and in all organs.

12.5.2 Development and growth processes

Temperature significantly influences plant development; thus the concept of thermal time is most commonly used. High temperatures generally increase leaf appearance rates. Leaf-elongation rates increase at high temperatures but leaf-elongation duration decreases. High temperature stress can result in significant increases in leaf numbers, particularly when reproductive development is arrested without any decrease in leaf photosynthetic rates (Prasad *et al.*, 2006a). Cell division and cell growth are the two primary processes involved in plant growth. High temperature stress can stimulate cell division and cell elongation rates. The impact of high temperature stress on leaf-area expansion and dynamics are less well understood.

High temperature stress alters the initiation and duration of developmental phases. In general, time from emergence to initiation of reproductive phase is shorter when temperatures increase occurs within the optimal range. At extreme high temperature stress, however, the duration to panicle initiation and duration from panicle initiation to anthesis can be delayed. For example in grain sorghum, severe high temperature stress (>40°C) during later vegetative phases inhibits the conversion of vegetative to reproductive phase and plants remain vegetative until the stress is relieved. Panicle initiation in sorghum was delayed by 28 days under high temperature stress (Prasad *et al.*, 2008a). High temperature stress can also cause the cessation of panicle development at any stages between panicle initiation and flowering. Severe temperature stress inhibits panicle exsertion and delayed flowering (Prasad *et al.*, 2006b). Once the panicle emerges, the rate of development is primarily a function of temperature, particularly when water and nutrients are not limited.

High temperature stress shortens the duration of the reproductive phase (period during which potential kernel or seed numbers are determined) and grain fill (during which the grain or seed weight are determined). High temperature stress during later stages of panicle or flower development can increase the time from flowering to seed-set (Prasad *et al.*, 1999). Long duration of spikelet development and high spike weight at anthesis is positively correlated with final grain yield in wheat under high temperature stress conditions. For cereal crops, longer periods of vegetative and reproductive development are often necessary to improve reproductive potential (number of productive tillers and kernels) and leaves and tillers to provide assimilate supply during grain fill.

Grain or seed fill duration is the time from seed-set to physiological maturity. For most crop species, particularly those with a physical restriction for growth of seeds, such as rice (which has a fixed pericarp) and legumes such as groundnut or soybean (which have fixed locule size for development of the seed), yield is mainly a function of seed numbers per unit area and seed fill duration. High temperature stress decreases the seed filling duration, leading to smaller seed size. High temperature stress following flowering is known to have little effect on seed filling rates, but seed fill duration is shortened, which leads to small seed size or seed yield. The increase in seed filling rate does not compensate for loss of duration, thus resulting in smaller seed size and seed yields in most grain crops.

12.5.3 Reproductive processes and yield components

High temperature stress has a major influence on plant reproduction. Reproductive processes, particularly those of microsporogensis and megasporogensis, anthesis, pollination, pollen tube growth, fertilization and early embryo development are all highly susceptible to high temperature stress. Failure of any of these processes decreases fertilization or increases early embryo abortion, leading to a lower number of grains and limiting yield.

Reproductive stages of crop development are differentially sensitive to environmental stresses. Stress just prior to anthesis and at anthesis causes a significant increase in floral abortion and lower seed numbers in many grain crops. Exposure to high temperature stress during flowering results in pollen sterility and loss of seed-set in legumes (cowpea, Ahmed *et al.*, 1992; dry bean, Prasad *et al.*, 2002; groundnut, Prasad *et al.*, 2003; soybean, Salem *et al.*, 2007) and cereals (wheat, Saini *et al.*, 1983; rice, Jagadish *et al.*, 2007). Lower seed-set under high temperature stress can be caused by poor anther dehiscence, hence low numbers of germinating pollen grains on the stigma, or by decreased pollen viability or ovule function. Short periods of high temperature stress can also influence pollen viability, seed-set and grain growth (Prasad *et al.*, 2006b). Exposure to temperature of >37°C for a period as brief as 1 h during the flowering stage decreases seed-set in rice (Matsui *et al.*, 2000; Jagadish *et al.*, 2010); similarly, exposure to temperature of >33°C for the first half of the day (6 h after anthesis) decreases pollen viability and thus seed-set in groundnut (Prasad *et al.*, 2000). Short episodes of temperature stress can cause sterility, and the timing of the episode of the high temperature relative to peak flowering will be critical to quantify the impact of high temperature stress, particularly to evaluate the phenotype and for modelling the impact of high temperature stress (Wheeler *et al.*, 2000). High temperature stress during flowering in soybean causes decreased pollen germination by degrading tapetal cells, as well as programmed cell death. Apart from this in sorghum, high temperature stress increases pollen reactive oxygen species production and membrane damage, eventually leading to lower pollen germination potential (Prasad and Djanaguiraman, 2011).

Yield is mainly a function of various components that can be broadly divided into the number of plants (germination and emergence), dry matter production (growth, tillers, potential reproductive sites), seed numbers (reproductive processes and seed-set), and seed-size (product of seed filling rate and seed filling duration). High temperature stress influences yield through seed numbers by influencing pollen or ovule function, which results in lower seed-set. High temperature stress directly influences seed fill duration by decreasing the grain fill duration, leading to smaller seed size and lower yields (Prasad *et al.*, 2002, 2003, 2006a,b, 2008a,b). High temperature stress can also decrease biomass production, particularly if the stress is high enough to cause decrease in photosynthesis. High temperatures in combination with low soil moisture can result in poor seedling emergence and early season vigour.

12.6 High Night-time Temperatures

Night-time temperatures are expected to increase at a faster rate than daytime temperatures under climate change (Alward *et al.*, 1999). Peng *et al.* (2004) ascribed year-to-year variation in rice grain yield to night-time temperature. High night-time temperature had a more pronounced negative effect than daytime temperatures on the yield of rice, with a 10% reduction in yield for every 1°C increase in night-time temperature and an overall increase in mean night-time temperature of 1.13°C from 1992 to 2003. Studies also show that high night-time temperature decreases pollen viability and spikelet fertility in rice (Mohammed and Tarpley, 2009). High night-time temperatures (22/34°C, day/night) are more detrimental to seed size (grain weight) in rice than high daytime temperatures (34/22°C) (Morita *et al.*, 2002). High temperatures increase the seed growth rate in the early ripening period but decrease the duration of seed growth, ultimately resulting in lower grain weight. Similarly, wheat yields decreased linearly with increasing night-time temperatures and, when compared with the control (14°C), grain fill duration decreased by 3 and 7 days at night temperatures of 20°C and 23°C, respectively (Prasad *et al.*, 2008b). High night-time temperature (>20°C) decreased spikelet fertility, grains per spike and grain size, and photosynthesis decreased at night temperatures >14°C. High night-time temperatures also significantly decreased yields in soybean and grain sorghum by increasing respiration rate and decreasing photosynthesis rate (Prasad and Djanaguiraman, 2011; Djanaguiraman *et al.*, 2013). High night-time temperature increased oxidative damage in leaves and pollen grains (Prasad and Djanaguiraman, 2011), leading to decreased pollen viability and germination, which eventually resulted in decreased pod/seed set.

12.7 Stages Sensitive to High Temperature Stress

The impacts of high temperature stress on crop growth and yield depend on the severity and duration of stress and the plant developmental stage at which the stress occurs. High temperature stress tolerance tends to be the greatest during early vegetative stages and decreases progressively during flowering and early seed filling stages. The most sensitive stages of development to high temperature stress generally are during panicle development and flowering in cereals and the period just prior to flowering and during flowering in legumes. In most cases, the reproductive growth and processes are more sensitive than vegetative growth and processes. Understanding the mechanisms that lead to improved reproductive success at high temperatures is critical to the development of high-temperature-tolerant crops. The stages sensitive to high temperature stress in selected crop species are given in Table 12.3.

12.8 Traits Used in Breeding for High Temperature Stress

12.8.1 Physiological processes

High temperature stress alters membrane functions by altering membrane fluidity. In plant cells, membrane function is especially important for crucial processes such as photosynthesis and respiration. Commonly used assays for testing membrane damage are the cell membrane stability assay and the chlorophyll *a* fluorescence assay. The loss of membrane integrity may be the cause of ion leakage from the cell; this could be caused by inhibition of membrane-bound enzyme activity under high temperature stress. Thylakoid membrane damage (F_o/F_m ratio) can be assessed by measuring ground state chlorophyll *a* fluorescence (F_o) and maximum chlorophyll *a* fluorescence (F_m) using a chlorophyll fluorometer. Thylakoid membranes and plasma membranes are prone to damage by high temperatures, and genotypes are known to differ in their response. These differential responses can be sensed by changes in chlorophyll fluorescence traits (Ristic *et al.*, 2008a,b).

High temperature stress decreases chlorophyll content (Prasad *et al.*, 2008a,b). Studies on sorghum, wheat and soybean in controlled environments revealed significant variability in photosynthetic rate and stomatal conductance when plants are exposed to high temperature stress, which is partly explained by differences in chlorophyll content (Reynolds *et al.*, 1994; Prasad *et al.*, 2006a; 2008b; Djanaguiraman *et al.*, 2011b).

Carbon isotope analysis integrates the ratio of stable isotopes of carbon (^{13}C:^{12}C) across the life of the analysed plant tissue.

Table 12.3. Stages sensitive to high temperature stress in selected crop species.

Crops	Stages	References
Rice	Early microspore following meiosis and flowering	Endo *et al.* (2009); Matsui and Omasa (2002)
Wheat	Floret formation and anthesis	Saini and Aspinall (1982); Wardlaw and Wrigley (1994)
Maize	Tasselling and seed set	Schoper *et al.* (1986, 1987)
Sorghum	Panicle initiation and anthesis	Prasad *et al.* (2008a)
Kidney bean	6–15 days before anthesis and anthesis	Monterroso and Wien (1990)
Common bean	6–8 days before anthesis and anthesis	Gross and Kigel (1994)
Cowpea	6–12 days before anthesis	Ahmed *et al.* (1992)
Chickpea	10 days before flowering, anthesis and pod development	Wang *et al.* (2006a)
Soybean	Flowering and pod set	Djanaguiraman *et al.* (2011b, 2013)
Groundnut	3 days before anthesis and 15 days after anthesis	Prasad *et al.* (1999)

Stable carbon isotope discrimination differences were positively associated with high temperature stress tolerance (Djanaguiraman *et al.*, 2011a). Saranga *et al.* (2001) observed QTLs associated with carbon isotope discrimination, which can help identify the contribution of stomatal conductance and photosynthetic capacity relationships under high temperature stress.

The canopy temperature depression (CTD) is highly correlated with yield in wheat and thus is identified as an escape mechanism (Cornish *et al.*, 1991; Reynolds *et al.*, 1994). On the other hand, evaluation of sorghum genotypes revealed both escape and true tolerance. Certain genotypes maintained elevated leaf temperature with no decreases in grain yield; that can be exploited for true high temperature tolerance (Mutava *et al.*, 2011). Such mechanisms should be exploited in other grain crops and both escape and tolerance mechanisms need to be exploited during breeding for high temperature stress.

12.8.2 Reproductive processes and partitioning

In rice, spikelet sterility caused by high temperature may be due to reduced pollen production or viability or number of germinated pollen grains on stigmatic surface or ovary fertility. Genotypes of grain crops are known to differ in their sensitivity to high temperature stress in terms of pollen viability, spikelet fertility and seed set (Kakani *et al.*, 2002; Craufurd *et al.*, 2003; Prasad *et al.*, 2006a,b). Cardinal temperatures for pollen grain germination (T_{min}, minimum temperature below which pollen does not germination; T_{opt}, optimum temperature at which pollen germination was maximum; and T_{max}, maximum temperature above which pollen grains fail to germinate) can be used for screening germplasm for high temperature stress tolerance, as observed in groundnut and cotton (Kakani *et al.*, 2002, 2005). In general, there is a strong positive relation between pollen viability or pollen number and seed-set; thus, genotypes with a higher T_{opt} (optimum) or T_{max} (ceiling temperature) may show greater tolerance to higher temperatures, particularly during reproductive stages of crop development. In addition, the occurrence of flowering early in the morning may be a useful trait that imparts high temperature tolerance by avoidance in rice (Jagadish *et al.*, 2007, 2010; Ishimaru *et al.*, 2010) and not true tolerance.

High temperature stress decreases the crop growing season by increasing the crop development rate, and a linear relationship has been observed between temperature and rate of development from base temperature to optimum temperature, above which the rate of development can decline. Photosynthate production is more likely to be limiting under high temperature stress conditions if the stress occurs during grain or seed filling stage. The reduction in photosynthate production eventually results in reduced transport of carbon to the developing seed, when the demand for assimilates is greatest. The total above ground biomass shows a stronger association with grain yield, which influences harvest index. Yield and yield components in stress conditions remain, however, the most effective tools for stress evaluation. Our experiments on rice, sorghum, wheat, groundnut and soybean showed that seed-set percentage or pod-set percentage, seed size, harvest index and seed dry weight are additional indicators of high temperature stress tolerance (Prasad *et al.*, 2008a,b; Djanaguiraman *et al.*, 2011b). Various traits of interest for high temperature stress and their association with yield are presented in Table 12.4.

12.9 Approaches for Developing High Temperature Stress Tolerant Genotypes

The traditional method for breeding for high temperature stress is to grow advanced breeding lines in a target environment and select those lines that have a yield advantage over the current locally adapted variety in that particular environment. Indirect selection using high-temperature-tolerant

Table 12.4. Examples of traits of interest for screening high temperature stress and its relationship with yield.

Traits	Relationship with yield	References
Cell membrane stability	Positive	Blum *et al.* (2001)
Canopy temperature depression	Positive	Reynolds *et al.* (1994)
Carbon isotope discrimination differences	Positive	Monneveux *et al.* (2003)
Leaf chlorophyll content	Positive	Ristic *et al.* (2008b)
Thylakoid membrane damage	Negative	Prasad *et al.* (2008a,b)
Photosynthesis	Positive	Prasad and Djanaguiraman (2011); Prasad *et al.* (2008b)
Stomatal conductance	Positive	Reynolds *et al.* (2007)
Leaf carbohydrates and related enzymes	Positive	Lafta and Lorenzen (1995)
Reactive oxygen species	Negative	Suzuki and Mittler (2006)
Antioxidant enzymes activity	Positive	Suzuki and Mittler (2006)
Protein synthesis elongation factor (EF-Tu), Rubisco activase (RCA)	Positive	Ristic *et al.* (2007); Ristic *et al.* (2008a)
Pattern of flowering, time of the day of flowering (early morning flowering)	Positive	Ishimaru *et al.* (2010); Jagadish *et al.* (2007)
Panicle exertion	Positive	Prasad *et al.* (2006b)
Pollen production	Positive	Prasad *et al.* (2006b)
Pollen dehiscence and sugar contents	Positive	Jain *et al.* (2007)
Pollen viability and germination; number of germinated pollen grains on stigma	Positive	Prasad *et al.* (2006a); Jagadish *et al.* (2010)
Seed-set percentage	Positive	Prasad *et al.* (2006a; 2008a)
Seed size	Positive	Prasad *et al.* (2006a; 2008a)
Stay green	Positive	Reynolds *et al.* (2001)

traits is the other efficient approach; it involves early generation selection for specific traits during a particular stage of crop development (e.g. vegetative, flowering or grain filling). The breeding approaches vary from crop to crop. Traditional backcross approaches, marker-assisted breeding involving QTLs and transgenic approaches are some of the breeding methods to improve high temperature stress tolerance in crops.

12.9.1 Traditional backcross approach

Rodriguez-Garay and Barrow (1988) developed high-temperature-tolerant cotton lines by the backcross method. A high-temperature-tolerant line (7456) of *Gossypium barbadense* L. was used as the donor parent and Paymaster 404 was used as the high-temperature-sensitive line. The crosses were made with pollen treated for 15 h at 35°C to generate F_1, F_2, and first and second backcross populations. Increased temperature tolerance was observed in plants from all populations. Similar observations were also noticed in controlled environment studies.

New rice varieties with high temperature stress tolerance at the milky stage (791 backcross introgression lines) were derived from the cross of high-temperature-tolerant rice line N22 and high-temperature-sensitive rice line Xieqingzao B. Through evaluation of phenotypic characteristics and comparison of genomic polymorphism, two backcross introgression rice lines, 703T (high temperature tolerant) and 704S (high temperature sensitive), with similar genetic backgrounds but significantly different high temperature tolerance at the milky stage, were selected and renamed as XN0437T (high temperature tolerant) and

XN0437S (high temperature sensitive) (Liao *et al.*, 2011).

In cowpea, high night-time temperatures were more deleterious to grain yield than high daytime temperatures. High night-time temperature caused male sterility and substantially decreased grain yield by increasing floral abscission and decreasing the number of pods m^{-2}. Breeding cowpea for high temperature tolerance involves exposing progenies to very high night temperatures and long daytime conditions, then selecting plants with the ability to produce more flowers and greater pod-set. Tolerance to high temperature stress during pod-set was more difficult to incorporate, even though it is mainly controlled by a single dominant gene, because the heritability was low (0.26) and the inheritance also depends on some minor genes. However, the electrolyte leakage from the leaf discs subjected to high temperature stress significantly correlated with pod set percentage (Hall, 2004). High-temperature-tolerant genes (or closely linked genes) also had a progressive dwarfing effect, which resulted in reduced plant height and increased pod-set percentage. Successful cowpea genotypes with improved high temperature tolerance have been developed (Hall, 2004).

Nine high-temperature-tolerant tomato breeding lines, four high-temperature-tolerant cultivars and four high-temperature-sensitive cultivars were evaluated for high temperature stress tolerance on the basis of fruit set, yield, fruit quality and seed production. Under high temperature conditions, the temperature-tolerant lines, the temperature-tolerant cultivars and the temperature-sensitive cultivars produced 186, 94 and 55 flowers per plant; and had 70%, 52% and 30% of fruit set per plant; and yield of 410, 173 and 11 g per plant, respectively. High temperature stress caused increased flower abscission and reduced fruit set and yield (Abdul-Baki, 1991). Alsadon *et al.* (2006) classified tomato cultivars into high temperature tolerant and susceptible on the basis of leaf electrolyte leakage. Temperature-stress-tolerant cultivars had an electrical conductivity (EC) of 27.75 to 41.69 μmhos cm^{-1} (Queen, Indian, Super Strain B, Pearson and Strain-B); moderately temperature-stress-tolerant cultivars had an EC of 43.80 to 50.70 μmhos cm^{-1} (Pacmore, Tnshet Star, Rocky, Shohba, Raad VF, VFN- 8, Moneymaker, Imperial and Pearson Improved), and temperature-stress-sensitive cultivars had an EC of 53.12 to 72.20 μmhos cm^{-1} (Malica, Castle Rock, Edkawi, Chico, Pakmore VF and Super Marmand). Pollen release and germination ability under high temperature were revealed to be the most important factors determining the fruit set ability and could be used in breeding programmes aiming for better fruit set under high temperatures in tomato.

In snap bean, a cross was made between a high-temperature-tolerant (Cornell 503) and a sensitive line (Majestic) to enhance tolerance using the abscission of reproductive organs as an indicator. The distribution of abscission scores in segregating generations (F_2 and backcrosses) indicated that a high rate of abscission in response to high temperature stress was controlled by a single recessive gene from Majestic (Rainey and Griffiths, 2005).

12.9.2 Quantitative trait loci approach

High temperature tolerance is not controlled by a single 'thermotolerant' gene in most crops. Different components of tolerance may be determined by different sets of genes at different stages of the crop growth. QTL analysis can be used to identify candidate genes underlying a trait. Once a region of DNA is identified as causal to a phenotype, it can be sequenced. The DNA sequence of any genes in this region can then be compared to a database of DNA for genes whose functions are already known. Studies of genetic diversity among cultivars, wild accessions and ecotypes are useful for discovery of novel QTLs and alleles that can be further exploited for the high temperature stress tolerance. Populations of wild species frequently harbour high intraspecies variation for tolerance traits that is absent in modern cultivars. A 'constitutive' QTL is detected consistently across most environments, whereas an 'adaptive' QTL is detected only in specific environmental conditions or increases in expression with the level of an environmental factor.

Rice genetic resources tolerant to high temperature were found in both *indica* and *japonica* subspecies. QTL mapping studies for high temperature tolerance have been conducted on various rice populations at booting (Zhao *et al.,* 2006) and flowering stages (Xiao *et al.,* 2011). Two QTLs underlying tolerance to high temperature stress were identified using recombinant inbred lines derived from a cross between high temperature stress tolerant rice cultivar 996 and sensitive cultivar 4628. Pollen fertility was used as an indicator of high temperature tolerance for the lines subjected to high temperature stress at the flowering stage. Two QTLs that affected pollen fertility, qPF4 and qPF6, were detected between RM5687 and RM471 on chromosome 4, and between RM190 and RM225 on chromosome 6, by using the composite interval mapping analysis. The two QTLs explained 15.1% and 9.31% of the total phenotypic variation in pollen fertility and increased the pollen fertility of the plants subjected to high temperature stress by 7.15% and 5.25%, respectively (Xiao *et al.*, 2011). QTLs for high temperature stress tolerance using the N22 rice variety were detected by Ye *et al.* (2012), who identified four chromosomal locations associated with spikelet fertility under high temperature conditions by genotyping a BC_1F_1 population and four putative QTLs by composite interval mapping of an F_2 population. The QTL on chromosome 1 was identified in both BC_1F_1 and F_2 populations. This QTL explained 12.6% of the variation of spikelet fertility under high temperature conditions. The QTL with the highest logarithm of the odds ratio score (LOD) (6.66) for spikelet fertility under high temperature was located on chromosome 4. This QTL explained 17.6% of the variation of spikelet fertility under high temperature conditions. The other QTL had a low LOD value. QTLs for rice high temperature stress tolerance at the flowering stage have been mapped on all chromosomes except chromosomes 6 and 7. Comparing the results from different studies clearly showed that QTLs on chromosomes 1, 2, 3, 4, 8, 9, 10 and 11 were common in at least two different studies.

QTLs for high temperature tolerance in wheat have been reported for different traits, such as grain filling duration, canopy temperature depression, and yield and senescence-related traits. In a recombinant inbred line (RIL) population of spring wheat, cellular membrane thermostability (CMS) was correlated with seedling thermotolerance and yield of RIL grown under hot summer conditions (Fokar *et al.*, 1998; Blum *et al.*, 2001). Mitochondrial respiration activity as measured by the triphenyl-tetrazolium chloride (TTC) reduction assay under high temperature stress did not correlate significantly (Fokar *et al.*, 1998). Involvement of heat shock proteins (HSPs) in temperature tolerance was suggested in a co-segregation analysis of a small subset of RILs of winter wheat (Joshi *et al.*, 1997). RILs were developed by crossing high-temperature-tolerant (NW1014) and susceptible (HUW468) cultivars. Composite interval mapping showed that significant genomic regions on 2B, 7B and 7D were associated with high temperature tolerance. Of these, two (2B and 7B) were co-localized QTLs and explained more than 15% phenotypic variation for heat susceptibility index of thousand-grain weight, heat susceptibility index of grain filling duration and canopy temperature depression (Paliwal *et al.*, 2012). A RIL population derived from the varieties Halberd and Cutter was evaluated for high temperature stress tolerance 10 days after pollination. QTL analysis identified 15 and 12 QTLs associated with HSI in two years of study. Five QTL regions were detected on chromosomes 1A, 2A, 2B and 3B. These same regions were commonly associated with QTLs for flag leaf length, width and visual wax content. Pleiotropic trade-offs between the maintenance of kernel number versus increasing single kernel weight under high temperature stress were present at some QTL regions (Mason *et al.*, 2010). Using composite interval mapping, cooler canopy temperatures and higher grain yields were associated with the Babax allele on linkage groups 1B, 3B and 4A under high temperature conditions. Cool canopy effects also were detected due to the Seri allele on linkage group 2B but with no effect on yield. The effects of these QTLs were to decrease canopy temperatures by 0.2 to 0.5°C (Pinto *et al.*, 2008).

QTL analysis has been used to dissect high temperature stress tolerance in maize RIL population (T232 x CM 37) by analysing CMS (Ottaviano *et al.* 1991). They detected six QTLs that accounted for 53% of the genetic variability of CMS. In the same RIL population, researchers identified five QTLs for pollen ability to germinate at high temperature and six QTLs for pollen tube growth at high temperature (Frova and Sari-Gorla, 1993). The analysis of the correlation between HSP and CMS revealed no significant association of the two traits. Maize QTLs (three) associated with tolerance to high seed drying temperature were identified and mapped to chromosomes 6 and 8, explaining 39% of the phenotypic variation of the trait with additive, dominance and over-dominance gene action (Salgado *et al.* 2008).

The great challenge in enhancing high temperature stress tolerance based on the QTL approach for exploiting the segregation of natural alleles is that QTLs often do not relate well across genetic backgrounds, frequently showing smaller adaptation effects or disappearing altogether in different backgrounds, even under the same growing conditions. The candidate gene approach is better suited for QTL cloning when genes are known to control the same or a similar trait in another species. Another possibility is to consider candidates on the basis of a specific biochemical activity rather than biological function. The candidate gene approach appears to be most successful when a specific biochemical activity underlying the tolerance trait is known. Transcriptome profiling from model or target crop species also can be used to identify candidate genes.

12.9.3 Transgenic approaches

Gene expression under stress conditions can be classified as: (i) genes encoding proteins with known enzymatic or structural functions; (ii) proteins with unknown functions; and (iii) regulatory proteins. Initial attempts to develop transgenics involved single action genes that would confer tolerance to stress. Stress-induced proteins with known functions such as water channel proteins, osmolyte biosynthesis (proline, betaine, sugars such as trehalose, and polyamines), detoxification enzymes and transport proteins were the initial targets; however, stress tolerance is mediated by many genes at a time, so single-gene tolerance is unlikely to be sustainable. Attempts to transform plants with stress-induced genes (regulatory proteins) have therefore emerged. Through these proteins, many genes involved in the stress response can be regulated simultaneously by a single gene encoding a stress-inducible transcription factor.

Heat shock protein (HSP) genes

The increased transcription of a set of genes in response to temperature is a highly conserved biological response in all organisms. The response is mediated by heat shock transcription factor (HSF), which is present in a monomeric, non-DNA binding form in unstressed cells and is activated by stress to a trimeric form that can bind to promoters of heat shock genes. The induction of genes encoding heat shock proteins is one of the most prominent responses observed at the molecular level of organisms exposed to high temperature (Vierling, 1991). High-temperature-stressed maize had more abundant HSP101 in tassel, endosperm and embryos than in roots and leaves (Young *et al.*, 2001). In elongating segments of primary roots in maize, HSP101 is produced and plays an important role in high temperature stress tolerance (Nieto-Sotelo *et al.*, 2002). Two tomato genotypes, *Lycopersicon esculentum* and *Lycopersicon pennellii*, differing in HSP synthesis, were crossed. The results showed that F_1 produced only half of the *L. esculentum*-specific HSPs and all of the *L. pennellii* HSPs. In barley, F_1 plants obtained from a cross between Onice × Georgie expressed HSP patterns more similar to Georgie than to Onice. F_1 also produced several unique HSPs, indicating that F_1-specific genetic backgrounds activate heat shock genes normally not expressed in both parental backgrounds (Marmiroli *et al.*, 1989). In maize, the F_1 hybrid from B73 × Mo17 produced all HSPs of both parents, indicating dominant inheritance. In F_2 progeny, alleles for six low molecular weight HSPs show typical 3:1 Mendelian ratios for dominant expression pattern (Fu *et al.*, 2012).

Protein synthesis elongation factor Tu (EF-Tu)

Protein synthesis elongation factor Tu (EF-Tu) is a protein that plays an important role in high temperature stress tolerance. Initially in maize, EF-Tu was reported as a 45-kDa HSP, which expressed at higher level under high temperature stress and acts as a molecular chaperone. The positive correlation between plastid EF-Tu accumulation and the high-temperature-tolerant phenotype was reported for several maize hybrids (Ristic *et al.*, 1991, 1996). A study in winter wheat showed that the EF-Tu gene expression is up-regulated under heat stress conditions, and wheat cultivars that accumulate more EF-Tu protein display better tolerance to high temperature stress (Ristic *et al.*, 2008b). High temperature stress-induced expression/accumulation of both EF-Tu and EF-1α has been observed in mature plants of spring high temperature cultivars (Bukovnik *et al.*, 2009). Fu and Ristic (2010) hypothesized that over-expression of an EF-Tu gene may enhance high-temperature tolerance in wheat and introduced a maize plastid EF-Tu gene (*Zmeftu1*) into two cultivars of hexaploid wheat (Bobwhite (BW) and Xinchun 9 (XC9)). The transgenic plants with increased levels of the plastid EF-Tu had reduced thylakoid membrane damage and enhanced photosynthetic rate and grain yield upon exposure to high temperature stress. The enhanced CO_2 fixation may be due to the EF-Tu factor's protection of photosynthetic membranes and photosynthesis-related enzymes under the high temperature stress.

Detoxification pathway genes

Overexpression of CBF1/DREB1B genes resulted in improved tolerance to temperature stress in crop plants such as rice, wheat and canola. Increase in biosynthesis of proline was achieved by co-expression of the *Escherichia coli* P5C biosynthetic enzymes gamma-glutamyl kinase 74 (GK74) and gamma-glutamylphosphate reductase (GPR) and the antisense transcription of proline dehydrogenase (ProDH) in *Arabidopsis* and tobacco. The transgenic plants had improved high temperature stress tolerance because of elevated accumulation of cell wall proline-rich proteins. Simultaneous co-expression of dehydroascorbate reductase (DHAR), glutathione reductase (GR) or glutathione-S-transferase (GST) and glutathione reductase (GR) in tobacco plants also resulted in the increased tolerance of temperature stress in the transgenic plants (Reguera *et al.*, 2012). Reactive oxygen species (ROS) are produced under high temperature stress conditions; these ROS are toxic molecules that cause oxidative damage to proteins, DNA and lipids. Overexpression of Mn-superoxide dismutase (Mn SOD) that mediates the conversion of O_2^- to H_2O_2, in wheat resulted in higher tolerance to temperature stress and improved yields under field conditions (Gusta *et al.*, 2009). Ascorbate peroxidases (APX) and catalases (CAT) are two important enzymes that participate in H_2O_2 detoxification. Expression of the cAPX gene in tomato improved tolerance to high temperature stress (Wang *et al.*, 2006b). The combined expression of antioxidant enzymes may be a promising strategy to enhance high temperature stress tolerance. Constitutive expression of the stress-response transcriptional coactivator multiprotein bridging factor 1c (MBF1c) in *Arabidopsis thaliana* enhances the tolerance of transgenic plants to bacterial infection, high temperature and osmotic stress. Moreover, the enhanced tolerance of transgenic plants to osmotic and high temperature stress was maintained even when these two stresses were combined. Transcriptome profiling and inhibitor studies suggest that MBF1c expression enhances the tolerance of transgenic plants to high temperature and osmotic stress by partially activating, or perturbing, the ethylene-response signal transduction pathway (Suzuki *et al.*, 2005). By overexpressing Athsp101 protein, Katiyar-Agarwal *et al.* (2003) generated a high-temperature-tolerant transgenic rice (cv. Pusa basmati 1) line. They observed that almost all the transgenic plants recovered after severe high temperature stress of 45–50°C and exhibited vigorous growth during the subsequent recovery at 28°C, whereas the untransformed plants could not recover to a similar extent. High-temperature-tolerant rice plants have also been produced by overexpressing a rice small HSP, sHSP17.7 (Murakami *et al.* 2004). Qi *et al.* (2011) have reported that mtHsp70 overexpression suppresses programmed cell death by maintaining mitochondrial membrane potential and preventing ROS signal amplification in rice protoplasts.

12.10 Conclusions

High temperature stress is one of the major environmental factors limiting productivity of grain crops. High temperature stress decreases grain yield by influencing several physiological, growth and yield processes. Improving tolerance of food grain crops to high temperature is critical to improving yield and food security. Breeding for high temperature tolerance is vital not only in current climates, but also in future projected climates with increased frequency of high temperature stress events. New sources for high temperature tolerance need to be exploited. Crop wild relatives and landraces have the potential to contribute high temperature tolerance to grain crops, but wild accessions and landraces that may provide the highest number of tolerance genes are under-represented in many germplasm collections. Only about 8% of accessions in the world crop germplasm collections are crop wild relatives, and some collections have little or no wild germplasm. Improvements in molecular technology are contributing to faster identification of traits in wild relatives, and innovative methods have helped develop new and improved varieties. Most of these resources remain untapped and are potentially precious. Wild genetic resources can thus provide unique opportunities for the introgression or cloning and transformation of important genes and QTLs from wild to cultivated crops by classical and modern techniques for advancing crop improvement. The principal bottleneck to the use of wild relatives at present is the paucity of material in our collections. Changing climates and climate variability highlight the need to discover more rapidly genetic variability and new sources of tolerance and to exploit all available genetic resources, identify new resources and conserve them for the future.

Note

* Contribution no. 13-160-B from Kansas Agricultural Experiment Station.

References

Abdul-Baki, A.A. (1991) Tolerance of tomato cultivars and selected germplasm to heat stress. *Journal of the American Society for Horticultural Science* 116, 1113–1116.

Ahmed, F.E., Hall, A.E. and DeMason, D.A. (1992) Heat injury during floral development in cowpea (*Vigna unguiculata*, Fabaceae). *American Journal of Botany* 79, 784–791.

Al-Khatib, K. and Paulsen, G.M. (1984) Mode of high temperature injury to wheat during grain development. *Physiologia Plantarum* 61, 363–368.

Alsadon, A.A., Wahb-allah, M.A. and Khalil, S.O. (2006) *In vitro* evaluation of heat stress tolerance in some tomato cultivars. *Journal of King Saud University* 19, 13–24.

Alward, R.D., Detling, J.K. and Milchunas, D.G. (1999) Grassland vegetation changes and nocturnal global warming. *Science* 283, 229–231.

Baskin, C.C. and Baskin, J.M. (1998) *Seeds – Ecology, Biogeography, and Evolution of Dormancy and Germination.* Academic Press, San Diego, California.

Blum, A., Klueva, N. and Nguyen, H.T. (2001) Wheat cellular thermotolerance is related to yield under heat stress. *Euphytica* 117, 117–123.

Bukovnik, U., Fu, J., Bennett, M., Prasad, P.V.V. and Ristic, Z. (2009) Heat tolerance and expression of protein synthesis elongation factors, EF-Tu and EF-1α, in spring wheat. *Functional Plant Biology* 36, 234–241.

Camejo, D., Jimenez, A., Alarcon, J.J., Torres, W., Gomez, J.M. *et al.* (2006) Changes in photosynthetic parameters and antioxidant activities following heat-shock treatment in tomato plants. *Functional Plant Biology* 33, 177–187.

Canci, H. and Toker, C. (2008) Evaluation of annual wild *Cicer* species for drought and heat resistance under field conditions. *Genetic Resources and Crop Evolution* 56, 1–6.

Coleman, W.K. (2008) Evaluation of wild *Solanum* species for drought resistance: 1. *Solanum gandarillasii* Cardenas. *Environmental and Experimental Botany* 62, 221–230.

Cornish, K., Radin, J.W., Turcotte, E.L., Lu, Z.M. and Zeiger, E. (1991) Enhanced photosynthesis and stomatal conductance of Pima cotton (*Gossypium barbadense* L.) bred for increased yield. *Plant Physiology* 97, 484–489.

Crafts-Brander, C. and Salvucci, M.E. (2002) Sensitivity to photosynthesis in the C_4 plant, maize to heat stress. *Plant Cell* 12, 54–68.

Craufurd, P.Q., Prasad, P.V.V., Kakani, V.G., Wheeler, T.R. and Nigam, S.N. (2003) Heat tolerance in groundnut. *Field Crops Research* 80, 63–77.

De Ronde, J.A.D., Cress, W.A., Kruger, G.H.J., Strasser, R.J. and Staden, J.V. (2004) Photosynthetic response of transgenic soybean plants containing an Arabidopsis P5CR gene, during heat and drought stress. *Journal of Plant Physiology* 61, 1211–1244.

Djanaguiraman, M., Prasad, P.V.V., Boyle, D.L. and Schapaugh, W.T. (2011a) High temperature stress and soybean leaves: leaf anatomy and photosynthesis. *Crop Science* 5, 2125–2131.

Djanaguiraman, M., Prasad, P.V.V. and Kassim, A. (2011b) Ethylene perception inhibitor 1-MCP decreases oxidative damage of leaves through enhanced antioxidant defense mechanisms in soybean plants grown under high temperature stress. *Journal of Environmental and Experimental Botany* 71, 215–223.

Djanaguiraman, M., Prasad, P.V.V., Boyle, D.L. and Schapaugh, W.T. (2013) Soybean pollen anatomy, viability, and pod set under high temperature stress. *Journal of Agronomy and Crop Science* 199, 171–177.

Egli, D.B., TeKrony, D.M. and Spears, J.F. (2005) High temperature stress and soybean seed quality: Stage of seed development. *Seed Technology* 33, 345–352.

Ellis, R.H. (2011) Rice seed quality development and temperature during late development and maturation. *Seed Science Research* 21, 95–101.

Endo, M., Tsuchiya, T., Hamada, K., Kawamura, S., Yano, K. *et al.* (2009) High temperatures cause male sterility in rice plants with transcriptional alterations during pollen development. *Plant Cell Physiology* 50, 1911–1922.

Fenner, M. (1991) The effects of the parent environment on seed germinability. *Seed Science Research* 1, 75–84.

Fischer, G., Shah, M. and van Velthuizen, H. (2002) Impacts of climate on agro-ecology. In: Fisher, G., Shah, M. and Velthuizen, H. (eds) *Climate Change and Agricultural Vulnerability*, IIASA, Vienna, pp. 38–91.

Fokar, M., Nguyen, H.T. and Blum, A. (1998) Heat tolerance in spring wheat: I. Estimating cellular thermotolerance and its heritability. *Euphytica* 104, 1–8.

Frova, C. and Sari-Gorla, M. (1993) Quantitative expression of maize HSPs: genetic dissection and association with thermotolerance. *Theoretical and Applied Genetics* 86, 213–220.

Fu, J. and Ristic, Z. (2010) Analysis of transgenic wheat (*Triticum aestivum* L.) harboring a maize (*Zea mays* L.) gene for plastid EF-Tu: segregation pattern, expression and effects of the transgene. *Plant Molecular Biology* 73, 339–347.

Fu, J., Momcilovic, I. and Prasad, P.V.V. (2012) Molecular bases and improvement of heat tolerance in crop plants. In: Josipovic, S. and Ludwig, E. (eds) *Heat Stress: Causes, Treatment and Prevention.* Nova Science, USA, pp. 185–214.

Gibson, L.R. and Mullen, R.E. (1996) Soybean seed quality reductions by high day and night temperature. *Crop Science* 36, 1615–1619.

Gross, Y. and Kigel, J. (1994) Differential sensitivity to high temperature of stages in the reproductive development in common bean (*Phaseolus vulgaris* L.). *Field Crops Research* 36, 201–212.

Guo, Y.P., Zhou, H.F. and Zhang, L.C. (2006) Photosynthetic characteristics and protective mechanisms against photooxidation during high temperature stress in two citrus species. *Scientia Horticulturae* 108, 260–267.

Gusta, L., Benning, N., Wu, G., Luo, X., Liu, X. *et al.* (2009) Superoxide dismutase: an all-purpose gene for agri-biotechnology. *Molecular Breeding* 24, 103–115.

Hajjar, R. and Hodgkin, T. (2007) The use of wild relatives in crop improvement: a survey of developments over the last 20 years. *Euphytica* 156, 1–13.

Hall, A.E. (2004) Breeding for adaptation to drought and heat in cowpea. *European Journal of Agronomy* 21, 447–454.

Heckathorn, S.A., Downs, C.A., Sharkey, T.D. and Coleman, J.S. (1998) The small, methionine-rich chloroplast heat-shock protein protects photosystem II electron transport during heat stress. *Plant Physiology* 11, 6439–6444.

Hede, A.R., Skovmand, B., Reynolds, M.P., Crossa, J., Vilhelmsen A.L. *et al.* (1999) Evaluating genetic diversity for heat tolerance traits in Mexican wheat landraces. *Genetic Resources and Crop Evolution* 46, 37–45.

Ishimaru, T., Hirabayashi, H., Ida, M., Takai, T., San-Oh, Y.A. *et al.* (2010) A genetic resource for early-morning flowering trait of wild rice *Oryza officinalis* to mitigate high temperature-induced spikelet sterility at anthesis. *Annals of Botany* 106, 515–520.

Jagadish, S.V.K., Craufurd, P.Q. and Wheeler, T.R. (2007) High temperature stress and spikelet fertility in rice (*Oryza sativa* L.). *Journal of Experimental Botany* 58, 1627–1635.

Jagadish, S.V.K., Muthurajan, R., Oane, R., Wheeler, T.R., Heuer, S. *et al.* (2010) Physiological and proteomic approaches to address heat tolerance during anthesis in rice (*Oryza sativa* L.). *Journal of Experimental Botany* 61, 143–156.

Jain, M., Prasad, P.V.V., Boote, K.J., Allen Jr., L.H. and Chourey, P.S. (2007) Effect of season-long high temperature growth conditions on sugar-to-starch metabolism in developing microspores of grain sorghum (*Sorghum bicolor* L. Moench). *Planta* 227, 67–79.

Jarvis, A., Lane, A. and Hijmans, R.J. (2008) The effect of climate change on crop wild relatives. *Agriculture, Ecosystems and Environment*, 126, 13–23.

Joshi, C.P., Klueva, N.Y., Morrow, K.J. and Nguyen, H.T. (1997) Expression of a unique plastid-localized heat-shock protein is genetically linked to acquired thermotolerance in wheat. *Theoretical and Applied Genetics* 95, 834–841.

Kakani, V.G., Prasad, P.V.V., Craufurd, P.Q. and Wheeler, T.R. (2002) Response of *in vitro* pollen germination and pollen tube growth of groundnut (*Arachis hypogaea* L.) genotypes to temperature. *Plant, Cell and Environment* 25, 1651–1661.

Kakani, V.G., Reddy, K.R., Koti, S., Wallace, T.P., Prasad, P.V.V. *et al.* (2005) Differences in *in-vitro* pollen germination and pollen tube growth of cotton cultivars in response to high temperature. *Annals of Botany* 96, 59–67.

Katiyar-Agarwal, S., Agarwal, M. and Grover, A. (2003) Heat tolerant basmati rice engineered by over-expression of hsp101. *Plant Molecular Biology* 51, 677–686.

Khanna-Chopra, R., and Sabarinath, S. (2004) Heat-stable chloroplastic Cu/Zn superoxide dismutase in *Chenopodium murale. Biochemical and Biophysical Research Communications* 320, 1187–1192.

Lafta, A.M. and Lorenzen, J.H. (1995) Effect of high temperature on plant growth and carbohydrate metabolism in potato. *Plant Physiology* 109, 637–643.

Lane, A. and Jarvis, A. (2007) Changes in climate will modify the geography of crop suitability: Agricultural biodiversity can help with adaptation. *SAT eJournal* 4, 1–12.

Liao, J., Zhang, H., Shao, X., Zhong, P. and Huang, Y. (2011) Identification for heat tolerance in backcross recombinant lines and screening of backcross introgression lines with heat tolerance at milky stage in rice. *Rice Science* 18, 279–286.

Marmiroli, N., Lorenzon, C., Stanca, A. M. and Terzi, V. (1989) Preliminary study of the inheritance of temperature stress proteins in barley (*Hordeum vulgare* L.). *Plant Science* 62, 147–156.

Mason, R.E., Mondal, S., Beecher, F.W., Pacheco, A., Jampala, B. *et al.* (2010) QTL associated with heat susceptibility index in wheat (*Triticum aestivum* L.) under short-term reproductive stage heat stress. *Euphytica* 174, 423–436.

Matsui, T. and Omasa, K. (2002) Rice (*Oryza sativa* L.) cultivars tolerant to high temperature at flowering: anther characteristics. *Annals of Botany* 89, 683–687.

Matsui, T., Omasa, K. and Horie, T. (2000) High temperature at flowering inhibit swelling of pollen grains, a driving force for thecae dehiscence in rice (*Oryza sativa* L.). *Plant Production Science* 3, 430–434.

Menendez, R., Gonzalez, A., Hill, J.K., Braschler, B., Willis, S. *et al.* (2006) Species richness changes lag behind climate change. *Proceedings of the Royal Society B: Biological Sciences* 273 (1593), 1465–1470.

Mohammed, A.R. and Tarpley, L. (2009) Impact of high nighttime temperature on respiration, membrane stability, antioxidant capacity, and yield of rice plants. *Crop Science* 49, 313–322.

Monneveux, P., Pastenes, C. and Reynolds, M.P. (2003) Limitations to photosynthesis under light and heat stress in three high-yielding wheat genotypes. *Journal of Plant Physiology* 160, 657–666.

Monterroso, V.A. and Wien, H.C. (1990) Flower and pod abscission due to heat stress in beans. *Journal of American Society of Horticultural Science* 115, 631–634.

Morales, D., Rodriguez, P., Dellamico, J., Nicolas, E., Torrecillas, A. *et al.* (2003) High-temperature preconditioning and thermal shock imposition affects water relations, gas exchange and root hydraulic conductivity in tomato. *Biologia Plantarum* 47, 203–208.

Morita, S., Shiratsuchi, H., Takanashi, J., and Fujita, K. (2002) Effect of high temperature on ripening in rice plants – comparison of the effects of high night temperatures and high day temperatures. *Japanese Journal of Crop Science* 71, 102–109.

Murakami, T., Matsuba, S., Funatsuki, H., Kawaguichi, K., Saruyama, H. *et al.* (2004) Over-expression of a small heat shock protein, sHSP17.7, confers both heat tolerance and UV-B resistance to rice plants. *Molecular Breeding* 13, 165–175.

Mutava, R.N., Prasad, P.V.V., Tuinstra, M.R., Kofoid, K.D. and Yu, J. (2011) Characterization of sorghum genotypes for traits related to drought tolerance. *Field Crops Research* 123, 10–18.

Nautiyal, P.C., Rajgopal, K. and Zala, P.V. (2008) Evaluation of wild *Arachis* species for abiotic stress tolerance: I. Thermal stress and leaf water relations. *Euphytica* 159, 43–57.

Nieto-Sotelo, J., Martinez, L.M., Ponce, G., Cassab, G.I., Alagon, A. *et al.* (2002) Maize HSP101 plays important roles in both induced and basal thermotolerance and primary root growth. *Plant Cell* 14, 1621–1633.

Ottaviano, E., Sari-Gorla, M., Pe, E. and Frova, C. (1991) Molecular markers (RFLPs and HSPs) for the genetic dissection of thermotolerance in maize. *Theoretical and Applied Genetics* 81, 713–719.

Paliwal, R., Roder, M.S., Kumar, U., Srivastava, J.P. and Joshi, A.K. (2012) QTL mapping of terminal heat tolerance in hexaploid wheat (*T. aestivum* L.) *Theoretical and Applied Genetics* 125, 561–575.

Peng, S., Huang, J., Sheehy, J., Laza, R., Visperas, R. *et al.* (2004) Rice yields decline with higher night temperature from global warming. *Proceedings of National Academy of Science USA* 101, 9971–9975.

Pinto, S., Chapman, S.C., McIntyre, C.L., Shorter, R. and Reynolds, M. (2008) QTL for canopy temperature response related to yield in both heat and drought environments. In: Appels, R., Eastwood, R., Lagudah, E., Langridge, P., Mackay, M. *et al.* (eds) *Proceedings of the 11th International Wheat Genetics Symposium,* 24–28 August. Brisbane, Australia. Sydney University Press, Sydney. Available at: http://ses.library.usyd.edu.au/bitstream/2123/3351/1/P172.pdf (accessed 14 January 2013).

Pradhan, G.P., Prasad, P.V.V., Fritz, A.K., Kirkham, M.B. and Gill, B.S. (2012) High temperature tolerance in *Aegilops* species and its potential transfer to wheat. *Crop Science* 52, 292–304.

Prasad, P.V.V. and Djanaguiraman, M. (2011) High night temperature decreases leaf photosynthesis and pollen function in grain sorghum. *Functional Plant Biology* 38, 993–1003.

Prasad, P.V.V., Craufurd, P.Q. and Summerfield, R.J. (1999) Sensitivity of peanut to timing of heat stress during reproductive development. *Crop Science* 39, 1352–1357.

Prasad, P.V.V., Craufurd, P.Q., Summerfield, R.J. and Wheeler, T.R. (2000) Effects of short episodes of heat stress on flower production and fruit-set of groundnut. *Journal of Experimental Botany* 51, 777–784.

Prasad, P.V.V., Boote, K.J., Allen Jr., L.H. and Thomas, J.M.G. (2002) Effects of elevated temperature and carbon dioxide on seed-set and yield of kidney bean (*Phaseolus vulgaris* L.). *Global Change Biology* 8, 710–721.

Prasad, P.V.V., Boote, K.J., Allen Jr., L.H. and Thomas, J.M.G. (2003) Super-optimal temperatures are detrimental to reproductive processes and yield of peanut under both ambient and elevated carbon dioxide. *Global Change Biology* 9, 1775–1787.

Prasad, P.V.V., Boote, K.J. and Allen Jr., L.H. (2006a) Adverse high temperature effects on pollen viability, seed-set, seed yield, and harvest index of grain-sorghum [*Sorghum bicolor* (L.) Moench] are more severe at elevated carbon dioxide due to higher tissue temperatures. *Agricultural and Forest Meteorology* 139, 237–251.

Prasad, P.V.V., Boote, K.J., Allen Jr., L.H., Sheehy, J.E. and Thomas, J.M.G. (2006b) Species, ecotype and cultivar differences in spikelet fertility and harvest index of rice in response to high temperature stress. *Field Crop Research* 95, 398–411.

Prasad, P.V.V., Pisipati, S.R., Mutava, R.N. and Tuinstra, M.R. (2008a) Sensitivity of grain sorghum to high temperature stress during reproductive development. *Crop Science* 48, 1911–1917.

Prasad, P.V.V., Pisipati, S.R., Ristic, Z., Bukovnik, U. and Fritz, A.K. (2008b) Impact of night time temperature on physiology and growth of spring wheat. *Crop Science* 48, 2372–2380.

Qi, Y., Wang, H., Zou, Y., Liu, C., Liu, Y. *et al.* (2011) Over-expression of mitochondrial heat shock protein 70 suppresses programmed cell death in rice. *FEBS Letters* 585, 231–239.

Rainey, K.M. and Griffiths, P.D. (2005) Inheritance of heat tolerance during reproductive development in Snap bean (*Phaseolus vulgaris* L.). *Journal of American Society of Horticultural Science* 130, 700–706.

Reguera, M., Peleg, Z. and Blumwald, E. (2012) Targeting metabolic pathways for genetic engineering abiotic stress-tolerance in crops. *Biochimica et Biophysica Acta* 1819, 186–194.

Rekika, D., Monneveux, P. and Havaux, M. (1997) The *in vivo* tolerance of photosynthetic membranes to high and low temperatures in cultivated and wild wheats of the *Triticum* and *Aegilops* genera. *Journal of Plant Physiology* 6, 734–738.

Reynolds, M.P., Balota, M., Delgado, M.I.B., Amani, I. and Fischer, R.A. (1994) Physiological and morphological traits associated with spring wheat yield under hot, irrigated conditions. *Australian Journal of Plant Physiology* 21, 717–730.

Reynolds, M.P., Nagarajan, S., Razzaque, M.A. and Ageeb, O.A.A. (2001) Heat tolerance. In: Reynolds, M.P., Ortiz-Monasterio, J.I. and McNab, A. (eds) *Application of Physiology in Wheat Breeding*. CIMMIT, Mexico, DF, pp. 124–135.

Reynolds, M.P., Pierre, C.S., Saad, A.S.I., Vargas, M. and Condon, A.G. (2007) Evaluating potential genetic gains in wheat associated with stress-adaptive trait expression in elite genetic resources under drought and heat stress. *Crop Science* 47, S-172–S-189.

Ristic, Z., Gifford, D.J. and Cass, D.D. (1991) Heat shock proteins in two lines of *Zea mays* L. that differ in drought and heat resistance. *Plant Physiology* 97, 1430–1434.

Ristic, Z., Williams, G., Yang, G., Martin, B. and Fullerton, S. (1996) Dehydration, damage to cellular membranes, and heat-shock proteins in maize hybrids from different climates. *Journal of Plant Physiology* 149, 424–432.

Ristic, Z., Momcilovic, I., Fu, J., Callegari, E. and DeRidder, B.P. (2007) Chloroplast protein synthesis elongation factor, EF-Tu, reduces thermal aggregation of rubisco activase. *Journal of Plant Physiology* 64, 1564–1571.

Ristic, Z., Bukovnik, U., Momcilovic, I., Fu, J. and Vara Prasad, P.V. (2008a) Heat-induced accumulation of chloroplast protein synthesis elongation factor, EF-Tu, in winter wheat. *Journal of Plant Physiology* 165, 192–202.

Ristic, Z., Bukovnik, U., Prasad, P.V.V. and West, M. (2008b) A model to predict heat stability of photosynthetic membranes. *Crop Science* 48, 1513–1522.

Rodriguez-Garay, B. and Barrow, J.R. (1988) Pollen selection for heat tolerance in cotton. *Crop Science* 28, 857–859.

Saini, H.S. and Aspinall, D. (1982) Abnormal sporogenesis in wheat (*Triticum aestivum* L.) induced by short periods of high temperature. *Annals of Botany* 49, 835–846.

Saini, H.S., Sedgley, M. and Aspinall, D. (1983) Effect of heat stress during floral development on pollen tube growth and ovary anatomy in wheat (*Triticum aestivum* L.) *Australian Journal of Plant Physiology* 10, 137–144.

Salem, M.A., Kakani, V.G., Koti, S. and Reddy, K.R. (2007) Pollen-based screening of soybean germplasm for high temperatures. *Crop Science* 47, 219–231.

Salgado, K.C.P.C., Von Pinho, E.V.R., Guimaraes, C.T., Von Pinho, R.G., Ferreira C.A. *et al.* (2008) Mapping of quantitative trait locus associated with maize tolerance to high seed drying temperature. *Genetics and Molecular Research* 7, 1319–1326.

Salvucci, M.E. and Crafts-Brandner, S.J (2004) Mechanism for deactivation of Rubisco under moderate heat stress. *Physiologia Plantarum* 122, 513–519.

Saranga, Y., Menz, M., Jiang, C.X., Robert, J.W., Yakir, D. *et al.* (2001) Genomic dissection of genotype x environment interactions conferring adaptation of cotton to arid conditions. *Genome Research* 11, 1988–1995.

Scafaro, A.P., Haynes, P.A. and Atwell, B.J. (2010) Physiological and molecular changes in *Oryza meridionalis* Ng., a heat-tolerant species of wild rice. *Journal of Experimental Botany* 61, 191–202.

Schoper, J.B., Lambert, R.J. and Vasilas, B.L. (1986) Maize pollen viability and ear receptivity under water and high temperature stress. *Crop Science* 26, 1029–1033.

Schoper, J.B., Lambert, R.J. and Vasilas, B.L. (1987) Pollen viability, pollen shedding, and combining ability for tassel heat tolerance in maize. *Crop Science* 27, 27–31.

Sharkey, T.D. (2005) Effects of moderate heat stress on photosynthesis: Importance of thylakoid reactions, rubisco deactivation, reactive oxygen species, and thermotolerance provided by isoprene. *Plant, Cell and Environment* 28, 269–277.

Spears, J.F., TeKrony, D.M. and Egli, D.B. (1997) Temperature during seed filling and soybean seed germination and vigour. *Seed Science and Technology* 25, 233–244.

Suzuki, N. and Mittler, R. (2006) Reactive oxygen species and temperature stresses: A delicate balance between signaling and destruction. *Physiologia Plantarum* 126, 45–51.

Suzuki, N., Rizhsky, L., Liang, H., Shuman, J., Shulaev, V. *et al.* (2005) Enhanced tolerance to environmental stress in transgenic plants expressing the transcriptional coactivator multiprotein bridging factor 1c. *Plant Physiology* 139, 1313–1322.

Thomas, C.D., Cameron, A., Green, R.E., Bakkenes, M., Beaumont, L.J. *et al.* (2004) Extinction risk from climate change. *Nature* 427, 145–148.

Thomas, J.M.G., Prasad, P.V.V., Boote, K.J. and Allen Jr., L.H. (2009) Seed composition, seedling emergence and early seedling vigour of red kidney bean seed produced at elevated temperature and carbon dioxide. *Journal of Agronomy and Crop Science* 195, 148–156.

van Ginkel, M and Ogbonnaya, F. (2007) Novel genetic diversity from synthetic wheats in breeding cultivars for changing production conditions. *Field Crop Research* 104, 86–94.

Vierling, E. (1991) The roles of heat shock proteins in plants. *Annual Review of Plant Physiology and Plant Molecular Biology* 42, 579–620.

Waines, J.G. (1994) High temperature stress in wild wheats and spring wheats. *Australian Journal of Plant Physiology* 21, 705–715.

Wang, J., Gan, Y.T., Clarke, F. and McDonald, C.L. (2006a) Response of chickpea yield to high temperature stress during reproductive development. *Crop Science* 46, 2171–2178.

Wang, Y., Wisniewski, M., Meilan, R., Cui, M. and Fuchigami, L. (2006b) Transgenic tomato (*Lycopersicon esculentum*) overexpressing cAPX exhibits enhanced tolerance to UV-B and heat stress. *Journal of Applied Horticulture* 8, 87–90.

Wardlaw, I.F. and Wrigley, C.W. (1994) Heat tolerance in temperate cereals: an overview. *Australian Journal of Plant Physiology* 21, 695–703.

Wheeler, T.R., Craufurd, P.Q., Ellis, R.H., Porter, J.R. and Prasad, P.V.V. (2000) Temperature variability and the yield of annual crops. *Agriculture, Ecosystems and Environment* 82, 159–167.

Wise, R.R., Olson, A.J., Schrader, S.M. and Sharkey, T.D (2004) Electron transport is the functional limitation of photosynthesis in field-grown Pima cotton plants at high temperature. *Plant Cell Environment* 27, 717–724.

Xiao, Y., Pan, Y., Luo, L., Deng, H., Zhang, G. *et al.* (2011) Quantitative trait loci associated with pollen fertility under high temperature stress at flowering stage in rice (*Oryza sativa*) *Rice Science* 18, 204–209.

Xu, W., Subudhi, P.K., Crasta, O.R., Rosenow, D.T., Mullet, J.E. *et al.* (2000) Molecular mapping of QTLs conferring stay-green in grain sorghum (*Sorghum bicolor* L. Moench). *Genome* 43, 461–469.

Yang, J., Sears, R.G., Gill, B.S. and Paulsen, G.M. (2002) Growth and senescence characteristics associated with tolerance of wheat-alien amphiploids to high temperature under controlled conditions. *Euphytica* 126, 185–193.

Ye, C., Argayoso, M.A., Redoña, E., Sierra, S.N., Laza, M.A. *et al.* (2012) Mapping QTL for heat tolerance at flowering stage in rice using SNP markers. *Plant Breeding* 131, 33–41.

Young, T.E., Ling, J., Geisler-Lee, C.J., Tanguay, R.L., Caldwell, C. *et al.* (2001) Developmental and thermal regulation of the maize heat shock protein, HSP101. *Plant Physiology* 127, 777–791.

Zanakis, G.N., Ellis, R.H. and Summerfield, R.J. (1994) A comparison of changes in vigour among three genotypes of soybean (*Glycine max*) during seed development and maturation in three temperature regimes. *Experimental Agriculture* 30, 157–170.

Zhao, Z.G., Jiang, L., Xiao, Y.H., Zhang, W.W., Zhai, H.Q. *et al.* (2006) Identification of QTLs for heat tolerance at the booting stage in rice (*Oryza sativa* L.). *Acta Agronomica Sinica* 32, 540–644.

13 Drought

Salvatore Ceccarelli

Formerly International Center for Agricultural Research in the Dry Areas (ICARDA), Aleppo, Syria (now retired)

13.1 Introduction

Today, few scientists question whether climate change is occurring or not and the discussion has shifted from whether it is happening to what to do about it. Using the results from formal economic models, it has been estimated that, in the absence of effective counteraction, the overall costs and risks of climate change will be equivalent to a 5% decrease in global gross domestic product (GDP) each year (Stern, 2007). If a wider range of risks and impacts is taken into account, the estimates of damage could rise to a 20% decrease in GDP or more, with a disproportionate burden and increased risk of famine in the poorest countries (Altieri and Koohafkan, 2003).

The most likely scenarios of the future climate are the following:

- Higher temperatures, which will reduce crop productivity, are certain.
- An increase in CO_2 concentration is certain with both direct and indirect effects.
- Increasing frequency of drought is highly probable.
- Increasing in the areas affected by salinity is highly probable.
- Increasing frequency of biotic stress is also highly probable.

Among these abiotic and biotic stresses, drought is the number one limitation to crop productivity in the world. As climate changes, the incidence and duration of drought and heat stress on our major crops will increase in many regions, negatively affecting crop yields and food security (Lauer *et al.*, 2012).

Drought is the most serious constraint to agricultural production and is still one of the most difficult challenges to agricultural scientists, as shown by the devastating effects of drought on the maize and soybean production in the USA in the summer of 2012 (US Drought Monitor, 2012) with, at the time of writing, 24% of the area experiencing extreme or severe drought conditions that are the worst in 65 years. It is likely that many current crop varieties will need replacement to enable them to better suit the new and changing agroenvironments (Luck *et al.*, 2011).

Drought is not easy to define because it means different things to different people in different places in different crops and therefore can be defined in several ways: for example, to a meteorologist it could mean a rare event (for example one of the 10 driest seasons per century); to an agronomist it means yield strongly limited by water; to a plant physiologist drought means plots not watered; and eventually to a biochemist/molecular biologist drought means plants or leaves left to dry on a lab bench (Passioura, 2006). The definition that I will use is the following: drought is that level of water availability below that needed for the crop to express its potential in terms of growth and production. Defined in this way it becomes clearer that drought can occur at different times of crop development, with a different intensities, depending on the amount of water available to the crop, over different lengths of time, and can occur one or more times during the life cycle of a plant. Farmers have to cope with all these different types of drought, which are largely unpredictable in the field. In a plastic house or in a growth chamber, where farmers usually do not plant their crops, drought is imposed artificially and hence is highly predictable. This is usually the type of drought scientists prefer to study. Whether the conclusions of these studies apply to the field drought remains to be demonstrated.

Scientists are also using, often almost as synonyms, drought tolerance and drought resistance when describing the responses of crops to limited water availability. Passioura (2006) suggested replacing these qualitative terms with water productivity, which can be quantified and expressed as units of amount of crop yield per volume of water supplied or used. The literature on drought indicates that this suggestion has not been generally followed.

Historically, drought has been the main factor limiting crop production. Water availability has been associated with the rise of several civilizations, whereas drought has caused the collapse of empires and societies, such as the Akkadian Empire (Mesopotamia, ~4200 calendar year BP), the Classic Maya (Yucatan Peninsula, ~1200 calendar year BP), the Moche IV–V Transformation (coastal Peru, ~1500 calendar year BP) (deMenocal, 2001) and the early bronze society in the southern part of the Fertile Crescent (Rosen, 1990).

Although in the past drought reached the front pages of the media only when it caused famine and death, in the past few years drought warnings have become more and more frequent. More people are now aware that drought is a permanent constraint to agricultural production in many developing countries and is increasingly causing crop losses in developed countries. The development of drought-resistant cultivars would be a major breakthrough (Ceccarelli *et al.*, 2004). From a genetic point of view, however, drought resistance is a very elusive trait because its occurrence, severity, timing and duration vary from location to location and in the same location from year to year. Cultivars successful in 1 dry year may fail in another, or cultivars resistant to late drought may not be resistant to intermittent drought or to drought occurring early in the season (Passioura, 2006). To make matters worse, drought seldom occurs in isolation; it often interacts with other abiotic (particularly temperature extremes) and biotic stresses (for example root diseases and nematodes) (Fresco and Baudoin, 2002). Moreover, areas with high risk of drought generally have low-input agriculture (Ceccarelli *et al.*, 1991) because the risk of crop failure or low yields is so high that farmers tend to limit investments to the bare minimum, i.e. seed and tillage. Thus, breeding for drought resistance is made more difficult by its interactions with other stresses. Eventually, and in relation to the complex syndrome of climate change, drought appears as a moving target because it is expected to gradually increase with time, and to be different from place to place, with consequences on breeding strategies including the choice of germplasm that we will discuss in the later sections of the chapter.

13.2 The Genetics of Drought Resistance

Drought resistance is commonly considered an extremely complex multigenic trait controlled by several genes with small effect (Parry *et al.*, 2005; Fleury *et al.*, 2010) and this complexity of drought resistance emerges from both phenotypic and genotypic studies.

Phenotypic studies have considered several traits as possible candidates to be indicators of drought resistance (leaf water retention of excised leaves, stomatal conductance, proline content, osmotic adjustment, reduced xylem vessel size in seminal roots, abscisic acid accumulation, leaf turgor, leaf waxiness, leaf area index, stomatal density, transpiration efficiency, early growth vigour, growth habit, accumulation and remobilization of stem reserves, Δ^{13} C discrimination, canopy temperature, relative water content, water use efficiency, flowering time, cell membrane stability, cell wall rheology, root characteristics and others). When we studied the association between a number of morphological traits and grain yield under stress over a number of years in barley (Ceccarelli *et al.*, 1991), we found, however, that individual traits play a different role in different years under different combinations of climatic stresses (Table 13.1).

In spite of large differences between the four cropping seasons (from near crop failure in 1986/87 to nearly 3 t ha^{-1} in 1987/88), high grain yield under drought was associated with earliness, large kernels and, mainly in six-row genotypes, with plant height under drought. In 1986/87, however, when a long period of low winter temperatures prevented the crop from growing during the rainy period, the correlation between grain yield and cold tolerance in two-row genotypes was almost as strong as with days to heading. In six-row genotypes, only the correlation coefficients with cold tolerance and growth vigour were significant. In that season, only the earliest of the cold tolerant two-row barleys and the most vigorous and cold tolerant six-row barleys produced harvestable yield. A similar pattern emerged in the six-row genotypes in 1988/89 cropping season also characterized by a long period of low winter temperatures coincidental with a long winter dry spell. In the two-row genotypes, the correlation between yield and cold tolerance was again as strong as with days to heading. In the environment used in this study (dryland Mediterranean with hot and dry summers and cold winters), earliness alone, despite its efficiency as an escape mechanism, is necessary but not sufficient to achieve maximum yield stability. The correlation coefficients between yield potential and yield in drought conditions were non-significant in five cases, negative in one case (two-row in 1988/89) and positive in two cases (two-row in 1985/86 and six-row in 1988/89). In these two cases, the variation of yield potential explained only 3% and 10%, respectively, of the sum of squares of yield in drought conditions. The multiple correlation coefficients (R) are not comparable across seasons (with the exception of 1985/86 and 1987/88) because some of the traits were not measured or scored each season. The comparison between two-row and six-row genotypes shows, however, that a given group of traits can have similar multiple correlation coefficients with yield in drought conditions in the same environment even though considerable differences exist between the correlation coefficients with individual traits.

Although the situation illustrated in Table 13.1 is certainly an oversimplification, it indicates that high grain yields in drought-stressed environments can be associated with different combinations of traits rather

Table 13.1. Simple and multiple correlation coefficients (R) between grain yield in stress conditions (YD in kg ha^{-1}) and days to heading (DH), plant height in drought conditions (PHD), growth vigour (GV), growth habit (GH), cold tolerance (CT), 1000 kernel weight (KW) and yield potential (YP) in two-row and six-row barley genotypes[a] in four cropping seasons (modified from Ceccarelli *et al.*, 1991).

YD	Two-row				Six-row			
versus	85/86	86/87	87/88	88/89	85/86	86/87	87/88	88/89
DH	−0.24**	−0.17**	−0.35**	−0.63**	−0.45**	−0.07	−0.35**	−0.76**
PHD	0.45**	n.a.[b]	0.02	0.02	0.42**	n.a.	0.38**	0.55**
GV	−0.12	−0.06	−0.03	−0.08	−0.19	−0.21*	−0.16	−0.29**
GH	−0.09	−0.01	−0.08	0.43**	−0.20*	−0.05	−0.20*	0.12
CT	n.a.	−0.16*	n.a.	−0.62**	n.a.	−0.21*	n.a.	−0.28**
KW	0.16*	0.21**	0.37**	0.31**	0.04	0.03	0.44**	0.61**
YP	0.17**	0.03	−0.04	−0.34**	−0.04	−0.03	−0.18	31**
R	0.56**	0.29**	0.49**	0.84**	0.55**	0.26*	0.60**	0.81**
YD	1602	74	2931	744	1514	45	2869	567

[a]236 two-row and 96 six-row in 85/86, 86/87 and 87/88; 203 two-row and 97 six-row in 88/89; [b]not available; *P<0.05; **P<0.01.

than to the same specific trait, and that within these combinations the role of individual traits changes with the type, the intensity and the timing of the drought stress, which vary from year to year. It is therefore possible for each of the traits considered in Table 13.1 to find genotypes with the same value of the trait, although widely different for grain yield in drought conditions. This indicates that within a given range of values for each trait, the final performance (grain yield in drought conditions), depends on the expression of other traits. Similar results were obtained by Condon *et al.* (2002) and by Reynolds *et al.* (2007) using a general model for drought adaptation in wheat.

Phenotyping for drought resistance has recently been made more precise by the use of techniques such as nuclear magnetic resonance, canopy spectral reflectance and infrared thermography, near-infrared spectroscopy, magnetic resonance imaging and positron emission tomography (Mir *et al.*, 2012).

Most of the evidence for the genetic complexity of drought resistance comes from the molecular dissection of drought tolerance using traditional quantitative trait loci (QTL) mapping and more recently with linkage disequilibrium (LD) based association mapping. These studies have identified QTLs for drought resistance/ tolerance defined as grain yield under stress and/or for characters associated with drought resistance in a number of crops; in the majority of these studies these QTLs are located on many chromosomal regions (see Table 1 in Mir *et al.*, 2012), each explaining only a small portion of the total phenotypic variance.

In gene expression studies it is not uncommon for hundreds if not thousands of genes to be found being up- or down-regulated under dehydration stress in the laboratory (Blum, 2011). A typical example of this complexity has been revealed by the analysis of the genes that are up-regulated or down-regulated by cold, drought and salinity stress in *Arabidopsis thaliana* (L.) Heynh (Seki *et al.*, 2002).

In this study, several genes were affected by drought and salinity (119 up-regulated and 31 down-regulated), and a smaller number by drought and cold and by cold and salinity. A relatively large number of genes (22 up-regulated and 17 down-regulated) were affected by all the three abiotic stresses (Seki *et al.*, 2002). As many as 128 and 26 genes were up- and down-regulated, respectively, only by drought. If these results can be extrapolated to agricultural crops and if we consider that drought and cold can occur at different developmental stages of the crop, for a variable length of time and with a variable intensity, the level of complexity of the molecular responses is likely to be higher than that shown in Fig. 13.1. In fact, data on barley show that the genes up-regulated and down-regulated at the given stress treatment are different from those up-regulated and down-regulated at a different stress treatment (Öztürk *et al.*, 2002).

A more recent example of the genetic complexity of drought resistance is the dissection of gene expression under drought stress in rice (*Oryza sativa* L. ssp. *indica*) in which the number of genes that were up-regulated in the shoot, flag leaf and panicle was 1020, 301 and 448, respectively. Further, 547, 614 and 1145 genes were down-regulated in the shoot, flag leaf and panicle, respectively (Zhou *et al.*, 2007).

However, although biotechnology has revolutionized the development of new varieties with genes being identified for engineering or marker-assisted selection (MAS) for a range of 'simple' traits, and despite a substantial investment in the study of more complex traits, there are not yet success stories based on the identification of specific genes and their utilization for drought resistance (Chapman, 2008). The only markers for traits associated with drought resistance that have found their way into practical plant breeding programmes are those for boron and aluminium tolerance (Gupta *et al.*, 2010). In general, MAS has contributed very little to the release of improved cultivars with greater tolerance to abiotic stresses, and this is perhaps partly

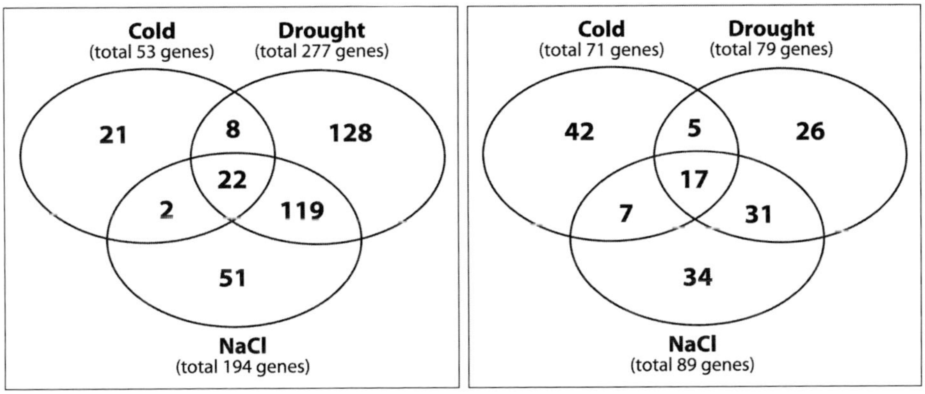

Fig.13.1. Up-regulated (left) and down-regulated (right) genes by cold, drought and salinity stress in *Arabidopsis thaliana* (from Ceccarelli *et al.*, 2004, based on Seki *et al.,* 2002).

because interactions between multiple stresses are seldom considered (Collins *et al.*, 2008).

Although QTL analysis and gene cloning have been used to investigate the same stress responses, the relationship between QTLs and stress-related sequences is still far from understood and will remain a challenge for the near future (Fleury *et al.*, 2010). This conclusion, which leaves the solutions to the near future, is common to most molecular studies.

13.3 Breeding for Drought Resistance

Two contrasting philosophies have been used to breed drought-resistant varieties. In the first, selection is done under optimum growing conditions assuming that an increased yield potential will have a carry-over effect in less favourable conditions. In the second, selection is done in the presence of drought, and can take two forms: (i) indirect selection for physiological or developmental traits (analytical breeding); and (ii) direct selection for grain yield (empirical or pragmatic breeding).

The first philosophy has been one of the main concepts upon which the 'Green Revolution' was based (Hazell, 2003), which in high input environments achieved large production increases but also decreased agricultural biodiversity and increased environmental degradation because the varieties it produced were able to express their superiority only if supplied with chemical inputs (fertilizers, pesticides and herbicides) (Keneni *et al.*, 2012). On the other hand, in the regions where farmers were too poor to afford chemical inputs and in those chronically affected by drought it failed to improve drought resistance.

Nearly one billion people in the world were still undernourished by 2010 (FAO, 2012) and drought continues to affect agricultural production severely, even in developed countries as mentioned earlier in the case of the USA in 2012 (US Drought Monitor, 2012). One corollary of this philosophy is that, because favourable environments tend to be similar in time and space, one or few varieties *widely adapted* can perform well in all of them, hence the reduction in agricultural biodiversity.

The second philosophy recognizes the complexity of drought and is based on breeding for *specific adaptation.* In the analytical approach, physiologists classify mechanisms for drought resistance into drought escape, drought avoidance and drought tolerance. As discussed earlier, several traits have been associated with those mechanisms.

The analytical approach has been very successful in understanding which traits are associated with drought resistance but much less successful in developing new cultivars with improved drought resistance under field conditions where drought varies in occurrence, severity, timing and duration. Therefore, it is the interaction among traits that determines the overall crop response to drought, rather than the expression of any specific trait (Ceccarelli *et al.*, 1991). One exception could be osmotic adjustment, which is effective in a specific drought environment (Munns and Richards, 2007).

Although breeding for drought resistance on the basis of direct selection for grain yield in the target environment (empirical or pragmatic breeding) appears intuitively as the most obvious solution, it has faced criticism because the chances of progress seem low and remote. Two issues with selection in stress environments are the precision of selection and the existence of several target environments.

On the first issue, a literature review (Ceccarelli, 1996a) has shown the absence of a consistent relationship between grain yield and the magnitude of heritability and the possibility of combining precision and relevance by conducting trials in the target environment even when it is a stress environment. The second issue, how to deal with the multitude of target environments, is intimately associated with broad and specific adaptation, two concepts that have been debated in plant breeding since the early 1920s and are still highly controversial (Ceccarelli and Grando, 2002). Because of the heterogeneity of the environments affected by drought, breeding for specific adaptation is essential to increase the yield of crops grown in these environments. Breeding for specific adaptation to drought conditions is often considered an undesirable breeding strategy, because it is usually associated with lower yields under favourable conditions and produces too many varieties for the seed companies to handle. Breeding for wide adaptation has been considered recently to be more cost-effective than selecting specifically adapted genotypes (Windhausen *et al.*, 2012). This is not necessarily true when a plant-breeding programme is organized in a participatory way (Mustafa *et al.*, 2006).

The issue of the heterogeneity and the multitude of the environments affected by drought has been addressed by developing an innovative type of breeding programme for drought prone areas based on the collaboration between farmers and scientists known as participatory plant breeding (PPB) (Ceccarelli *et al.*, 2007, 2009). PPB is based on breeding for specific adaptation to both the climatic and social environment and for wide adaptation over time (also called stability or dependability) and is conducted independently in areas that are consistently different. PPB has been shown to be able to reconcile yield increases with increased biodiversity in space and time, adaptation to climate changes and enhanced rate of adoption of new varieties (Ceccarelli *et al.*, 2009).

Genomic selection, a form of MAS that simultaneously estimates all locus, haplotype or marker effects across the entire genome to calculate genomic estimated breeding values (Heffner *et al.*, 2009), is expected to revolutionize breeding in the next decade. For the time being, however, genomic selection falls short of the expectations because its accuracy is limited by inefficiencies in the prediction of phenotype from genotype (Morrell *et al.*, 2012).

A fundamental component of a breeding strategy is the choice of the right type of germplasm and the next sections will illustrate examples of the importance of landraces and wild relatives in breeding for drought resistance.

13.4 The Value of Genetic Resources under Drought

Today there is a contradiction between the importance of biodiversity and the dangers associated with its decline, which is recurrently emphasized by the major scientific journals (e.g. see Cardinale *et al.*, 2012; Hooper *et al.*, 2012), and the vulnerability of our crops associated with the continuous decline of agrobiodiversity (Esquinas-Alcázar, 2005; Keneni *et al.*, 2012). Modern agriculture has required the development of uniform varieties, which has been met by both public and private plant breeders with a limited number of standard, high-yielding varieties: this has characterized the Green Revolution, which has made it possible to boost food production at the price of the loss of innumerable heterogeneous traditional farmers' varieties (Esquinas-Alcázar, 2005).

13.4.1 Landraces

In many developing countries, and for crops grown in low input conditions and in stress environments, landraces are still the backbone of agricultural production (Ceccarelli, 1984; Flower, 1996; Kelley *et al.*, 1996; Sthapit *et al.*, 1996; Grando *et al.*, 2001). The reasons why farmers prefer to grow only landraces or continue to grow landraces even after partial adoption of modern cultivars are many, often including quality attributes such as food and feed quality (Kelley *et al.*, 1996), and seed storability (Ceccarelli and Grando, 1999). Landraces are often able to produce some yield even in difficult conditions where modern varieties are less reliable. For example, where farmers have adopted modern cultivars they have kept the landraces in the most unfavourable areas of the farm (Cleveland *et al.*, 2000). The value of landraces as sources of drought tolerance has been well documented in the case of barley in Syria (Ceccarelli *et al.*, 1995; Grando *et al.*, 2001) and in several other crops elsewhere (Brush, 1999). The comparison between barley landraces and modern cultivars under a range of conditions from severe stress (low input and low rainfall) to moderately favourable conditions (high inputs and high rainfall) has consistently indicated that:

1. On average, landraces yield more than modern cultivars under low-input and stress conditions (see Table 13.2).
2. Within landraces, there is considerable variation for grain yield under low input and drought stress conditions, but all the landrace-derived lines yield something, whereas some modern cultivars fail (see Table 13.3).
3. Landraces are responsive to both inputs and rainfall and the yield potential of some lines is high, though not as high as modern cultivars (see Table 13.3).
4. The superiority of landraces is not associated with mechanisms of escape to drought stress, as shown by their heading date (see Table 13.4).
5. It is possible to find modern cultivars which under low-input and stress conditions yield almost as well as landraces, but their frequency is very low.

The data in Table 13.2 suggest that selection conducted only in high-input conditions or in intermediate conditions is likely to miss most of the lines that would have performed well under low input conditions. The assumption of most breeding programmes that landraces are genetically inferior is based on work conducted in research stations. Even those breeding programmes addressing target environments that have a low yield potential because of the combination of biotic and abiotic stresses have rarely challenged this assumption.

An example of the potential value of landraces in marginal environments is given in Table 13.3 where barley breeding lines, unrelated to local barley landraces, were compared with 77 pure lines extracted from Syrian landraces in two contrasting environments, one affected by drought stress represented by two locations with 186 mm and 185 mm rainfall, respectively (YS), and the second not affected by drought and represented by three locations with 504 mm, 234 mm and 317 mm, respectively.

The landraces have an average yield advantage of 60% under stress, whereas the modern cultivars have an average yield advantage of 14% in the absence of stress. In

Table 13.2. Grain yield (kg ha^{-1}) of pure lines derived from Syrian landraces and modern cultivars at three levels of stress[a] in northern Syria (from Ceccarelli, 1996b).

Environment	Landraces (n = 44)	Modern (n = 206)	Difference	P[b]
Stress	1038	591	447	<0.01
Intermediate	3105	3291	186	NS
Non-stressed	4506	6153	1647	<0.01

[a] as defined by average precipitation and soil fertility (Ceccarelli, 1996b); [b] based on *t*-tests for groups of unequal size; NS, not significant.

Table 13.3. Grain yield (kg ha^{-1}) under stress (YS) and grain yield under non-stress (YNS) of barley breeding lines classified according to the germplasm type (from Ceccarelli, 1994).

Type of germplasm	YS[b]			YNS[c]	
	N[a]	Yield	Range	Yield	Range
Non-landraces	155	488	0–893	3901	2310–4981
Landraces[d]	77	788	486–1076	3413	2398–4610
Best check		717		4147	

[a]Number of entries; [b]average of two stress sites; [c]average of three non-stress sites; [d]pure lines obtained by pure line selection within landraces.

Table 13.4. Means of morphological, developmental and agronomic traits in 1041 modern (unrelated to Syrian landraces) barley genotypes compared with 322 pure lines extracted from Syrian landraces[a] (from Ceccarelli, 1994).

Traits	Non-landraces (n = 1041)	Landraces (n = 322)
Early growth vigour (1 = good; 5 = poor)	2.5 b	3.2 a
Growth habit (1 = erect; 5 = prostrate)	2.8 b	4.0 a
Cold tolerance (1 = tolerant; 5 = susceptible)	3.0 a	1.3 b
Days to heading (from emergence)	117.9 b	121.2 a
Grain filling duration (days)	39.3 a	35.5 b
Yield potential (kg ha^{-1})	4398.0 a	3293.0 b
Yield under stress (kg ha^{-1})	483.1 b	984.0 a

[a]Means followed by the same letter are not significantly (P<0.05) different based on *t*-test for samples of unequal size.

addition to the mean performance of the two types of germplasm, the interval of variation is very informative. All 77 lines from landraces yielded something under stress, whereas some of the modern cultivars failed; the best modern cultivars yielded almost as much as the best landraces. Under non-stress conditions, it was interesting to find that the yield of some landraces was not significantly inferior to that of the best modern cultivars.

In the attempt to understanding the basis of the adaptation of landraces to drought stress conditions, we evaluated a number of morphological and developmental traits in landrace and non-landrace material hoping to find a relatively simple explanation to the superiority of the landraces under drought stress. The choice of the traits was limited to those that are easy, cheap and quick to measure because these are essential requisites for use in a breeding programme. The results were discouraging (Table 13.4). Landraces differ from non-landraces in a number of traits that together appear to form an adaptive complex.

If the genetic structure of landraces is considered as an evolutionary approach to survival and performance under arid and semi-arid conditions (Schulze, 1988), then it appears that during millennia of cultivation under adverse conditions, natural and artificial selection have not been able to identify either an individual genotype possessing 'a trait' associated with its superior performance or an individual genotype with a specific architecture of different traits. On the contrary, the combined effects of natural and artificial selection have led to an architecture of genotypes representing different combinations of traits each able to respond to different types of drought and different combinations of various abiotic and biotic stresses.

The value of landraces and their contribution to plant breeding has been recently summarized in the case of cereals by Newton *et al.* (2010). In the case of legumes, there is also quite a considerable body of evidence suggesting the presence of useful genes both in landraces and wild relatives (Teshome *et al.*, 2010). In these reviews, examples of the contribution of landraces to plant breeding include contributing useful genes both for resistance to abiotic stresses, particularly to drought and high temperatures, and for resistance or tolerance to biotic stresses. The latter has probably been the most common and widespread contribution of landraces that in several crops have been used as donors of resistance genes to virtually all the most important diseases. This is also highly relevant in the case of climate change because one of the expected side effects is an increase in the frequency of both pest and disease epidemics.

13.4.2 Crop wild relatives

Crop wild relatives have particular value because of their potential to contribute beneficial traits to crops, such as biotic and abiotic resistance, leading to improved yield and stability (Maxted *et al.*, 2006). Crop wild relative species tend to contain greater genetic variation than crops because they have not passed through the genetic bottleneck of domestication that has reduced the diversity available to breeders, and so they are expected to present a reservoir of useful variation for crop improvement (Vollbrecht and Sigmon, 2005).

Crop wild relatives have contributed significantly to improving food production and have been used particularly in barley (Lakew *et al.*, 2011, 2012, 2013), cassava (Nassar, 2003), potato (Bradshaw *et al.*, 2006), rice (Brar, 2005), tomato (Rick and Chetelat, 1995), lentil (Erskine *et al.*, 1994), chickpea (Singh and Ocampo, 1997) and wheat (Valkoun, 2001; Zaharieva *et al.*, 2001). Wild relatives have been mostly used as sources of pest and disease resistance genes (80% of the traits according to Hajjar and Hodgkin, 2007, 56% of the papers according to Maxted *et al.*, 2012), much less as sources of genes for drought tolerance (17% of the references cited abiotic stress as an objective in Maxted *et al.*, 2012) although several studies in a number of crops have identified useful genes for drought resistance in the corresponding wild relatives.

Barley breeders have long recognized the inherent value of wild species for the improvement of important agronomic, disease and quality traits. For example, the introgression of germplasm from wild barley, *Hordeum vulgare spp. spontaneum* (hereafter *H. spontaneum*), has been considered useful to increase the genetic variation for characters that contribute to drought tolerance in barley (Nevo, 1992; Forster *et al.*, 1997; Ivandic *et al.*, 2000; Grando *et al.*, 2001; Baum *et al.*, 2003; Ceccarelli *et al.*, 2004). One example of using wild relatives in a breeding programme was ICARDA's barley breeding programme that routinely used *H. spontaneum* for drought stressed environments. One of the reasons for choosing *H. spontaneum* was its ability to maintain a long stem even under severe drought. Plant height in very dry years is very important to farmers for two reasons: firstly, it allows mechanical harvesting, which is much cheaper than hand harvesting, and second it is associated with higher straw yield and the price of straw in dry years can nearly match the price of grain. Lodging susceptibility associated with tall plants is seldom a problem because of the prevailing low input and dry conditions.

The decisive year for including *H. spontaneum* in ICARDA's breeding programme for stress environments was 1987 when a *H. spontaneum* collection was evaluated at a very dry site in Syria. That year rainfall was only 176.2 mm and there was a crop failure (Ceccarelli, 1994) with virtually no breeding lines in the entire programme able to produce grain. The only lines able to head and to produce some grains were two accessions of *H. spontaneum*. In these, some photosynthetic activity was found between 0700 and 0800 hours, the average net photosynthesis was 2.7±0.4 μmol CO_2 m^{-2} s^{-1} and stomatal conductance was 0.060±0.003 mol m^{-2} s^{-1}. At the same time, the stomata of the locally adapted landrace, Arabi Aswad, considered to be drought resistant, were closed. By midday, the stomatal conductance of the *H. spontaneum* accessions decreased to 0.025 μmol CO_2 m^{-2} s^{-1} at a leaf water potential of −4.6 Mpa while the leaf water potential of Arabi Aswad was −3.5 Mpa and the stomatal conductance was zero. These lines combined earliness with an acceptable level of cold resistance, they were able to maintain good plant height under drought conditions and, despite the very dry conditions, they were also able to extrude part of the last internode outside the flag leaf (Grando *et al.*, 2001). The two *H. spontaneum* accessions have been used routinely in the crossing programme ever since.

Recently Lakew *et al.* (2011, 2012, 2013) used barley lines with various levels of introgression of the same two *H. spontaneum* accessions described above with the objective of identifying chromosomal regions in the two wild barleys responsible for increasing barley yields in marginal environments characterized by severe moisture stress. The lines were tested in a range of locations/years from highly drought stressed (about 0.7 t ha^{-1}) to favourable conditions (4.5 t ha^{-1}) (Fig. 13.2). The top yielding lines in the two driest environments (BR04 and KH05) were derived mostly from crosses with one of the two *H. spontaneum* accessions, whereas most of the improved varieties (group 8 with no *H. spontaneum* introgression) showed a positive genotype by environment (GE) interaction with the highest yielding environments (TH04, TH05 and TH06). The results of the field experiments indicated that there was variation for grain yield under drought stress among barley genotypes, and that some of the lines derived from *H. spontaneum* had consistently superior specific adaptation to the range of severe stress conditions used in this study even if the study was unable to associate this superiority with specific chromosomal regions. The usefulness of *H. spontaneum* in breeding programmes for stress conditions is likely to increase in view of the predicted increase in the occurrence of high temperatures and droughts.

13.5 A Dynamic Response to Climate Change

As landraces and wild relatives are likely to become important sources of useful genetic variation when trying to cope with increased levels of drought resistance in the varieties needed to cope with climate change, the emphasis will be on where to access this type of germplasm.

The two most obvious sources of novel genetic variation are the gene banks and/or the farmers' fields. The major difference between the two sources of genetic variation is that the first is static, in the sense that it represents the genetic variation available in the collection sites at the time the collection was made, whereas the second is dynamic because landraces and wild relatives are heterogeneous populations and, as such, they evolve and can generate continuously novel genetic variation. The difference can be relevant as shown recently by Nevo *et al.*, (2012) who found that accessions of wild wheat and barley relatives collected in Israel in 2008 were significantly earlier than the accessions collected in the same locations 28 years earlier and kept in a gene bank. The finding is important because it confirms that even populations of predominantly self-pollinated plants can evolve, as also shown by Morran *et al.* (2009).

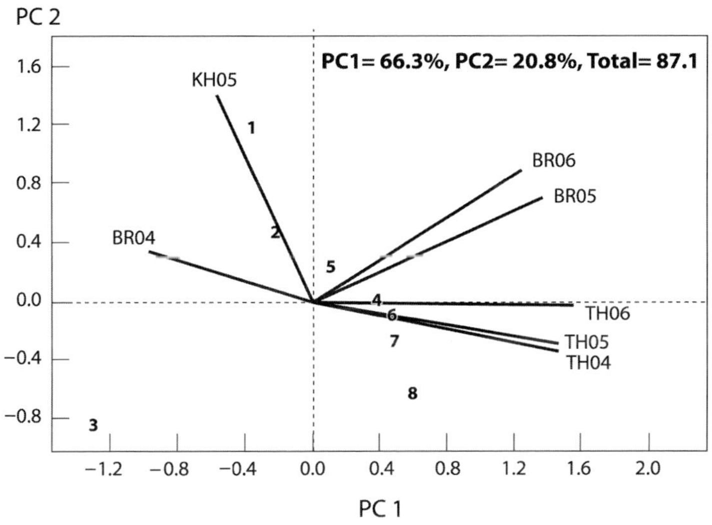

Fig. 13.2. GGE biplot for grain yield data in seven year × location combinations of eight groups of barley genotypes differing for the percentage of observed introgression with *H. spontaneum* (group 1 = 44.5%, group 2 = 24.2%, group 3 = 22.6%, group 4 = 45.4%, group 5 = 19.0%, group 6 = 15.5%, group 7 = 11.4%, group 8 = 0%). Groups 1, 2, 3 and 6 are introgressed with one accession of *H. spontaneum*, and groups 4, 5 and 7 with a different accession (from Lakew *et al.*, 2011).

A strategy to cope with drought resistance, defined earlier as a moving target in the case of climate change, is to use population (evolutionary) breeding which produces heterogeneous varieties better able to cope with the year-to-year variability expected to increase with climate change. Evolutionary plant breeding can also produce uniform varieties capable of responding to market demand by superimposing artificial selection within the evolving population.

Evolutionary breeding (Suneson, 1956) is based on populations with large genetic variability, such as mixtures of F_2s, which can be handled directly by farmers in a multitude of environments while they slowly evolve and adapt. At ICARDA, evolutionary breeding was combined with the participatory programmes for barley and wheat and implemented in Syria, Jordan, Iran, Eritrea, Algeria, Ethiopia, and recently in Italy as a component of a European project on organic agriculture. The aim is to increase the probability of recombination within a population, which is constituted deliberately to harbour a very large amount of genetic variation.

In the case of barley such a population consisted of a mixture of nearly 1600 F_2s (Ceccarelli *et al.*, 2010), whereas, in the case of durum and bread wheat, the population consisted of a mixture of slightly more than 700 crosses and more than 2500 F_2-derived F_3 and F_4 generations, respectively. The barley population was planted at 19 locations in five countries, whereas the durum and bread wheat populations have been planted at five locations. The three populations have been left evolving under the pressure of changing climate conditions with the expectation of an increase in the frequency of genotypes with adaptation to the conditions (climate, soil, agronomic practices, abiotic and biotic stresses) of the location where each year the population is grown. The simplest and cheapest way of implementing evolutionary breeding is for the farmers to plant and harvest in the same location. The breeder and the farmers can, however, superimpose artificial selection with criteria that may change from location to location and with time. While the population is evolving, lines can be derived and tested as pure lines in the participatory breeding programmes, or can be used as multilines, or a sub-sample of the population can be directly used for cultivation exploiting the advantages of intra-population genetic diversity described earlier. The key aspect of the method is that, while the lines are continuously extracted, the population is left evolving for an indefinite amount of time, thus becoming a unique source of continuously better adapted genetic material directly in the hands of the farmers. This approach generates a large agro-biodiversity both in space and in time. In all the countries where the evolutionary populations are currently grown (including some developed countries), the farmers share the excess seed with others so that the populations are rapidly spreading across the farmers' communities. This guarantees that the improved material will be readily available to farmers without the bureaucratic and inefficient systems of variety release and formal seed production.

13.6 Conclusions

Drought is a complex problem that affects the livelihoods of millions of poor every year and most likely will become more frequent as the climate changes. Current knowledge on the molecular basis of drought resistance is not yet sufficient to develop more drought resistant crops or varieties. Therefore, for the time being, conventional breeding based on specific adaptation and evolutionary strategies, and with the collaboration of farmers seems to be the only realistic avenue to develop in the short-term drought-resistant varieties quickly available on the ground. As our knowledge of the genetic basis of drought resistance improves, biotechnological techniques and genomic tools might offer the possibility of introducing new traits and regulatory mechanisms that would improve the efficiency of selection. As this is now admittedly recognized as an avenue for the future, there is a need to re-address the allocation of resources, which are currently grossly unbalanced in favour of approaches that are not likely to accelerate the production of new varieties in the timeframe of climate change.

References

Altieri, M.A. and Koohafkan, P. (2003) *Enduring Farms: Climate Change, Smallholders and Traditional Farming Communities.* Third World Network, 131 Jalan Macalister, 10400 Penang, Malaysia, p. 72.

Baum, M., Grando, S., Backes, G., Jahoor, A., Sabbagh, A. *et al.* (2003) QTLs for agronomic traits in the Mediterranean environment identified in recombinant inbred lines of the cross 'Arta'×*H. spontaneum* 41 - 1. *Theoretical Applied Genetics* 107, 1215–1225.

Blum, A. (2011) Drought resistance – is it really a complex trait? *Functional Plant Biology* 38, 753–757.

Bradshaw, J.E., Bryan, G.J., and Ramsay, G. (2006) Genetic resources (including wild and cultivated S*olanum* species) and progress in their utilisation in potato breeding. *Potato Research* 49, 49–65.

Brar, D. (2005) Broadening the gene pool and exploiting heterosis in cultivated rice. In: Toriyama, K., Heong, K.L. and Hardy, B. (eds) *Rice is Life: Scientific Perspectives for the 21st Century.* Proceedings of the World Rice Research Conference, Tokyo and Tsukuba, Japan, 4–7 November 2004, pp. 157–160.

Cardinale, B.J., Duffy, J.E., Gonzalez, A., Hooper, D.U., Perrings, C. *et al.* (2012) Biodiversity loss and its impact on humanity. *Nature* 486, 59–67.

Ceccarelli, S. (1984) Utilization of landraces and *Hordeum spontaneum* in barley breeding for dry areas. *Rachis* 3, 8–11.

Ceccarelli, S. (1994) Specific adaptation and breeding for marginal conditions. *Euphytica* 77, 205–219.

Ceccarelli, S. (1996a) Adaptation to low/high input cultivation. *Euphytica* 92, 203–214.

Ceccarelli, S. (1996b) Positive interpretation of genotype by environment interactions in relation to sustainability and biodiversity. In: Cooper, M. and Hammers, G.L. (eds) *Plant Adaptation and Crop Improvement.* CAB International, Wallingford, UK, ICRISAT, Andra Pradesh, India, and IRRI, Manila, Philippines, pp. 467–486.

Ceccarelli, S, and Grando, S. (1999) Barley landraces from the Fertile Crescent: a lesson for plant breeders. In: Brush, S.B. (ed.) *Genes in the Field: On-Farm Conservation of Crop*

Diversity. IPGRI, IDRC and Lewis Publishers, Boca Raton, Florida, pp. 51–76.

Ceccarelli, S. and Grando, S. (2002) Plant breeding with farmers requires testing the assumptions of conventional plant breeding: Lessons from the ICARDA barley program In: Cleveland, D.A. and Soleri, D. (eds) *Farmers, Scientists and Plant Breeding: Integrating Knowledge and Practice.* CAB International, Wallingford, UK, pp. 297–332.

Ceccarelli, S., Acevedo, E. and Grando, S. (1991) Breeding for yield stability in unpredictable environments: single traits, interaction between traits, and architecture of genotypes. *Euphytica* 56, 169–185.

Ceccarelli, S., Grando, S., and van Leur, J.A.G. (1995) Barley landraces of the Fertile Crescent offer new breeding options for stress environments. *Diversity* 11, 112–113.

Ceccarelli, S., Grando, S., Baum, M. and Udupa, S.M. (2004) Breeding for drought resistance in a changing climate. In: Rao, S.C. and Ryan, J. (eds) *Challenges and Strategies for Dryland Agriculture.* CSSA Spec. Publ. 32. ASA and CSSA, Madison, Wisconsin, pp. 167–190.

Ceccarelli, S., Grando, S. and Baum, M. (2007) Participatory plant breeding in water-limited environments. *Experimental Agriculture* 43, 411–435.

Ceccarelli, S., Guimaraes, E.P. and Weltzien, E. (eds) (2009) *Plant Breeding and Farmer Participation.* FAO, Rome, Italy.

Ceccarelli, S., Grando, S., Maatougui, M., Michael, M., Slash, M. *et al.* (2010) Plant breeding and climate changes. *Journal of Agricultural Science* 148, 627–638.

Chapman, S.C. (2008) Use of crop models to understand genotype by environment interactions for drought in real world and simulated plant breeding trials. *Euphytica* 161, 195–208.

Cleveland, D.A., Soleri, D. and Smith, S.E. (2000) A biological framework for understanding farmers' plant breeding. *Economic Botany* 54, 377–390.

Collins, N.C., Tardieu, F. and Tuberosa, R. (2008) Quantitative trait loci and crop performance under abiotic stress: where do we stand? *Plant Physiology* 147, 469–486.

Condon, A.G., Richards, R.A., Rebetzke, G.J. and Farquhar, G. (2002) Improving intrinsic water-use efficiency and crop yield. *Crop Science* 42, 122–131.

deMenocal, P.B. (2001) Cultural responses to climate change during the late holocene. *Science* 292, 667–673.

Erskine, W., Tufail, M., Russell, A., Tyagi, M.C., Rahman, M.M. *et al.* (1994) Current and future strategies in breeding lentil for resistance to biotic and abiotic stresses. *Euphytica* 73, 127–135.

Esquinas-Alcázar, J. (2005) Protecting crop genetic diversity for food security: political, ethical and technical challenges. *Nature Reviews Genetics* 6, 946–953.

Food and Agriculture Organization of the United Nations (2012) Hunger. Available at: www.fao.org/hunger/en/ (accessed 21 November 2012).

Fleury, D., Jefferies, S., Kuchel, H. and Langridge, P. (2010) Genetic and genomic tools to improve drought tolerance in wheat. *Journal of Experimental Botany* 61, 3211–3222.

Flower, D.J. (1996) Physiological and morphological features determining the performance of the sorghum landraces of northern Nigeria. *Experimental Agriculture* 32, 129–141.

Forster, B.P., Russell, J.R., Ellis, R.P., Handley, L.L., Robinson, D. *et al.* (1997) Locating genotypes and genes for abiotic stress tolerance in barley: a strategy using maps, markers and the wild species. *New Phytologist* 137, 141–147.

Fresco, L.O. and Baudoin, W.O. (2002) Food and nutrition security towards human security. In: *ICV Souvenir Paper*, International Conference on *Vegetables: World Food Summit Five Years Later.* 11–13 June 2002, Rome, Italy.

Grando, S., von Bothmer, R. and Ceccarelli, S. (2001) Genetic diversity of barley: use of locally adapted germplasm to enhance yield and yield stability of barley in dry areas. In: Cooper, H.D., Spillane, C. and Hodgin, T. (eds) *Broadening the Genetic Base of Crop Production.* CABI, New York, FAO, Rome, and IPGRI, Rome, pp. 351–372.

Gupta, P., Langridge, P. and Mir, R. (2010) Marker-assisted wheat breeding: present status and future possibilities. *Molecular Breeding* 26, 145–161.

Hajjar, R. and Hodgkin, T. (2007) The use of wild relatives in crop improvement: a survey of developments over the last 20 years. *Euphytica* 156, 1–13.

Hazell, P.B.R. (2003) The Green Revolution. In: Mokyr, J. (ed.) *Encyclopaedia of Economic History.* Oxford University Press, Oxford, UK, pp. 478–480.

Heffner, E.L., Sorrells, M.E. and Jannink, J-L. (2009) Genomic selection for crop improvement. *Crop Science* 49, 1–12.

Hooper, D.U., Adair, E.C., Cardinale, B.J., Byrnes, J.E.K, Hungate, B.A. *et al.* (2012) A global synthesis reveals biodiversity loss as a major driver of ecosystem change. *Nature* 486, 105–108.

Ivandic, V., Hackett, C.A., Zhang, Z.J., Staub, J.E., Nevo, D. *et al.* (2000) Phenotypic responses of wild barley to experimentally imposed water stress. *Journal of Experimental Botany* 51, 2021–2029.

Kelley, T.G., Parthasarathy Rao, P., Weltzien, E. and Purohit, M.L. (1996) Adoption of improved cultivars of pearl millet in an arid environment: straw yield and quality considerations in western Rajasthan. *Experimental Agriculture* 32, 161–171.

Keneni, G., Bekele, E., Imtiaz, M. and Dagne, K. (2012) Genetic vulnerability of modern crop cultivars: causes, mechanism and remedies. *International Journal of Plant Research* 2, 69–79.

Lakew, B., Eglinton, J., Henry, R.J., Baum, M., Grando, S. *et al.* (2011) The potential contribution of wild barley (*Hordeum vulgare* spp. *spontaneum*) germplasm to drought resistance of cultivated barley (*Hordeum vulgare* spp. *vulgare*). *Field Crops Research* 120, 161–168.

Lakew, B., Henry, R.J., Eglinton, J., Baum, M., Ceccarelli, S. *et al.* (2012) SSR analysis of introgression of drought tolerance from the genome of *Hordeum spontaneum* into cultivated barley (*Hordeum vulgare* ssp. *vulgare*). *Euphytica* DOI 10.1007/s10681-012-0795-9.

Lakew, B., Henry, R.J., Ceccarelli, S., Grando, S., Eglinton, J. *et al.* (2013) Genetic analysis and phenotypic associations for drought tolerance in *Hordeum spontaneum* introgression lines using SSR and SNP markers. *Euphytica* 189, 9–29.

Lauer, J.G., Bijl, C.G., Grusak, M.A., Baenziger, P.S., Boote, K. *et al.* (2012) The scientific grand challenges of the 21st century for the Crop Science Society of America. *Crop Science* 52, 1003–1010.

Luck, J., Spackmand, M., Freemand, A., Trębicki, P., Griffiths, W. *et al.* (2011) Climate change and diseases of food crops. *Plant Pathology* 60, 113–121.

Maxted, N., Ford-Lloyd, B.V., Jury, S.L., Kell, S.P. and Scholten, M.A. (2006) Towards a definition of a crop wild relative. *Biodiversity Conservation* 15, 2673–2685.

Maxted, N., Kell, S., Ford-Lloyd, B., Dulloo, E. and Toledo, A. (2012) Toward the systematic conservation of global crop wild relative diversity. *Crop Science* 52, 774–785.

Mir, R.R., Zaman-Allah, M., Sreenivasulu, N., Trethowan, R. and Varshney, R.K. (2012) Integrated genomics, physiology and breeding approaches for improving drought tolerance in crops. *Theoretical Applied Genetics* 125, 625–645.

Morran, L.T., Parmenter, M.D. and Phillips, P.C. (2009) Mutation load and rapid adaptation favour outcrossing over self-fertilization. *Nature* 462, 350–352.

Morrell, P.L., Buckler, E.S., Ross-Ibarra, J. (2012) Crop genomics: advances and applications. *Nature Reviews Genetics* 13, 85–96.

Munns, R. and Richards, R.A. (2007) Recent advances in breeding wheat for drought and salt stresses. In: Jenks, M.A.P., Hasegawa, M. and Jain, S.M. (eds) *Advances in Molecular Breeding toward Drought and Salt Tolerant Crops.* Springer-Verlag, Dordrecht, the Netherlands, pp. 565–586.

Mustafa, Y., Grando, S. and Ceccarelli, S. (2006) *Assessing the Benefits and Costs of Participatory and Conventional Barley Breeding Programs in Syria.* ICARDA, Aleppo, Syria.

Nassar, N.M.A. (2003) Cassava, *Manihot esculenta* Crantz genetic resources: VI Anatomy of a diversity center. *Genetics and Molecular Research* 2, 214–222.

Nevo, E. (1992) Origin, evolution, population genetics and resources for breeding of wild barley, *Hordeum spontaneum*, in the Fertile Crescent. In: Shewry, R.P. (ed.), *Barley: Genetics, Biochemistry, Molecular Biology and Biotechnology.* CAB International, Wallingford, UK, pp. 19–43.

Nevo, E., Fu,Y.-B., Pavlicek, T., Khalifa, S., Tavasi, M. *et al.* (2012) Evolution of wild cereals during 28 years of global warming in Israel. *Proceedings of the National Academy of Sciences USA* 109, 3412–3415.

Newton, A.C., Akar, T., Baresel, J.P., Bebeli, P.J., Bettencourt, E. *et al.* (2010) Cereal landraces for sustainable agriculture. A review. *Agronomy for Sustainable Development* 30, 237–269.

Öztürk, Z.N., Talamé, V., Michalowski, C.B., Gozukirmizu, N., Tuberosa, R. *et al.* (2002) Monitoring large-scale changes in transcript abundance in drought- and salt-stressed barley. *Plant Molecular Biology* 48, 551–573.

Parry, M.A.J., Flexas, J. and Medrano, H. (2005) Prospects for crop production under drought: research priorities and future directions. *Applied Biology* 147, 211–226.

Passioura, J.B. (2006) Increasing crop productivity when water is scarce – from breeding to field management. *Agricultural Water Management* 80, 176–196.

Reynolds, M., Dreccer, F. and Trethowan, R. (2007) Drought-adaptive traits derived from wheat wild relatives and landraces. *Journal of Experimental Botany* 58, 177–186.

Rick, C. and Chetelat, R. (1995) Utilization of related wild species for tomato improvement. *Acta Horticulturae* 412, 21–38.

Rosen, A.M. (1990) Environmental change at the

end of early Bronze Age Palestine. In: De Miroschedji, P. (ed.) *L'Urbanisation de la Palestine à l'Âge du Bronze Ancient.* BAR International, Oxford, UK, pp. 247–255.

Schulze, E.D. (1988) Adaptation mechanisms of non cultivated arid-zone plants: useful lesson for agriculture? In: Bidinger, F.R. and Johansen, C. (eds) *Drought Research Priorities for the Dryland Tropics.* ICRISAT, Hyderabad, India, pp. 159–177.

Seki, M., Narusaka, M., Ishida, J., Nanjo, T., Fujita, M. *et al.* (2002) Monitoring the expression profiles of 7000 *Arabidopsis* genes under drought, cold and high-salinity stresses using a full-length cDNA microarray. *Plant Journal* 31, 279–292.

Singh, K.B. and Ocampo, B. (1997) Exploitation of wild *Cicer* species for yield improvement on chickpea. *Theoretical and Applied Genetics* 95, 418–423.

Sthapit, B.R., Joshi, K.D. and Witcombe, J.R. (1996) Farmer participatory crop improvement. III. Participatory plant breeding, a case study for rice in Nepal. *Experimental Agriculture* 32, 479–496.

Stern, N. (2007) *The Economics of Climate Change: The Stern Review.* Cambridge University Press, Cambridge. Also available at: www.hm-treasury.gov.uk/independent_revie ws/stern_review_economics_climate_change/stern_review_Report.cfme (accessed 20 November 2012).

Suneson, C.A. (1956) An evolutionary plant breeding method. *Agronomy Journal* 48, 188–191.

Teshome, A., Brown, A.H.D. and Hodgkin, T. (2010) Diversity in landraces of cereal and legume crops. *Plant Breeding Reviews* 21, 221–261.

US Drought Monitor (2012) Available at: www.droughtmonitor.unl.edu/ (accessed 21 November 2012).

Valkoun, J. (2001) Wheat pre-breeding using wild progenitors. *Euphytica* 119, 17–23.

Vollbrecht, E. and Sigmon, B. (2005) Amazing grass: developmental genetics of maize domestication. *Biochemical Society Transactions* 33, 1502–1506.

Windhausen, V.S., Wagener, S., Magorokosho, C., Makumbi, D., Vivek, B. *et al.* (2012) Strategies to subdivide a target population of environments: results from the CIMMYT-led maize hybrid testing programs in Africa. *Crop Science* 52, 2143–2152.

Zaharieva, M., Gaulin, E., Havaux, M., Acevedo, E. and Monneveux, P. (2001) Drought and heat responses in the wild wheat relative *Aegilops geniculata* Roth: potential interest for wheat improvement. *Crop Science* 41, 1321–1329.

Zhou, J., Wang, X., Jiao, Y., Qin, Y., Liu, X. *et al.* (2007) Global genome expression analysis of rice in response to drought and high-salinity stresses in shoot, flag leaf, and panicle. *Plant Molecular Biology* 63, 591–608.

14 Salinity

William Erskine,[1] Hari D. Upadhyaya[2] and Al Imran Malik[3]

[1]*International Centre for Plant Breeding Education and Research (ICPBER) and Centre for Legumes in Mediterranean Agriculture (CLIMA), The University of Western Australia, Crawley, Australia;* [2]*International Crops Research Institute for the Semi-Arid Tropics (ICRISAT), Patancheru, India; and* [3]*Centre for Legumes in Mediterranean Agriculture (CLIMA) and Institute of Agriculture, The University of Western Australia, Crawley, Australia*

14.1 Introduction

Plants that tolerate high levels of salinity have been recognized for hundreds of years and are termed halophytes. Although the definition of halophytes is equivocal, they represent approximately 1% of the world's flora (Flowers and Colmer, 2008). Their very existence highlights the genetic diversity of responses to soil salinity within plants. In striking contrast, however, salinity is a key abiotic stress that affects the productivity of our salt-sensitive (glycophytic) crops worldwide.

Soil salinization refers to when the salt content and salinity rises. This can occur from natural processes, called primary salinity, such as the retreat of seawater and from the mineralization of rocks that weather to release soluble chloride salts. Secondary salinity, induced by the action of humans, is caused by processes such as irrigation (irrigation salinity) and the clearing of land for agriculture (dryland salinity). Both types of secondary salinity are caused by water-table rises that mobilize salt in the soil. Salinity can be due to the ions K^+, Na^+, Ca^{2+}, Mg^{2+} and Cl^-. Saline soils have a high concentration of soluble salts and are classed as saline when the electrical conductivity (EC_e) is ≥ 4 dS m^{-1} according to the USDA Salinity Laboratory (USSL, 2005). A conductivity of 4 dS m^{-1} is equivalent to 40 mM NaCl. (Seawater has an EC_e of ~50 dS m^{-1} and 550 mM NaCl.) This chapter focuses on salinity caused by the Na^+ and Cl^- ions.

14.2 Salinity Today and in 2050

Today, although salt-affected soils are found on all continents and under most climatic conditions, they are more extensive in arid and semi-arid regions where evaporation is highest (Abrol *et al.*, 1988). The distribution of salt-affected soils worldwide was given by Abrol *et al.* (1988) – these figures are indicative – and Rhoades *et al.* (1992) reported the following countries with notable salinity problems: Australia, China, Egypt, India, Iraq, Mexico, Pakistan, the republics of the former Soviet Union, Syria, Turkey and the USA. FAO (2000) estimated that about 20–30 million hectares (ha) of irrigated land are seriously damaged by the build-up of salts and 0.25–0.5 million hectares are estimated to be lost from production every year as a result of salt build-up. In a more recent FAO expert consultation to assess the status of biosaline agriculture, although the inconsistencies of the data on the extent of the problem were emphasized, there was agreement that the problem is extensive and growing (FAO, 2009).

Looking ahead to 2050, there is *high confidence* that many semi-arid areas (e.g. the Mediterranean Basin, western USA, southern Africa and north-eastern Brazil) will suffer a decrease in water resources owing to climate change (IPCC, 2007). Drought-affected areas are projected to increase in extent. So, large increases in irrigation water demand as a result of climate changes are projected regionally (IPCC, 2007). This will exacerbate the irrigation salinity problems that exist today. Additionally, in coastal areas sea level rise will aggravate water resource constraints due to increased salinization of groundwater supplies (IPCC, 2007). In Australia, regarding dryland salinity, it is estimated that by 2050 the area of regions with a high risk of salinity may increase from the 5.7 million ha now affected to 17 million ha (ADSA, 2000).

There is major diversity among the various crop/soil salinity scenarios depending on soil structural composition, ion content, and rainfall pattern and timing. This complex is hard to manage and breeding for salinity tolerance is often an approach of choice – complementary to improved agronomic practices and landscape management. As mentioned above, most of our crops are salt sensitive, despite the widespread occurrence of halophytes. Although terrestrial halophytes come from a wide spectrum of families, two families are particularly important, the Chenopodiaceae and the Poaceae. Members of these families provide models through which the physiology of salt tolerance can be understood (Flowers and Flowers, 2005).

14.3 Physiology of Salinity Tolerance

There are two distinct phases in which plants encounter soil salinity stress. The presence of excessive salt in the soil solution makes it strenuous for plant roots to take up water (i.e. water stress), and then a toxic concentration of salt can be accumulated within plants from taking up excess salt from the soil solution (i.e. ion specific stress) (Greenway and Munns, 1980; Munns and Termaat, 1986). Roots face the immediate effect of a high salt concentration in the soil solution that reduces plant growth; however, over time salt accumulates to toxic concentrations in shoots to affect the whole plant function, causing necrosis.

The mechanisms of salinity tolerance in plant species are classified into three: osmotic stress tolerance, ion exclusion and tissue tolerance (Colmer *et al.*, 2005; Munns and Tester, 2008). Osmotic stress reduces cell expansion, loses cell turgor and also closes stomata. Plants respond to the stress by osmotic adjustment – compartmentation of ions into vacuoles and the synthesis of organic solutes to restore cell turgor (Wyn Jones *et al.*, 1977). Ion exclusion restricts the entry of potentially toxic ions – Na^+ and/or Cl^- – into the shoots. Failure to do so results in the accumulation of ions in the leaves to a concentration that is toxic and results in the premature death of older leaves (Munns and James, 2003; Tester and Davenport, 2003). Tissue tolerance is the ability of the plant to compartmentalize toxic ions, Na^+ and/or Cl^-, into vacuoles to keep the enzyme function active in the cytosol and/or organelles (Yeo and Flowers, 1983). The physiological traits associated with these diverse mechanisms of salinity tolerance are being documented (Colmer *et al.*, 2005; Munns and Tester, 2008).

14.4 Variation Among Crops to Salinity Tolerance

As previously mentioned, wide variation to salinity tolerance is observed among plants (Fig. 14.1). Among cereal crops, rice *(Oryza sativa* L.) is the most sensitive and barley (*Hordeum vulgare* L.) is the most tolerant; bread wheat (*Triticum aestivum* L.) and durum wheat (*Triticum turgidum* L. subsp. *durum*) lie in between (see review in Munns and Tester, 2008). Sea barley grass (*Hordeum marinum* L.), a wild relative of barley and tall wheatgrass (*Thinopyrum ponticum* (Podp.) Barkworth & D.R. Dewey), and a wheat wild relative can grow at salt concentrations approaching that of sea water (Colmer *et al.*, 2005; Garthwaite *et al.*, 2005). Although most legume crops are more salt sensitive

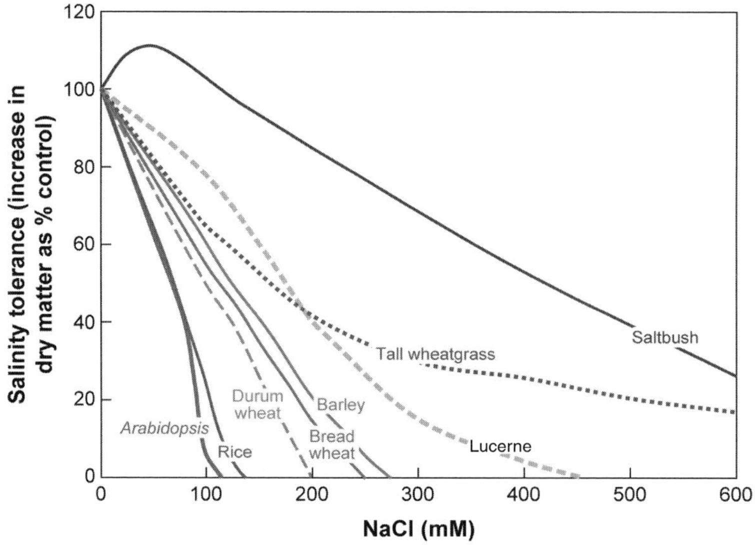

Fig. 14.1. Diversity in the salt tolerance of various species, shown as increases in shoot dry matter after growth in solution or sand culture containing NaCl for at least 3 weeks, relative to plant growth in the absence of NaCl. Data are for rice (*Oryza sativa*), durum wheat (*Triticum turgidum* subsp. *durum*), bread wheat (*Triticum aestivum*), barley (*Hordeum vulgare*), tall wheatgrass (*Thinopyrum ponticum*), Arabidopsis (*Arabidopsis thaliana*), lucerne (*Medicago sativa*) and saltbush (*Atriplex amnicola*). Reprinted from Munns and Tester (2008).

than rice (Lauchli, 1984), lucerne (*Medicago sativa* L.) is relatively tolerant to salinity (Al-Khatib *et al.*, 1992).

14.5 Inheritance of Salinity Tolerance

With this genetic variability both among and within crops in salinity tolerances, the complexity in factors affecting salinity suggests that it depends on the action of many genes. Indeed salinity tolerance shows all the characteristics of a multigenic trait, and quantitative trait loci (QTLs) have been identified in barley (Zhou *et al.*, 2012), chickpea (*Cicer arietinum* L.) (Vadez *et al.*, 2012), rice (Lee *et al.*, 2007), soybean (*Glycine max* L.) (Lee *et al.*, 2004), tomato (*Lycopersicon esculentum* L.) (Villalta *et al.*, 2007) and wheat (Ma *et al.*, 2007; De Leon *et al.*, 2011). In barley a total of five QTLs for salinity tolerance were identified on chromosomes 1H, 2H, 5H, 6H and 7H (Zhou *et al.*, 2012). In rice a QTL for salt tolerance was found to encode a sodium transporter (Ren *et al.*, 2005). QTLs for salt tolerance, however, differ depending on growth conditions and on the stage of growth and may only be expressed in particular bi-parental cross combinations (Flowers and Flowers, 2005), leading to the limited usefulness of linked DNA markers.

14.6 What is Needed to Develop Salt-tolerant Cultivars?

As a precursor to developing a screening method and embarking on selection for crop improvement for salinity, the target field scenario of salinity needs to be clearly understood. What ions are involved in the field salinity problem? What is the timing/duration (transience) of the salinity and during which growth stage does it occur in the field? What is the structural composition of the affected soil?

In the soil, salinity levels are notoriously variable, which makes field screening frustratingly imprecise. A recent review by

Bazihizina *et al.* (2012) has emphasized that even within the root-zone of single plants the salinity of the soil solution can vary widely. The review showed that water uptake by roots from the least saline part of the soil is the key factor driving shoot growth. Because of soil salinity heterogeneity and a lack of large-scale screening facilities in the field (Arzani, 2008), screenings for salinity tolerance are usually carried out under more controlled conditions, but screening in the field remains crucial for confirming controlled environment results. An example of methodology development for such controlled environment screening is given by Munns and James (2003) for durum wheat.

With a screening method in hand, then the identification of useful genetic variation for salinity tolerance should begin among available cultivars, then – if necessary – proceed to landraces, then – again if necessary – to wild relatives and finally to genetic modification. It must be emphasized that the identification of useful variability for salt tolerance is just the start of the process of exploiting the identified tolerance in crop improvement, and the proposed search scheme is to minimize the time from the identification of useful levels of tolerance to the release of tolerant cultivars to growers.

14.7 Salinity Tolerance within the Domesticates

Significant genetic variation for salinity tolerance has been found among the landraces of many crop species. Progress in breeding for salt tolerance in rice was reviewed by Gregorio *et al.* (2002) and salt-tolerant landrace rice varieties such as Pokkali and Nona Bokra originated in coastal areas of India and are the source of salt tolerance in most rice breeding programmes. Similarly, a review of salinity tolerance in chickpea (Flowers *et al.*, 2010) indicated that screenings of diverse germplasm showed sufficient variation to enable improvement in yield under saline conditions via breeding. There is equivalent information on screening landrace populations for salinity tolerance in other species and a summary is given in Table 14.1 for chickpea, pigeonpea, pearl millet, groundnut, sorghum, finger millet and foxtail millet.

14.7.1 Mini-core approach to screen germplasm collections for salt tolerance

Salinity tolerance as a trait is determined by several sub-traits, any (or many) of which can in turn be determined by several genes (Yamaguchi and Blumwald, 2005), creating great diversity within germplasm collections. Although large germplasm collections have been assembled worldwide for different crops, identification of genotypes that are highly diverse in general or in specific combinations that can be used in crop improvement is lacking. The huge size of germplasm collections makes it impractical and expensive to assess their genetic worth, a hurdle to select specific lines by breeders for their use. Frankel (1984) suggested use of representative core collections (10% of entire collection) in such cases. In the case of large collections (tens of thousands) such as those maintained at the CGIAR centers, however, even a core collection will be unwieldy for extensive evaluation to identify trait-specific germplasm. To overcome this, Upadhyaya and Ortiz (2001) postulated a mini core (10% of core or 1% of entire collection), adequately representing variability in the core and entire collections. At ICRISAT mini-core collections of chickpea, pigeonpea, groundnut, sorghum, pearl millet, finger millet and foxtail millet (Upadhyaya and Ortiz, 2001; Upadhyaya *et al.*, 2002, 2006, 2009, 2010, 2011a, b) have been developed. A mini-core collection (35–242 accessions) can be easily evaluated for different quantitative (including biotic and abiotic stress tolerance) and quality traits at both the phenotypic and genetic level. The highly diverse genetic resources identified thus are useful as donors not only to pyramid stress tolerance genes but also to broaden the genetic base of cultivars. For example, phenotypic diversity (based on salinity tolerance and multi-environment agronomic evaluation) and genetic diversity (based on SSR marker data of salinity tolerant lines)

Table 14.1. Genetic resources of chickpea, pigeonpea, pearl millet, groundnut, sorghum, finger millet and foxtail millet: responses to salt stress.

Crop	Botanical name	Concentration of NaCl (mM)	Stage of stress imposition	Duration of stress	Percentage growth reduction versus control	Reference	Remarks
Chickpea	*Cicer arietinum (L.)*	80 mM	Sowing and 12 DAS[a] in two half doses	Entire crop duration	40–60	Krishnamurthy *et al.*, 2011	Yield reduction primarily due to reduction in pod and seed number
Pigeonpea	*Cajanus cajan (L) Millsp.*	1.01g NaCl kg^{-1} soil	Sowing to 10 DAS in three equal doses	7–69 DAS	15–52	Srivastava *et al.*, 2006	The mini-core collection provided a larger range of variation for salinity tolerance than accessions originating from putative saline areas; a large number of tolerant accessions were from Bangladesh
Groundnut	*Arachis hypogaea (L)*	100 mM in three equal doses	Sowing to 10 DAS in 3 equal doses	Entire crop duration	30	Srivastava, 2010 ICRISAT, 2010	The mini-core provided sixfold variation for pod yield under saline conditions
Pearl millet	*Pennisetum glaucum (L) R.Br.*	100 mM	Sowing	Sowing to 39 DAS	51	Kulkarni *et al.*, 2006 Krishnamurthy *et al.*, 2007	Wide variation for grain as well as stover yield
Sorghum	*Sorghum bicolor (L) Moench*	200 mM	Sowing	Sowing to 39 DAS	38	ICRISAT, 2008	Wide variation for grain as well as stover yield
Finger millet	*Eleusine coracana (L) Gaertn.*	100 mM	Sowing and 10 DAS in two equal half doses	Entire crop duration	30	ICRISAT, 2012	Wide variation for biomass and grain yield
Foxtail millet	*Setaria italica (L.) P. Beauv*	100 mM	Sowing and 10 DAS in two equal half doses	Entire crop duration	37	ICRISAT, 2012	Wide variation for grain as well as stover yield

[a]DAS, days after sowing.

from the mini-core set of accessions was assessed and a large number of highly diverse pairs of accessions were identified in chickpea, finger millet and foxtail millet at ICRISAT, India. These highly diverse lines identified systematically on the basis of a number of economically important traits and molecular markers enable breeders to select source lines to incorporate into breeding.

Recent alternative approaches to screen for salt tolerance in cultivated rice have included somaclonal variants (Gregorio *et al.*, 2002). Overall, however, the level of variation within crop germplasm has been found in most cases as inadequate to be useful in the field and, among crops, there are rather remarkably few released cultivars with salinity tolerance. A review by Ashraf and Akram (2009) listed only examples of three lucerne cultivars AZ-97MEC and AZ-97MEC-ST (Dobrenz *et al.*, 1989) and AZ-germ salt-II (Al-Doss and Smith, 1998), and two cultivars in bread wheat – S24 (Ashraf and O'Leary, 1996) and KRL1-4 (Hollington, 2000) based on the landrace Kharchia 65 – all evaluated on natural salt-affected soils. Given this limited genetic diversity among landraces for a salinity response, many scientists have taken the approach to identify suitable genetic variation for salt tolerance among wild relatives of crops.

14.8 Salinity Tolerance in Crops through Wide Hybridization

Domestication created genetic bottlenecks in the emerging crops, and crop wild relatives (and progenitor) represent a major under-exploited genetic resource with great potential (Tanksley and McCouch, 1997). The agronomic performance of crops has been improved using wild relatives (wheat, Mujeeb-Kazi *et al.*, 1996; rice, Brar, 2005). But wild relatives, in general, being rather poor in agronomic traits such as yield, are little used to breed for improved yield (Hajjar and Hodgkin, 2007); and are primarily exploited for pest and disease resistance (Rick and Chetelat, 1995; Brar and Khush, 1997; Hajjar and Hodgkin, 2007). Evidence of a wide range of variation among crop wild relatives for abiotic stress tolerance such as salinity tolerance (e.g. wheat: Garthwaite *et al.*, 2005; Colmer *et al.*, 2006; barley: Qiu *et al.*, 2011; rice: Flowers *et al.*, 1990; tomato: Tal and Shannon, 1983; Bolarn *et al.*, 1991; bean: Bayuelo-Jiménez *et al.*, 2002) is well documented.

With these sources of salinity tolerance identified in crop wild relatives, attempts have been made to exploit them by breeding for salinity tolerance. In tomato, Tal and Shannon (1983) evaluated three wild relatives (*Lycopersicon cheesmanii* subsp. *minor* (Hook.) C.H. Mull., *Lycopersicon peruvianum* subsp. *puberulum* (Phil.) Luckwill and *Solanum pennellii* Corr.) along with two tomato cultivars and F_1 hybrids. The study demonstrated the salinity tolerance in the F_1 hybrids. In another attempt, a cultivar (*Lycopersicon esculentum* Mill. cv. Walter) was crossed with *L. cheesmanii*. Plants selected from the F_2 and successive backcrosses to 'Walter' survived and produced fruit when irrigated with up to 70% seawater in the sandy soil culture trials, whereas 'Walter' did not survive (Rush and Epstein, 1980). In sunflower, salinity tolerance traits from *Helianthus paradox* Heiser, a wild relative of sunflower, were successfully introduced and five maintainer lines have been released (Miller and Seiler, 2003). In Brassica, improved salinity tolerance has been reported in the amphiploids *Brassica napus* L., *Brassica juncea* (L.) Czern. and *Brassica carinata* A. Braun compared with the diploids *Brassica campestris L.* and *Brassica nigra* (L.) W.D.L. Koch. Presumably, salinity tolerance in the amphiploids has been accumulated from the diploid relatives (Ashraf and McNeilly, 2004; Kumar *et al.*, 2009). In rice, a comparison has been made between sources of salinity tolerance from among landraces (Yeo and Flowers, 1986; Yeo *et al.*, 1990; Lisa *et al.*, 2004, 2011) and some wild relatives (Akbar *et al.*, 1987). In both screenings, salinity tolerances have been reported at different developmental stages; however, in some cases, landraces showed better and/or similar tolerance compared to wild relatives. For example, the *indica* landrace Nona Bokra (Indian origin) and *Oryza glaberrima* Steud. (African origin) showed better tolerance at a 2–3 leaf stage compared with wild rice; and wild rice, *Oryza rufipogon* Griff., had similar tolerance to landrace Pokkali (Gregorio *et al.*, 2002). By contrast, the distant wild relative of rice *Porteresia coarctata* (Roxb.) Tateoka, which needs a high salt level to produce grain, could not be included in the breeding programme owing to instability in F_1s. It has been suggested, however, that modern crop biotechnology tools may help to transfer desired traits of salinity tolerance of *P. coarctata* to cultivated rice (Latha *et al.*, 2004). In barley, a wide range of variation in salinity tolerance has been reported in germplasm of the progenitor *Hordeum vulgare* subsp. *spontaneum* (K. Koch) Thell and was in focus to improve salinity tolerance (Gorham *et al.*, 1991; Nevo *et al.*, 1993; Forster *et al.*, 1994; Ellis *et al.*, 2000). But, as in the previous rice comparison, there was little difference observed between the least damaged classes of cultivated accessions of barley and those of the close wild relative (in this case the progenitor) *spontaneum* accessions. In wheat, wild relatives have been used in attempts to increase salinity tolerance by wide hybridization. For example, the tall wheatgrass species *Agropyron elongatum* (syn. *Lophopyrum elongatum* syn. *Thinopyrum elongatum*), *Thinopyrum bessarabicum* (Dvořák and Ross, 1986; Omielan *et al.*, 1991; Mahmood and Quarrie, 1993; Colmer *et al.*, 1995) and sea barley grass (*Hordeum marinum* L.) (Islam *et al.*, 2007) have been crossed with bread wheat and amphiploids have shown considerable improvement in salt tolerance compared with wheat. Despite the presence of salinity tolerance traits in the wild relatives, owing to the complex nature of the inheritance in only a few cases have crops reached the stage of registered germplasm with genetic contributions for salinity tolerance from wild relatives (e.g. sunflower (*Helianthus annuus* L.) (Miller and Seiler 2003)), and to the authors' knowledge there are no such cultivar releases.

14.9 Genetic Modification Approach

There has been massive investment in salt tolerance research. A search in August 2012 of US, European and Australian patent databases revealed 17,692 patents granted that invoke plant salinity tolerance (Patentlens, 2012). Most of these patents involve specific genes. One approach being pursued to improve salinity tolerance is genetic transformation. Transgenic plants are genetically altered by the insertion of desired genes into their cells. Transgenic crops now cover 160 million ha worldwide with herbicide tolerance and insect resistance as the predominant transformed traits (James, 2011). Clearly transgenic crops are a major commercial success. For salinity tolerance there are many promising assays of plants transformed with specific genes that have led to enhanced tolerance under laboratory and controlled environment conditions (Munns, 2005; Ashraf and Akram, 2009). The genes involved may be grouped into ion transporters, compatible organic solutes and enhanced antioxidant production, as reviewed by Ashraf and Akram (2009). There is only one study by Xue *et al.* (2004) in which the performance of a bread wheat transgenic line overexpressing the vacuolar Na^+/H^+ antiporter gene *AtNHX1* from Arabidopsis was evaluated under both greenhouse and natural field conditions. With this one notable exception, the testing of transgenic plants for salinity tolerance under field conditions is very much the exception rather than the rule. A licence (DIR053/2004) to field-trial genetically modified salt-tolerant wheat on saline land was granted by the Office of the Gene Technology Regulator in Australia in 2004 (OGTR, 2012). There are as yet no released transgenic salt-tolerant crop cultivars. Field results differ from laboratory assays for several reasons: field environments have salinity levels that vary in time and space; the level of tolerance can vary through plant development; and laboratory tests are often only done with NaCl whereas many soils contain additional salts. Most importantly salinity tolerance is a complex multigenic trait and the chance of finding a single gene to act as a silver bullet is low.

14.10 New Domestications for Salinity

Dicotyledonous halophytes are able to grow at similar rates to salt-sensitive plants (glycophytes) even though they use energy in the accumulation and compartmentalization of the ions required for osmotic adjustment (Flowers and Colmer, 2008). Unlike the situation with xerophytes (plants adapted to life in a dry habitat) that survive drought with very low growth rates (see Ceccarelli, Chapter 13, this volume), which makes them unsuitable as crops, halophytes can grow rapidly under saline conditions (Flowers and Colmer, 2008). We now give four examples of ongoing domestications for saline environments:

1. Domestication of the halophyte *Salicornia bigelovii* Torr.
2. Use of woody biomass crops such as oil mallee eucalypts for dryland salinity.
3. The domestication of a new pasture plant, *Melilotus siculus* (Turra) B. D. Jacks., for areas that are both salted and waterlogged.
4. Domestications for bio-amelioration.

14.10.1 *Salicornia bigelovii*

The domestication of halophytes was proposed for salt-affected arable land (O'Leary, 1984; Glenn *et al.*, 1999) and there are several reflections of the potential of halophytes as crops within the literature (Glenn *et al.*, 1991, 1999; Troyo-Dieguez *et al.*, 1994; Zurayk and Baalbaki, 1996). Although the domestication of some halophytes as fodder crops has been evaluated (O'Leary, 1984), the high salt concentration in the shoots diminished success as possible alternatives in the cropping system (Malcolm, 1969, 1974; Zahran, 1994; Ashour *et al.*, 1997; Glenn *et al.*, 1998; Brown and Glenn, 1999). *Salicornia bigelovii* is a halophyte with good potential to become a vegetable crop owing to its high biomass (18 t ha^{-1} in 200 day cycle irrigated with sea water) and seed production (2 t ha^{-1}); and seed protein (31%) and oil (28%) content. Moreover, the oil is high in polyunsaturated fatty acids, particularly linoleic acid (74% of total) (Glenn *et al.*, 1991, 1999). *S. bigelovii* is now cultivated as a speciality vegetable for US and European fresh produce markets (OASE, 2009).

14.10.2 Woody biomass crops – oil mallee eucalypts

In Australia dryland salinity, with the need to control recharge and the recognition that carbon emissions are an important national issue, are now the key drivers in the domestication of deep-rooted, short-production cycle, woody crops (Hobbs *et al.*, 2009). Woody perennial systems can accumulate and store significant quantities of carbon in both living plant biomass and soil profiles and provide offsets as an alternative feedstock for energy and transport fuel production. A range of Australian native species with potential for domestication as new biomass industry crops have been evaluated and the primary focus is now on the genetic improvement of the blue mallee *Eucalyptus polybractea* R.T. Baker as a potential woody species suited to mixed farm systems as part of the overall agricultural landscape. This approach with mallee species, being taken by the Cooperative Research Centre Future Farm Industries, reduces saline seepage through the deep roots of the woody species and is a form of salinity mitigation, but it does not represent direct breeding for salt tolerance per se (Sudmeyer and Goodreid, 2007). Screening of eucalypts directly for saline lands is also under way (Zohar *et al.*, 2010).

14.10.3 *Melilotus siculus* – salt- and waterlogging-tolerant pasture

Areas affected by salinity often experience periods of waterlogging. In such conditions, plants are exposed to root-zone oxygen deficiency in addition to salinity. This combined stress makes the plant more susceptible to waterlogging compared to the salinity stress alone (Barrett-Lennard, 2003; Malik *et al.*, 2009). In the cropping system, a healthy legume component is desired for good forage quality and soil health. But forage legumes are sensitive to salinity with the exception of lucerne (Rogers *et al.*, 2005). Salinity and waterlogging tolerance, separately and in combination, of forage legumes has been documented (Rogers *et al.*, 2011). Among forage legumes, the *Melilotus* genus arose as a research priority because of its variation in tolerance to salinity and waterlogging and the stresses in combination (Rogers *et al.*, 2008). The genus *Melilotus* contains 25 species (Allen and Allen, 1981), which are moderately winter hardy and drought resistant, and can be grazed as pasture plants (Stevenson, 1969). Within the genus, the species *Melilotus siculus* is now the focus of domestication efforts in West Australia (Rogers *et al.*, 2011).

Bio-amelioration or biological reclamation, using plants tolerant to salinity, sodicity and waterlogging to improve soil health, is the most economical way of bringing these soils back to productivity. Biological reclamation using highly salt-tolerant halophytes is an alternative option to ameliorate alkali soils as well as produce fuel wood, forage, medicinal and other products of economic use. Some halophytes are able to rehabilitate degraded soils through desalination and stabilization, as well as heavy-metal extraction (phytoremediation) and fertility improvement, thus creating micro-habitats favourable for the development of several salt-sensitive annuals (Chedly *et al.*, 2006).

14.10.4 Growing salt mining species/cultivars

In India an improved rice cultivar CSR 10 is recommended to bio-ameliorate highly deteriorated sodic soils having stress up to pH 10.2 without any soil amendment, reducing the soil pH to 9.5 and exchangeable sodium percentage (ESP) from 85 to 40, along with improvement in the soil physical characteristics, after the third crop. Continuous cultivation of this variety without any soil/chemical amendment reclaimed alkali soils in 3 years to the extent that other crops such as wheat, barley and mustard could then be grown (Mishra *et al.*, 2003).

In Pakistan, Kallar grass (*Leptochloa fusca* L.) and *Sesbania bispinosa* (Jacq.) W.F. Wight have been used with or without added gypsum to ameliorate saline soils. Kallar grass thrives on low-fertility saline soils, fixes significant amounts of N (Malik *et al.*, 1997) and maintains P uptake under salinity stress (Mahmood *et al.*, 1995). Organic matter added to the soil by the crop and the acidifying action of the roots makes soluble calcium available. *Sesbania* is also excellent for reclamation because of its high fodder yield, vigorous and deep root system, and ability to fix N for subsequent crops (Qadir *et al.*, 1997). Three halophytic species of the Chenopodiaceae family, *Haloxylon recurvum* Bunge ex Boiss, *Suaeda nudiflora* Moq. and *Salsola baryosma* (Schult.) Dandy, were found to be effective for bioremediation of a secondary salinized soil and their use as 'primers' for enhancing the process of reclamation of saline soils, beside biomass production, was recommended by Shekhawat *et al.* (2006).

Rehabilitation of degraded lands with non-conventional salt-tolerant trees is feasible, though over a longer period. For example, in India, 20-year-old plantations of *Prosopis julifera* (Sw.) DC., *Acacia nilotica* (L.) Delile, *Eucalyptus tereticornis* Sm., *Albizia lebbeck* (L.) and *Terminalia arjuna* (Roxb.) Wight & Arn. could ameliorate an alkali soil (pH >10) by adding organic matter through litter to the extent that arable crops could be grown successfully on this soil (Dagar, 2009). At another site, after 7 years of plantation, *Prosopis julifera* and *Acacia nilotica* reduced the soil pH, EC_e and ESP from 8.8, 4 dS m^{-1} and 35, respectively, to 8.5, 1.3 dS m^{-1} and 10–14 (Dagar, 2009).

14.11 Future for Genetic Resources and Breeding for Salinity Tolerance

A brief glance at the vegetation growing in a salt marsh or a mangrove swamp demonstrates the ready availability of abundant genetic diversity for salinity tolerance; but these halophytes are not domesticated. Some species growing in salt marshes are, however, wild relatives of existing crop species (e.g. *H. marinum*). The exploitation of the genes for salt tolerance in these wild relatives through wide crossing with domesticates is currently limited by fertility issues, which will be the focus of future research. It is likely with the increasing importance of salinity anticipated as a consequence of climate change that efforts towards the domestication of new halophytic species will continue either for new food, pasture/forage and bioremediation uses or to mitigate dryland salinity, for example the oil mallees in southern Australia.

Within crop species we have also seen that attempts to improve crop salt tolerance have so far met with very limited success because of the complexity of the trait, both genetically and physiologically. The systematic screening of structured samples of germplasm collections (mini-core) has revealed some new levels of salinity tolerance. The problem of salinity is compounded by associated mineral deficiencies (Zn, P), toxicities (Fe, Al and organic acids), drought and submergence. The varying magnitude and interactions of these soil stresses over time and place influence the stability of tolerance of a cultivar, making long-term adaptability dependent on its ability to withstand all these correlated stresses. Thus, multiple stress tolerance traits must be considered for breeding for saline environments (Gregorio *et al.*, 2002). Furthermore, salt tolerance in plants seems to be a developmentally regulated process because tolerance varies with growth stage, generally increasing with the age of the plant (Richards, 1996). Given these diverse components of salinity tolerance, the future screening of genetic resources can focus on identifying the extremes for the individual components, which breeders can then recombine.

References

Abrol, I.P., Yadav, J.S.P. and Massoud, F.I. (1988) *Salt-Affected Soils and their Management.* FAO Soils Bulletin 39, Food and Agriculture Organization of the United Nations (FAO), Rome, Italy. Available at: www.fao.org/docrep/x5871e/x5871e00.htm#Contents (accessed 13 November 2012).

ADSA (Australian Dryland Salinity Assessment) (2000) *Australian Natural Resources Atlas.* Available at: www.anra.gov.au/topics/salinity/pubs/national/salinity_summary.html (accessed 13 November 2012).

Akbar, M., Jena, K.K. and Seshu, D.V. (1987) Salt tolerance in wild rices. *International Rice Research News* 12, 15.

Al-Doss, A.A. and Smith, S.E. (1998) Registration of AZ-97MEC and AZ-97MEC-ST very non-dormant alfalfa germplasm pools with increased shoot weight and differential response to saline irrigation. *Crop Science* 38, 568.

Al-Khatib, M., McNeilly, T. and Collins, J.C. (1992) The potential of selection and breeding for improved salt tolerance in lucerne (*Medicago sativa* L.). *Euphytica* 65, 43–51.

Allen, O.N. and Allen, E.K. (1981) *The Leguminosae. A Source Book of Characteristics, Uses and Nodulation.* University of Wisconsin Press, Madison, Wisconsin.

Arzani, A. (2008) Improving salinity tolerance in crop plants: a biotechnological view. *In vitro Cellular Developmental Biology-Plant* 44, 373–383.

Ashour, N.I., Serag, M.S., Abd El-Haleem, A.K. and Mekki, B.B. (1997) Forage production from three grass species under saline irrigation in Egypt. *Journal of Arid Environments* 37, 299–307.

Ashraf, M. and Akram, N.A. (2009) Improving salinity tolerance of plants through conventional breeding and genetic engineering: an analytical comparison. *Biotechnology Advances* 27, 744–752.

Ashraf, M. and O'Leary, J.W. (1996) Responses of some newly developed salt-tolerant genotypes of spring wheat to salt stress. 1. Yield components and ion distribution. *Journal of Agronomy and Crop Science* 176, 91–101.

Ashraf, M. and McNeilly, T. (2004) Salinity tolerance in *Brassica* oilseeds. *Critical Reviews in Plant Sciences* 23, 157–174.

Barrett-Lennard, E.G. (2003) The interaction between waterlogging and salinity in higher plants: causes, consequences and implications. *Plant and Soil* 253, 35–54.

Bayuelo-Jiménez, J.S., Debouck, D.G. and Lynch, J.P. (2002) Salinity tolerance in *Phaseolus* species during early vegetative growth. *Crop Science* 42, 2184–2192.

Bazihizina, N., Barrett-Lennard, E.G. and Colmer, T.D. (2012) Plant growth and physiology under heterogeneous salinity. *Plant and Soil* 354, 1–19.

Bolarn, M.C., Fernandez, F.G., Cruz, V. and Cuartero, J. (1991) Salinity tolerance in four wild tomato species using vegetative yield salinity response curves. *Journal of American Society for Horticultural Science* 116, 286–290.

Brar, D. (2005) Broadening the gene pool and exploiting heterosis in cultivated rice. In: Toriyama, K., Heong, K.L. and Hardy B. (eds) *Rice is Life: Scientific Perspectives for the 21st Century. Proceedings of the World Rice Research Conference*. Tokyo and Tsukuba, Japan, pp. 157.

Brar, D. and Khush, G. (1997) Alien introgression in rice. *Plant Molecular Biology* 35, 35–47.

Brown, J.J. and Glenn, E.P. (1999) Reuse of highly saline aquaculture effluent to irrigate a potential forage halophyte, *Suaeda estoroa*. *Aquaculture Engineering* 20, 91–111.

Chedly, A., Barhoumi, Z., Ghnaya, T., Debez, A., Hamed, K.B. *et al.* (2006) Potential utilisation of halophytes for the rehabilitation and valorisation of salt-affected areas in Tunisia. In*:* Ozturk, M., Waisel, Y., Ajmal Khan, M. and Gork, G. (eds) *Biosaline Agriculture and Salinity Tolerance in Plants.* Birkhauser Verlog, Basel, Switzerland.

Colmer, T.D., Epstein, E. and Dvořák, J. (1995) Differential solute regulation in leaf blades of various ages in salt-sensitive wheat and a salt-tolerant wheat × *Lophopyrum elongatum* (Host) A. Love amphiploid. *Plant Physiology* 108, 1715–1724.

Colmer, T.D., Munns, R. and Flowers, T.J. (2005) Improving salt tolerance of wheat and barley: future prospects. *Australian Journal of Experimental Agriculture* 45, 1425–1443.

Colmer, T.D., Flowers, T.J. and Munns, R. (2006) Use of wild relatives to improve salt tolerance in wheat. *Journal of Experimental Botany* 57, 1059–1078.

Dagar, J.C. (2009) Opportunities for alternate land uses in salty and water scarcity areas. *International Journal of Ecology and Environmental Sciences* 35, 53–66.

De Leon, J.L.D., Escoppinichi, R., Geraldo, N., Castellanos, T., Mujeeb-Kazi, A. *et al.* (2011) Quantitative trait loci associated with salinity tolerance in field grown bread wheat. *Euphytica* 181, 371–383.

Dobrenz, A.K., Robinson, D.L., Smith. S.E. and Poteet, D.C. (1989) Registration of AZ-germ salt-II nondormant alfalfa germplasm. *Crop Science* 29, 493.

Dvořák, J. and Ross, K. (1986) Expression of tolerance of Na^+, K^+, Mg^{2+}, Cl^- and SO_{2-4} ions and sea water in the amphiploid of *Triticum aestivum* × *Elytrigia elongata*. *Crop Science* 26, 658–660.

Ellis, R.P., Forster, B.P., Robinson, D., Handley, L.L., Gordon, D.C. *et al.* (2000) Wild barley: a source of genes for crop improvement in the 21st century? *Journal of Experimental Botany* 51, 9–17.

FAO (Food and Agriculture Organization) (2000) *Crops and Drops: Making the Best Use of Water for Agriculture.* FAO Advance Edition, FAO, Rome.

FAO (Food and Agriculture Organization) (2009) *Advances in the Assessment and Monitoring of Salinization and Status of Biosaline Agriculture.* Report of expert consultation held in Dubai, United Arab Emirates, 26–29 November 2007. World Soil Resources Reports No. 104. FAO, Rome.

Flowers, T.J. and Colmer, T.D. (2008) Salinity tolerance in halophytes. *New Phytologist* 179, 945–963.

Flowers, T.J. and Flowers, S.A. (2005) Why does salinity pose such a difficult problem for plant breeders? *Agricultural Water Management* 78, 15–24.

Flowers, T.J., Flowers, S.A., Hajibagheri, M.A. and Yeo, A.R. (1990) Salt tolerance in the halophytic wild rice, *Porteresia coarctata* Tateoka. *New Phytologist* 114, 675–684.

Flowers, T.J., Gaur, P.M., Gowda, C.L.L., Krishnamurthy, L., Samineni, S. *et al.* (2010) Salt sensitivity in chickpea. *Plant, Cell and Environment* 33, 490–509.

Forster, B.P., Pakniyat, H., Macaulay, M., Matheson, W., Phillips, M. *et al.* (1994) Variation in the leaf sodium content of the *Hordeum vulgare* (barley) cultivar Maythorpe and its derived mutant cv. Golden Promise. *Heredity* 73, 249–253.

Frankel, O.H. (1984) Genetic perspective of germplasm conservation. In: Arber, W., Limensee, K., Peacock, W.J. and Stralinger, P. (eds) *Genetic Manipulations: Impact on Man and Society.* Cambridge University Press, Cambridge, pp. 161–470.

Garthwaite, A.J., von Bothmer, R. and Colmer, T.D. (2005) Salt tolerance in wild *Hordeum* species is associated with restricted entry of Na^+ and Cl^- into the shoots. *Journal of Experimental Botany* 56, 2365–2378.

Glenn, E., O'Leary, J., Watson, M., Thompson, T.L. and Kuehl, R. (1991) *Salicornia bigelovii* Torr.:

an oilseed halophyte for seawater irrigation. *Science* 251, 1065–1067.

Glenn, E., Tanner, R., Miyamoto, M., Fitzsimmons, K. and Boyer, J. (1998) Water use productivity and forage quality of the halophyte *Atriplex nummularia* grown on saline waste water in a desert environment. *Journal of Arid Environment* 38, 45–62.

Glenn, E.P., Brown, J.J. and Blumwald, E. (1999) Salt tolerance and crop potential of halophytes. *Critical Reviews in Plant Sciences* 18, 227–255.

Gorham, J., Bristol, A., Young, E.M. and Wyn Jones, R.G. (1991) The presence of the enhanced K/Na discrimination trait in diploid *Triticum* species. *Theoretical and Applied Genetics* 82, 729–736.

Greenway, H. and Munns, R. (1980) Mechanisms of salt tolerance in non-halophytes. *Annual Review of Plant Physiology* 31, 149–190.

Gregorio, G.B., Senadhira, D., Mendoza, R.D., Manigbas, N.L., Roxas, J.P. *et al.* (2002) Progress in breeding for salinity tolerance and associated abiotic stresses in rice. *Field Crops Research* 76, 91–101.

Hajjar, R. and Hodgkin, T. (2007) The use of wild relatives in crop improvement: a survey of developments over the last 20 years. *Euphytica* 156, 1–13.

Hobbs, T.J., Bartle, J. and Bennell, M. (2009) *Reviews of High Priority Species for Woody Biomass Crops in Lower Rainfall Southern Australia.* FLORASEARCH 3B Publication No. 09/044 Rural Industries Research and Development Corporation, Canberra, Australia.

Hollington, P.A. (2000) Technological breakthroughs in screening/breeding wheat varieties for salt tolerance. In: Gupta, S.K., Sharma, S.K. and Tyagi, N.K. (eds) *Proceedings of the National Conference Salinity Management in Agriculture.* Central Soil Salinity Research Institute, India, pp. 273–289.

ICRISAT Archival Report (2008) *Sustaining Biodiversity of Sorghum, Pearl Millet, Small Millets, Groundnut, Pigeonpea and Chickpea for Current and Future Generations.* Hyderabad, India.

ICRISAT Archival Report (2010) *Sustaining Biodiversity of Sorghum, Pearl Millet, Small Millets, Groundnut, Pigeonpea and Chickpea for Current and Future Generations.* Hyderabad, India.

ICRISAT (2012) *Sustainable Conservation and Utilization of Genetic Resources of Two Underutilized Crops – Finger Millet and Foxtail Millet – to Enhance Productivity, Nutrition and Income in Africa and Asia.* Final Report of the project. Hyderabad, India.

IPCC (Intergovernmental Panel on Climate Change) (2007) *Climate Change 2007: Synthesis Report.* IPCC, Geneva, Switzerland.

Islam, S., Malik, A.I., Islam, A.K.M.R. and Colmer, T.D. (2007) Salt tolerance in a *Hordeum marinum–Triticum aestivum* amphiploid, and its parents. *Journal of Experimental Botany* 58, 1219–1229.

James, C. (2011) *Global Status of Commercialized Biotech/GM Crops.* ISAAA Brief 43-2011. ISAAA, Ithaca, New York.

Krishnamurthy, L., Serraj, R., Rai, K.N., Hash, C.T, and Dakheel, A.J. (2007) Identification of pearl millet [*Pennisetum glaucum* (L.) R. Br.] lines tolerant to salinity. *Euphytica* 158,179–188.

Krishnamurthy, L., Turner, N.C., Gaur, P.M., Upadhyaya, H.D., Varshney, R.K. *et al.* (2011) Consistent variation across soil types in salinity resistance of a diverse range of chickpea (*Cicer arietinum* L.) genotypes. *Journal of Agronomy and Crop Science* 197, 214–227.

Kulkarni, V.N., Rai, K.N., Dhakheel, A.J., Ibrahim, M., Hebbara, M. *et al.* (2006) Pearl millet germplasm adapted to saline conditions. *International Sorghum and Millet Newsletter* 47, 103–106.

Kumar, G., Purty, R.S., Sharma, M.P., Singla-Pareek, S.L. and Pareek, A. (2009) Physiological responses among *Brassica* species under salinity stress show strong correlation with transcript abundance for SOS pathway-related genes. *Journal of Plant Physiology* 166, 507–520.

Latha, R., Srinivas, R.C., Sr Subramaniam, H.M., Eganathan, P. and Swaminathan, M.S. (2004) Approaches to breeding for salinity tolerance – a case study on *Porteresia coarctata. Annals of Applied Biology* 144, 177–184.

Lauchli, A. (1984) Salt exclusion; an adaptation of legumes for crops and pastures under saline conditions. In: Staples, R.C. (ed.) *Salinity Tolerance in Plants: Strategies for Crop Improvement.* Wiley, New York, pp. 171–187.

Lee, G.J., Boerma, H.R., Villagarcia, M.R., Zhou, X., Carter Jr., T.E. *et al.* (2004) A major QTL conditioning salt tolerance in S-100 soybean and descendent cultivars. *Theoretical and Applied Genetics* 109, 1610–1619.

Lee, S.Y., Ahn, J.H., Cha, Y.S., Yun, D.W., Lee, M.C. *et al.* (2007) Mapping QTLs related to salinity tolerance of rice at the young seedling stage. *Plant Breeding* 126, 43–46.

Lisa, L.A., Seraj, Z.I., Fazle, E.C.M., Das, K.C., Biswas, K. *et al.* (2004) Genetic variation in microsatellite DNA, physiology and morphology of coastal saline rice (*Oryza sativa* L.) landraces of Bangladesh. *Plant and Soil* 263, 213–228.

Lisa, L.A., Elias, S.M., Rahman, M.S., Shahid, S., Iwasaki, T. *et al.* (2011) Physiology and gene expression of the rice landrace Horkuch under salt stress. *Functional Plant Biology* 38, 282–292.

Ma, L., Zhou, E., Huo, N., Zhou, R., Wang, G. *et al.* (2007) Genetic analysis of salt tolerance in a recombinant inbred population of wheat (*Triticum aestivum* L.). *Euphytica* 153, 109–117.

Mahmood, A. and Quarrie, S.A. (1993) Effects of salinity on growth, ionic relations and physiological traits of wheat, disomic addition lines from *Thinopyrum bessarabicum*, and two amphiploids. *Plant Breeding* 110, 265–276.

Mahmood, K, Van der Deelen, J. and Baert, L. (1995) Phosphorus uptake by Kallar grass (*Leptochloa fusca* (L.) Kunth) from different depths in saline and non-saline soil. *European Journal of Agronomy* 4, 245–251.

Malcolm, C.V. (1969) Use of halophytes for forage production on saline wastelands. *Journal of the Australian Institute of Agricultural Science* 35, 38–49.

Malcolm, C.V. (1974) Forage production from shrubs on saline land. *Journal of Agriculture Western Australia* 15, 68–73.

Malik, A.I., English, J.P. and Colmer, T.D. (2009) Tolerance of *Hordeum marinum* accessions to O_2 deficiency, salinity and these stresses combined. *Annals of Botany* 103, 237–248.

Malik, K.A., Bilal, R., Mehnaz, S., Rasul, G., Mirza, M.S. *et al.* (1997) Association of nitrogen-fixing, plant-growth-promoting rhizobacteria (PGPR) with kallar grass and rice. *Plant and Soil* 194, 37–44.

Miller, J.F. and Seiler, G.J. (2003) Registration of five oilseed maintainer (HA 429–HA 433) sunflower germplasm lines. *Crop Science* 43, 2313–2314.

Mishra, B., Singh, R.K. and Senadhira, D. (2003) Advances in breeding salt tolerant rice varieties. In: Khush, G.S., Brar, D.S. and Hardy, B. (eds) *Advances in Rice Genetics. Supplement to Rice Genetics IV. Proceedings of the Fourth International Rice Genetics Symposium*, IRRI, Los Baños, Philippines, pp. 5–7.

Mujeeb-Kazi, A., Rosas, V. and Roldan, S. (1996) Conservation of the genetic variation of *Triticum tauschii* (Coss.) Schmalh. *Aegilops squarrosa* auct. non L.) in synthetic hexaploid wheats (*T. turgidum* L. s. lat.3T. tauschii; 2n=6x=42, AABBDD) and its potential utilization for wheat improvement. *Genetic Resources and Crop Evolution* 43, 129–134.

Munns, R. (2005) Genes and salt tolerance: bringing them together. *New Phytologist* 167, 645–663.

Munns, R. and Termaat, A. (1986) Whole-plant responses to salinity. *Functional Plant Biology* 13, 143–160.

Munns, R. and James, R.A. (2003) Screening methods for salinity tolerance: a case study with tetraploid wheat. *Plant and Soil* 253, 201–218.

Munns, R. and Tester, M. (2008) Mechanisms of salinity tolerance. *Annual Review of Plant Biology* 59, 651–681.

Nevo, E., Krugman, T. and Beiles, A. (1993) Genetic resources for salt tolerance in the wild progenitors of wheat (*Triticum dicoccoides*) and barley (*Hordeum spontaneum*) in Israel. *Plant Breeding* 110, 338–341.

OASE Foundation (2009) *Biosaline Agriculture.* Amsterdam, the Netherlands.

OGTR (Office of the Gene Technology Regulator) (2012) Table of applications and licences for Dealings involving Intentional Release (DIR) into the environment. Available at: www.ogtr.gov.au/internet/ogtr/publishing.nsf/Content/ir-1 (accessed 13 November 2012).

O'Leary, J.W. (1984) The role of halophytes in irrigated agriculture. In: Staples, R.C. and Toenniessen, G.H. (eds) *Salinity Tolerance in Plants: Strategies for Crop Improvement.* Wiley, New York, pp. 285–300.

Omielan, J., Epstein, E. and Dvořák, J. (1991) Salt tolerance and ionic relations of wheat as affected by individual chromosomes of salt-tolerant *Lophopyrum elongatum. Genome* 34, 961–974.

Patentlens (2012) *Patent Lens.* Cambia, Canberra, Australia.

Qadir, M., Qureshi, R.H. and Ahmad, N. (1997) Nutrient availability in a calcareous saline-sodic soil during vegetative bioremediation. *Arid Soil Research Rehabilitation* 11, 343–352.

Qiu, L., Wu, D., Ali, S., Cai, S., Dai, F. *et al.* (2011) Evaluation of salinity tolerance and analysis of allelic function of *HvHKT1* and *HvHKT2* in Tibetan wild barley. *Theoretical and Applied Genetics* 122, 695–703.

Ren, Z.-H., Gao, J.-P., Li, L.-G., Cai, X.L., Huang, W. *et al.* (2005) A rice quantitative trait locus for salt tolerance encodes a sodium transporter. *Nature Genetics* 37, 1141–1146.

Rhoades, J.D., Kandiah, A. and Mashali, A.M. (1992) *The Use of Saline Waters for Crop Production.* FAO Irrigation and Drainage paper 48. Food and Agriculture Organization of the United Nations, Rome.

Richards, R.A. (1996) Defining selection criteria to improve yield under drought. *Plant Growth Regulation* 20,157–166.

Rick, C.M. and Chetelat, R.T. (1995) Utilization of related wild species for tomato improvement. *Acta Horticulture* 412, 21–38.

Rogers, M.E., Craig, A.D., Munns, R., Colmer, T.D., Nichols, P.G.H. *et al.* (2005) The potential for developing fodder plants for the salt-affected areas of southern and eastern Australia: an overview. *Australian Journal of Experimental Agriculture* 45, 301–329.

Rogers, M., Colmer, T.D., Frost, K., Henry, D., Cornwall, D. *et al.* (2008) Diversity in the genus *Melilotus* for tolerance to salinity and waterlogging. *Plant and Soil* 304, 89–101.

Rogers, M.E., Colmer, T.D., Nichols, P.G.H., Hughes, S.J., Frost, K. *et al.* (2011) Salinity and waterlogging tolerance amongst accessions of messina (*Melilotus siculus*). *Crop and Pasture Science* 62, 225–235.

Rush, P.W. and Epstein, E. (1980) Breeding and selection for salt tolerance by the incorporation of wild germplasm into a domestic tomato. *Journal of the American Society for Horticultural Science* 106, 699–704.

Shekhawat, V.P.S., Kumar, A. and Neumann, K-H. (2006) Bioreclamation of secondary salanized soils using halophytes. In: Ozturk, M., Waisel, Y., Ajmal Khan, M. and Gork, G. (eds) *Biosaline Agriculture and Salinity Tolerance in Plants.* Birkhauser Verlog, Basel, Switzerland, pp.147–154.

Srivastava, N. (2010) Identification of constricting parental material for salinity tolerance based on morphological, physiological, and biochemical study and molecular diversity by SSR molecular markers in groundnut (*Arachis hypogaea* L.). PhD thesis. Jawaharlal Nehru Technological University, Hyderabad, India.

Srivastava, N., Vadez, V., Upadhyaya, H.D. and Saxena, K.B. (2006) Screening for intra and inter specific variability for salinity tolerance in pigeonpea (*Cajanus cajan*) and its related wild species. *Journal of SAT Agricultural Research* 2, 1–12.

Stevenson, G. (1969) An agronomic and taxonomic review of the genus *Melilotus* Mill. *Canadian Journal of Plant Science* 49, 1–20.

Sudmeyer, R.A. and Goodreid, A. (2007) Short-rotation woody crops: a prospective method for phytoremediation of agricultural land at risk of salinisation in southern Australia? *Ecological Engineering* 29, 350–361.

Tal, M. and Shannon, M. (1983) Salt tolerance in the wild relatives of the cultivated tomato: responses of *Lycopersicon esculentum, L. cheesmanii, L. peruvianum, Solanum pennellii* and F_1 hybrids to high salinity. *Functional Plant Biology* 10, 109–117.

Tanksley, S.D. and McCouch, S.R. (1997) Seed banks and molecular maps: unlocking genetic potential from the wild. *Science* 277, 1063–1066.

Tester, M. and Davenport, R. (2003) Na+ tolerance and Na+ transport in higher plants. *Annals of Botany* 91, 503–527.

Troyo-Dieguez, E., Ortega-Rubio, A., Maya, Y. and Leon, J.L. (1994) The effect of environmental conditions on the growth and development of the oilseed halophyte *Salicornia bigelovii* Torr. in arid Baja California Sur, Mexico. *Journal of Arid Environments* 28, 207–213.

Upadhyaya, H.D. and Ortiz, R. (2001) A mini core subset for capturing diversity and promoting utilization of chickpea genetic resources in crop improvement. *Theoretical and Applied Genetics* 102, 1292–1298.

Upadhyaya, H.D., Bramel, P.J., Ortiz, R. and Singh, S. (2002) Developing a mini core of peanut for utilization of genetic resources. *Crop Science* 42, 2150–2156.

Upadhyaya, H.D., Reddy, L.J., Gowda, C.L.L., Reddy, K.N. and Singh, S. (2006) Development of a mini core subset for enhanced and diversified utilization of pigeonpea germplasm resources. *Crop Science* 46, 2127–2132.

Upadhyaya, H.D., Pundir, R.P.S., Dwivedi, S.L., Gowda, C.L.L., Reddy, G. *et al.* (2009) Developing a mini core collection of sorghum for diversified utilization of germplasm. *Crop Science* 49, 1769–1780.

Upadhyaya, H. D., Sharma, N.D.R.K., Ravishankar, C.R., Albrecht, T., Narasimhudu *et al.* (2010) Developing a mini-core collection in finger millet using multilocation data. *Crop Science* 50, 1924–1931.

Upadhyaya, H.D., Ravishankar, C.R., Narsimhudu, Y., Sharma, N.D.R.K., Singh, S.K. *et al.* (2011a) Identification of trait-specific germplasm and developing a mini core collection for efficient use of foxtail millet genetic resources in crop improvement. *Field Crops Research* 124, 459–467.

Upadhyaya, H.D., Yadav, D., Reddy, K.N., Gowda C.L.L. and Singh, S. (2011b) Development of pearl millet mini core collection for enhanced utilization of germplasm. *Crop Science* 51, 217–223.

USSL (United States Salinity Laboratory) (2005) George E. Brown Jr. Salinity Laboratory. Riverside USDA-ARS California, USA.

Vadez, V., Krishnamurthy, L., Thudi, M., Anuradha, C., Colmer, T.D. *et al.* (2012) Assessment of ICCV 2 × JG 62 chickpea progenies shows sensitivity of reproduction to salt stress and reveals QTL for seed yield and yield components. *Molecular Breeding* 30, 9–21.

Villalta, I., Bernet, G.P., Carbonell, E.A. and Asins, M.J. (2007) Comparative QTL analysis of salinity tolerance in terms of fruit yield using two *Solanum* populations of F_7 lines. *Theoretical and Applied Genetics* 114, 1001–1017.

Wyn Jones, R.G., Storey, R., Leigh, R.A., Ahmed, N. and Pollard, A. (1977) A hypothesis on cytoplasmic osmoregulation. In: Marré, E. and Cifferi, O. (eds) *Regulation of Cell Membrane Activities in Plants*. Elsevier, Amsterdam, the Netherlands, pp. 121–136.

Xue, Z.-Y., Zhi, D.-Y., Xue, G.-P., Zhang, H., Zhao, Y.-X. *et al.* (2004) Enhanced salt tolerance of transgenic wheat (*Triticum aestivum* L.) expressing a vacuolar Na+/H+ antiporter gene with improved grain yields in saline soils in the field and a reduced level of leaf Na+. *Plant Science* 167, 849–859.

Yamaguchi, T. and Blumwald, E. (2005) Developing salt-tolerant crop plants: challenges and opportunities. *Trends in Plant Science* 10, 615–620.

Yeo, A.R. and Flowers, T.J. (1983) Varietal differences in the toxicity of sodium ions in rice leaves. *Physiologia Plantarum* 59, 189–195.

Yeo, A.R. and Flowers, T. (1986) Salinity resistance in rice (*Oryza sativa* L.) and a pyramiding approach to breeding varieties for saline soils. *Functional Plant Biology* 13, 161–173.

Yeo, A.R., Yeo, M.E., Flowers, S.A. and Flowers, T.J. (1990) Screening of rice (*Oryza sativa* L.) genotypes for physiological characters contributing to salinity resistance, and their relationship to overall performance. *Theoretical and Applied Genetics* 79, 377–384.

Zahran, M.A. (1994) *Juncus and Kochi Fiber-Producing and Fodder-Producing Halophytes under Salinity and Aridity Stress.* Marcel Dekker, New York, pp. 505–528.

Zhou, G., Johnson, P., Ryan, P.R., Delhaize, E. and Zhou, M. (2012) Quantitative trait loci for salinity tolerance in barley (*Hordeum vulgare* L.) *Molecular Breeding* 29, 427–436.

Zohar, Y., Di Stefano, J. and Bartle, J. (2010) Strategy for screening eucalypts for saline lands. *Agroforestry Systems* 78, 127–137.

Zurayk, R. and Baalbaki, R. (1996) *Inula crithmoides*: a candidate plant for saline agriculture. *Arid Soil Research and Rehabilitation* 10, 213–223.

15 Response to Flooding: Submergence Tolerance in Rice

Abdelbagi M. Ismail[1] and David J. Mackill[2]

[1]*International Rice Research Institute (IRRI), Los Baños, Philippines and* [2]*Department of Plant Sciences, University of California, Davis, USA*

15.1 Introduction

Floods cause considerable damage to crops worldwide. For dryland cereals such as maize, wheat, rye and barley, excessive water in the root zone is sufficient to cause major yield reductions. It has been estimated that about 10% of all irrigated lands are seriously affected by waterlogging, causing up to 20% reduction in crop yields in some areas. Besides irrigated areas, vast rainfed areas are prone to temporary waterlogging and floods, particularly in low-lying fields (Setter and Waters, 2003; Jackson, 2004). Several factors contribute to waterlogging of farmlands. These include excessive rainfall or irrigation with poor internal drainage or impeded runoff, and seasonal floods from surface water bodies during periods of high rainfall. Excessive irrigation with poor drainage can increase the incidences of soil waterlogging through rise in the water table and, in drier regions, it also causes the build-up of soil salinity and/or alkalinity, as in the Sindh Province in the Indus valley of Pakistan and the Indo-Gangetic Plains of northern India.

Waterlogging causes various adverse conditions in the root zone that affect root growth and function. Oxygen becomes scarce hindering root respiration, while other gases such as CO_2 and ethylene build up. Low oxygen causes a reduction in root growth and function, reducing nutrient and water uptake and ultimately leading to plant death. Phytotoxic substances such as reduced iron (Fe^{+2}) and manganese (Mn^{+2}) and hydrogen sulphide (H_2S), as well as intermediates in anaerobic carbon metabolism like butyric and acetic acids, accumulate to toxic levels as a consequence of the activity of anaerobic microbes. These toxic substances cause serious damage when absorbed in excessive quantities by plant roots. Furthermore, changes in soil pH alter nutrient availability and uptake (Drew and Lynch, 1980). Apparently, all these undesirable changes are associated with low oxygen in the root zone, and crop plants and varieties with the mechanisms to overcome these problems can survive and grow well in waterlogged soils. Genetic diversity in waterlogging tolerance has been observed in several crops including wheat, barley and rye, but little progress has been made in developing tolerant varieties (Drew, 1997; Setter and Waters, 2003).

Unlike dryland cereals, rice is peculiar in its ability to survive and flourish in waterlogged and submerged soils. This is because of its extensive aerenchyma system that facilitates aeration of the roots and the rhizosphere, alleviating the stresses associated with low oxygen (Setter *et al.*, 1997; Jackson and Ram, 2003). Despite its tolerance of waterlogging, however, rice is sensitive to prolonged partial (more than 25 cm to a few metres) or transient complete inundation, common in rainfed lowlands and flood prone areas. These floods cause considerable losses in rice production each year, and rural poverty and food insecurity are widespread in these areas because fewer opportunities are available for farmers. In these areas, rice is the predominant food and occasionally a cash crop; but its yields are low, only 0.5 to 2.0 t ha^{-1}, and unstable because farmers use few inputs and still rely on traditional landraces that are relatively more adapted to different types of floods than modern varieties. The enormous diversity and plasticity in adaptation to contrasting hydrological conditions made rice one of the crops most amenable for genetic manipulation to develop varieties suitable for excess water conditions. Breeding such varieties will also help to keep up with the progressively increasing flood incidences and severity in the tropics and subtropics, frequently attributed to climate change. This chapter will focus on rice as a model crop where progress was made in developing flood-tolerant varieties and where opportunities exist for further improvements through mining the enormous genetic diversity available in preserved collections worldwide.

Only two species of rice, *Oryza sativa* and *Oryza glaberrima*, are cultivated (Jackson *et al.*, 1997). *O. glaberrima* was domesticated in West Africa and is still grown there on a limited scale, whereas the Asian species *O. sativa* is widely grown on about 144 million ha worldwide. The geographical distribution of rice-growing areas extends from 50°N in northern China to 35°S in New South Wales, Australia, as well as in Argentina. Rice is also grown across a wide range of altitudes, from 3 m below sea level in Kerala, India, to as high as 3000 m in the uplands of Nepal and Bhutan (Australia: Office of the Gene Technology Regulator, 2005). Within *O. sativa*, two major categories have been identified:

- The *japonica* type that includes temperate *japonica*, confined mostly to temperate regions because of its better tolerance of low temperature, and tropical *japonica*, mostly grown in the uplands of both tropical and subtropical areas.
- The *indica* type grown in tropical and subtropical areas.

Rice is also adapted to a variety of soil types, but is mostly grown in medium- and heavy-textured soils, and to variable hydrological conditions and water regimes, from fully aerated soils in uplands to flooded soils in irrigated and rainfed lowlands, and to long-duration partial submergence in flood-prone areas.

This considerable plasticity in the cultivated rice species has resulted in substantial diversity within the cultivated gene pool. The T.T. Chang Genetic Resources Center (hereafter, TTC-GRC) of the International Rice Research Institute (IRRI) currently holds more than 117,000 accessions of mostly landraces and wild relatives collected worldwide and some varieties and breeding lines. Out of these accessions, 1657 are *O. glaberrima*, 4465 accessions are wild relatives, 15 accessions are *Oryza*-related genera and the rest belong to *O. sativa* (N.R. Sackville-Hamilton and F.R. Hay, Los Baños, 2012, personal communication). This diversity in adaptation to various ecologies also made rice one of the most widely grown crops over an extreme range of habitats, broadly classified into four different categories primarily based on hydrology: upland, irrigated lowland, rainfed lowland and flood-prone ecosystems (Maclean *et al.*, 2002). Rice in uplands is grown in aerated soils, with variable topography, from undulating and steep sloping lands to well-drained flat fields. Soils in uplands vary in texture and water-holding capacity, and in fertility, from poor highly leached soils of West Africa to fertile soils in some areas in South-east Asia. Uplands currently constitute about 13% of the world rice area, and are generally characterized by low yields of about 1 t ha^{-1} and poor farming communities, except in areas where new high-yielding varieties and better cropping systems are adopted as in Yunnan, China, and in Brazil. Rice varieties adapted to uplands are mostly short maturing, with deeper roots to avoid drought, and have higher tolerance of acid soils.

In lowlands, rice is either irrigated or rainfed. The irrigated ecosystem covers about 55% of the world rice area and provides about 75% of world rice production. Fields in irrigated areas have an assured water supply with standing shallow water for most of the season. Dwarf high-yielding varieties are predominant, and yields are usually high, averaging over 5 t ha^{-1}. Rainfed lowlands, on the other hand, constitute one quarter of rice area and contribute about 18% of rice production. These areas are generally prone to different abiotic stresses such as drought and submergence because of the lack of water control, and adverse soils resulting from salt stress and other soil problems in both coastal and inland areas. Local landraces with yields of less than 2 t ha^{-1} still dominate in most rainfed areas but, recently, new high-yielding varieties tolerant of drought, submergence and salt stress are becoming available and gradually replacing farmers' old landraces.

Flood-prone rice ecosystems are frequently affected by either transient flash floods causing complete submergence or longer-term partial floods of up to 0.5 m in stagnant-flood (SF) areas to over 5.0 m in deepwater areas. More than 20 million ha in South and South-east Asia are annually affected by uncontrolled floods, with low yields averaging less than 1.5 t ha^{-1}; yet, these areas support more than 100 million people. Traditional varieties are more dominant in rainfed lowlands and flood-affected areas; however, varieties that tolerate complete submergence were recently developed through the marker-assisted incorporation of the *SUB1A* gene. These varieties can tolerate more than 2 weeks of complete submergence and have considerable yield benefits (Mackill *et al.*, 2012).

The erratic floods being experienced in rainfed and flood-affected areas are usually caused by heavy rain and overflowing of rivers. They can also sometimes be driven by tidal movements in coastal areas. These floods cause serious problems for rice and other crops because the topography of the land in most cases prevents fast water movement to drain flooded fields. Flooding is therefore considered a major challenge for rice production in South and Southeast Asia, the home for the majority of the world's rice farmers. Evidently, rice is the only crop that can grow well in flood-affected areas because they are too wet for other crops, providing opportunities to use lands that are predominantly or seasonally wet for food production. However, rice typically grows well in shallow water (paddy fields), and most varieties die after a few days of complete submergence or grow poorly when partially flooded (Singh *et al.*, 2011; Mackill *et al.*, 2012). Developing rice varieties that produce well when partially flooded or that can withstand transient complete submergence will lighten the undesirable effects of floods in many rice-producing areas and help cope with the adversities of climate change in areas that are or will be receiving more rainfall and floods in the future.

15.2 Climate Change and Floods

Economic losses from flood damage in rice are not accurately known, but some estimates in the 1990s indicated that they were US$600 million to US$1 billion a year in Asia (Herdt, 1991; Dey and Upadhyaya, 1996). These losses are expected to mount considerably with the worsening flooding incidences caused mostly by human-induced global warming, projected as heavier precipitation and increased frequency of natural disasters such as coastal storms, cyclones and typhoons. In 2007 the expected rise in sea level by 2100 was estimated at 0.18 to 0.59 m (Metz *et al.*, 2007) but more recent estimates put the value much higher (NRC-NAS, 2010), a consequence of expansion of ocean water and melting of glaciers, thus seriously threatening coastal areas with floods and erosion. Other human activities are contributing to and exacerbating the severity of specific regional floods as a consequence of climate change, such as deforestation causing excessive runoff from uplands, draining of wetlands that act as reservoirs for excess water, for expansion of agriculture and other uses, and development activities that interfere with natural drainage such as building of roads and other structures.

15.2.1 Projected increase in flooding incidence caused by global warming

Numerous models predicted serious negative impacts of climate change on food production and food security in several regions of the world (Lobell *et al*., 2008). Among the serious consequences, flood incidence is expected to increase in most regions of the tropics due to increases in heat waves, higher intensity of tropical storms, precipitation extremes and sea-level rise (Kemp *et al*., 2011; Coumou and Rahmstorf, 2012). Worldwide variation in sea level was reported to be strongly associated with global mean temperatures in scales of decades to centuries (Vermeer and Rahmstorf, 2009). Global warming could enhance incidences of extreme rainfall beause warmer air can hold more moisture, leading to heavier rains when conditions are conducive, and the increased latent energy of the atmosphere when holding higher moisture could drive catastrophic storms. In agreement with these studies, Metz *et al*. (2007) concluded that global warming is highly likely to be associated with future increases in the frequency of heat waves and extreme precipitation, with a probability of more than 90%.

Sea-level rise is probably the most certain consequence of climate change for coastal zones, as predicted from changes in sea-surface and atmospheric temperatures. Tropical storms and cyclones will probably be more frequent and intense, with stronger winds and heavier rainfall with the current trends in global warming (Metz *et al*., 2007). A modest increase in incidences of moderate to severe storms was noticed in recent years in some countries, with a few devastating storms, such as cyclone Nargis in Myanmar in May 2008, with a surge of over 4 m, destroying more than 1.7 million ha of rice crop (USDA-FAS, 2008). In 2007, Cyclone Sidr devastated more than 640,000 ha of cropland in southern Bangladesh, where rice is the main crop during the monsoon or *aman* season, when it struck (Hossain *et al*., 2008). Besides these incidences, numerous abnormal occurrences of catastrophic floods and other unprecedented extreme weather events have been witnessed in other areas such as in Pakistan, Rajasthan (India), eastern Australia and numerous other areas worldwide (Coumou and Rahmstorf, 2012); however, the extent to which these incidences are linked to climate change still needs to be determined. The evidence is that global warming will probably be associated with more floods in the tropics, where rice farming is the predominant agricultural activity, especially in the coastal zones of South and South-east Asia.

15.2.2 Impacts of climate change on rice in flood-prone areas

The challenges facing rice production in flood-prone areas are becoming ever more complex with the long-term adverse effects of climate change, and the areas affected by flooding are expected to increase considerably as a consequence of extreme weather (Bates *et al*., 2008). Impacts on land-use activities in the tropics are already being seen in highly vulnerable areas in recent years, as in the Mekong River Delta of southern Vietnam, the Ayeyarwaddy delta in southern Myanmar and the Ganges-Brahmaputra delta of southern Bangladesh. These deltas are highly populated and each provides from 30% to 70% of the total rice produced in these countries. Any changes in flooding incidence and intensity in these areas will have substantial effects on food security (Ismail and Tuong, 2009; Wassmann *et al*., 2009).

Historically, these deltas have been the main targets for ensuring a progressive increase in rice production to meet the increasing demand in these countries because of their fertile soils and availability of water resources, and recent statistics show that they still maintain the increasing trend in production figures throughout Asia. However, continuance of this trend is now in question. The land in most of these large deltas is also subsiding because of sediment compaction resulting from pumping of oil, gas and water, coupled with trapping of sediments in upstream reservoirs in most basins and deposition along rivers because of the slower water flow (Syvitski *et al.*, 2009), making these deltas even more vulnerable. Any decline in rice production in these coastal zones will have a considerable impact on local food security as well as on international rice trade.

A general consensus among most predictions suggests that climate change will increase global average rainfall beyond coastal areas to some inlands in mid and higher latitudes, causing more frequent floods and submergence. Efforts are therefore needed to halt any further decline in productivity of these areas, and even to maintain a steady increase in rice production in both coastal and inland areas to cope with the global increase in food demand. This will ultimately contribute to the efforts for coping with the consequences of global warming and the anticipated food shortages. Developing rice varieties tolerant of the abiotic stresses likely to increase in the future, such as submergence, drought and salinity, will provide a step forward in these efforts (Ismail *et al.*, 2008; Ismail and Tuong, 2009). Without sufficient adaptation measures, these areas could experience serious negative effects on the production of the major crops, including rice, that are important for a large number of already food-insecure human populations.

15.3 Mining for Adaptation to Flood-prone Areas

The early domestication of the *Oryza* species, dating back to approximately 9000 BC in China and about 3500 BC in Africa (Molina *et al.*, 2011), and the extensive geographical spread of these species over extremely diverse climates and ecological niches generated a considerable amount of diversity in rice. This diversity has been extensively used for germplasm improvement over decades. There are now more than 750,000 genebank accessions of *Oryza* species kept in several genebanks around the world, most of them comprising the Asian rice *O. sativa* (Hay *et al.*, 2013). The largest of these collections is held at the TTC-GRC, containing about 20% of the rice germplasm conserved worldwide, donated from over 110 countries (Jackson, 1997). This collection provides a significantly rich source of diverse genetic resources for exploitation through breeding.

Earlier efforts for improving rice adaptation to flood-prone areas at IRRI started in the early 1970s through IRRI's Genetic Evaluation and Utilization (GEU) Program. Approximately 18,000 accessions were screened through this programme but only about 0.5% of them showed tolerance (Palada and Vergara, 1972; Khush and Coffman, 1977; Setter and Laureles, 1996), suggesting that tolerance of submergence in rice is relatively rare. In late 1974, an offshoot of this programme started in Thailand – the Thai–IRRI Collaborative Deepwater Rice Research Project – conducted on 6 ha of naturally flooded lands, with 11 artificially constructed ponds for controlled evaluation of genetic material from various sources. Moreover, considerable research efforts were carried out in other countries such as India and Japan (Yamada, 1959; Richharia and Misro, 1960; IRRI, 1975; Vergara and Mazaredo, 1975).

Rice landraces tolerant of complete submergence or having the ability to elongate at a pace commensurate with the increase in water depth in deepwater areas were probably known to farmers in flood-prone areas for centuries but were not sufficiently recognized until the 1950s (Richharia and Misro, 1960). Systematic screening to identify tolerant germplasm for both transient submergence tolerance for flash-flood areas and elongation ability for deepwater areas began at IRRI during the early 1970s (Vergara and Mazaredo, 1975; HilleRisLambers and Vergara, 1982). Screening under controlled submergence in tanks, using 10-day-old seedlings completely submerged for 10 days followed by 5 days of recovery, allowed scientists to evaluate thousands of germplasm accessions. Several landraces were identified that had higher tolerance of complete submergence. Most of these varieties were from Sri Lanka and a few were from India and other countries (Table 15.1). These landraces were, however, found to be unsuitable for direct use by farmers because of their low yield and poor

Table 15.1. Rice landraces identified during the 1970s and 1980s as tolerant of transient complete submergence.

Variety	Origin	TTC-GRC accessions	Remarks
Jamis Wee	Sri Lanka	**15290, 15294**, 40811	
Kurkaruppan	Sri Lanka	**15449**	High tolerance, used in earlier crosses
Thavalu	Sri Lanka	**15314, 15325**	High tolerance, used in earlier crosses
Goda Heenati	Sri Lanka	**15419, 31393**, 120958	High tolerance, used in several studies and in crosses
Madael	Sri Lanka	7722, **7727**, 8970, 8979, 11701, 15234, 15426, **36310**, 117525	
Weli Handrian	Sri Lanka	11717,**15433**, 15649, 31559, 31560, 40889, 40890, 50957	
Kaharamana	Sri Lanka	11961, 15273, 15306, 15307, 15379, **31420**, 40812	
Lumbini	Sri Lanka	47719, 31465, 31464, **15319**, **15291**, 11947, 121627	
Kalukanda	Sri Lanka	**15405,** 15430, 15754, 15759, 31424, 31425, 31426, 31427, 36271, 36272, 36273, 36274	
Kanni Murunga	Sri Lanka	**15432,15594**, 47379, 121010	
Vellai	Sri Lanka	8997	
Kaluwee	Sri Lanka	11986, **12009, 15425, 15439**, **15597, 36283, 8898, 8971**	
Kottamalli	Sri Lanka	15424, 15435	
Ratawee	Sri Lanka	**11991**, **15185**, **15338**, **31521**, **31522**, **36361**, 8914, **8920**, **9000**	
Periya Karuppan	Sri Lanka	15460, **15692**, 15697	
Kalu Gries	Sri Lanka	**15330**	
Buruma Thavalu	Sri Lanka	**15335**	
Madabaru	Sri Lanka	**15333**	
Sudu Gries	Sri Lanka	**15331**	
Pokura Samba (15337)	Sri Lanka	15245, 15385, 15227, 47721, **15337**	
FR13A	India	**6144**, 8887, 122454, 117267, 124390	Highly tolerant, most widely used in crosses and mechanistic studies, source of *SUB1A-1* allele
FR43B	India	**6143**, 8886	
Padi Ewang Janggut	Indonesia	24758	Adapted to swampy areas
SML Temerin	Surinam	46685	
Nang Dum To	Vietnam	**231**	
T442-57	India	51221	
Nam Sagui 19	Thailand	40009	High to moderate tolerance, used as check in earlier screening

Landraces with bold accession numbers are also classified as deepwater types in the genebank database (Vergara and Mazaredo, 1975; Mackill *et al.*, 1996; TTC-GRC database).

grain quality. They were generally photoperiod sensitive and tall, making them susceptible to lodging and not compatible with high-input farming. None the less, the discovery of these tolerant landraces made it possible for breeders to transfer tolerance into high-yielding genetic backgrounds using conventional breeding methods.

More intensive hybridization efforts started in the late 1970s (Mohanty and Khush, 1985; Mohanty and Chaudhary, 1986; Haque *et al.* 1989) and, by the 1990s, submergence-tolerant, semi-dwarf and high-yielding breeding lines became available (Mackill *et al.*, 1993). In most crosses, FR13A, a pure-line selection from the traditional variety Dhalputtia (Richharia and Govindaswami, 1966), was used as a donor parent because of its consistent performance in numerous trials and regions (HilleRisLambers and Vergara, 1982). A few of the breeding lines developed from these crosses were released as varieties, but they were not widely grown by farmers because they still possessed some undesirable traits inherited from the landraces used as donors, such as FR13A.

Tolerance of flooding during germination is another important agronomic trait for direct-seeded rice in flood-prone and other rainfed ecosystems. Farmers in these areas often face uncertainty of rainfall, preventing the large-scale adoption of direct seeding as a method for crop establishment, mainly because of the high sensitivity of rice to flooding during germination and the early stages of seedling growth (Yamauchi *et al.*, 1993; Ismail *et al.*, 2009, 2012; Angaji *et al.*, 2010). Heavy rains or early floods pose serious problems even for transplanted rice in rainfed areas, and early crop establishment through direct seeding could help avoid further damage caused by floods just after transplanting. Moreover, direct seeding is more effective in rainfed areas when the monsoon is delayed or drought occurs early in the season. Flooding fields after direct seeding in irrigated ecosystems can also effectively suppress most weeds and reduce the need for extensive hand weeding or the use of herbicides (Ismail *et al.*, 2012). Breeding cultivars with higher tolerance of flooding during germination and early growth will therefore be important in both rainfed and irrigated systems.

Earlier attempts to breed varieties tolerant of flooding during germination met with limited success, mainly because donors with sufficient tolerance were not identified from the relatively smaller set of germplasm evaluated (Yamauchi *et al.*, 1993; Yamauchi and Winn, 1996; Biswas and Yamauchi, 1997). More recently, more than 8000 accessions of different origins were screened; most of them were from the TTC-GRC collected from flood-prone areas. Several lines that can germinate and establish in flooded soils were identified (Angaji *et al.*, 2010). Furthermore, some of these lines were characterized for the mechanisms associated with tolerance of anaerobic conditions or low oxygen during germination and early growth (Ismail *et al.*, 2009). Numerous crosses were also made and tolerant breeding lines, for both irrigated and rainfed conditions and that can successfully germinate in fields with standing shallow water, were developed. In addition, some of these early crosses were used to develop mapping populations, and to identify quantitative trait loci (QTLs) associated with tolerance for use in marker-assisted breeding and for cloning of the functional genes involved in tolerance traits.

15.4 Progress in Breeding Flood-tolerant Rice Cultivars

The extensive efforts devoted during the past five decades to collecting and mining genetic resources for sources of tolerance of different types of flooding successfully identified a large number of tolerant landraces. These landraces were then used extensively for physiological and genetic studies, substantially enhancing our understanding of the mechanisms associated with tolerance of different flood conditions. These studies also highlighted the tremendous genetic diversity that evolved within cultivated rice species to adapt to various hydrological and climatic conditions. Knowledge of the mechanisms associated with tolerance of these different types of floods was sufficiently advanced over the past few decades, as briefly summarized in Table 15.2. Here, we summarize the progress made in exploiting the genetic variation in these traits for rice germplasm improvement to develop high-yielding varieties for flood-affected areas and for areas where floods are expected to worsen with climate change.

15.4.1 Tolerance of flooding during germination

Rice varieties suited to direct seeding in lowlands will be adopted on a large scale by farmers if these varieties can tolerate flooding during germination and seedling establishment. Breeding of such varieties began at IRRI in 2002–2003, when several crosses were made between the most tolerant landraces identified by then, such as *Khaiyan*, *Khao Hlan On* and *Ma-Zhan Red* (Angagi *et al.*, 2010), and several popular varieties adapted to both rainfed and irrigated ecosystems. Stable breeding lines were developed and are currently being evaluated. Several of the tolerant landraces were also used to develop mapping populations. One population was derived from a cross between *Khao Hlan On*, a landrace from Myanmar, and IR64, a popular lowland variety. Six QTLs derived from *Khao Hlan On* were identified (Angaji *et al.*, 2010). The strongest of these QTLs was mapped on the long arm of chromosome 9 (*qAG-9-2*), with a LOD score of 20.3 and phenotypic variance of 33.5%. This major QTL was subsequently fine-mapped to a 58-kb region and several putative candidate genes were annotated in the locus. These genes are now being studied using complementation and silencing to identify the gene(s) responsible for this QTL (Septiningsih *et al.*, 2013, unpublished results). Fine-mapping of this QTL also facilitated its use in marker-assisted back-cross (MABC) breeding to transfer it into popular varieties and new breeding material.

Another mapping population was developed using *Ma-Zhan Red*, a tolerant landrace from China, crossed with IR42, a rainfed lowland variety sensitive to all types of floods. Six significant QTLs were identified in this population, on chromosomes 2, 5, 6 and 7, and all their tolerant alleles were contributed by *Ma-Zhan Red*. The largest QTL was identified on the short arm of chromosome 7 (called '*qAG7.1*'), with a LOD score of 14.5 and a phenotypic variance of 31.7% in this population. This QTL is being fine-mapped for use in breeding and for cloning (Septiningsih *et al.*, 2013). The current emphasis is to develop a MABC system to combine both *qAG7.1* and *qAG-9-2* into high-yielding varieties to enhance their tolerance for flooding during germination, for use in direct-seeding systems in both irrigated and rainfed ecosystems.

Besides these two, additional QTLs from other donors are being targeted that could contribute to tolerance. Several mechanisms seem to be associated with tolerance; some of them are independent (Ismail *et al.*, 2009) and this underlies the basis of a pyramiding strategy. Other groups have identified additional QTLs associated with tolerance (Jiang *et al.*, 2004, 2006) and these will also be explored. Combining multiple QTLs is necessary to obtain higher tolerance because the expression of this trait seems to be affected by several climatic, seed and seedbed conditions (Ella *et al.*, 2010, 2011). New varieties combining several of these QTLs will probably have sufficient and more stable tolerance under field conditions, and will then help in expanding the adoption of direct-seeded rice in flood-prone areas and in areas that are expected to receive more floods with climate change. Moreover, these varieties will be suitable for direct seeding in irrigated systems, where flooding of the field with a shallow layer of water after drilling dry seeds can be used as an effective measure for weed suppression, considerably reducing the cost of hand weeding and the hazards and costs associated with the use of herbicides (Ismail *et al.*, 2012).

Table 15.2. Major types of floods affecting rice in different ecosystems, major traits associated with tolerance, and major QTL and genes identified for tolerance

Type of flood	Traits required for tolerance	Major QTLs/genes identified	References
Transient complete submergence (a few days to 18 days)			
During germination and early seedling growth	Fast germination and seedling growth, fast coleoptile elongation	*qAG7.1*	Ismail *et al.*, 2009
	Ability to break down and use starch under low O_2	*qAG-9-2*	Angaji *et al.*, 2010
	Anaerobic respiration to generate energy under low oxygen		Septiningsih *et al.*, 2013
During vegetative stage	Reduced elongation and low carbohydrate consumption during submergence	*SUB1A-1*	Jackson and Ram, 2003
			Sarkar *et al.*, 2006
	Underwater photosynthesis		Xu *et al.*, 2006
	Chlorophyll retention underwater		Mackill *et al.*, 2012
	Fast recovery		
Longer-term partial floods (a few weeks to a few months)			
Stagnant, medium-deep (30–50 cm)	Partial, slow elongation of the shoot	Not genetically characterized	Singh *et al.*, 2011
	High tillering ability underwater		
	Strong culms, resistant to lodging		
	Sufficient leaf area above water		
	Large fertile panicles		
Deepwater (>50 cm to over 5 m)	Fast internode elongation with rising water (> 20 cm per day)	*SNORKEL1*	Catling, 1992
	Sufficient leaf area above water	*SNORKEL2*	Hattori *et al.* 2007, 2009
	Kneeing ability when water recedes		
	Large fertile panicles		

15.4.2 Tolerance of longer-term partial submergence

Partial submergence of rice plants for a few weeks to a few months during the monsoon season is common in rainfed and flood-prone ecosystems. Affected areas are broadly classified into two main categories on the basis of the depth of water and the type of varieties adapted to these areas. Areas with stagnant flood (SF) are characterized by water depths of 30–50 cm for most of the season, sometimes referred to as medium-deep or semi-deep areas. Deepwater (DW), on the other hand, refers to areas where water depth is greater than 50 cm, and can reach several metres at times. In both SF and DW areas, the plants must retain the ability to elongate to keep up with rising water for aeration. Grain yields of these areas are typically very low for numerous reasons, including the lack of suitable modern varieties. Farmers' landraces still predominate, with lower yields and, in some cases, low grain quality. The low yield of these varieties is associated with low tillering ability, small panicles and susceptibility to lodging (Table 15.2). Efforts to breed high-yielding varieties for DW areas started several decades ago, but had relatively limited progress, in part because varieties suitable for these areas need to invest a high amount of energy into vegetative growth at the expense of their grain yield (HilleRisLambers and Vergara, 1982; Catling, 1992; Mackill *et al.*, 2010).

Tolerance of stagnant flooding

Modern rice varieties are sensitive to stagnant floods, and can suffer substantial yield penalties when grown under conditions with water depth exceeding 30 cm in the field. In cases when complete draining of water is not possible following a period of flash floods that causes complete submergence, the effects of stagnant flood become even more severe and such events are common in flood-prone areas (Singh *et al.*, 2011). Breeding varieties with sufficient tolerance of SF received relatively little attention in the past despite the enormous yield losses caused by this type of stress in large areas in South and South-east Asia. In recent years, we screened several hundred diverse rice accessions collected from areas of recurrent floods in the TTC-GRC. These sets also included released and obsolete varieties and elite lines developed through various breeding programmes. Numerous accessions were identified that have high survival and produce many tillers with high numbers of filled grains when partially flooded with a water depth of about 50 cm through maturity. The majority of the modern varieties tested showed high mortality, fewer tillers and poor yield under SF, which might explain why farmers in affected areas still rely on growing local landraces that have lower yield but are moderately tolerant. Tolerant genotypes also vary in the extent of elongation of the internodes in a manner dependent on water depth.

More recently, breeding lines with relatively higher yields under SF have been developed at IRRI through standard breeding methods (Mackill *et al.*, 2010). These lines are being evaluated in target areas in South and South-east Asia, as well as in a few countries in sub-Saharan Africa. Tolerance of SF seems more complex than tolerance of other types of floods that affect rice, and numerous QTLs/genes are expected to be involved. This is because the traits associated with tolerance, such as the ability to produce more tillers and sturdy culms, and elongate in a need-based manner with rising water, are probably independently controlled. Our recent studies using a larger set of tolerant lines further showed that tolerant lines are either inherently tall or are capable of partial elongation in response to submergence (Y. Kato and A. Ismail, unpublished results). Mapping populations are now being developed to identify loci associated with these various tolerance traits for subsequent use in breeding.

Deepwater rice

Prolonged floods of 50 cm to 1 m depth are most common in DW areas; however, in some of these areas, deeper flood can surpass 5 m and stagnate for several months. Water levels generally rise faster in DW areas than in SF areas. Therefore, fast elongation of leaves, leaf sheaths and internodes is an important characteristic of DW rice varieties. This is to keep up with the fast-rising water to escape complete submergence and to maintain oxygen supply to roots via aerenchyma tissue (Richharia and Misro, 1960; Catling, 1992; Setter *et al.*, 1997). Some traditional cultivars adapted to DW ecosystems can elongate by more than 20 cm per day based on the depth and rate by which floodwater rises in the field, and produce numerous adventitious and aquatic roots from submerged nodes, probably enabling the plant to obtain oxygen and nutrients from floodwater and to remain floating in deeper waters, hence the name 'floating rice'. Because of their excessive vegetative growth, rice varieties in DW areas generally have low yields and are susceptible to lodging. In addition, these varieties have low tillering capacity because tiller formation is usually suppressed under flooded conditions because of diminished meristematic activity under low oxygen.

Apparently, survival of DW rice varieties is dependent on rapid underwater internode elongation to ensure that upper leaves maintain contact with air, both for aeration of underwater plant parts, including roots, and for photosynthesis (Vergara *et al.*, 1976; Catling, 1992). This rapid elongation is induced by ethylene and gibberellic acid in a manner opposite to the strategy of rice varieties adapted to areas affected by complete submergence. Several studies using different DW rice varieties have identified QTLs associated with internode elongation as well as with variation in the number of elongating internodes, on chromosomes 1, 3 and 12, and most of these QTLs were common across several mapping populations (Sripongpangkul *et al.*, 2000; Hattori *et al.*, 2007, 2009; Kawano *et al.*, 2008). More recently, the large-effect QTL on chromosome 12 was recognized as the major determinant of rapid internode elongation in DW rice. Cloning of this QTL identified two functional genes, *SNORKEL1* (*SK1*) and *SNORKEL2* (*SK2*), responsible for internode elongation. Both genes are members of the ERF (Ethylene Responsive Factor) genes of subgroup VII, which also includes the *SUB1A* gene responsible for tolerance of transient complete submergence (Xu *et al.*, 2006; Hattori *et al.*, 2009).

Breeding DW rice varieties probably started earlier than for other types of flooding. Richharia and Misro (1960) outlined the early breeding strategies as selection among germplasm collected from DW areas and hybridization, even between *O. sativa* and *O. rufipogon* (formerly known as *O. perennis*, and suggested as the origin of DW rice, Oka, 1975) because it can cross easily with *O. sativa* and retain most of the traits of DW rice, including poor tillering, long internodes, swollen nodes and growth of nodal shoots. Several cultivars were developed by pure-line selection from DW varieties, and these were introduced in some areas with partial success. However, later breeding work emphasized hybridization between these tall, elongating DW varieties and high-yielding semi-dwarfs. The objective was to transfer the elongation trait into higher-yielding varieties. This was indeed achieved with some improved varieties such as RD19 in Thailand; it has shorter stature under shallow conditions but can elongate when water is deeper. Progress in developing high-yielding DW varieties has not, however, been fully successful, and areas under DW rice are now shrinking because of low productivity. In most countries, these areas are being replaced by dry-season rice with the availability of irrigation facilities from surface-water bodies and shallow tube wells, as in Bangladesh, India and Vietnam. However, with the worsening of weather patterns in recent years and the increase in incidence of abnormally heavy rains and floods, more areas are expected to be affected by deeper and protracted floods. Renewed emphasis is therefore needed to develop high-yielding varieties for this system, especially with the recent progress in identifying major genes controlling elongation under DW conditions.

15.4.3 Tolerance of transient complete submergence during the vegetative stage

The genetic variation in submergence tolerance in rice revealed through the Asia-wide screening programmes begun by IRRI in the 1970s marked the initial steps for a formal breeding programme at IRRI and elsewhere (Palada and Vergara, 1972; Khush and Coffman, 1977). The numerous tolerant landraces identified through this programme were also used extensively in genetic and physiological studies that revealed the bases of tolerance and helped in setting effective breeding strategies. The most significant of these findings is the discovery and cloning of the *SUB1* gene, identified from the Indian variety FR13A, and its deployment to farmers in the background of popular varieties (Jackson and Ram, 2003; Sarkar *et al.*, 2006; 2009; Bailey-Serres *et al.*, 2010; Mackill *et al.*, 2012). A summary of the historical events for the discovery and deployment of this gene is provided in Table 15.3 and discussed in detail below.

Use of the SUB1 gene in the development of submergence-tolerant varieties

After the identification of traditional varieties tolerant of submergence (Table 15.1), breeders began using them to develop tolerant high-yielding cultivars. Early breeding efforts suggested that the trait had high heritability, and semi-dwarf breeding lines with strong tolerance were identified. Many of them had low yield and poor grain quality, however. Ultimately, it was possible to break the association between lower yield and tolerance (Mackill *et al.*, 1993). Advanced breeding lines were evaluated in a number of countries but these varieties were not found to be the equal of the high-yielding varieties already available to farmers, even though they had much higher submergence tolerance.

A number of varieties with high yield, good adaptation to the rainfed lowland regions of South and South-east Asia and good grain quality began to spread rapidly among farmers, and attained the status of mega-varieties, covering one million to a few million hectares. These became reference varieties that were well accepted by farmers, millers and consumers. These varieties have become very difficult to replace, even though they are not tolerant of many biotic and abiotic stresses. The identification of the *SUB1* gene provided a means to improve these mega-varieties for submergence tolerance without compromising their desirable characteristics. By using the MABC approach (Tanksley *et al.*, 1989), the *SUB1* locus could be transferred into the mega-varieties without any undesirable characteristics of the donor parent and ensuring that the subsequent varieties retained all the desirable features of the recurrent parent. Six varieties were originally converted in two to three backcrosses: IR64 from IRRI-Philippines; Swarna, Samba Mahsuri and Sabitri from India; BR11 from Bangladesh; and TDK1 from Laos (Neeraja *et al.*, 2007; Septiningsih *et al.*, 2009; Iftekharuddaula *et al.*, 2011). Background markers were used to prevent the transfer of unlinked genes and markers flanking the *SUB1* locus were used to limit the size of the chromosome fragment possessing the gene. The resulting varieties were estimated to be at least 98% genetically identical to the original varieties based on genome-wide marker analysis.

Evaluation of the varieties under field conditions confirmed that they were phenotypically identical to the original mega-varieties but that they possessed the strong submergence tolerance trait of the donor variety. The only exception was that two varieties that originally had brown-coloured hulls (Swarna and TDK1) were converted to the straw hull trait. This was because of the tight linkage between *SUB1* and the gene *IBF1* (Inhibitor of Brown Furrows 1) on chromosome 9 (Shao *et al.*, 2012). This change in hull colour was not considered undesirable and has been advantageous in distinguishing between seeds of Swarna and Swarna-Sub1. The Sub1 varieties were compared with the original parents in both control and submerged conditions. Under control conditions without submergence, the grain yield of the Sub1 varieties was the same as that of the original varieties. Under

Table 15.3. Historical roadmap for the development of Sub1 rice varieties.

Milestone	Year	Effort	Location/Institution	References
Screening of genebank accessions at IRRI for submergence tolerance and development of hybridization programmes	1970s	~10 years	IRRI	Vergara and Mazaredo, 1975 Khush and Coffman, 1977
Improved breeding lines developed using tolerant donors	1980s	15 years	IRRI	Mackill *et al.*, 1993, 1996
Mapping of the *SUB1* QTL on chromosome 9 and markers developed for selection	1995	5 years	UCD	Xu and Mackill, 1996 Xu *et al.*, 2000, 2004
MABC programme to develop mega varieties with *SUB1*	2005	3 years	IRRI	Neeraja *et al.*, 2007 Septiningsih *et al.*, 2009
Cloning of *SUB1* gene and identification of improved markers	2006	10 years	UCD, UCR, IRRI	Xu *et al.*, 2006
Testing of Sub1 varieties in India and Bangladesh	2006	3 years	CRRI and NDUAT; BRRI	Mackill *et al.*, 2012
Swarna-Sub1 released as the first Sub1 variety in India	2009	2 years	IRRI, CRRI and NDUAT, India	Reddy *et al.*, 2010
Sub1 varieties commercialized in different Asian countries	2012	3 years	India (2), Bangladesh (3), Nepal (2), Indonesia (4), Philippines (2) and Myanmar (1)	Mackill *et al.*, 2012 Ismail *et al.*, 2013

IRRI, International Rice Research Institute; UCD, University of California, Davis; UCR, University of California, Riverside; CRRI, Central Rice Research Institute, Cuttack, India; NDUAT, Narenda Dev University of Agriculture and Technology, Faizabad, Uttar Pradesh, India; BRRI, Bangladesh Rice Research Institute, Gazipur, Bangladesh.

various degrees of submergence stress, usually 12 to 18 days of complete submergence, the Sub1 varieties outyielded the original varieties and the yield advantage generally ranged from 1 to 3.5 t ha^{-1} above that of the intolerant varieties (Sarkar *et al.*, 2009; Singh *et al.*, 2009; Mackill *et al.*, 2012). This yield advantage was mostly a consequence of the higher survival and faster recovery of the Sub1 varieties after submergence. However, even under milder submergence stress, under which most intolerant plants recovered, the Sub1 varieties consistently had a 1 t ha^{-1} or more yield advantage (Mackill *et al.*, 2012).

Impacts of Sub1 varieties in farmers' fields

Crosses were made at IRRI in 2003–2004 to develop the first six mega-varieties and the MABC-derived varieties were isolated after only 2–3 years. Swarna-Sub1 was the first variety developed and it was evaluated and underwent seed increase in 2006. This allowed initiation of testing in farmers' fields by 2007. Over two seasons, the results from comparisons between Swarna-Sub1 and Swarna validated the consistent yield superiority of the Sub1 variety under natural submergence events in farmers' fields. Yield advantages of 1–3.5 t ha^{-1} were observed and in many cases the intolerant variety was completely destroyed, whereas Swarna-Sub1 yielded 3 t ha^{-1} or more.

Similar results were consistently obtained from other Sub1 varieties evaluated in thousands of trials conducted by farmers in India, Bangladesh, Nepal, the Philippines and Indonesia since 2008. The superior performance of these varieties was consistent when flooding occurred any time from the early seedling stage to about a week before heading. Sub1 varieties also helped eliminate the cost incurred for re-transplanting damaged fields, which usually caused a delay in harvest, with a further reduction in yield and deterioration in grain quality. The success of these varieties also triggered most national programmes to adjust their varietal release guidelines, thus facilitating faster commercialization in India, Bangladesh and Indonesia (Mackill *et al.*, 2012; Ismail *et al.*, 2013).

The benefits of Sub1 varieties in areas affected by transient submergence have stimulated considerable global interest in them, and several breeding programmes (such as at IRRI and in India) are now considering introducing the gene in all lowland rice varieties likely to be affected by floods. This success has also stimulated substantial international support for evaluation, large-scale seed production and delivery to farmers, including support from the Bill & Melinda Gates Foundation (BMGF) through a project called 'Stress-Tolerant Rice for Africa and South Asia (STRASA)' and from the United States Agency for International Development (USAID) and the International Fund for Agricultural Development (IFAD). Additional resources have also been allocated by several national programmes to produce and distribute sufficient seeds of Sub1 varieties to farmers in flood-affected areas in Asia. Some of these varieties have already been commercialized in India, Bangladesh, Nepal, Indonesia and the Philippines (Mackill *et al.*, 2012; Ismail *et al.*, 2013). The Sub1 varieties were also evaluated in flood-prone areas in sub-Saharan Africa, where they performed similarly as in Asia, and efforts were begun to introduce *SUB1* into several rice varieties popular in rainfed areas in Africa.

15.5 Future Directions in Flood-tolerance Research in Rice to Cope with Climate Change

Rainfed lowlands and flood-prone rice areas are more susceptible to variations in rainfall patterns and flood incidence. Several climate change models predicted floods to increase because of increases in the intensity of tropical storms, precipitation extremes and sea-level rise (Coumou and Rahmstorf, 2012); however, it is still not clear from current trends whether flood events are becoming more frequent or severe. Future breeding programmes need to keep up with any further deterioration in natural resources to avoid serious food shortages. These efforts should recognize the importance of the collection and preservation of additional germplasm resources, especially from current hotspots, to be used for mining new tolerance traits and genes for different types of floods and other climate disturbances likely to worsen with climate change.

SUB1 can provide protection for 4 to 18 days of submergence; however, some areas are witnessing flash floods that persist for 22 to 25 days, suggesting the need for additional tolerance genes. Furthermore, *SUB1* can protect the plants when they are completely submerged but it is not effective in situations when plants are partially flooded for prolonged durations, as in stagnant-flood and deepwater areas. Efforts are needed to develop varieties that contain *SUB1* but that can also survive and yield well when flooding is partial and long-lasting. More research is needed to decipher the traits and genes associated with tolerance of stagnant floods to facilitate developing these varieties. Tolerance of floods during germination is another important feature in both rainfed and irrigated rice production areas, as rice crop establishment shifts from transplanting seedlings into puddled fields to sowing dry seed in soil. Tolerance of both stagnant flooding and submergence during germination is independent of submergence tolerance based on *SUB1* but all three traits are compatible and can potentially be combined (Mackill *et al.*, 2010, 2012; Singh *et al.*, 2011). This combination will provide varieties that can be adopted on a larger scale in rainfed and flood-prone areas.

In nearly all flood-prone areas, other abiotic stresses are also encountered during different times within the season. For example, both drought and submergence can occur in rainfed lowlands and mostly within the same season (Mackill *et al.*, 1996), and varieties combining tolerance of both stresses will be very useful in these areas. Several QTLs associated with drought tolerance have recently been identified and are being transferred into popular varieties using MABC, and some of these varieties also have *SUB1* (Verulkar *et al.*, 2010; Mackill *et al.*, 2012). In coastal areas, both salinity and submergence are encountered during the monsoon season. Salinity is normally high during crop establishment and submergence can occur any time during the season (Ismail *et al.*, 2008). Varieties tolerant of both stresses are being developed by combining *SUB1* with *Saltol*, a major QTL conferring moderate to high salinity tolerance at the seedling stage (Thomson *et al.*, 2010), and good progress was made in identifying additional QTL targets that can further be combined with *Saltol* for higher tolerance. Besides tolerance of these major abiotic stresses, these varieties need to have resistances to major pests and diseases. The availability of such resilient varieties will further encourage farmers to use more fertilizer and better crop management practices to ensure higher and more stable yields. In addition, most of the current varieties are late maturing, and farmers are demanding earlier maturing varieties with multiple stress tolerance. These varieties can provide more flexibility to adjust and optimize planting time to generate opportunities for growing multiple crops of rice or additional crops, thus increasing crop diversity and annual productivity.

15.6 Conclusions

Rice is the most versatile cereal crop in its adaptability to a variety of climatic and edaphic factors. It is currently grown under a wide range of hydrological conditions, from aerobic soils in uplands to deepwater areas, where water depth can fluctuate from 50 cm to more than 5 metres for most of the growing season. This enormous versatility and adaptive plasticity made rice one of the crops most liable to genetic manipulation to cope with climate adversities. Submergence, drought, salinity and high temperature are already causing severe losses in global food production and are considered major threats to food security worldwide. The current diversity in rice germplasm could, however, potentially help develop varieties capable of coping with the consequences of climate change if effectively exploited, especially when combined with effective mitigation strategies to halt further deterioration.

Exploring genetic resources to breed stress-tolerant varieties started in the early 1970s at IRRI, and considerable progress was made in identifying and characterizing sources of tolerance and using them in breeding. The impact of Sub1 varieties in flood-prone areas clearly demonstrated the significant role that germplasm resources could have if effectively exploited using recent technological interventions. Significant progress was also made in developing varieties tolerant of other types of floods, such as submergence during germination and prolonged stagnant floods during the vegetative stage. Varieties combining genes for tolerance of a particular stress and/or for different stresses are being developed and are expected to be adopted on a larger scale by farmers. Success in developing more adapted varieties to keep up with changing climates is therefore plausible, and considerable efforts are needed to arrest further deterioration in rice production.

15.7 Acknowledgements

The research presented in this chapter was funded by several donors. We would like particularly to acknowledge the support of the Bill & Melinda Gates Foundation, the German Federal Ministry for Economic Cooperation and Development (BMZ), the European Commission (EC) and the International Fund for Agricultural Development (IFAD) for funding the development and evaluation of Sub1 varieties. We also thank all our research and development partners and farmers in South and South-east Asia for their enormous support and assistance during the evaluation of Sub1 varieties. We thank all colleagues who contributed to the work summarized in this chapter, and Bill Hardy for his excellent editorial contribution.

References

Angaji, S., Septiningsih, E.M., Mackill, D.J. and Ismail, A.M. (2010) QTLs associated with tolerance of anaerobic conditions during germination in rice (*Oryza sativa* L.). *Euphytica* 172, 159–168.

Australia: Office of the Gene Technology Regulator (2005) The biology and ecology of rice (*Oryza sativa* L.) in Australia. Office of the Gene Technology Regulator, Woden, ACT. Available at: www.ogtr.gov.au/internet/ogtr/publishing.nsf/content/rice-3/$FILE/biologyrice1.pdf (accessed on 27 November 2012).

Bailey-Serres, J., Fukao, T., Ronald, P.C., Ismail, A.M., Heuer, S. *et al.* (2010) Submergence tolerant rice: *SUB1*'s journey from landrace to modern cultivar. *Rice* 3, 138–147.

Bates, B.C., Kundzewicz, Z.W., Wu, S. and Palutikof, J.P. (eds) (2008) *Climate Change and Water.* IPCC Secretariat, Geneva, Switzerland.

Biswas, J.K. and Yamauchi, M. (1997) Mechanism of seedling establishment of direct-seeded rice (*Oryza sativa* L.) under lowland conditions. *Botanical Bulletin Academia Sinica* 38, 29–32.

Catling, D. (1992) *Rice in Deep Water.* International Rice Research Institute, Los Baños, Philippines.

Coumou, D. and Rahmstorf, S. (2012) A decade of weather extremes. *Nature Climate Change* 2, 491–496.

Dey, M.M. and Upadhyaya, H.K. (1996) Yield loss due to drought, cold and submergence in Asia. In: Evenson, R.E., Herdt, R.W. and Hossain, M. (eds) *Rice Research in Asia: Progress and Priorities.* CAB International, Wallingford, UK, pp. 291–303.

Drew, M.C. (1997) Oxygen deficiency and root metabolism: injury and acclimation under hypoxia and anoxia. *Annual Review of Plant Physiology and Plant Molecular Biology* 48, 223–250.

Drew, M.C. and Lynch, J.M. (1980) Soil anaerobiosis, microorganisms and root function. *Annual Review of Phytopathology* 18, 37–66.

Ella, E.S., Dionisio-Sese, M.L. and Ismail, A.M. (2010) Proper management improves seedling survival and growth during early flooding in contrasting rice (*Oryza sativa* L.) genotypes. *Crop Science* 50, 1997–2008.

Ella, E.S., Dionisio-Sese, M.L. and Ismail, A.M. (2011) Seed pre-treatment in rice reduces damage, enhances carbohydrate mobilization and improves emergence and seedling establishment under flooded conditions. *Annals of Botany PLANTS* 2011, 10.1093/aobpla/plr007.

Haque, Q.A., HilleRisLambers, D., Tepora, N.M. and dela Cruz, Q.D. (1989) Inheritance of submergence tolerance in rice. *Euphytica* 41, 247–251.

Hattori, Y., Miura, K., Asano, K., Yamamoto, E., Mori, H. *et al.* (2007) A major QTL confers rapid internode elongation in response to water rise in deepwater rice. *Breeding Science* 57, 305–314.

Hattori, Y., Nagaki, K., Furukawa, S., Song, X.L., Kawano, R. *et al.* (2009) The ethylene response factors *SNORKEL1* and *SNORKEL2* allow rice to adapt to deep water. *Nature* 460, 1026–1030.

Hay, F.R., Sackville Hamilton, N.R., Furman, B.J., Upadhyaya, H.D., Reddy, K.N. *et al.* (2013) Cereals. In: Normah, M.N., Chin, H.F. and Reed, B.M. (eds) *Conservation of Tropical Plant Species.* Springer, New York, pp. 293–315.

Herdt, R.W. (1991) Research priorities for rice biotechnology. In: Khush, G.S. and Toenniessen, G.H. (eds) *Rice Biotechnology.* CAB International, UK, pp. 19–54.

HilleRisLambers, D. and Vergara, B.S. (1982) Summary results of an international collaboration on screening methods for flood tolerance. In: *Proceedings of the 1981 International Deepwater Rice Workshop.* IRRI, Los Baños, Philippines, pp. 347–353.

Hossain, M.Z., Islam, M.T., Sakai, T. and Ishida, M. (2008) Impact of tropical cyclones on rural infrastructures in Bangladesh. *Agriculture Engineering International: the CIGR Ejournal*, Invited Overview No. 2, Vol. X, April 2008. Available at: http://www.cigrjournal.org/index.php/Ejounral/article/view/1036/1029 (accessed 2 December 2012).

Iftekharuddaula, K., Newaz, M., Salam, M., Ahmed, H., Mahbub, M. *et al.* (2011) Rapid and high-precision marker assisted backcrossing to introgress the *SUB1* QTL into BR11, the rainfed lowland rice mega variety of Bangladesh. *Euphytica* 178, 83–97.

IRRI (1975) Genetic Evaluation and Utilization (GEU) program: deep water and flood tolerance. IRRI Annual Report for 1975. IRRI, Los Baños, Philippines, pp. 171–177.

Ismail, A.M. and Tuong, T.P. (2009) Brackish water coastal zones of monsoon tropics: challenges and opportunities. In: Haefele, S. and Ismail, A.M. (eds) *Natural Resource Management for Poverty Reduction and Environmental Sustainability in Fragile Rice-Based Systems.* Limited Proceedings No. 15. IRRI, Los Baños, Philippines, pp. 113–121.

Ismail, A.M., Thomson, M.J., Singh, R.K., Gregorio, G.B. and Mackill, D.J. (2008) Designing rice varieties adapted to coastal areas of south and Southeast Asia. *Journal of the Indian Society of Coastal Agricultural Research* 26, 69–73.

Ismail, A.M., Ella, E.S., Vergara, G.V. and Mackill, D.J. (2009) Mechanisms associated with tolerance to flooding during germination and early seedling growth in rice (*Oryza sativa*). *Annals of Botany* 103, 197–209.

Ismail, A.M., Johnson, D.E., Ella, E.S., Vergara, G.V. and Baltazar, A.M. (2012) Adaptation to flooding during emergence and seedling growth in rice and weeds, and implications for crop establishment. *Annals of Botany PLANTS* 2012: pls019; doi:10.1093/aobpla/pls019.

Ismail, A.M., Singh, U.S., Singh, S., Dar, M. and Mackill, D.J. (2013) The contribution of submergence-tolerant (*Sub1*) rice varieties to food security in flood-prone areas. *Field Crops Research.* Available at http://dx.doi.org/10.1016/j.far.2013.01.007

Jackson, M.B. (2004) The impact of flooding stress on plants and crops. Available at: www.plantstress.com/Articles/index.asp (accessed 8 January 2013).

Jackson, M.B. and Ram, P.C. (2003) Physiological and molecular basis of susceptibility and tolerance of rice plants to complete submergence. *Annals of Botany* 91, 227–241.

Jackson, M.T. (1997) Conservation of rice genetic resources: the role of the International Rice Genebank at IRRI. *Plant Molecular Biology* 35, 61–67.

Jackson, M.T., Loresto, G.C., Appa Rao, S., Jones, M., Guimaraes, E.P. *et al.* (1997) Rice. In: Fuccillo, D., Sears, L. and Stapleton, P. (eds) *Biodiversity in Trust.* Cambridge University Press, Cambridge, UK, pp. 273–291.

Jiang, L., Ming-yu, H.W., Ming, C. and Jian-min, W. (2004) Quantitative trait loci and epistatic analysis of seed anoxia germinability in rice (*Oryza sativa* L.). *Rice Science* 11, 238–244.

Jiang, L., Liu, S., Hou, M., Tang, J., Chen, L. *et al.* (2006) Analysis of QTLs for seed low temperature germinability and anoxia germinability in rice (*Oryza sativa* L.). *Field Crops Research* 98, 68–75.

Kawano, R., Doi, K., Yasui, H., Mochizuki, T. and Yoshimura, A. (2008) Mapping of QTLs for floating ability in rice. *Breeding Science* 58, 47–53.

Kemp, A.C., Horton, B.P., Donnelly, J.P., Mann, M.E., Vermeer, M. *et al.* (2011) Climate related sea-level variations over the past two millennia. *Proceedings of the National Academy of Sciences USA* 108, 11017–11022.

Khush, G. and Coffman, W.R. (1977) Genetic Evaluation and Utilization (GEU) program: the rice improvement program of the International Rice Research Institute. *Theoretical and Applied Genetics* 51, 97–110.

Lobell, D.B., Burke, M.B., Tebaldi, C., Mastrandrea, M.D., Falcon, W.P. *et al.* (2008) Prioritizing climate change adaptation needs for food security in 2030. *Science* 319, 607–610.

Mackill, D.J., Amante, M.M., Vergara, B.S. and Sarkarung, S. (1993) Improved semidwarf rice lines with tolerance to submergence of seedlings. *Crop Science* 33, 749–753.

Mackill, D.J., Coffman, W.R. and Garrity, D.P. (1996) *Rainfed Lowland Rice Improvement.* International Rice Research Institute, Los Baños, Philippines.

Mackill, D.J., Ismail, A.M., Pamplona, A.M., Sanchez, D.L., Carandang, J.J. *et al.* (2010) Stress tolerant rice varieties for adaptation to a changing climate. *Crop, Environment and Bioinformatics* 7, 250–259.

Mackill, D.J., Ismail, A.M., Singh, U.S., Labios, R.V. and Paris, T.R. (2012) Development and rapid adoption of submergence-tolerant (Sub1) rice varieties. *Advances in Agronomy* 115, 303–356.

Maclean, J.L., Dawe, D.C., Hardy, B. and Hettel, G.P. (eds) (2002) *Rice Almanac.* International Rice Research Institute, Los Baños, Philippines.

Metz, B., Davidson, O.R., Bosch, P.R., Dave, R. and Meyer, L.A. (eds) (2007) *Contribution of Working Group III to the Fourth Assessment Report of the Intergovernmental Panel on Climate Change, 2007.* Cambridge University Press, Cambridge, UK and New York.

Mohanty, H.K. and Chaudhary, R.C. (1986) Breeding for submergence tolerance in rice in India. In: *Progress in Rainfed Lowland Rice.* IRRI, Los Baños, Philippines, pp. 191–200.

Mohanty, H.K. and Khush, G.S. (1985) Diallel analysis of submergence tolerance in rice, *Oryza sativa* L. *Theoretical and Applied Genetics* 70, 467–473.

Molina, J., Sikora, I.M., Garud, I.N., Flowers, J.M., Rubinstein, S. *et al.* (2011) Molecular evidence for a single evolutionary origin of domesticated rice. *Proceedings of the National Academy of Sciences USA* 108, 8351–8356.

National Research Council of the National Academies (2010) *Advancing the Science of Climate Change.* National Academies Press. Available at: www.nap.edu/catalog.php?record_id=12782 (accessed 8 January 2013).

Neeraja, C., Maghirang-Rodriguez, R., Pamplona, A., Heuer, S., Collard, B. *et al.* (2007) A marker-assisted backcross approach for developing submergence-tolerant rice cultivars. *Theoretical and Applied Genetics* 115, 767–776.

Oka, H.I. (1975) Prospects for the improvement of floating rice varieties. In: *Proceedings of the International Seminar on Deepwater Rice*, August 1974. Bangladesh Rice Research Institute, Dhaka, Bangladesh, pp. 71–76.

Palada, M.C. and Vergara, B.S. (1972) Environmental effects on the resistance of rice seedlings to complete submergence. *Crop Science* 12, 209–212.

Reddy, J.N., Sarkar, R.K., Patnaik, S.S.C., Singh, D.P., Singh, U.S. *et al.* (2010) Improvement of rice germplasm for rainfed lowlands of eastern India. SABRAO 13th International Congress. Conference abstract available at: http://open.irri.org/sabrao/images/stories/conference/site/papers/apb09final00211.pdf.

Richharia, R.H., and Govindaswami, S. (1966) *Rices of India.* Scientific Book Company, Patna, India.

Richharia, R.H. and Misro, B. (1960) Flood and deep water rices and the future plan of their improvement. *Indian Agriculturist* 4, 135–143.

Sarkar, R.K., Reddy, J.N., Sharma, S.G. and Ismail, A.M. (2006) Physiological basis of submergence tolerance in rice and implications for crop improvement. *Current Sciences* 91, 899–906.

Sarkar, R.K., Panda, D., Reddy, J.N., Patnaik, S.S.C., Mackill, D.J. *et al.* (2009) Performance of submergence tolerant rice (*Oryza sativa*) genotypes carrying the *Sub1* quantitative trait locus under stressed and non-stressed natural field conditions. *Indian Journal of Agricultural Sciences* 79, 876–883.

Septiningsih, E.M., Pamplona, A.M., Sanchez, D.L., Neeraja, C.N., Vergara, G.V. *et al.* (2009) Development of submergence tolerant rice cultivars: the *Sub1* locus and beyond. *Annals of Botany* 103, 151–160.

Septiningsih, E.M., Ignacio, J.C.I., Sendon, P.M.D., Sanchez, D.L., Ismail, A.M. (2013) QTL mapping and confirmation for tolerance of anaerobic conditions during germination derived from the rice landrace Ma-Zhan Red. *Theoretical and Applied Genetics* 126, 1357–1366.

Setter, T.L. and Laureles, E.V. (1996) The beneficial effect of reduced elongation growth on submergence tolerance of rice. *Journal of Experimental Botany* 47, 1551–1559.

Setter, T.L. and Waters, I. (2003) Review of prospects for germplasm improvement for waterlogging tolerance in wheat, barley and oats. *Plant and Soil* 253, 1–34.

Setter, T.L., Ellis, M., Laureles, E.V., Ella, E.S., Senadhira, D. *et al.* (1997) Physiology and genetics of submergence tolerance in rice. *Annals of Botany* 79, 67–77.

Shao, T., Qian, Q., Tang, D., Chen, J., Li, M. *et al.* (2012) A novel gene *IBF1* is required for the inhibition of brown pigment deposition in rice hull furrows. *Theoretical and Applied Genetics* 125, 381–390.

Singh, S., Mackill, D.J. and Ismail, A.M. (2009) Responses of *SUB1* rice introgression lines to submergence in the field: yield and grain quality. *Field Crops Research* 113, 12–23.

Singh, S., Mackill, D.J. and Ismail, A.M. (2011) Tolerance of longer-term partial stagnant flooding is independent of the *SUB1* locus in rice. *Field Crops Research* 121, 311–323.

Sripongpangkul, K., Posa, G.B.T., Senadhira, D.W., Brar, D., Huang, N. *et al.* (2000) Genes/QTLs affecting flood tolerance in rice. *Theoretical and Applied Genetics* 101, 1074–1081.

Syvitski, J.P.M., Kettner, A.J., Overeem, I., Hutton, E.W.H., Hannon, M.T. *et al.* (2009) Sinking deltas due to human activities. *Nature Geoscience* 2, 681–686.

Tanksley, S.D., Young, N.D., Paterson, A.H. and Bonierbale, M.W. (1989) RFLP mapping in plant breeding: new tools for an old science. *Nature Biotechnology* 7, 257–264.

Thomson, M.J., de Ocampo, M., Egdane, J., Rahman, M.A., Sajise, A.G. *et al.* (2010) Characterizing the *Saltol* quantitative trait locus for salinity tolerance in rice. *Rice* 3, 148–160.

USDA-FAS (2008) BURMA: Widespread cyclone damage in major rice production regions. Available at: www.pecad.fas.usda.gov/highlights/2008/05/Burma_Cyclone_Nargis_Rice_Impact.htm (accessed 3 December 2012).

Vergara, B.S. and Mazaredo, A. (1975) Screening for resistance to submergence under greenhouse conditions. In: *Proceedings of the International Seminar on Deepwater Rice*, August 1974. Bangladesh Rice Research Institute, Dhaka, Bangladesh, pp. 67–70.

Vergara, B.S., Jackson, M.B. and De Datta, S.K. (1976) Deepwater rice and its response to deep-water stress. In: *Climate and Rice.* IRRI, Los Baños, Philippines, pp. 301–319.

Vermeer, M. and Rahmstorf, S. (2009) Global sea level linked to global temperature. *Proceedings of the National Academy of Sciences USA* 106, 21527–21532.

Verulkar, S.B., Mandal, N.P., Dwivedi, J.L., Singh, B.N., Sinha, P.K. *et al.* (2010) Breeding resilient and productive genotypes adapted to drought-prone rainfed ecosystem of India. *Field Crops Research* 117, 197–208.

Wassmann, R., Jagadish, S.V.K., Sumfleth, K., Pathak, H., Howell, G. *et al.* (2009) Regional vulnerability of climate change impacts on Asian rice production and scope for adaptation. *Advances in Agronomy* 102, 91–133.

Xu, K. and Mackill, D.J. (1996) A major locus for submergence tolerance mapped on rice chromosome 9. *Molecular Breeding* 2, 219–224.

Xu, K., Xu, X., Ronald, P.C. and Mackill, D.J. (2000) A high-resolution linkage map in the vicinity of the rice submergence tolerance locus Sub1. *Molecular and General Genetics* 263, 681–689.

Xu, K.N., Deb, R. and Mackill, D.J. (2004) A microsatellite marker and a codominant PCR-based marker for marker-assisted selection of submergence tolerance in rice. *Crop Science* 44, 248–253.

Xu, K., Xia, X., Fukao, T., Canlas, P., Maghirang-Rodriguez, R. *et al.* (2006) *Sub1A* is an ethylene response factor-like gene that confers submergence tolerance to rice. *Nature* 442, 705–708.

Yamada, N. (1959) Physiological basis of resistance of rice plant against overhead flooding. *Bulletin of the National Institute of Agricultural Sciences, Series D* (*Plant Physiology, Genetics and Crops in General*) 8, 1–112 (in Japanese with extensive English summary and captions).

Yamauchi, M. and Winn, T. (1996) Rice seed vigor and seedling establishment in anaerobic soil. *Crop Science* 36, 680–686.

Yamauchi, M., Aguilar, A.M., Vaughan, D.A. and Seshu, D.V. (1993) Rice (*Oryza sativa* L.) germplasm suitable for direct sowing under flooded soil surface. *Euphytica* 67, 177–184.

16 Effects of Climate Change on Plant–Insect Interactions and Prospects for Resistance Breeding Using Genetic Resources

Jeremy Pritchard,[1] Colette Broekgaarden[2] and Ben Vosman[2]

[1]School of Biosciences, University of Birmingham, Birmingham, UK and [2]Wageningen UR Plant Breeding, Wageningen, the Netherlands

16.1 Introduction

Elevated CO_2 is having profound impacts on climate. When considering plants and the interaction with their pests and pathogens, climate change can be sub-divided into three main components:

- The elevated CO_2 itself.
- Elevated temperature.
- The indirect alteration in rainfall patterns that will impose drought or excessive rainfall depending on the area.

Of these, it is generally considered that only temperature will have a direct effect on pest herbivores of plants (Bale, 1988). By far the greatest effects on herbivores are arguably indirect as the components of climate change alter the plants that form their diets (Douglas, 2006; Pritchard *et al.*, 2007). Elevated CO_2 potentially provides more carbon, temperature alters plant metabolism and development, while drought alters both the partitioning and composition of nutrients in the plant. In addition, the competing requirements within plants for efficient carbon assimilation and water economy may reduce stomatal opening, producing plants that are more drought tolerant rather than of increased biomass and altered composition (Schroeder *et al.*, 2001). Such changes in water use efficiency (WUE) may affect plant range and thus their associated herbivores.

Layered over this simplified scenario, however, insect herbivores exist in different feeding guilds, broadly represented by: (i) the chewers such as lepidopteron caterpillars; (ii) cell content feeders such as spider mites; and (iii) phloem feeders including plant hoppers, aphids and whiteflies. Consideration of feeding guild is important because plants are not homogeneous in their response to abiotic stresses such as elevated temperature or drought, and different organs and cell types respond in different ways and to different degrees (Pritchard *et al.*, 2007). In addition, plants respond differently depending on the attacking herbivore (Stout *et al.*, 1998; Kawazu *et al.*, 2012), Thus changes in whole leaf composition and defence experienced by a caterpillar, for example, may be very different from the changes occurring in the sieve element (SE), the food source of an aphid, on the same plant.

To predict the impact of climate change on herbivores, a description of the effect on plant composition (diet) and the relationship between herbivore performance and diet is required in addition to a consideration of any direct effects on the pest insect. A good understanding of these components at both the physiological and underlying molecular level will aid strategies to improve crops using a wide range of genetic resources – in particular, landraces and crop wild relatives.

16.2 Components of Climate Change: Direct Effects on Herbivores

Of the three main components of climate change, elevated CO_2 and drought are likely to have little direct effect on herbivores. Temperature is, however, known to have large effects, both in terms of lethality at high and low temperature, and in affecting both phenology and distribution. Bale and Hayward identify cold tolerance, diapause (a suspension of development in response to regular periods of cold) and high temperature as the three most important components (Bale and Hayward, 2010). A general elevation of global temperature under climate change will reduce low temperature mortality and increase that at high temperature.

Climate change models predict that changes in temperature will alter the distribution pattern of pest insects (Aragon and Lobo, 2012). Susceptibility to low temperature is affected by ontogeny (Bowler and Terblanche, 2008). For example, aphid eggs are far more cold tolerant than anholocyclic adults (only producing asexual morphs), so that the potential for pest outbreak depends on the preceding winter temperature. Insects have a number of strategies to deal with low temperatures and these differ at the species and genus levels, and so prediction of general effects is difficult. Negative effects of changes in temperature may be seen when strategies such as diapause are affected. Incubation at 20°C rather than 15°C led to a reduced proportion of insects undergoing diapause and shorter duration for those individuals that did (McWatters and Saunders, 1998). As temperature increases, insects can completely abort diapause (Fatinou and Kagkou, 2000). Even if changes in temperature only alter the timing, there can still be large effects on subsequent insect populations if the diapause disengages life cycle from the onset of cold stress. Such disengagement of insects from their host plant can have severe effects because there is often an existing mismatch between herbivore and plant that climate change will only exacerbate (Singer and Parmesan, 2010). Temperature elevation may result in changes in pest distribution, either due to direct migration or local extinction and selection. Insects in tropical areas are probably close to their maximum temperature and so small changes will have potentially magnified effects (Piyaphongkul *et al.*, 2012).

16.3 Components of Climate Change: Indirect Effects on Herbivores

Although the direct effects of climate change, in particular temperature changes, are straightforward to study, in terms of crop protection the indirect effects are arguably as important if not more so. The effects of CO_2, drought and temperature will have significant effects on plant composition, i.e. the insects' diet. Prediction of outcomes is, however, currently severely limited by the absence of a detailed understanding of how elevated CO_2 and drought will affect plant composition and how the diverse range of herbivores will respond to this change in diet and how they themselves more directly respond to the stresses of climate change.

16.3.1 Indirect effects of climate change on defence

Plant defence and climate change

Plants employ a range of defences, both physical and chemical, against herbivores. The components of climate change will potentially alter these defences both qualitatively and quantitatively as resource partitioning priorities are affected. Theoretically, if carbon becomes more available during climate change, plants may allocate more to defence, whereas resources such as N are diluted and therefore become more at a premium then defence could be compromised. Current information suggests that a consistent effect on front-line defence allocation is unlikely, perhaps not surprising given the diverse life history strategies of plants. For example, growth of *Asclepias syriaca* (the common milkweed) in elevated CO_2 led to an increase in terpenes and latex but a decrease in allocation to chemical defences (Vannette and Hunter, 2011). The increase in C availability under elevated CO_2 was accompanied by an increase in carbohydrate rich tannins but a decrease in N rich alkaloids, consistent with allocation to defence following an increase in the C/N ratio (Coley, 2000). A decrease in glucosinolates was seen under elevated CO_2 in *Brassica napus* (Himanen *et al.*, 2008), whereas a drought-induced increase in tissue glucose concentration was accompanied by a decrease in phenolics (Gutbrodt *et al.*, 2012). Grown under elevated CO_2, tobacco plants produced more secondary metabolites and had increased resistance to the green peach aphid (Fu *et al.*, 2010). In soybean, an enhanced resistance to herbivorous beetles was implied as defence-related genes showed increased expression under elevated CO_2 (Casteel *et al.*, 2008). Interestingly, flavonoids increased in sorghum grown under elevated CO_2, an effect that has a potential benefit to herbivores by providing additional antioxidants in their diet (O'Neill *et al.*, 2010). Although adaptation to drought may divert resources from defences, a comparison of two populations of *Acacia* adapted to different water regimes did not identify any changes in plant defence associated with drought (Ward *et al.*, 2012).

Although the effects of climate change on front-line defence appear diverse, the effects on regulatory networks may be more predictable, providing a more robust target for breeding strategies. Herbivores can be considered as a specific type of pathogen; an overlap between genes induced by hemipteran herbivores and micropathogens has been predicted (Kaloshian and Walling, 2005). Jasmonic acid (JA), salicylic acid (SA), ethylene and even small RNA are thought to regulate induced and constitutive front-line defences such as protease inhibitors or nicotine (Thompson and Goggin, 2006; Pandey and Baldwin, 2007). Responses can be subtly different; different feeding guilds on tomato represented by chewers, thrips and leaf miners induced qualitatively different patterns of JA, SA and ethylene gene induction (Kawazu *et al.*, 2012).

Recent emerging data are indicating that elevated CO_2 may alter regulation of plant defence in a predictable way. In a range of soybean genotypes grown under elevated CO_2 herbivore damage generally decreased JA while SA was increased (Casteel *et al.*, 2012). In tomato, a similar general pattern was seen with an increase in SA and decrease in JA under elevated CO_2 accompanying a CO_2 induced increase in C/N (Huang *et al.*, 2012). A study examining tomato defence against root knot nematodes (*Meloidogyne incognita*) found that SA defence was increased and JA defence decreased (Sun *et al.*, 2012). Tomato grown under elevated CO_2 had reduced resistance to cotton bollworm with a concomitant decrease in levels of JA and defence compounds including proteinase inhibitors (Guo *et al.*, 2012). The apparently consistent and contrasting effects of elevated CO_2 on SA and JA pathways may lead to differential effects on herbivores because biotrophic pathogens are regarded as regulated by the SA pathway, whereas necrotropic pathogens are affected by the JA route (Glazebrook *et al.*, 2003).

Elevated CO_2 may also have an effect on indirect defences. Lima beans (*Phaseolus lunatus*) grown under elevated CO_2 had elevated emissions of volatile organic compounds (VOCs), which could affect the ability of parasite predators of herbivore pests (Ballhorn *et al.*, 2011). The variability of the response of defence regulation in plants to elevated CO_2 is highlighted by the different responses of individual soybean genotypes, leading to the suggestion that genetic variation in regulation of defence is responsible for the observed variation in the effect of elevated CO_2 on plant–insect interactions (Casteel *et al.*, 2012).

A review of the response of microbial pathogen defence pathways in plants suggests that common features of defence may be predictable under climate change but that the complexity of the stresses and responses make individual predictions difficult, suggesting a systems biology approach as a potential solution (Newton *et al.*, 2012). The common feature of the effects of climate change on regulation of defence are encouraging in terms of crop improvement, but further work is needed to determine whether there is a consistent effect on front-line defence and a consequent improvement in resistance of plants to herbivory.

Leaf tissue as diet – the effect on chewer performance

It is generally accepted that elevated CO_2 increases the C/N ratio of plant tissue as more carbon is available for photosynthesis, boosting plant biomass. A positive effect of elevated CO_2 is suggested to be greater for C_3 plants because they lack the CO_2 concentrating mechanism possessed by C_4 plants (Delucia *et al.*, 2012). *Brassica rapa* grown under elevated CO_2 had a 16% reduction in nitrogen content and an associated 20% increase in C/N (Karowe and Grubb, 2011). Since additional fixed carbon effectively dilutes other nutrients, including nitrogen, positive effects on plant growth will be limited in the absence of sufficient additional nitrogen. As a consequence, plants with the ability to fix nitrogen will also do better under elevated atmospheric CO_2 with subsequent consequences for their herbivores.

Because most herbivores are limited by dietary nitrogen (Douglas, 2006), an increase in C/N is predicted to increase the impact of chewing herbivores as they consume more in compensatory feeding and concomitantly have reduced performance on a less ideal diet. The chewing pests *Achaea janata* (the castor semilooper) and *Spodoptera litura* (cutworm) both increased feeding on castor (*Ricinus communis*) grown under elevated CO_2 but, presumably because of the reduced quality of the diet, had reduced relative growth (RGR) rate (Rao *et al.*, 2009). Qualitatively similar results were obtained in a long-term field study on a range of tree species where elevated CO_2 led to a consistent elevation in C/N of 26% and a reduction in leaf N where herbivores (represented by leaf miners) increased consumption, reduced RGR and had reduced population density on a range of oak tree species (Stiling and Cornelissen, 2007). A meta-analysis of 42 studies suggested that the elevation of C/N is a general phenomenon following elevated CO_2 (Zvereva and Kozlov, 2006). The interaction between CO_2 and N availability is clearly demonstrated in N-fixing legumes where elevated CO_2 did not increase C/N ratio and chewing pests displayed increased performance on this material (Karowe and Migliaccio, 2011). Thus an increase in tissue C/N is a generally observed phenomenon of elevated CO_2.

Phloem sap as diet – the effect on sap-feeding herbivores

Chewing herbivores consume the whole plant tissue making the interaction with a host plant relatively simple to analyse compared with sap-feeding pests such as aphids, whiteflies and plant hoppers, whose diet consists mainly of phloem (sieve element) sap. Sieve element sap is not representative of plant material as a whole, being nutritionally imbalanced with an already high C/N ratio and a high osmotic pressure, presenting the herbivore simultaneously with location, nutritional and osmotic challenges (Douglas, 2006). Climate change will have a significant effect on C/N ratio, sieve element osmotic pressure and consequently on sap feeding pests. The sieve element forms the central part of the plant transport system for photosynthates and nitrogen in the form of amino acids (Van Bel, 2003). Although sieve element N is largely in the form of amino acids, it is deficient in many of the essential amino acids required by herbivores. Additionally, phloem composition is regulated by the plant with high sucrose concentrations achieved as the phloem is loaded by source-located sucrose transporters (Kuhn *et al.*, 1999), whereas amino acids are regulated by a range of amino acid transporters such as the amino acid permeases (AAPs; Hunt *et al.*, 2010). Elevated CO_2 and osmoregulation are both predicted to increase plant C/N ratios and so reduce N availability in the sieve element. The generally observed changes in bulk tissue levels of C and N make it reasonable to expect associated changes in supplying sieve elements but there is currently little direct evidence of changes in sieve element composition under elevated CO_2. Sun *et al.* (2009), however, reported reductions in phloem amino acid content in cotton grown under elevated CO_2, suggesting an increase in sieve element C/N.

The importance of N to sap-feeding pests is demonstrated because aphids perform better on plants grown under high nitrogen. For example, *Aphis gossypi* had shorter developmental time on cotton grown under high nitrogen (Hosseini *et al.*, 2010). Similarly, *Rhopalosiphum padi* on barley with supplemental N had a higher rate of reproduction, found the phloem more quickly and fed longer (Ponder *et al.*, 2000). Theories of plant defence predict that an increase in nutrient availability should lead to an increase in herbivore performance (Stamp, 2003). Consistent with this, the aphid *Sitobium avenae* performed better on a range of barley genotypes with supplementary nutrients (Rowntree *et al.*, 2010). A correlation between sucking pest performance and N is, however, often not seen as clearly as for chewing pests. This may not only be partly due to the plants' ability to regulate phloem N in the face of externally varying levels of N, but also to the ability of the aphid to compensate for the non-ideal spectrum of amino acids by upgrading non-essential to essential amino acids through the symbiotic bacterium *Buchnera* (Douglas, 1998). Consistent with this, sieve element amino acid concentration and composition was altered in loss of function mutants of AAPs in *Arabidopsis*; but despite the change in aphid diet there was no significant effect on subsequent performance (Hunt *et al.*, 2010). A similar lack of correlation of aphid performance and phloem N was noted in cotton grown under elevated CO_2, which had lower phloem amino acids but performance of the aphid *A. gossypii* was unchanged (Sun *et al.*, 2009). The role of the symbiont *Buchnera* in compensating for altered diet is confirmed when the bacteria are removed by antibiotic treatment, and an association between phloem N and aphid performance can then be seen (Kemp, 2011).

Perhaps as a result of the influence of the *Buchnera* symbiont there is little or no consistent correlation in the literature between sap feeders and plants grown under elevated CO_2. The aphid *Myzus persicae* increased in abundance at elevated CO_2 despite a lack of change in foliar N (Bezemer *et al.*, 1998). A recent survey of studies on wheat and cotton in China reported that two species of aphid showed increased performance, no effect on the sap feeding whitefly but a detrimental effect on the chewer *Helicoverpa armigera* (Sun *et al.*, 2011). A study of a diverse group of aphids on their respective host plants found no consistent patterns of aphid performance at high CO_2 despite increases in host plant C/N ratio of bulk tissue and concluded that responses were species specific and mediated by changes in aphid behaviour (Hughes and Bazzaz, 2001). Thus any effects of climate change on sap feeding pests through changes in the N aspects of diet may be ameliorated and blurred by the *Buchnera* symbiont, making diet a less attractive target for potential control.

16.3.2 Components of climate change: indirect effects of drought

A major effect of climate change will be altered rainfall patterns and increased drought in many important crop-growing areas. Plants as a group are generally well adapted to dealing with water shortage and, unlike animals where internal homeostasis is the key, plants respond to water deficit by decreasing their water potential to maintain uptake. This strategy requires significant changes in intracellular composition, usually by osmoregulation through active accumulation of solutes (Morgan, 1984). Osmoregulation will have significant effects on plant composition and consequent implications for the herbivores who exploit them as diet. In addition, herbivores will be faced with a diet of increased osmotic pressure, a greater risk of dehydration and so higher cost of maintaining an osmotic internal environment.

Consequence of phloem osmoregulation for sap feeders

The osmotic challenge of the plant as diet is often overlooked, a problem that can be predicted to increase under climate change and may provide a novel target for control. Phloem sap has a high osmotic pressure ranging from 0.6 to over 3.0MPa and varies between and within plants (Downing, 1978; Fisher and Cash-Clark, 2000). Sap-feeding herbivores are efficient osmoregulators because the osmotic pressure of ingested phloem sap is hyperosmotic to haemolymph, whereas haemolymph and egested honeydew are iso-osmotic (Wilkinson *et al.*, 1997). An elevation in sieve element osmotic pressure during drought will elevate the cost of osmoregulation and therefore compound the challenge of increasing C/N ratios.

During drought, phloem osmotic pressure increases to the greatest extent of any plant compartment (Pritchard *et al.*, 2007) with an associated increase in amino acid concentration (Hale *et al.*, 2003) contrasting with a decrease in sieve element N under elevated CO_2 (Sun *et al.*, 2011). Aphids ingested less sap and had reduced performance on droughted plants with higher SE osmotic pressure and phloem nitrogen, highlighting the conflict between phloem N levels and the need for osmoregulation (Hale *et al.*, 2003). The response may be subtle and depend on the level of drought, because aphid performance increased under moderate drought (80% relative water content) but was decreased at higher levels (60% relative water content) (Tariq *et al.*, 2012). Aphids performed equally or better when reared on potato with a lower SE sucrose concentration highlighting the cost of osmoregulation. This study also indicated that responses to changes in phloem composition can be highly species-specific for sap feeders, because *M. persicae* did better on a lower sucrose diet, whereas *Macrosiphum euphorbia* did worse on the same plant material (Pescod *et al.*, 2007).

To deal with the osmotic challenge of ingesting phloem sap, aphids have a range of potential solutions that may present novel targets for control (Douglas *et al.*, 2006). One strategy is rehydration from the xylem (Pompon *et al.*, 2011). Aphids dehydrated for 24 h spent 26% of their time xylem feeding when placed back onto a host plant, whereas those starved for only 1 h displayed no xylem feeding (Spiller *et al.*, 1990). If xylem feeding is important for osmoregulation, and if xylem becomes unfavourable, rehydration by sap feeders will be restricted and performance negatively affected (Daniels *et al.*, 2009).

Reducing the osmotic pressure of ingested phloem sap is an alternative to dilution by xylem feeding. Polymerization of ingested sucrose into oligosaccharides of lower osmotic pressure can make phloem sap iso-osmotic to haemolymph. Under high artificial dietary sucrose concentrations the honeydew oligosaccharides significantly increase (Wilkinson *et al.*, 1997), pea aphid (*Acyrthosiphon pisum*) fed on artificial diets showed sucrose polymerization resulting in a 34% reduction in sucrose concentration of the honeydew (Rhodes *et al.*, 1997). In *A. pisum,* sucrase activity has been localized to the gut distal to the stomach (Ashford *et al.*, 2000). Sucrase inhibitors fed to aphids reduced ability to regulate haemolymph water potential (Karley *et al.*, 2005). Gradients of solutes originating in the phloem have been implicated as cues for phloem location (Hewer *et al.*, 2011) and these are altered by both elevated CO_2 and drought, thus affecting the ability of herbivores to find the phloem. Thus both phloem location and phloem acceptance may be affected by the various components of climate change and may provide novel strategies for controlling sap-feeding pests.

Consequence for leaf chewers

Chewing pests are also affected by plant composition (diet) and acceptance. Lepidopteran larvae feeding on drought-stressed host plants showed differences in performance consistent with changes in diet but also altered feeding preferences, indicating that in the short term host plant acceptance may be as important as plant composition (Gutbrodt *et al.*, 2012). *Spodoptera littoralis* (cotton leafworm) feeding on the drought-stressed grass species *Holcus lanatus* took longer to develop but had higher emergence rates, whereas the plant tissue had reduced N but increased carbohydrate levels (Walter *et al.*, 2012). The elevated carbohydrate levels in the previous study are consistent with plant osmo-regulation under drought.

Components of climate change therefore affect many aspects of herbivore interaction with their host plant, from host location and acceptance, to the quality of the diet that drives the insect's growth and reproduction. For sap feeders in particular an increase in the concentration of their diet may provide an additional challenge. Understanding the molecular and physiological basis of the interaction of a herbivore with its host may provide additional routes to generating resistant crops additional to conventional breeding.

16.4 Genetic Resources Important for Increasing Pest/Disease Resistance in Crop Plants

Breeding new varieties of crops with resistance against pest insects and diseases depends on the availability of novel sources of genes and alleles. Wild plant species closely related to crops (crop wild relatives) and traditional, locally adapted landraces are vital sources of such genes. Within these resources there is considerable genetic variation for resistance traits that has most likely been maintained by differential selection pressures acting in the natural environment (Anderson and Mitchell-Olds, 2011). Besides variation among wild plant populations, artificial selection during domestication has also resulted in variation in resistance (Diamond, 2002; Broekgaarden *et al.*, 2012). Natural variation among crop wild relatives, landraces or modern cultivars can be explored as sources of resistance, which can be introduced into susceptible crop plants (Broekgaarden *et al.*, 2011). For this, fast and sophisticated methods to screen large collections of germplasm are needed.

The Focused Identification of Germplasm Strategy (FIGS) is an approach to target accessions more likely to possess specific traits sought by breeders (Mackay and Street, 2004). The idea behind this strategy is that plants survive in environments with a high selection pressure because they developed adaptive traits to cope with this pressure. Environments where a particular trait is advantageous are expected to harbour populations with high levels of resistance. This method is dependent on the existence of known trait evaluation data in addition to climate and environmental data. The FIGS approach has so far been successfully applied to identify resistance in wheat to stem rust (Endresen *et al.*, 2011; Bari *et al.*, 2012), powdery mildew (Bhullar *et al.*, 2009), Russian wheat aphids (El Bouhssini *et al.*, 2011) and Sunn pest (El Bouhssini *et al.*, 2009). It should be noted that the FIGS approach can only be applied successfully when detailed information on the environmental conditions or past disease/insect pressure is available. Such information is, however, often lacking.

Screening large germplasm collections is essential to identify effective sources of resistance. Nowadays this phenotyping is the limiting step in breeding and there is a demand for high-throughput phenotyping methods. Most phenotyping studies involve tedious counting thereby limiting the number of plants that can be handled in a specific time period. One way to overcome this is to use image processing techniques by using the visible spectrum to detect damage caused by insect feeding (Kloth *et al.*, 2012). Such image analyses also have the potential to measure insect performance parameters. In addition, it is also possible to determine resistance levels in an indirect manner. By exploiting the ability of sap feeding insects to transmit plant viruses, resistance can be related to the success of virus transmission by measuring virus concentrations in the plant after infestation by virus-infected insects (Chen *et al.*, 2012).

Once identified, the molecular and

genetic basis of the source of resistance should be elucidated. Proteomic and transcriptomic analysis is leading to an emerging picture of changes in gene expression induced by herbivore feeding. For sucking pests, one strategy is to examine the phloem-specific expression of proteins or genes and to see if putative functions can be matched to defence. Enriched protein libraries for the sieve element are often enriched with annotations suggesting biotic interactions (Barnes *et al.*, 2004; Kehr, 2006) providing potential targets for further study. A related strategy is to examine gene lists from phloem-specific (or enriched) cDNA libraries. Useful candidate genes are often seen including protease inhibitors, amino acid transporters and aquaporins (Divol *et al.*, 2005; Doering-Saad *et al.*, 2006). Such phloem-focused studies have so far only looked at constitutive expression of phloem-specific genes. Changes in gene expression in response to herbivore pressure have so far only been sought in a whole-tissue approach. One such study in *Arabidopisis* fed on by the aphid *Myzus persicae* identified a range of induced genes, including those involved in cell wall modification, oxidative stress, glucosyl transferase and lectin production (Couldridge *et al.*, 2007). Although such lists may well be species specific, some insight into the generality of the responses may be obtained by comparing changes in expression between different plant and herbivore/pathogen combinations. The sucking pest whitefly feeding on *Arabidopsis* also induced genes in common with those induced by the aphid including glucosyl transferase and the SA/JA pathway regulator PAD4 (Kempema *et al.*, 2007). There were also some pathogen-induced genes common between *M. persicae* and the fungus *Alternaria brassicicola* (Saskia *et al.*, 2003) and the bacterium *Pseudomonas syringae* (Thilmony *et al.*, 2006).

A challenge in identifying genetic resources to contribute to herbivore control generally is to determine genes in common with a large number of interactions, while continued study of individual plant–herbivore interactions may identify specific individual solutions. In both cases progress is hampered by the diversity of experimental systems, treatments and approaches making direct comparisons of the range of studies difficult.

Constitutive gene expression may also play an important role in identifying genetic resources for breeding programmes. A transcriptomics analysis of constitutive gene expression across 12 rice genotypes was interrogated with resistance data produced by Electrical Penetration Graph (EPG) analysis, in which the insect and the plant are made part of an electric circuit to monitor insect feeding behaviour, of brown planthopper (BPH) feeding (Ghaffar *et al.*, 2011) and revealed 239 genes contributing to resistance/susceptibility (Ghaffar, 2012). The annotation of these candidate genes allowed a range of hypotheses about potential gene function to be proposed, with putative modes of action in both phloem location and phloem acceptance. Similar data sets across genotypes with diverse resistance can be used to inform models identifying novel plant genetic resources, a strategy that may be more efficient than the study of individual candidate genes.

Only in a limited number of cases resistance genes towards pest insects have been identified. These include quantitative trait loci (QTL) type of resistance as well as R-gene based resistance. Examples of the QTL type of resistance are specific secondary metabolites (acyl sugars, methyl ketones), which are often produced in glandular trichomes. In tomato such resistances are active against a range of insects and are durable. The R-gene based resistances are based on highly specific interactions, often between a particular biotype of pest insect and an R-gene present in the plant. Such resistances are often not durable, as was shown for Hessian fly, Russian wheat aphid and BPH resistance (Broekgaarden *et al.*, 2011).

16.5 Strategies for Introduction of Pest and Disease Resistance

Nowadays there is a plethora of tools available that facilitate resistance breeding. These include both traditional and biotechnological solutions as well as tools that make use of biotechnology but where the end product is not genetically modified. As mentioned, novel genes conferring resistance to pests and diseases are often not present in a crop species itself but need to be introduced from its wild relatives. Traditional breeding with crop wild relatives is not always an easy task because of linkage drag, and negative traits must be removed before varieties with improved resistances can be put into the market. To accomplish this, use can be made of marker-assisted selection (Mohan *et al.*, 1997; Collard and Mackill, 2008) which is a well-established and widely used strategy for crop improvement. It has proven to be effective in backcrossing, gene pyramiding and recurrent parent selection, among others (Huang *et al.*, 1997; Moose and Mumm, 2008). With the development of novel genomics tools like next generation sequencing (Tang *et al.*, 2006; Imelfort *et al.*, 2009) and marker platforms that allow the analysis of hundreds of thousands of markers in a large number of genotype samples (Anithakumari *et al.*, 2010; Shirasawa *et al.*, 2010) it is now feasible in almost any crop.

For the production of hybrid varieties a new strategy has been developed, which was termed 'reverse breeding' (Dirks *et al.*, 2009; Wijnker *et al.*, 2012). In a traditional breeding process hybrid seeds are produced by crossing two inbred lines that contain favourable complementary characteristics. In reverse breeding, a hybrid plant with favourable characteristics is selected from a population of offspring derived from a cross. In that plant meiotic recombination is suppressed, for instance by silencing the DMC1 gene (Wijnker *et al.*, 2012), which results in the segregation of non-recombined parental chromosomes. Next, doubled haploid plants are produced from unfertilized ovules or from microspore and anther culture (Jain *et al.*, 1996), which can be genotyped to select those plants that have complementary sets of chromosomes to reconstitute the original hybrid (Dirks *et al.*, 2009). The whole procedure has been demonstrated using *Arabidopsis thaliana* (Wijnker *et al.*, 2012) and can, in principle, be applied to any crop, although its efficiency drops with increasing chromosome number.

There are several mutation-based techniques that may be used to create insect-resistant plant material. For several crops TILLING (Targeting Induced Local Lesions in Genomes) populations have been produced using EMS (ethyl methanesulfonate) (Gady *et al.*, 2009) that can be evaluated for mutations conferring increased resistance. Similarly, activation tag libraries (Marsch-Martinez *et al.*, 2002) have been used and mutants with increased aphid resistance identified (Levy *et al.*, 2005; Chen *et al.*, 2012). An activation tag library effectively produces gain of function mutants; e.g. genes are expressed at a higher level in a different tissue, or at a different stage than in wild-type plants. Next to these non-targeted mutation techniques, strategies have been developed to create, in a highly specific way, mutations in genes that may increase the resistance towards insects. For these techniques, candidate genes involved in insect resistance have to be known. Techniques to direct these mutations include oligonucleotide directed mutagenesis (Oh and May, 2001; Dekker *et al.*, 2003) and the use of zinc-finger nucleases (Lloyd *et al.*, 2005; Wright *et al.*, 2005).

Genetically modified organism (GMO)-based approaches have shown to be very effective in the control of insects as well, and examples include the use of Bt genes and RNAi constructs. The use of Bt genes to combat insects in plants had started in the mid-1980s and has proven to be very successful. The total global area on which Bt crops were grown in 2009 was more than 50 million hectares (Sanahuja *et al.*, 2011). The use of RNAi constructs is more recent and still in the experimental stage, although results look promising (Baum *et al.*, 2007; Mao *et al.*, 2007; Buranda and Hunter, 2012; Zhang *et al.*, 2013). In this approach a double-stranded (ds)RNA molecule is used to silence an essential gene in the pest insect. Such a construct can be delivered to the insect by feeding on a transgenic plant producing it. Which delivery method is the most effective is, however, still the subject of many studies (Huvenne and Smagghe,

2010; Yu *et al.*, 2013). When the dsRNA molecule is applied transiently it can be considered as a highly specific insecticide.

16.6 Conclusions

Climate change can have profound effects on plant–insect interactions, potentially disturbing interactions directly and indirectly. It can change population development and distribution patterns of the pest insect as well as of its natural enemies. Society has to deal with this and adapt plant varieties by introducing resistance genes. Such genes are often available in crop wild relatives. Also the means to introduce them are available. A developing and increasingly detailed understanding of the mechanisms that underlie resistance and susceptibility can highlight new targets for control (e.g. osmoregulation and the JA/SA pathways). In addition, such information can indicate the level at which studies must be pursued. For example, although sucking pests show no clear patterns of response to climate change as a feeding guild, general responses are emerging in chewing pests. Thus the former requires species-specific solutions, whereas the latter guild may be controlled by more overarching measures. Guided by this information, a range of biotechnological tools is now being developed that may be used to increase food security more sustainably under a changing climate by reducing the need to use insecticides. The availability of and access to genetic variation in genebanks are currently significant components of the response to insect pests and their importance will increase in the coming years.

16.7 Acknowledgements

J.P. and B.V would like to acknowledge funding from the EU PGR Secure project and C.B. acknowledges funding from the STW Perspectief Program: Learning from Nature to protect crops (LFN).

References

Anderson, J.T. and Mitchell-Olds, T. (2011) Evolutionary ecology of plant defences. Ecological genetics and genomics of plant defences: evidence and approaches. *Functional Ecology* 25, 312–324.

Anithakumari, A.M., Tang, J., van Eck, H.J., Visser, R.G.F., Lounissen, J.A.M. *et al.* (2010) A pipeline for high throughput detection and mapping of SNPs from EST databases. *Molecular Breeding* 26, 65–75.

Aragon, P. and Lobo, J.M. (2012) Predicted effect of climate change on the invasibility and distribution of the Western corn root-worm. *Agricultural and Forest Entomology* 14, 13–18.

Ashford, D.A., Smith, W.A. and Douglas, A.E. (2000) Living on a high sugar diet: the fate of sucrose ingested by a phloem-feeding insect, the pea aphid *Acyrthosiphon pisum. Journal of Insect Physiology* 46, 335–341.

Bale, J.S. (1980) Insect cold hardiness – freezing and supercooling – an ecophysiological perspective. *Journal of Insect Physiology* 33, 899–908.

Bale, J.S. and Hayward, S.A.L. (2010) Insect overwintering in a changing climate. *Journal of Experimental Biology* 213, 980–994.

Ballhorn, D.J., Reisdorff, C. and Pfanz, H. (2011) Quantitative effects of enhanced CO_2 on jasmonic acid induced plant volatiles of lima bean (*Phaseolus lunatus* L.) *Journal of Applied Botany and Food Quality* 84, 65–71.

Bari, A., Street, K., Mackay, M., Endresen, D.T.F., De Pauw, E. *et al.* (2012) Focused identification of germplasm strategy (FIGS) detects wheat stem rust resistance linked to environmental variables. *Genetic Resources and Crop Evolution* 59, 1465–1481.

Barnes, A., Bale, J.S., Constantinidou, C., Ashton, P., Jones, A.C. and Pritchard, J. (2004) Determining protein identity from sieve element sap in *Ricinus communis* L. by quadrupole time of flight (Q-TOF) mass spectrometry. *Journal of Experimental Botany* 55, 1473–1481.

Baum, J.A., Bogaert, T., Clinton, W., Heck, G.R., Feldmann, P. *et al.* (2007) Control of coleopteran insect pests through RNA interference. *Nature Biotechnology* 25, 1322–1326.

Bezemer, T.M., Hefin Jones, T. and Knight, K.J. (1998) Long-term effects of elevated CO_2 and temperature on populations of the peach potato aphid *Myzus persicae* and its parasitoid *Aphidius matricariae. Oecologia* 116, 128–135.

Bhullar, N.K., Street, K., Mackay, M., Yahiaoui, N. and Keller, B. (2009) Unlocking wheat genetic resources for the molecular identification of previously undescribed functional alleles at the Pm3 resistance locus. *Proceedings of the National Academy of Sciences USA* 106, 9519–9524.

Bowler, K. and Terblanche, J.S. (2008) Insect thermal tolerance: what is the role of ontogeny, ageing and senescence? *Biological Reviews* 83, 339–355.

Broekgaarden, C., Snoeren, T.A.L., Dicke, M. and Vosman, B. (2011) Exploring natural variation to identify insect-resistance genes. *Plant Biotechnology Journal* 9, 819–825.

Broekgaarden, C., Riviere, P., Steenhuis, G., Cuenca, M., Kos, M. *et al.* (2012) Phloem-specific resistance in *Brassica oleracea* against the whitefly *Aleyrodes proletella*. *Entomologia Experimentalis et Applicata* 142, 153–164.

Buranda, J.P and Hunter, W.B. (2012) RNAi: Future in insect management. *Journal of Invertebrate Pathology* http://dx.doi.org/10.1016/j.jip.2012.07.012.

Casteel, C.L., O'Neill, B.F., Zavala, J.A., Bilgin, D.D., Berenbaum, M.R. *et al.* (2008) Transcriptional profiling reveals elevated CO_2 and elevated O3 alter resistance of soybean (*Glycine max*) to Japanese beetles (*Popillia japonica*). *Plant, Cell and Environment* 31, 419–434.

Casteel, C.L., Segal, L.M., Niziolek, O.K., Berenbaum, M.R. and DeLucia, E.H. (2012) Elevated carbon dioxide increases salicylic acid in *Glycine max. Environmental Entomology* 41, 1435–1442.

Chen, X., Vosman, B., Visser, R.G.F., van der Vlugt, R.A.A. and Broekgaarden, C. (2012) High throughout phenotyping for aphid resistance in large plant collections. *Plant Methods* 8, 33.

Coley, P.D. (2000) Possible effects of climate change on plant/herbivore interactions in moist tropical forests. *Climatic Change* 39, 455–572.

Collard, B.C.Y. and Mackill, D.J. (2008) Marker assisted selection: an approach for precision plant breeding in the twenty-first century. *Philosophical Transactions of the Royal Society B: Biological Sciences* 363, 557–572.

Couldridge, C.E.C., Newbury, H.J., Ford-Lloyd, B., Bale, J.S. and Pritchard, J. (2007) Exploring plant responses to aphid feeding using a full *Arabidopsis* microarray reveals a small number of genes with significantly altered expression. *Bulletin of Entomological Research* 97, 523–532.

Daniels, M., Bale, J.S., Newbury, H.J., Lind, R.J. and Pritchard, J. (2009) A sublethal dose of thiamethoxam causes a reduction in xylem feeding by the bird cherry-oat aphid (*Rhopalosiphum padi*), which is associated with dehydration and reduced performance. *Journal of Insect Physiology* 55, 758–765.

Dekker, M., Brouwers, C. and Te Riele, H. (2003) Targeted gene modification in mismatch-repair-deficient embryonic stem cells by single-stranded DNA oligonucleotides. *Nucleic Acids Research* 31, e27.

DeLucia, E.H., Nabity, P.D., Zavala, J.A. and Berenbaum, M.A. (2012) Climate change: resetting plant–insect interactions. *Plant Physiology* 160, 1677–1685.

Diamond, J. (2002) Evolution, consequences and future of plant and animal domestication. *Nature* 418, 700–707.

Dirks, R., van Dun, K., de Snoo, C.B., van den Berg, M., Lelivelt, C.L. *et al.* (2009) Reverse breeding: a novel breeding approach based on engineered meiosis. *Plant Biotechnology Journal* 7, 837–845.

Divol, F., Vilaine, F., Thibivilliers, S., Amselem, J., Palauqui, J.C. *et al.* (2005) Systemic response to aphid infestation by *Myzus persicae* in the phloem of *Apium graveolens. Plant Molecular Biology* 57, 517–540.

Doering-Saad, C., Newbury, H.J., Couldridge, C.E., Bale, J.S. and Pritchard, J. (2006) A phloem-enriched cDNA library from *Ricinus*: insights into phloem function. *Journal of Experimental Botany* 55, 3183–3193.

Douglas, A.E. (1998) Nutritional interactions in insect–microbial symbioses: aphids and their symbiotic bacteria. *Buchnera* Annual *Review of Entomology* 43, 17–37.

Douglas, A.E. (2006) Phloem-sap feeding by animals: problems and solutions. *Journal of Experimental Botany* 57, 747–754.

Douglas, A.E., Price, D.R.G., Minto, L.B., Jones, E., Pescod, K.V. *et al.* (2006) Sweet problems: insect traits defining the limits to dietary sugar utilisation by the pea aphid, *Acyrthosiphon pisum. Journal of Experimental Biology* 209, 1395–1403.

Downing, N. (1978) Measurements of the osmotic concentrations of stylet sap, haemolymph and honeydew from an aphid under osmotic stress. *Journal of Experimental Biology* 77, 247–250.

El Bouhssini, M.E., Street, K., Joubi, A., Ibrahim, Z. and Rihawi, F. (2009) Sources of wheat resistance to Sunn pest, *Eurygaster integriceps* Puton, in Syria. *Genetic Resources and Crop Evolution* 56, 1065–1069.

El Bouhssini, M.E., Street, K., Amri, A., Mackay, M., Ogbonnaya, F.C. *et al.* (2011) Sources of resistance in bread wheat to Russian wheat aphid (*Diuraphis noxia*) in Syria identified using the focused identification of germplasm strategy (FIGS). *Plant Breeding* 130, 96–97.

Endresen, D.T.F., Street, K., Mackay, M., Bari, A., Amri, A. *et al.* (2011) Sources of resistance to stem rust (Ug99) in bread wheat and durum wheat identified using focused identification of germplasm strategy. *Crop Science* 52, 764.

Fatinou, A.A. and Kagkou, E.A. (2000) Effect of thermoperiod on diapauses induction of *Sesamia nonagriodes* (Lepidoptera-Noctuidae). *Environmental Entomology* 29, 489–494.

Fisher, D.B. and Cash-Clark, C.E. (2000) Gradients in water potential and turgor pressure along the translocation pathway during grain filling in normally watered and water-stressed wheat plants. *Plant Physiology* 123, 139–147.

Fu, X., Ye, L., Kang, L. and Ge, F. (2010) Elevated CO_2 shifts the focus of tobacco plant defences from cucumber mosaic virus to the green peach aphid. *Plant Cell and Environment* 33, 2056–2064.

Gady, A.L.F., Hermans, F.W.K., Van De Wal, M.H.B.J., Van Loo, E.N., Visser R.G.F. *et al.* (2009) Implementation of two high through-put techniques in a novel application: detecting point mutations in large EMS mutated plant populations. *Plant Methods* 5, 13.

Ghaffar, M.B.A.B. (2012) Transcriptomic analysis of phloem-feeding insect resistance in rice germplasm. PhD thesis, University of Birmingham, Birmingham, UK.

Ghaffar, M.B.A.B., Pritchard, J. and Ford-Lloyd, B. (2011) Brown planthopper (*N. lugens* Stal) feeding behaviour on rice germplasm as an indicator of resistance. PLOS ONE 6, e22137.

Glazebrook, J., Chen, W.J., Estes, B., Chang, H.S., Nawrath, C. *et al.* (2003) Topology of the network integrating salicylate and jasmonate signal transduction derived from global expression phenotyping. *Plant Journal* 34, 217–228.

Guo, H., Sun, Y., Ren, Q., Zhu-Salzman, K., Kang, L. *et al.* (2012) Elevated CO_2 reduces the resistance and tolerance of tomato plants to *Helicoverpa armigera* by suppressing the JA signaling pathway. *PLoS One* 7, e41426.

Gutbrodt, B., Dorn, S. and Mody, K. (2012) Drought stress affects constitutive but not induced herbivore resistance in apple plants. *Arthropod-Plant Interactions* 6, 171–179.

Hale, B.K., Bale, J.S., Pritchard, J., Masters, G.J. and Brown, V.K. (2003) Effects of host plant drought stress on the performance of the bird cherry-oat aphid, *Rhopalosiphum padi* (L.): a mechanistic analysis. *Ecological Entomology* 28, 666–677.

Hewer, A., Becker, A. and van Bel, A.J.E. (2011) An aphid's Odyssey – the cortical quest for the vascular bundle. *Journal of Experimental Biology* 22, 3868–3879.

Himanen, S.J., Nissinen, A., Auriola, S., Poppy, G.M., Stewart, C.N. *et al.* (2008) Constitutive and herbivore-inducible glucosinolate concentrations in oilseed rape (*Brassica napus*) leaves are not affected by Bt Cry1Ac insertion but change under elevated atmospheric CO_2 and O-3. *Planta* 227, 427–437.

Hosseini, M., Ashouri, A., Enkegaard, A., Goldansaz, S.H., Mahalati, M.N. *et al.* (2010) Performance and population growth rate of the cotton aphid, and associated yield losses in cucumber, under different nitrogen fertilization regimes. *International Journal of Pest Management* 56, 127–135.

Huang, L., Ren, Q., Sun, Y., Ye, L., Cao, H. *et al.* (2012) Lower incidence and severity of tomato virus in elevated CO_2 is accompanied by modulated plant induced defence in tomato. *Plant Biology* 6 905–913.

Huang, N., Angeles, E.R., Domingo, J., Magpantay, G., Singh, S. *et al.* (1997) Pyramiding of bacterial blight resistance genes in rice: marker-assisted selection using RFLP and PCR. *Theoretical and Applied Genetics* 95, 313–320.

Hughes, L. and Bazzaz, F.A. (2001) Effects of elevated CO_2 on five plant-aphid interactions. *Entomologia Experimentalis et Applicata* 99, 87–96.

Hunt, E., Gattolin, S., Newbury, H.J., Bale, J.S., Tseng, H.-M. *et al.* (2010) A mutation in amino acid permease *AAP6* reduces the amino acid content of the *Arabidopsis* sieve elements but leaves aphid herbivores unaffected. *Journal of Experimental Botany* 61, 55–64.

Huvenne, H. and Smagghe, G. (2010) Mechanisms of dsRNA uptake in insects and potential of RNAi for pest control: a review. *Journal of Insect Physiology* 56, 227–235.

Imelfort, M., Duran, C., Batley, J. and Edwards, D. (2009) Discovering genetic polymorphisms in next-generation sequencing data: review article. *Plant Biotechnology Journal* 7, 312–317.

Jain, S., Sopory, S.K. and Veilleux, R.E. (eds) (1996) In Vitro *Haploid Production in Higher Plants. Volume 1: Fundamental Aspects and Methods.* Kluwer Academic Publishers, Dordrecht, the Netherlands.

Kaloshian, I. and Walling, L.L. (2005) Hemipterans as plant pathogens. *Annual Review of Phytopathology* 43, 491–521.

Karley, A.J., Ashford, D.A., Minto, L.M., Pritchard, J. and Douglas, A.E. (2005) The significance of gut sucrase activity for osmoregulation in the pea aphid, *Acyrthosiphon pisum. Journal of Insect Physiology* 51, 1313–1319.

Karowe, D.N. and Grubb, C. (2011) Elevated CO_2 increases constitutive phenolics and trichomes, but decreases inducibility of phenolics in *Brassica rapa* (Brassicaceae). *Journal of Chemical Ecology* 37, 1332–1340.

Karowe, D.N. and Migliaccio, A. (2011) Performance of the legume-feeding herbivore, *Colias philod* (Lepidoptera: Pieridae) is not affected by elevated CO_2. *Arthropod–Plant Interactions* 5, 107–114.

Kawazu, K., Mochizuki, A., Sato, Y., Sugeno, W., Murata, M. *et al.* (2012) Different expression profiles of jasmonic acid and salicylic acid inducible genes in the tomato plant against herbivores with various feeding modes. *Arthropod–Plant Interactions* 6, 221–230.

Kehr, J. (2006) Phloem sap proteins: their identities and potential roles in the interaction between plants and phloem-feeding insects. *Journal of Experimental Botany* 57, 767–774.

Kemp, H. (2011) Investigating the effect of plant amino acid transporters AtAAP1 and AtAAP2 on aphid–plant interactions. Ph.D. thesis, University of Birmingham, Birmingham, UK.

Kempema, L.A., Cui, X., Holzer, F.M. and Walling, L.L. (2007) Arabidopsis transcriptome changes in response to phloem-feeding silverleaf whitefly nymphs. Similarities and distinctions in responses to aphids. *Plant Physiology* 143, 849–865.

Kloth, K.J., Thoen, M.P.M., Bouwmeester, H.J., Jongsma, M.A. and Dicke, M. (2012) Association mapping of plant resistance to insects. *Trends in Plant Science* 17, 311–319.

Kuhn, C., Barker, L., Burkle, L. and Frommer, W.B. (1999) Update on sucrose transport in higher plants. *Journal of Experimental Botany* 50, 935–953.

Levy, M., Wang, Q., Kaspi, R., Parrella, M.P. and Abel, S. (2005) Arabidopsis IQD1, a novel calmodulin-binding nuclear protein, stimulates glucosinolate accumulation and plant defense. *Plant Journal* 43, 79–96.

Lloyd, A., Plaisier, C.L., Carroll, D. and Drews, G.N. (2005) Targeted mutagenesis using zinc-finger nucleases in Arabidopsis. *Proceedings of the National Academy of Sciences USA* 102, 2232–2237.

Mackay, M.C. and Street, K. (2004) Focused identification of germplasm strategy—FIGS. In: Black, C.K., Panozzo, J.F. and Rebetzke, G.J. (eds) *Proceedings of the 54th Australian Cereal Chemistry Conference and the 11th Wheat Breeders' Assembly.* Royal Australian Chemical Institute, Melbourne, Australia, pp 138–141.

Mao, Y.B., Cai, W.J., Wang, J.W., Hong, G.J., Tao, X.Y. *et al.* (2007) Silencing a cotton bollworm P450 monooxygenase gene by plant-mediated RNAi impairs larval tolerance of gossypol. *Nature Biotechnology* 25, 1307–1313.

Marsch-Martinez, N., Greco, R., Van Arkel, G., Herrera-Estrella, L. and Pereira, A. (2002) Activation tagging using the En-I maize transposon system in *Arabidopsis. Plant Physiology* 129, 1544–1556.

McWatters, H.G. and Saunders, D.S. (1998) Maternal temperature has different effects on the photoperiodic response and duration of larval diapause in blow fly (*Calliphora vicina*) strains collected at two latitudes. *Physiological Entomology* 23, 369–375.

Mohan, M., Nair, S., Bhagwat, A., Krishna, T.G., Yano, M. *et al.* (1997) Genome mapping, molecular markers and marker-assisted selection in crop plants. *Molecular Breeding* 3, 87–103.

Moose, S.P. and Mumm, R.H. (2008) Molecular plant breeding as the foundation for 21st century crop improvement. *Plant Physiology* 147, 969–977.

Morgan, J.M. (1984) Osmoregulation and water-stress in higher-plants. *Annual Review of Plant Physiology and Plant Molecular Biology* 35, 299–319.

Newton, A.C., Torrance, L., Holden, N., Toth I.K.,Cooke, D.E.L. *et al.* (2012) Climate change and defense against pathogens in plants. In: Sariaslani, S. and Gadd, G.M. (eds) *Advances in Applied Microbiology* 81, 89–132.

Oh, T.J. and May, G.D. (2001) Oligonucleotide-directed plant gene targeting. *Current Opinion in Biotechnology* 12, 169–172.

O'Neill, B.F., Zangerl, A.R., Dermody, O., Bilgin, D.D., Casteel, C.L. *et al.* (2010) Impact of elevated levels of atmospheric CO_2 and herbivory on flavonoids of soybean (*Glycine max* Linnaeus). *Chemical Ecology* 36, 35–45.

Pandey, S.P. and Baldwin, I.T. (2007) RNA-directed RNA polymerase 1 (RdR1) mediates the resistance of *Nicotiana attenuata* to herbivore attack in nature. *Plant Journal* 50, 40–53.

Pescod, K.V., Quick, W.P. and Douglas, A.E. (2007) Aphid responses to plants with genetically manipulated phloem nutrient levels. *Physiological Entomology* 32, 253–258.

Piyaphongkul, J., Pritchard, J. and Bale, J. (2012) Can tropical insects stand the heat? A case study with the brown planthopper *Nilaparvata lugens* (Stål). PLoS ONE 7(1), e29409. doi:10.1371/journal.pone.0029409.

Pompon, J., Quiring, D., Goyer, C., Giordanengo, P. and Pelletier, Y. (2011) A phloem-sap feeder mixes phloem and xylem sap to regulate osmotic potential. *Journal of Insect Physiology* 57, 1317–1322.

Ponder, K.L., Pritchard, J., Harrington, R. and Bale, J.S. (2000) Difficulties in location and acceptance of phloem sap combined with reduced concentration of phloem amino acids explain lowered performance of the aphid *Rhopalosiphum padi* on nitrogen deficient barley (*Hordeum vulgare*) seedlings. *Entomologia Experimentalis et Applicata* 97, 203–210.

Pritchard, J., Griffiths, B. and Hunt, E.J. (2007) Can the plant-mediated impacts on aphids of elevated CO_2 and drought be predicted? *Global Change Biology* 13, 1616–1629.

Rao, M.S., Srinivas, K., Vanaja, M., Rao, G.G.S.N., Venkateswarlu, B. *et al.* (2009) Host plant (*Ricinus communis* Linn.) mediated effects of elevated CO_2 on growth performance of two insect folivores. *Current Science* 97, 1047–1054.

Rhodes, J.D., Croghan, P.C. and Dixon, A.F.G. (1997) Dietary sucrose and oligosaccharide synthesis in relation to osmoregulation in the pea aphid, *Acyrthosiphon pisum*. *Physiological Entomology* 22, 373–379.

Rowntree, J.K., McVennon, A. and Preziosi, R.F. (2010) Plant genotype mediates the effects of nutrients on aphids. *Oecologia* 163, 675–679.

Sanahuja, G., Banakar, R., Twyman, R.M., Capell, T. and Christou, P. (2011) *Bacillus thuringiensis*: a century of research, development and commercial applications. *Plant Biotechnology Journal* 9, 283–300.

Saskia, C.M. van Wees, Chang, H.S., Zhu, T. and Glazebrook, J. (2003) Characterization of the early response of Arabidopsis to *Alternaria brassicicola* infection using expression profiling. *Plant Physiology*, 132, 606–617.

Schroeder, J.I., Kwak, J.M. and Allen, G.J. (2001) Guard cell abscisic acid signalling and engineering drought hardiness in plants. *Nature* 410, 327–330.

Shirasawa, K., Isobe, S., Hirakawa, H., Asamizu, E., Fukuoka, H. *et al.* (2010) SNP discovery and linkage map construction in cultivated tomato. *DNA Research* 17, 381–391.

Singer, M.C. and Parmesan, C. (2010) Phenological asynchrony between herbivorous insects and their hosts: signal of climate change or pre-existing adaptive strategy? *Philosophical Transactions of the Royal Society B – Biological Sciences* 365, 3161–3176.

Spiller, N.J., Koenders, L. and Tjallingii, W.F. (1990) Xylem ingestion by aphids – a strategy for maintaining water-balance. *Entomologia Experimentalis et Applicata* 55, 101–104.

Stamp, N. (2003) Out of the quagmire of plant defense hypotheses. *Quarterly Review of Biology* 78, 23–55.

Stiling, P. and Cornelissen, T. (2007) How does elevated carbon dioxide (CO_2) affect plant–herbivore interactions? A field experiment and meta-analysis of CO_2-mediated changes on plant chemistry and herbivore performance. *Global Change Biology* 13, 1823–1842.

Stout, M.J., Workman, K.V., Bostock, R.M and Duffey, S.S. (1998) Specificity of induced resistance in the tomato, *Lycopersicon esculentum*. *Oecologia* 113, 74–81.

Sun, Y.C., Jing, B.B. and Ge, F. (2009) Response of amino acid changes in *Aphis gossypii* (Glover) to elevated CO_2 levels. *Journal of Applied Entomology*.133, 189–197.

Sun, Y.C., Yin, J., Chen, F.J., Wu, G. and Ge, F. (2011) How does atmospheric elevated CO_2 affect crop pests and their natural enemies? Case histories from China. *Insect Science* 18, 393–400.

Sun, Y., Yin, J., Cao, H., Li. C.Y., Kang, I. *et al.* (2012) Elevated CO_2 influences nematode-Induced defense responses of tomato genotypes differing in the JA pathway. *Plos One* 6, e19751.

Tang, J., Vosman, B., Voorrips, R.E., van der Linden, C.G. and Leunissen, J.A.M. (2006) QualitySNP: a pipeline for detecting single nucleotide polymorphisms and insertions/deletions in EST data from diploid and polyploid species. *BMC Bioinformatics* 7, 438.

Tariq, M., Wright, D.J., Rossiter, J.T. and Staley, J.T. (2012) Aphids in a changing world: testing the plant stress, plant vigour and pulsed stress hypotheses. *Agricultural and Forest Entomology* 14, 177–185.

Thilmony, R., Underwood, W. and He, S.Y. (2006) Genome-wide transcriptional analysis of the *Arabidopsis thaliana* interaction with the plant pathogen *Pseudomonas syringae* pv. tomato DC3000 and the human pathogen *Escherichia coli* O157:H7. *Plant Journal* 46, 34–53.

Thompson, G.A. and Goggin, F.L. (2006) Transcriptomics and functional genomics of plant defence induction by phloem-feeding insects. *Journal of Experimental Botany* 57, 755–766.

Van Bel, A.J.E. (2003) The phloem, a miracle of ingenuity. *Plant, Cell and Environment* 26, 125–149.

Vannette, R.L. and Hunter, M.D. (2011) Genetic variation in expression of defense phenotype may mediate evolutionary adaptation of *Asclepias syriaca* to elevated CO_2. *Global Change Biology* 17, 1277–1288.

Walter, J., Hein, R., Auge, H., Beierkuhnlein, C., Loffler, S. *et al.* (2012) How do extreme drought and plant community composition affect host plant metabolites and herbivore performance? *Arthropod–Plant Interactions* 6, 15–25.

Ward, D., Shrestha, M.K. and Golan-Goldhirsh, A. (2012) Evolution and ecology meet molecular genetics: adaptive phenotypic plasticity in two isolated Negev desert populations of *Acacia raddiana* at either end of a rainfall gradient. *Annals of Botany* 109, 247–255.

Wilkinson, T.L., Ashford, D.A., Pritchard, J. and Douglas, A.E. (1997) Honeydew sugars and osmoregulation in the pea aphid *Acyrthosiphon pisum. Journal of Experimental Biology* 200, 2137–2143.

Wijnker, E., van Dun, K., de Snoo, C.B., Lelivelt, C.L.C., Keurentjes, J.J.B. *et al.* (2012) Reverse breeding in *Arabidopsis thaliana* generates homozygous parental lines from a heterozygous plant. *Nature Genetics* 44, 467–470.

Wright, D.A., Townsend, J.A., Winfrey, R.J., Irwin, P.A., Rajagopal, J. *et al.* (2005) High-frequency homologous recombination in plants mediated by zinc-finger nucleases. *Plant Journal* 44, 693–705.

Yu, N., Christiaens, O., Liu, J., Niu, J., Cappelle, K. *et al.* (2013) Delivery of dsRNA for RNAi in insects: an overview and future directions. *Insect Science* 20, 4–14.

Zhang, H., Li, H.-C. and Miao X.-X. (2013) Feasibility, limitation and possible solutions of RNAi-based technology for insect pest control. *Insect Science* 20, 15–30.

Zvereva, E.L. and Kozlov, M.V. (2006) Consequences of simultaneous elevation of carbon dioxide and temperature for plant–herbivore interactions: a metaanalysis. *Global Change Biology* 12, 27–41.

Index

Aerosols, 40–41, 45, 47
A European Genebank Integrated System (AEGIS), 26, 28
Agriculture, 78–92
  adaptation strategy
    cropping system, 89–90
    crop production, 91
    germplasm, 90
  climate change impact, 78–81, 167
    CO_2 concentrations, 81
    pests and diseases, 82
    water availability impacts, 81–82
  liberalization and lower economic growth, 64
  rice (*see* Rice)
  wheat (*see* Wheat)
  world prices, 70–72
Alternaria brassicicola, 277
Aphis gossypi, 274
Arabidopsis thaliana, 161, 169, 173–176, 191–194, 224, 225, 278
Asclepias syriaca, 272

Breeder Toolbox component of CottonDB, 162, 163

Climate change
  agriculture (*see* Agriculture)
  crop wild relatives (*see* Crop wild relatives (CWRs))
  drought
    abiotic and biotic stresses, 221
    chewing pests, 275–276
    costs and risks, 221
    evolutionary plant breeding, 231–232
    famine, 221
    gene banks/farmers' fields, 230
    sap feeders, phloem osmoregulation, 275
    water shortage, 274
  elevated CO_2, 270
  elevated temperature, 270, 271
  feeding guilds, 270
  flood, rice production
    coastal areas, 255
    economic losses, 253
    food shortages, 255
    global warming, 254
    land-use activities, impacts on, 254
    research, 264–265
  food production (*see* Food production)
  on-farm conservation (*see* On-farm conservation, crop landraces)
  pest/disease resistance, in crop plants
    crop wild relatives, 278
    gene pyramiding and recurrent parent selection, 278
    genetic resources, 276–277
    marker-assisted selection, 278
    mutation-based techniques, 278
    reverse breeding, 278
  plant defence, indirect effects on
    chewer performance, 273
    front-line defence, 272
    microbial pathogen defence pathways, 272–273
    physical and chemical, 271–272
    sap-feeding herbivores, 273–274
    tomato defence, 272

Climate projections, 38–59
  abrupt climate change, 55–58
  climate variations and changes
    externally forced natural variability, 39
    greenhouse gases changes, 39–40
    internal variability, 39
    land-cover change, 41
    natural and human-made aerosols, 40–41
    solar irradiance, 39
    timescales, 39
  CO_2 concentration, 38–39
  Earth's surface, 38
  forecasts and projections
    aerosol emission, 44
    butterflies, 43–44
    carbon sinks, 44–45
    global climate sensitivity, 45–46
    greenhouse gas, 44
    land-use change, 44
    Pooh Sticks, 44
    predictions of, 43
    regional climate responses, 46
  GCM, 41–42
  long-term projections
    CO_2 concentrations, 54
    near-term projections, 55
    precipitation, 50–52
    regional climate change, 52–53
    temperature, 49–50
    weather extremes, changes in, 54–55
  near-term forecasting, 58–56
  SCMs, 42
  solar energy, 38
  testing climate model projections
    climate models vs. observations, 47–49
    observations, climate change, 46–47
Consultative Group on International Agricultural Research (CGIAR), 27, 99, 104, 109–111, 120, 151
Convention on Biological Diversity (CBD), 16, 17, 26, 98, 102–103
Coupled Model Intercomparison Project phase 5 (CMIP5), 82
Crop wild relatives (CWRs), 114–130
  climate change impacts, 116, 117
  conservation
    ex situ, 127
    in situ, 127–130
    and use, 115–116
  crop improvement, 118
  definition, 114
  drought resistance
    genetic variation, 229
    ICARDA's barley breeding programme, 229–230
  economic value, 115
  food security, 115
  genetic diversity
    Euro-Mediterranean region, 118, 120
    genetic erosion, 118
    Global Crop Diversity Trust project, 120
    IUCN Red List assessment, 120–122
    risk of extinction modelling, 122–123
    Sampled Red List Index for Plants project, 118
  major crops, 114, 115
  methods of
    climate change susceptibility assessment, 125–126
    IUCN Red List categories and criteria, 123–125
    species distribution modelling, 126–127
  pest/disease resistance, in crop plants, 278
  potential climate change mitigation traits, 118, 119

Double strand breaks (DSBs), 192
Drought, 9, 11, 147, 221–232
  abiotic and biotic stresses, 222
  Africa, 80
  agricultural production, 221, 222
  agrobiodiversity, decline of, 226–227
  breeding, 222, 225–226
  climate change
    abiotic and biotic stresses, 221
    chewing pests, 275–276
    costs and risks of, 221
    evolutionary plant breeding, 231–232
    famine, 221
    gene banks/farmers' fields, 230
    sap feeders, phloem osmoregulation, 275
    water shortage, 274
  crop production, 222
  crop wild relatives
    genetic variation, 229
    ICARDA's barley breeding programme, 229–230
  definition of, 221
  drought-resistant cultivars, 222
  empires and societies, collapse of, 222
  genetic complexity of
    LD based association mapping, 224
    marker-assisted selection, 224–225
    QTLs, 224
    up-and down-regulated genes, 224–225
  grain yield and cold tolerance, 223
  individual traits, climatic stresses, 223–224
  landraces
    agricultural production, 227

barley landraces vs. modern cultivars, 227–228
genetic structure, 229
plant breeding, 229
limited water availability, 81, 222
maize and soybean production, effects on, 221
morphological traits and grain yield, 222
North Island, 52
southern Australia, 52
terminal drought, 84, 85
wheat, 224
yield potential, 223

Early morning flowering (EMF), 203–204, 210
Electrical Penetration Graph (EPG) analysis, 277
EURISCO web catalogue, 157–159
European central crop database, 154–158
Europe's regional collaborative programme (ECPGR), 26, 28

Floods, 255–266
in deltas and lowlying regions, 167
dryland cereals, 251
rice crop
climate change, Impacts of, 254–255
deepwater rice varieties, 260–261
economic losses, 253
germination, 257–259
global warming, 254
hybridization, 257
in irrigated lowland, 252–253
Oryza species, 252
rainfed lowlands, 253
research, 264–265
rural poverty and food insecurity, 251–252
in South and Southeast Asia, 253
stagnant floods, tolerance of, 260
submergence tolerant landraces, 255–257
submergence-tolerant varieties, *SUB1* gene, 262–263
Sub1 varieties, in farmers' fields, 263–264
in uplands, 252
Focused Identification of Germplasm Strategy (FIGS), 276
Food production, 19, 61–76
agroecological zones, 72
CO_2 concentrations, 73–74
$2\times CO_2$ scenarios, climate models
agricultural adaptation, 62, 64
full trade liberalization, 64
lower economic and population growth rates, 64–65
unmitigated emissions of, 61–62
yield effects, 62, 63
crop production, biophysical effects, 68–70
crop wild relatives, 229
development pathways
risk of hunger, 67, 68
yield effects, 65, 67
droughts, 167
Green Revolution, 227
higher resolution climate models, 65, 66
IFPRI, 74–75
land use, 72
macroeconomic approach, 72
tropical and equatorial regions, 73
world prices, 70–72
Food security, 190, 215, 221, 254, 255, 265, 279
climate change, impacts of, 76
CWRs, 115, 130
definition of, 2
GHI, 2
increased production and productivity, 88
PGRFA, contributions of, 17–18, 107, 109
rice, 3
sorghum-based technologies, 89
Trust's mission, 27
Frankenstein foods, 190

General circulation models (GCM), 41–42, 48, 62, 65, 116, 142
Genetically modified organism (GMO), 278
Genetic Evaluation and Utilization (GEU) Program, 255
Genetic resources, 3–8, 147
conservation and use, 98–112
benefit-sharing fund, 111
CBD, 102–103
CGIAR centres, 104, 109–111
FAO Commission, 102
Global Crop Diversity Trust, 109
International Undertaking on Plant Genetic Resources, 100–102
ITPGRFA (*see* International Treaty on Plant Genetic Resources for Food and Agriculture (ITPGRFA))
Nagoya Protocol, 103–104
UPOV Convention, 100
omics technology (*see* Omics technology)
pest/disease resistance, 276–277
Genome-wide association studies (GWAS), 171
agronomic traits, 179
phenotyping datasets, 180

Germplasm conservation
  biodiversity decade, 16–17
  ex situ genebanks
    AEGIS context, 26
    CBD, 26
    climate change, 20
    core collections, 23–24
    crop descriptors and documentation, 25–26
    ECPGR, 26
    genebank collections, 20
    genebank standards, 24–25
    in vitro conservation, 22–23
    seed collections, 21–22
  GPA2, 18–19
  in situ conservation
    adaptive evolution, climate change, 29
    definition, 29
    genetic reserves, 30–31
    landraces, 31–32
  molecular biology, 19–20
  rational conservation approaches
    AEGIS, 28
    ECPGR, 28
    Global Crop Diversity Trust, 27
    GRIN-Global project, 28–29
    Svalbard Global Seed Vault, 27–28
  SOW2, 17–18
  use of, 32–33
Germplasm databases and informatics, 151–164
  genebanks, 151
  genome databases, 161–163
  international germplasm databases
    Crop Wild Relatives Global Portal, 160
    EURISCO web catalogue, 157–159
    European central crop databases, 156–158
    GENESYS portal, 159–160
  national germplasm databases and portals
    India, 160
    Japan, 160, 162
    national multi-crop databases, 160, 161
    USA, 161
  plant biodiversity informatics, 151
  plant genetic resources
    characterization and evaluation descriptors, 153, 156
    passport descriptors and MCPDs, 152, 154, 155
  World Wide Web, 151
Germplasm Resources Information Network (GRIN-Global project), 28–29
Global Crop Diversity Trust (GCDT), 27–28, 105, 109, 120, 129, 159
Global Hunger Index (GHI), 2
Global Plan of Action (GPA2), 17–19, 27, 102, 105, 108
GrainGenes, 162
Halophytes, 236, 243–245
Harnessing meiotic recombination, crop varieties, 190–198
  Arabidopsis thaliana, 191
  barley, 191–192
  blunt tool, 190
  Brassicas, 191
  crop species
    B. napus, 196
    changing environmental conditions, 197
    chromatin organization, 197
    genome sequencing and molecular marker techniques, 196
    plant breeders, 196
    targeting gene expression, 197
    transferring novel genes, 197–198
  meiosis
    chromosome segregation, 192
    CO control, polyploids, 195–196
    CO formation, barley and cereals, 194–195
    DSBs, 192, 193
    frequency and distribution, 194
    hotspots, 194
    meiotic homologous recombination, 193
    meiotic prophase I, 192
    reproducing organisms, 192
    synaptonemal complex (SC), 192
Heat shock proteins (HSPs), 204, 212–214
Helicoverpa armigera, 274
High temperature stress, 201–215
  breeding for
    physiological processes, 208–209
    reproductive processes and partitioning, 209, 210
  crops and wild relatives, 201–202
  genotypes
    detoxification pathway genes, 214–215
    EF-Tu, 214
    genes encoding proteins, 213
    HSP genes, 213–214
    QTL, 211–213
    stress-induced proteins, 213
    traditional backcross approach, 210–211
  impacts of, 208
  landraces and crop wild relatives, 202–204
  leaf chlorophyll content, 204–205
  night-time temperatures, 207
  photosystem (PS) II, 205
  plant growth and development, 205–206
  plant reproduction processes and yield components, 206–207
  rubisco catalytic activity, 205
  species fitness, 202
Holcus lanatus, 276

Integrated Assessment Models (IAMs), 44, 79
International Rice Genome Sequencing Project, 6
International Rice Research Institute (IRRI), 3, 21, 167, 203, 252
International Treaty on Plant Genetic Resources for Food and Agriculture (ITPGRFA), 111–112
  CBD, 106
  Contracting Parties, 109
  Farmers' Rights, 106
  Global Information System, 108
  Global Plan of Action, 108
  Governing Body, 107
  Multilateral System, 106, 107
  networks, 108
  SMTA, 106
IUCN Red List Categories and Criteria, 123–125

Linkage disequilibrium (LD), 144, 180, 224

Maize Genetics and Genomics Database (MaizeGDB), 162, 163
Marker-assisted selection (MAS), 24, 162, 190, 224–225, 278
Material Transfer Agreement (MTA), 104, 152
Melilotus siculus, 243–244
Multiparent Advanced Generation Inter-Cross (MAGIC), 6
Myzus persicae, 274, 275, 277

National Institute of Agrobiological Sciences (NIAS), 160, 173

Omics technology, 166–182
  climate change, agriculture, 167
  epigenome, 172–173
  exploratory tools
    allele flow, 180
    allele interactions, 181
    coalescent modelling, 180–181
    genotype x phenotype interactions, 181–182
    germplasm stocks/FIGS, 181
    recombination breakpoints, 182
    targeted selection, 181
  genome-wide association studies (GWAS)
    agronomic traits, 179
    phenotyping datasets, 180
  genomics
    DNA sequence(s), 169
    next generation sequencing, 170–171
    Sanger sequencing, 170
    third generation sequencing technology, 171–172
  germplasm use, 167–168
  informatics
    BigData, 177
    BioHDF5, 177–178
    cloud computing, 177
    controlled vocabularies and ontologies, 178
    databases and tools, 178–179
  ionome, 175
  metabolome, 174–175
  metagenome, 176–177
  molecular technologies, 168–169
  proteome, 174
  systems biology and interactome, 175–176
  transcriptome, 173
On-farm conservation, crop landraces, 137–147
  agricultural adaptation, 138
  biological and genetic diversity, 141
  community and farmer level, 145–146
  crop diversity, 141
  crop evolution, 139
  crop genetic resources, 138
  evolutionary service, 139, 143–145
  ex situ conservation, 137
  genetic and phenotypic diversity, 138
  genetic erosion, 140
  institutions, 141
  land suitability, 142–143
  maize, 147
  natural selection, 138–139
  poverty, 140
  replacement of, 140
  risk management, 147–148
  seed systems, 139–140
  small-scale farmers, 141
  socio-biological systems, 140

Pacific Biosciences system, 171
Participatory plant breeding (PPB), 226
Plant genetic resources for food and agriculture (PGRFA)
  GPA2, 18–19
  SOW2 report, 17–18
Protein synthesis elongation factor Tu (EFTu), 214
Pseudomonas syringae, 277

Quantitative trait loci (QTLs), 9, 24, 87, 175, 197
  drought resistance, 224
  flood-tolerant rice, 258–259
  high temperature stress, 211–213
  pest insects, resistance genes, 277
  salt tolerance, 238

Representative Concentration Pathways (RCPs), 78–79
Rhopalosiphum padi, 274
Rice, 3–13, 255–266
  Asia
    abiotic and biotic stresses, 86
    agroecological conditions, 86
    climate change adaptation, 87
    pest and diseases, 82–84, 86
    production losses, 86–87
  breeding
    irrigation and fertilizer, 11
    IRRI's experience, 12
    mega-varieties, 12
    rice production targets, 11
    rice seed market, 12
    seed industry, 11–12
    source, 13
    SRP platform, 12–13
  climate change
    drought, 9, 11
    flash flooding, 9–10
    heat, 8–9
    pest and disease complex, 11
    salinity tolerance, 10–11
  floods (*see* Floods)
  food security, 3
  genetic resources
    breeders, 6
    indica, 3
    IR8, 4, 5
    IR64, 4
    IR72, 4, 5
    japonica, 4
    Oryza rufipogon, 3, 4
    traditional rice varieties, 4
    wild species, 6
  prices of, 1
  temperatures, 83
Rice Biotechnology Programme, 6
Risk of hunger
  effects on, 67, 68
  impacts on, 75–76
  implications for, 73–74

Salicornia bigelovii, 243
Simple climate models (SCMs), 42, 48
Sitobium avenae, 274
Soil salinity, 6, 8, 10–11, 86, 91, 118, 236–245
  in arid and semi-arid regions, 236
  biosaline agriculture, 236
  Chenopodiaceae and Poaceae, 237
  in coastal areas, 237
  domestications
    Melilotus siculus, 243–244
    Salicornia bigelovii, 243
    salt mining species/cultivars, 244
    woody biomass crops, 243
  dryland salinity, 236, 237
  glycophytic crops, 236
  halophytes, 236
  ions, 236
  irrigation salinity, 236, 237
  primary salinity, 236
  salinity tolerance, in crops
    cereal crops, 237, 238
    genetic modification approach, 242–243
    genetic resources and breeding, 245
    germplasm collections, 239–241
    ion exclusion, 237
    landrace rice varieties, 239
    legume crops, 237–238
    osmotic stress tolerance, 237
    QTLs, 238
    screening method, 238–239
    tissue tolerance, 237
    toxic concentrations, 237
    wide hybridization, 241–242
  worldwide distribution of, 236
Sorghum production, 88–89, 91
Special Report on Emissions Scenarios (SRES), 44, 65
Spodoptera littoralis, 276
Standard Material Transfer Agreement (SMTA), 26, 104, 106, 107, 110
State of the World's Plant Genetic Resources, 17–18, 102
Stress-Tolerant Rice for Africa and South Asia (STRASA), 264
Sustainable Rice Production (SRP) platform, 12–13
Synaptonemal complex (SC), 192

Targeting Induced Local Lesions in Genomes (TILLING), 278
Trade-related Aspects of Intellectual Property Rights (TRIPS), 98, 100
Traditional backcross approach, 210–211
T.T. Chang Genetic Resources Center (TTC-GRC), 252

Union for the Protection of New Varieties of Plants (UPOV), 100
United States Department of Agriculture (USDA), 3

Wheat
  CIMMYT, 171
  growing season climates, 83

pests and diseases, 82
prices of, 1
proteomics, 174
south-west eastern Australia
climate change and impacts on, 84–85
conservation agriculture, 85
growing conditions in, 84
physiology and genetics, 84
production value, 84
wheat varieties, 85
temperatures, 83
World Wide Web, 151, 177